Nanofillers for Sustainable Applications

Nanofillers for Sustainable Applications provides an in-depth review of the wide-ranging applications of nanofillers. It explores both synthetic and natural nanofillers and focuses on their use as reinforcement and active fillers in composite structures.

Covering various aspects of nanofillers, including synthesis methods, characteristics, properties, and compatibility, this book highlights the potential of nanofillers as functional materials for different applications and offers a collection of comparative studies to showcase their efficacy. It emphasizes sustainability, intelligent design, and high-end applications in fields such as packaging, pulp and paper, aerospace, automotive, medicine, chemical industry, biodiesel, and chemical sensors. This book is organized into several sections, covering topics such as synthetic nanomaterials, nanosafety, natural nanofillers, polymer composites, metal nanofillers, nanofillers in various industries, nanofillers in renewable energy, nanofillers in biomedical sectors, and nanofillers in automotive and aerospace industries.

This book will be a useful reference for undergraduate and graduate students and academic researchers in the fields of materials science, nanomaterials, and polymer composites.

Key features:

- Focuses on the fabrication approaches used for nanofillers in nanocomposites.
- Covers materials selection, design solutions, manufacturing techniques, and structural analysis, highlighting their potential as functional materials in different applications.
- Explores the positive environmental impact and material property improvements resulting from increased composite utilization across diverse industries.
- Discusses other types of nanofillers like nanocellulose, metal-based, graphene, and wood-based materials.
- Includes case studies from leading industrial and academic experts.

Emerging Materials and Technologies

Series Editor: Boris I. Kharissov

The *Emerging Materials and Technologies* series is devoted to highlighting publications centered on emerging advanced materials and novel technologies. Attention is paid to those newly discovered or applied materials with potential to solve pressing societal problems and improve quality of life, corresponding to environmental protection, medicine, communications, energy, transportation, advanced manufacturing, and related areas.

The series takes into account that under present strong demands for energy, material, and cost savings, as well as heavy contamination problems and worldwide pandemic conditions, the area of emerging materials and related scalable technologies is a highly interdisciplinary field, with the need for researchers, professionals, and academics across the spectrum of engineering and technological disciplines. The main objective of this book series is to attract more attention to these materials and technologies and invite conversation among the international R&D community.

For more information about this series, please visit: www.routledge.com/Emerging-Materials-and-Technologies/book-series/CRCEMT

Nanofillers for Sustainable Applications

Edited by
N. M. Nurazzi, E. Bayraktar, M. N. F. Norrrahim, H. A. Aisyah, N. Abdullah, and M. R. M. Asyraf

CRC Press
Taylor & Francis Group
Boca Raton London New York

CRC Press is an imprint of the
Taylor & Francis Group, an **informa** business

Designed cover image: Shutterstock

First edition published 2024
by CRC Press
2385 NW Executive Center Drive, Suite 320, Boca Raton FL 33431

and by CRC Press
4 Park Square, Milton Park, Abingdon, Oxon, OX14 4RN

CRC Press is an imprint of Taylor & Francis Group, LLC

© 2024 selection and editorial matter, N. M. Nurazzi, E. Bayraktar, M. N. F. Norrrahim, H. A. Aisyah, N. Abdullah, and M. R. M. Asyraf; individual chapters, the contributors

Library of Congress Cataloging-in-Publication Data
Names: N. M. Nurazzi, editor. | E. Bayraktar, editor. | M. N. F. Norrrahim, editor. |
H. A. Aisyah, editor. | N. Abdullah, editor. | M. R. M. Asyraf, editor.
Title: Nanofillers for sustainable applications / edited by N. M. Nurazzi, E. Bayraktar,
M. N. F. Norrrahim, H. A. Aisyah, N. Abdullah, M. R. M. Asyraf.
Description: Boca Raton : CRC Press, 2024. | Series: Emerging materials and technologies |
Includes bibliographical references and index.
Identifiers: LCCN 2023028373 (print) | LCCN 2023028374 (ebook) |
ISBN 9781032510798 (hardback) | ISBN 9781032510804 (paperback) |
ISBN 9781003400998 (ebook)
Subjects: LCSH: Nanostructured materials. | Fillers (Materials)
Classification: LCC TA418.9.N35 N25266 2024 (print) | LCC TA418.9.N35 (ebook) |
DDC 620.1/15–dc23/eng/20231002
LC record available at https://lccn.loc.gov/2023028373
LC ebook record available at https://lccn.loc.gov/2023028374

ISBN: 978-1-032-51079-8 (hbk)
ISBN: 978-1-032-51080-4 (pbk)
ISBN: 978-1-003-40099-8 (ebk)

DOI: 10.1201/9781003400998

Typeset in Times
by codeMantra

Contents

Chapter 2   The Characterization Techniques of Nanomaterials 44

Mohd Ridhwan Adam, Muhammad Hakimin Shafie,
Mohd Saiful Shamsudin, Siti Khadijah Hubadillah,
Mohd Riduan Jamalludin, Mohd Haiqal Abd Aziz,
and Atikah Mohd Nasir

Chapter 3   Nanosafety: Exposure, Detection, and Toxicology 59

Nusrat Tara and Mukul Pratap Singh

Chapter 7

*Soleha Mohamat Yusuff, Mohd Nor Faiz Norrrahim, and
Mohd Nurazzi Norizan*

Chapter 8

*M. M. Harussani, S. M. Sapuan, Gwyth Muosso,
and Shahriar Ahmad Fahim*

Chapter 16   Nanofillers in Food Packaging 309

*A. Nazrin, R. M. O. Syafiq, S. M. Sapuan, M. Y. M. Zuhri,
I. S. M. A. Tawakkal, and R. A. Ilyas*

Chapter 17   Design of Recycled Aluminium (AA 7075+AA1050 Fine
Chips)-Based Composites Reinforced with Nano-SiC Whiskers,
Fine Carbon Fiber for Aeronautical Applications 319

*Özgür Aslan, Olga Klinkova, Dhurata Katundi,
Ibrahim Miskioglu, and Emin Bayraktar*

Chapter 18   State-of-Art Review on Nanofiller in Biodiesel Applications 329

*M. S. M. Misenan, M. S. Ahmad Farabi, N. A. Zulkipli,
M. A. Mohd Saad, and A. H. Shaffie*

Editors

N. M. Nurazzi is a senior lecturer at the School of Industrial Technology, Universiti Sains Malaysia, Penang, 11800 Malaysia. Before joining Universiti Sains Malaysia, he was a Postdoctoral Fellow at the Centre for Defence Foundation Studies, National Defence University of Malaysia, under the Newton Research Grant for the study on "Role of Intermolecular Interaction in Conductive Polymer Wrapped MWCNT as Organophosphate Sensing Material Structure". He obtained a Diploma in Polymer Technology from Universiti Teknologi MARA (UiTM) in 2009, a Bachelor of Science (BSc.) in Polymer Technology from UiTM in 2011, and a Master of Science (MSc.) from UiTM in 2014 under a Ministry of Higher Education (MOHE) Malaysia scholarship. In 2018, he was awarded a PhD from Universiti Putra Malaysia (UPM) in Materials Engineering under the MOHE Malaysia scholarship. His main research interest includes materials engineering, polymer composites and characterizations, natural fiber composites, and carbon nanotubes for chemical sensors. To date, he has authored and co-authored more than 100 citations indexed in journals on polymer composites, natural fiber composites, and materials science-related subjects; 30 book chapters; 2 edited book; 1 authored book; more than 20 conference proceedings/ seminars; and 2 guest editor of journal special issues.

Emin Bayraktar (Prof. Emeritus, Habil., Dr (Ph.D.), DSc—Doctor of Science) is an academic and research staff member in Mechanical and Manufacturing engineering at SUPMECA/Paris, France. His research areas include manufacturing techniques of new materials (basic composites—hybrid), metal forming of thin sheets (Design + test + FEM), static and dynamic behavior and optimization of materials (experimental and FEM—utilization and design of composite-based metallic and non-metallic, powder metallurgy, and energetic material aeronautical applications), metallic-based and non-metallic materials, powder metallurgy and metallurgy of steels, welding, and heat treatment, as well as the processing of new composites, sintering techniques, sinter–forging, thixoforming, etc. He has authored more than 200 publications in the International Journals and International Conference Proceedings and has also authored more than 90 research reports (European = Steel Committee projects, Test + Simulation). He has advised 32 Ph.D. and 120 MSc theses and is currently advising 7. He is a Fellow of World Academy of Science in Materials and Manufacturing Engineering (WAMME), Editorial Board—Member of *International Journal of Achievement in Materials and Manufacturing Engineering*, Advisory board member of AMPT—2009 (Advanced Materials Processing Technologies), and Reviewer Board and Guess Editor for Polymers. He was Visiting Professor at Nanyang Technology University, Singapore, in 2012, Xi'an Northwestern Technical University, Aeronautical Engineering, in 2016, University of Campinas, UNICAMP-Brazil in 2013 until 2023. He is a recipient of the Silesian University Prix pour "FREDERIK STAUB Golden Medal-2009" by the Academy of WAMME, "World Academy of Science"—Poland, materials science section, and a recipient of the William Johnson International Gold Medal—2014, AMPT academic association.

M. N. F. Norrrahim working as research officer in the Research Center for Chemical Defence, National Defence University of Malaysia. His research interests in nanotechnology, composites, materials science, polymers, chemical and biological defense, and biotechnology. He has conducted various studies on the benefits of nanocellulose and its applications. During his doctoral studies, he improved the nanocellulose production from oil palm biomass using mechanical treatment. Now, his research is focused on using nanocellulose for applications such as composites, adsorbents, and military. His expertise in this field has been proven by the high-impact publications he has produced related to nanocellulose research. To date, he has authored or co-authored more than 100 articles, including 40 book chapters in renowned journals on nanotechnology, materials science, chemistry, and biotechnology-related subjects. He also published one book in Elsevier entitled *Industrial Application of Nanocellulose and its Nanocomposites*. He has received several local and international innovation awards. His current H-index is Google Scholar is 34 and Scopus 29. Besides that, he has presented his research findings at several local and international conferences. He also received several innovation awards.

H. A. Aisyah is a research officer in the Engineering & Processing Research Division, Malaysian Palm Oil Board. Previously, she worked as a Postdoctoral Researcher Fellow in Department of Mechanical and Manufacturing Engineering, Faculty of Engineering, UPM. She holds a Master in Biocomposite Technology and Design from UPM in 2013 under the MOHE Malaysia scholarship. She attended graduate school at the same university in the Biocomposite Technology, graduating with his Ph.D. in 2019. From 2014, she was a postgraduate student at the Institute of Tropical Forestry and Forest Products. Her doctoral research focused on the development of textile polymer composite from natural fiber and characterization of the composite. Her research interests include biocomposite manufacturing technology, polymer composite, natural fiber and lignocellulosic based composite, and textile composite. She has authored and co-authored 40 citation indexed journals on composite and materials science-related subjects. She also authored 15 book chapters as well as 20 conference proceedings.

N. Abdullah is currently lecturer of chemistry at the National Defence University of Malaysia, Malaysia, since 2012. She received her Master of Philosophy degree in Advanced Materials from the University of Malaya (UM), Malaysia, in 2004. She worked as a research assistant at Fritz Haber Institut, The Max Plank Society, Germany, from 2002 until 2003 as part of the internship to complete her master's degree. She received scholarship under the MOHE to pursue Doctor of Philosophy study at UM in the field of Nanomaterials. She continued her interest in research by doing postdoctoral studies at University College London, England, in 2016 working in the field of heterogeneous catalysis. She was awarded with several research grants and as PI for an International grant Newton Fund in 2019. Overall, her research output includes more than 50 publications in respected journals and 10 abstracts in national and international conferences. Her areas of interests include material science focused on the synthesis and characterization of carbon-based materials and their application as sensing material and development of supported nanoparticles as heterogeneous catalysts for photocatalytic reaction and hydrogenation reaction.

M. R. M. Asyraf is a Senior Lecturer at the School of Mechanical Engineering, Faculty of Engineering, Universiti Teknologi Malaysia (UTM). Previously, he worked as a Postdoctoral Research Fellow in the Institute of Energy Infrastructure, Universiti Tenaga Nasional from 2021 to 2022. He has a Bachelor of Engineering (Mechanical) with first class honors from UPM, Malaysia, and Ph.D. in Material Engineering from UPM, Malaysia. Currently, he registered as a graduate engineer, a graduate technologist, and a corporate member of *Persatuan Pembangunan dan Industri Enau Malaysia* (PPIEM). He has been appointed as a member of international community project 2021 organized by MyOHUN-USAID, entitled "Society awareness on recycling of COVID-19 PPE waste toward environmental sustainable green technology". In 2019, he has been appointed as a member of editorial boards in *Journal of Advanced Research in Fluid Mechanics and Thermal Sciences* (Scopus Indexed Journal). Currently, he has been appointed as Guest Editor in Special Issue entitled *Lignocellulosic Fibre-based Composites in Forests* (Q1), MDPI. Besides that, he served as an Academic Editor in *International Journal of Polymer Sciences* (Q2) and Hindawi and Review Editor in *Frontiers of Materials* (Q2). He also is an editorial board members of *Journal of Natural Fibre Polymer Composites*. To date, he has authored and co-authored more than 100 publications including over 66 journal articles, 2 books, 30 chapters in book, and other publications. Currently, his H-Index is 23 based on Scopus database. Besides, he has delivered more than 10 international and local presentations from various conferences and seminars. He reviewed over 40 journal papers such as high-impact journals such as *Polymers* (Q1), *Materials* (Q1), *Gels* (Q1), *Mathematics* (Q1), MDPI; *Cellulose* (Q1), *Biomass Conversion and Biorefinery* (Q1), *Journal of Polymers and the Environment* (Q1), *Fibers and Polymers* (Q1), *Scientific Reports* (Q1), Springer; *Journal of Natural Fibers*, Taylor and Francis (Q1); *E-Polymers Letters* (Q1); *Applied Sciences* (Q2), *Coatings* (Q2), MDPI; *Journal of Vibration and Control* (Q2), and *Science Progress* (Q3), SAGE; *Scientia Iranica* (Q3) and *Functional Composites and Structures*, IOP (Scopus); and *World Journal of Engineering*, Emerald (Scopus). He was awarded Institute of Malaysia (IKM) Research Prize in Polymer and Material Sciences 2021 from Chemical IKM and Excellent Graduate Award 2019 from Yayasan Bank Rakyat, Malaysia. He also received the Royal Education Award from Conference of Rulers, Malaysia, and Excellent Leadership Award by Koperasi, UPM, Malaysia. Furthermore, he also demonstrated excellence in non-academic performance, among them are through his contribution as a committee member for the Seminar on Scientific Writing and Publications using KSI Technique in 2019 organized by the Department of Aerospace Engineering, UPM. He also is committee member for the Sugar Palm Seminar 2019 and 5th International Conference on Computational Methods in Engineering and Health Sciences (ICCMEH2019). Moreover, he was appointed as Deputy Chairman in MyOHUN-USAID International Webinar Series 2020 and MyOHUN-USAID One Stop Centre & Exhibition. In 2021, he has served as Deputy Secretary and adjucator in the International Conference on Sugar Palm and Allied Fibre Polymer Composites (SAPC 2021). He has been invited as Invited Speaker in the International Conference on Materials Science & Engineering 2022 as well as an evaluating panel committee in the International Symposium of Polymeric Materials 2022 (ISPM 2022).

Preface

Nanofillers for Sustainable Applications is a comprehensive and insightful review that explores the wide-ranging applications of nanofillers. This book offers a detailed examination of both synthetic and natural derived nanofillers, specifically focusing on their utilization as reinforcement and active fillers in composite structures. By delving into the synthesis methods, characteristics, properties, and compatibility of nanofillers, the book highlights their potential as functional materials for a diverse range of applications. The primary objective of this book is to provide an extensive collection of comparative studies, encompassing both past and recent research, to shed light on the efficacy of nanofillers in various selected applications. It offers a comprehensive review of natural and synthetic nanofillers, with a particular emphasis on their role in fostering sustainability, intelligent design, and high-end applications. Notably, the book explores their application in areas such as packaging, pulp and paper, aerospace, automotive, medicine, chemical industry, biodiesel, and chemical sensors, elucidating the ways in which nanofillers contribute to advancements in these fields. In addition to the above, the book also touches upon other types of nanofillers, including nanocellulose, metal-based, graphene, and wood-based materials. While providing a concise summary of these materials, the book primarily focuses on the various approaches employed in the fabrication of nanofillers for nanocomposites. Furthermore, it explores the positive environmental impact and improvements in material properties resulting from the increased utilization of composites across diverse industries.

This book covers topic of (1) Overviews of Synthetic Nanomaterials, Synthesis Methods, Characteristics and Recent Progress; (2) The Characterization Techniques of Nanomaterials; (3) Nanosafety: Exposure, Detection and Toxicology; (4) Natural Nanofillers: Preparation and Properties; (5) Compatibility Study of Nanofillers Based Polymer Composites; (6) Inclusion of Nano-Fillers in Natural Fibre Reinforced Polymer Composites: Overviews and Applications; (7) Metal Nanofillers in Composite Structure; (8) Bio-Oils as The Precursor for Carbon Nanostructure Formation; (9) Nanoplastics in Environment: Environmental Risk, Occurrence, Characterization, And Identification; (10) Nanofillers in Pulp and Paper; (11) Design of Recycled Aluminium (AA 7075) based Composites Reinforced with Nano filler Ni-Al Intermetallic and Nano Niobium Powder Produced with Vacuum Arc Melting for Aeronautical Applications; (12) Performance Evaluation of Nano-Lignin in Polymer Composites; (13) Natural Nanofillers in Biopolymer Based Composites: A Review; (14) Effect of Dispersion and Interfacial Functionalization of Multiwalled Carbon Nanotubes in Epoxy Composites: Structural and Thermogravimetric Analysis Characteristics; (15) Natural Nanofillers in Polyolefins-Based Composites: A Review; (16) Nanofillers in Food Packaging; (17) Design of Recycled Aluminium (AA 7075+AA1050 fine chips) Based Composites Reinforced with Nano SiC Whiskers, Fine Carbon Fibre for Aeronautical Applications; (18) State of Art Review on Nanofiller in Biodiesel Applications; (19) Recent Progress of Advanced Nanomaterials in Renewable Energy; (20) Emerging Development on Nanocellulose and Its Composites in Biomedical Sectors; (21) Cinnamon Pickering Emulsions as a Natural Disinfectant: Protection Against Bacteria and SARS-CoV-2; as well as (22) Nanofillers in Automotive and Aerospace Industry.

By combining a comprehensive overview of nanofillers with their applications, *Nanofillers for Sustainable Applications* serves as an invaluable resource for researchers, professionals, and students seeking a deeper understanding of the potential and practicality of nanofillers in numerous fields.

We are incredibly thankful to all authors who are experts in Nanofillers for Sustainable Applications who contributed book chapters in this edited book and supported it by providing valuable ideas and knowledge. We are also grateful to CRC Press, the supporting team, especially Kyra and Sonia, for helping us in administrating and their valuable advice in finalizing this book.

Acknowledgements

Editors express their highest gratitude to all the dedicated and excellent authors, and my mentor Prof. Ir. Dr. Sapuan Salit (S. M. Sapuan) and Dr. Ahmad Ilyas Rushdan (R. A. Ilyas). Not to forget my wife and son, Nadzirah and Nuh for their help and endless support.

Contributors

Ummi Hani Abdullah
Department of Forest Production,
    Faculty of Forestry
Universiti Putra Malaysia
Selangor, Malaysia

Norli Abdullah
Department of Chemistry and Biology,
    Centre for Defence Foundation Studies
Universiti Pertahanan Nasional
    Malaysia
Kuala Lumpur, Malaysia

Falah Abu
Eco-Technology Programme, School
    of Industrial Technology, Faculty of
    Applied Sciences
UiTM Shah Alam
Selangor, Malaysia
and
Nanocomposite Materials and
    Processing Research Group, Faculty
    of Applied Sciences
UiTM Shah Alam
Selangor, Malaysia

Mohd Ridhwan Adam
School of Chemical Sciences
Universiti Sains Malaysia
Pulau Pinang, Malaysia

Mohd Ridhwan Adam
School of Chemical Sciences
Universiti Sains Malaysia
Pulau Pinang, Malaysia

So'bah Ahmad
Department of Food Science and
    Technology, School of Industrial
    Technology, Faculty of Applied
    Sciences
UiTM Shah Alam
Shah Alam, Malaysia

Syeed Saifulazry Osman Al-Edrus
Institute of Tropical Forestry and Forest
    Product
Universiti Putra Malaysia
Selangor, Malaysia

Azirah Akbar Ali
School of Biological Sciences
Universiti Sains Malaysia
Penang, Malaysia

Yoshito Andou
Department of Biological Functions
    Engineering, Graduate School
    of Life Science and Systems
    Engineering
Kyushu Institute of Technology
Fukuoka, Japan
and
Collaborative Research Centre for
    Green Materials on Environmental
    Technology
Kyushu Institute of Technology
Fukuoka, Japan

Özgür Aslan
Atilim University
Computational Mechanics
Ankara, Turkey

M. R. M. Asyraf
Engineering Design Research Group
    (EDRG), Faculty of Mechanical
    Engineering
Universiti Teknologi Malaysia
Johor, Malaysia
and
Centre for Advanced Composite
    Materials (CACM)
Universiti Teknologi Malaysia
Johor, Malaysia

Azman Azid
Faculty of Bioresources and Food
    Industry
Universiti Sultan Zainal Abidin, Besut
    Campus
Terengganu, Malaysia

Mohd Haiqal Abd Aziz
Department of Chemical Engineering
    Technology, Faculty of Engineering
    Technology
Universiti Tun Hussein Onn Malaysia
Johor, Malaysia

B. F. A. Bakar
Institute of Tropical Forestry and Forest
    Products (INTROP)
Universiti Putra Malaysia
Selangor, Malaysia
and
Department of Wood and Fiber
    Industries, Faculty of Forestry and
    Environment
Universiti Putra Malaysia
Selangor, Malaysia

Rennan Felix Da Silva Barbosa
Engineering, Modeling, and Applied
    Social Sciences Center (CECS)
Federal University of ABC
Santo André, Brazil

Mohd Salahuddin Mohd Basri
Department of Process and Food
    Engineering, Faculty of Engineering
Universiti Putra Malaysia, Serdang
Selangor, Malaysia
and
Laboratory of Halal Science Research,
    Halal Products Research Institute
Universiti Putra Malaysia (UPM)
Serdang, Malaysia
and
Laboratory of Biopolymer and
    Derivatives, Institute of Tropical
    Forestry and Forest Products
    (INTROP)
Universiti Putra Malaysia
Selangor, Malaysia

Emin Bayraktar
School of Mechanical and
    Manufacturing Engineering
ISAE-Supmeca
Paris, France

C. H. Lee
Institute of Tropical Forestry and Forest
    Products (INTROP)
Universiti Putra Malaysia
Selangor, Malaysia

Ivana Barros De Campos
Santo André Regional Center
Adolfo Lutz Institute
Santo André, Brazil

E. S. Zainudin
Advanced Engineering Materials
    and Composites Research Centre
    (AEMC), Department of Mechanical
    and Manufacturing Engineering,
    Faculty of Engineering
Universiti Putra Malaysia
Selangor, Malaysia

F. A. B. Balkis
Institute of Tropical Forestry and Forest
 Products (INTROP)
Universiti Putra Malaysia
Selangor, Malaysia
and
Faculty of Forestry and Environment
Universiti Putra Malaysia
Selangor, Malaysia

Shahriar Ahmad Fahim
Department of Transdisciplinary
 Science and Engineering, School of
 Environment and Society
Tokyo Institute of Technology
Tokyo, Japan

M. S. Ahmad Farabi
Department of Chemistry, Faculty of
 Science
Universiti Putra Malaysia
Selangor, Malaysia

Ismail M. Fareez
School of Biology, Faculty of Applied
 Sciences
Universiti Teknologi MARA, Shah
 Alam Campus
Selangor, Malaysia
and
Collaborative Drug Discovery Research
 (CDDR), Faculty of Pharmacy
Universiti Teknologi MARA, Selangor
 Campus
Selangor, Malaysia

Mohammed Abdillah Ahmad Farid
Department of Biological Functions
 Engineering, Graduate School of Life
 Science and Systems Engineering
Kyushu Institute of Technology
Fukuoka, Japan

Rafaela Reis Ferreira
Engineering, Modeling, and Applied
 Social Sciences Center (CECS)
Federal University of ABC
Santo André, Brazil

Fabio Gatamorta
UNICAMP
University of Campinas, FEM
SP-Campinas, Brazil

Aizat Ghani
Faculty of Tropical Forestry
Universiti Malaysia Sabah
Kota Kinabalu, Malaysia

H. A. Aisyah
Advanced Engineering Materials
 and Composites Research Centre
 (AEMC), Department of Mechanical
 and Manufacturing Engineering,
 Faculty of Engineering
Universiti Putra Malaysia
Selangor, Malaysia
and
Biomass Technology Unit, Engineering
 and Processing Division, Malaysian
 Palm Oil Board (MPOB)
Persiaran Institusi
Selangor, Malaysia

Mohd Idham Hakimi
Department of Bioprocess Technology,
 Faculty of Biotechnology and
 Biomolecular Sciences
Universiti Putra Malaysia
Selangor, Malaysia

M. M. Harussani
Advanced Engineering Materials
and Composites Research Centre
(AEMC), Department of Mechanical
and Manufacturing Engineering
Universiti Putra Malaysia
Selangor, Malaysia
and
Department of Transdisciplinary
Science and Engineering, School of
Environment and Society
Tokyo Institute of Technology
Tokyo, Japan

S. A. Hassan
Centre for Advanced Composite
Materials (CACM)
Universiti Teknologi Malaysia
Johor, Malaysia

K. Z. Hazrati
German Malaysian Institute, Jalan
Ilmiah
Taman Universiti
Selangor, Malaysia

Lee Seng Hua
Department of Wood Industry, Faculty
of Applied Sciences
Universiti Teknologi MARA (UiTM),
Cawangan Kampus Jengka
Bandar Tun Razak, Malaysia

Siti Khadijah Hubadillah
School of Technology Management and
Logistics
Universiti Utara Malaysia
Kedah, Malaysia

R. A. Ilyas
School of Chemical and Energy
Engineering, Faculty of Engineering
Universiti Teknologi Malaysia
Johor, Malaysia

and
Centre for Advanced Composite
Materials (CACM)
Universiti Teknologi Malaysia (UTM)
Johor, Malaysia

Khairul Anwar Ishak
Centre for Fundamental and Frontier
Sciences in Nanostructure Self-
Assembly, Department of Chemistry,
Faculty of Science
Universiti Malaya
Kuala Lumpur, Malaysia

Siti Hasnawati Jamal
Department of Chemistry and Biology,
Centre for Defence Foundation
Studies
Universiti Pertahanan Nasional
Malaysia
Kuala Lumpur, Malaysia
and
Centre for Tropicalisation
Universiti Pertahanan Nasional
Malaysia
Kuala Lumpur, Malaysia

Mohd Riduan Jamalludin
Faculty of Mechanical Engineering
Technology
Universiti Malaysia Perlis (UniMAP),
Kampus Alam UniMAP
Perlis, Malaysia

Ainil Hawa Jasni
Department of Science in Engineering,
Faculty of Engineering
International Islamic University of
Malaysia
Kuala Lumpur, Malaysia

Mohd Azwan Jenol
Faculty of Biotechnology and
    Biomolecular Sciences
Universiti Putra Malaysia
Selangor, Malaysia

Naveen Jesuarockiam
School of Mechanical Engineering
Vellore Institute of Technology
Vellore, India

Nor Syaza Syahirah Amat Junaidi
Fakulti Perubatan dan Kesihatan
    Pertahanan
Universiti Pertahanan Nasional
    Malaysia (UPNM)
Kuala Lumpur, Malaysia

Siti Hasnah Kamarudin
Eco-Technology Programme, School
    of Industrial Technology, Faculty of
    Applied Sciences
UiTM Shah Alam
Selangor, Malaysia
and
Nanocomposite Materials and
    Processing Research Group, Faculty
    of Applied Sciences
UiTM Shah Alam
Selangor, Malaysia

Cagatay Kasar
Atilim University
Computational Mechanics
Ankara, Turkey

Norherdawati Kasim
Department of Chemistry and Biology,
    Centre for Defence Foundation
    Studies
Universiti Pertahanan Nasional
    Malaysia
Kuala Lumpur, Malaysia

Noor Azilah Mohd Kasim
Department of Chemistry and Biology,
    Centre for Defence Foundation
    Studies
Universiti Pertahanan Nasional
    Malaysia
Kuala Lumpur, Malaysia

M. H. M. Kassim
Bioresource Technology Division,
    School of Industrial Technology
Universiti Sains Malaysia
Penang, Malaysia
and
Green Biopolymer, Coatings &
    Packaging Cluster, School of
    Industrial Technology
Universiti Sains Malaysia
Penang, Malaysia

Maurício Maruo Kato
Engineering, Modeling, and Applied
    Social Sciences Center (CECS)
Federal University of ABC
Santo André, Brazil

Dhurata Katundi
School of Mechanical and
    Manufacturing Engineering
ISAE-Supmeca
Paris, France

Olga Klinkova
School of Mechanical and
    Manufacturing Engineering
ISAE-Supmeca
Paris, France

Victor Feizal Knight
Research Centre for Chemical Defence
Universiti Pertahanan Nasional
    Malaysia
Kuala Lumpur, Malaysia

H. M. Mohammed
Bioresource Technology Division,
    School of Industrial Technology
Universiti Sains Malaysia
Penang, Malaysia
and
Green Biopolymer, Coatings &
    Packaging Cluster, School of
    Industrial Technology
Universiti Sains Malaysia
Penang, Malaysia

Mohd Nurazzi Norizan
Bioresource Technology Division,
    School of Industrial Technology
Universiti Sains Malaysia
Penang, Malaysia
and
Green Biopolymer, Coatings &
    Packaging Cluster, School of
    Industrial Technology
Universiti Sains Malaysia
Penang, Malaysia

M. S. M. Misenan
Department of Chemistry, College of
    Art and Science
Yildiz Technical University, Davutpasa
    Campus
Istanbul, Turkey

Ibrahim Miskioglu
ME-EM Department
Michigan Technological University
Houghton, Missouri

Gwyth Muosso
Advanced Engineering Materials
    and Composites Research Centre
    (AEMC), Department of Mechanical
    and Manufacturing Engineering
Universiti Putra Malaysia
Selangor, Malaysia

Syed Umar Faruq Syed Najmuddin
Faculty of Science and Natural Resources
Universiti Malaysia Sabah
Sabah, Malaysia

Atikah Mohd Nasir
Centre for Diagnostic, Therapeutic and
    Investigative Studies (CODTIS),
    Faculty of Health Sciences
Universiti Kebangsaan Malaysia
Kuala Lumpur, Malaysia

A. Nazrin
Laboratory of Biocomposite Technology,
    Institute of Tropical Forestry and
    Forest Products (INTROP)
Universiti Putra Malaysia
Selangor, Malaysia

M. N. F. Norrrahim
Centre for Defence Foundation Studies
Universiti Pertahanan Nasional
    Malaysia
Kuala Lumpur, Malaysia

Mohd Nor Faiz Norrrahim
Research Centre for Chemical Defence
Universiti Pertahanan Nasional
    Malaysia
Kuala Lumpur, Malaysia

Syaiful Osman
Eco-Technology Programme, School
    of Industrial Technology, Faculty of
    Applied Sciences
UiTM Shah Alam
Selangor, Malaysia
and
Nanocomposite Materials and
    Processing Research Group, Faculty
    of Applied Sciences
UiTM Shah Alam
Selangor, Malaysia

R. A. Ilyas
Faculty of Chemical and Energy
  Engineering
Universiti Teknologi Malaysia
Johor, Malaysia

M. Rafidah
Department of Civil Engineering,
  Faculty of Engineering
Universiti Putra Malaysia
Selangor, Malaysia

Nurul Latiffah Abd Rani
Pusat Asasi STEM
Universiti Malaysia Terengganu
Terengganu, Malaysia

Mohd Saiful Asmal Rani
Department of Physics, Faculty of
  Science
Universiti Putra Malaysia
Selangor, Malaysia
and
Institute of Tropical and Forest Products
  (INTROP)
Universiti Putra Malaysia
Selangor, Malaysia

Ahmad Rashedi
School of Mechanical and Aerospace
  Engineering
Nanyang Technological University
Singapore, Singapore

N. F. M. Rawi
Bioresource Technology Division,
  School of Industrial Technology
Universiti Sains Malaysia
Penang, Malaysia
and
Green Biopolymer, Coatings &
  Packaging Cluster, School of
  Industrial Technology
Universiti Sains Malaysia
Penang, Malaysia

Derval Dos Santos Rosa
Santo André Regional Center
Adolfo Lutz Institute
Santo André, Brazil

S. H. Lee
Institute of Tropical Forestry and Forest
  Products (INTROP)
Universiti Putra Malaysia
Selangor, Malaysia

M. A. Mohd Saad
Faculty of Science and Technology
Universiti Sains Islam Malaysia
Negeri Sembilan, Malaysia

Suresh Sagadevan
Nanotechnology & Catalysis Research
  Centre
University of Malaya
Kuala Lumpur, Malaysia

K. M. Salleh
Bioresource Technology Division,
  School of Industrial Technology
Universiti Sains Malaysia
Penang, Malaysia
and
Green Biopolymer, Coatings &
  Packaging Cluster, School of
  Industrial Technology
Universiti Sains Malaysia
Penang, Malaysia

MohdSaiful Samsudin
Environmental Technology Division,
  School of Industrial Technology
Universiti Sains Malaysia
Penang, Malaysia
and
Renewable Biomass Transformation
  Cluster, School of Industrial
  Technology
Universiti Sains Malaysia
Penang, Malaysia

S. M. Sapuan
Advanced Engineering Materials
  and Composites Research Centre
  (AEMC), Department of Mechanical
  and Manufacturing Engineering
Universiti Putra Malaysia
Selangor, Malaysia

Luiz Fernando Grespan Setz
Engineering, Modeling, and Applied
  Social Sciences Center (CECS)
Federal University of ABC
Santo André, Brazil

A. H. Shaffie
Halal Action Laboratory, Kolej
  GENIUS Insan
Universiti Sains Islam Malaysia
Negeri Sembilan, Malaysia

Muhammad Hakimin Shafie
Analytical Biochemistry Research
  Centre (ABrC), Universiti Sains
  Malaysia
University Innovation Incubator
  Building, SAINS@USM Campus
Penang, Malaysia

Noor Aisyah Ahmad Shah
Department of Chemistry and Biology,
  Centre for Defence Foundation
  Studies
Universiti Pertahanan Nasional
  Malaysia
Kuala Lumpur, Malaysia

**Nik Noorul Shakira Mohamed
Shakrin**
Fakulti Perubatan dan Kesihatan
  Pertahanan
Universiti Pertahanan Nasional
  Malaysia (UPNM)
Kuala Lumpur, Malaysia

and
Centre for Tropicalization (CENTROP)
Universiti Pertahanan Nasional
  Malaysia (UPNM)
Kuala Lumpur, Malaysia

Mohd Saiful Shamsudin
Department of Environmental
  Technology, School of Industrial
  Technology
Universiti Sains Malaysia
Pulau Pinang, Malaysia

Intan Juliana Shamsudin
Department of Chemistry and Biology,
  Centre for Defence Foundation
  Studies
Universiti Pertahanan Nasional
  Malaysia
Kuala Lumpur, Malaysia

S. Sharma
Mechanical Engineering Department,
  University Center for Research &
  Development
Chandigarh University
Mohali, India

D. D. C. Vui Sheng
Applied Mechanics Research and
  Consultancy Group (AMRCG),
  Faculty of Mechanical Engineering
Universiti Teknologi Malaysia
Johor, Malaysia

Muhammad Faizan Abdul Shukor
Research Centre for Chemical Defence
Universiti Pertahanan Nasional
  Malaysia
Kuala Lumpur, Malaysia

Nurul Naqirah Shukor
Pusat Pengajian Siswazah
Universiti Pertahanan Nasional
    Malaysia
Kuala Lumpur, Malaysia

Mukul Pratap Singh
GL Bajaj Institute of Technology and
    Management
Greater Noida, India

Alana Gabrieli De Souza
Engineering, Modeling, and Applied
    Social Sciences Center (CECS)
Federal University of ABC
Santo André, Brazil

Siti Norasmah Surip
Eco-Technology Programme, School
    of Industrial Technology, Faculty of
    Applied Sciences
UiTM Shah Alam
Selangor, Malaysia
and
Nanocomposite Materials and
    Processing Research Group, Faculty
    of Applied Sciences
UiTM Shah Alam
Selangor, Malaysia

R. M. O. Syafiq
Laboratory of Biocomposite
    Technology, Institute of Tropical
    Forestry and Forest Products
    (INTROP)
Universiti Putra Malaysia
Selangor, Malaysia

A. Syamsir
Institute of Energy Infrastructure
Universiti Tenaga Nasional, Jalan
    IKRAM-UNITEN
Selangor, Malaysia

Mohamad Nurul Azman
Mohammad Taib
Interdisciplinary Research Center for
    Advanced Materials
King Fahd University of Petroleum and
    Minerals
Dhahran, Saudi Arabia

Nusrat Tara
Environmental Chemistry Research
    Laboratory, Department of
    Chemistry
Jamia Millia Islamia
New Delhi, India

Intan Syafinaz Mohamed Amin
Tawakkal
Department of Process and Food
    Engineering, Faculty of Engineering
Universiti Putra Malaysia, Serdang
Selangor, Malaysia

L. Y. Tee
Civil Engineering Department, College
    of Engineering
Universiti Tenaga Nasional, Jalan
    IKRAM-UNITEN
Selangor, Malaysia

Zaharah Wahid
International Islamic University of
    Malaysia
Department of Science in Engineering
Kulliyyah of Engineering International
    Islamic University, Kuala Lumpur,
    Malaysia

Tengku Arisyah Tengku
Yasim-Anuar
Nextgreen Pulp & Paper Sdn. Bhd.,
    Menara LGB
Kuala Lumpur, Malaysia

Eliana Della Coletta Yudice
Santo André Regional Center
Adolfo Lutz Institute
Santo André, Brazil

Soleha Mohamat Yusuff
Industrial Technology Division
Malaysian Nuclear Agency
Selangor, Malaysia

M. A. F. M. Zaki
Civil Engineering Department, College
   of Engineering
Universiti Tenaga Nasional, Jalan
   IKRAM-UNITEN
Selangor, Malaysia

Ramli M. Zaki
Faculty of Pharmacy and Health
   Sciences
Universiti Kuala Lumpur Royal College
   of Medicine Perak
Perak, Malaysia

M. Y. M. Zuhri
Advanced Engineering Materials
   and Composites Research Centre
   (AEMC), Department of Mechanical
   and Manufacturing Engineering
Universiti Putra Malaysia
Selangor, Malaysia

N. A. Zulkipli
Department of Chemical and
   Environmental Engineering, Faculty
   of Engineering
Universiti Putra Malaysia
Selangor, Malaysia

1 Overviews of Synthetic Nanomaterials, Synthesis Methods, Characteristics, and Recent Progress

Siti Hasnah Kamarudin
UiTM Shah Alam

Mohd Salahuddin Mohd Basri
Universiti Putra Malaysia

Falah Abu, Siti Norasmah Surip, and Syaiful Osman
UiTM Shah Alam

1.1 INTRODUCTION

The insufficiency of non-renewable raw materials, as well as the improper disposal of solid waste in the environment, drives society towards more sustainable materials which honour nature and the environment. Nanotechnology really does have a significant role to play in global sustainability efforts. Nanotechnology has a lot of potential and thus is thought to be a crucial technology for the twenty-first century. It is a linear or enabling technology that, in the medium term, will permeate all industrial sectors. As a matter of fact, nanomaterials have surfaced as a fascinating renewable class of materials with a wide range of practical applications. New materials for protective coating, computer systems, textile materials, cosmetics, packaging, and medicines are just a few examples of product types that can benefit from nanomaterials in terms of function and quality.

Organic and synthetic nanoparticles are classified based on their chemical composition, size, shape, and surface functionalization. Organic nanoparticles are made up of natural or synthetic polymers, lipids, and proteins. They can be designed to have specific properties such as biocompatibility, biodegradability, and drug delivery. Synthetic nanoparticles, on the other hand, are typically made up of inorganic materials such as metals, metal oxides, and carbon-based materials. They can be engineered to have unique properties such as high surface area, conductivity, and magnetic properties. Some examples of synthetic nanoparticles include quantum dots, carbon nanotubes, gold nanoparticles, and silica nanoparticles. The type and category of nanoparticle utilized depends on the desired properties and the specific

DOI: 10.1201/9781003400998-1

1

application in which they will be used. Synthetically produced nanoparticles are important in nanotechnology and are used in a variety of applications. They are a diverse class of materials with lengths ranging from 1 to 100 nm in all three dimensions. Because of their unique properties, synthetic nanoparticles are crucial aspects of nanotechnology. They are manufactured industrially and utilized in products or processes. Depending on the application and the type of product, the nanomaterials used are either in a more or less tightly bound form.

One significant advantage of nanoparticles and nanomaterials is their large surface area-to-volume ratio. At the nanoscale, materials have a much larger surface area per unit mass than the same material in bulk form. This property allows for increased reactivity and interactions with other materials, making them highly useful in applications such as catalysis, sensing, and drug delivery. Additionally, the unique physical, chemical, and biological properties of nanoparticles and nanomaterials can be tailored by controlling their size, shape, and surface functionalization, providing a high degree of flexibility and versatility in their use.

1.2 TYPES OF SYNTHETIC NANOMATERIALS

1.2.1 Carbon Nanotubes and Carbon Nanofibres

Carbon-based materials of synthetic nanomaterials', for example, have numerous applications in industry and science. Carbon nanotubes (CNTs) are composed of carbon atoms arranged in a cylindrical structure with a high aspect ratio (length-to-diameter ratio). They can be either single-walled or multi-walled, depending on the number of concentric tubes. CNTs exhibit excellent mechanical strength, high thermal and electrical conductivity, and unique optical properties, making them suitable for a wide range of applications (Dresselhaus et al., 2001).

Carbon nanomaterials (CNTs) and carbon nanofibres (CNFs) are made of sp^2 carbon atoms with one-dimensional (1D) structures. CNTs are made from graphene nanofoils that have hollow coils of atoms that are arranged in a honeycomb pattern. Single-layer CNTs and multi-layered CNTs have diameters as small as 0.7 and 100 nm, respectively, and lengths that typically range from a few micrometres to several millimetres. The ends may be closed by half-fullerene molecules or may be hollow (Jana et al., 2021). The rolled sheets are referred to as single-walled (SWNTs), double-walled (DWNTs), or multi-walled carbon nanotubes because they can have one, two, or multiple walls (MWNTs). It is common to synthesize carbon precursors by deposition, particularly atomic carbon precursors. Using a laser or an electric arc, carbons are vaporized from graphite and deposited on metal particles. They were recently created using the chemical vapour deposition (CVD) method (Tehrani & Khanbolouki, 2018).

CNTs have unique electronic properties that make them attractive for various electronics applications, such as field-effect transistors (FETs), interconnects, and memory devices. CNTs can be used as high-performance channel materials in FETs due to their excellent electron transport properties. They can also be used as interconnects in integrated circuits due to their high current-carrying capacity and low resistance

(Avouris et al., 2007). Furthermore, they have been studied extensively for their potential applications in energy-related fields, such as energy storage, conversion, and harvesting. They can be used as electrode materials in supercapacitors and batteries due to their high surface area, high electrical conductivity, and fast charge transfer kinetics. CNTs can also be used as catalysts for fuel cells and as light-absorbing materials in solar cells (Kaempgen et al., 2009).

CNTs have shown potential for various biomedical applications due to their unique physical and chemical properties. They can be used as drug delivery vehicles, biosensors, and tissue engineering scaffolds. CNTs can be functionalized with targeting ligands and therapeutic agents to selectively target cancer cells and improve drug delivery efficiency. They can also be used as biosensors for detecting biomolecules and cells due to their high sensitivity and specificity (Kam & Dai, 2005). Moreover, CNTs can be used as reinforcement materials in composites to improve their mechanical and electrical properties. CNTs can be dispersed in polymers, metals, and ceramics to enhance their stiffness, strength, and electrical conductivity. CNT-reinforced composites have potential applications in the aerospace, automotive, and sports industries (Coleman et al., 2006).

Overall, the unique physical and chemical properties of CNTs have led to a wide range of applications in various fields, including electronics, energy, biomedical, and composites. As research in this field continues, more applications are expected to emerge.

Carbon nanofibre is created in a similar way as graphene nanofoils, as well as carbon nanotubes, are. The difference is that instead of regular cylindrical tubes, it is wound into a cone shape (Foong et al., 2020). Furthermore, CNFs have cylindrical nanostructures with various graphene sheet stacking arrangements, such as stacked platelet, ribbon, or herringbone (Serp & Figueiredo, 2009).

The extraordinary mechanical, electrical, thermal, and electrochemical properties of CNF and CNTs have generated a great deal of activity in a majority of science and engineering fields (Vamvakaki et al., 2021). CNFs are cylindrical carbon fibres with diameters ranging from a few nanometres to tens of nanometres and lengths up to several micrometres. They have a graphitic structure and exhibit excellent mechanical properties, such as high strength, stiffness, and toughness. CNFs also have high electrical and thermal conductivity and can be functionalized to modify their surface properties. CNFs have shown potential applications in energy storage, such as supercapacitors and batteries, due to their high surface area and high electrical conductivity. CNFs can be used as electrode materials in supercapacitors and can achieve high specific capacitance and high power density. CNFs can also be used as anode materials in lithium-ion batteries and can achieve high specific capacity and cycling stability (Luo et al., 2013).

CNFs can be used as catalyst supports due to their high surface area and good electrical conductivity. They can be functionalized with various metal or metal oxide nanoparticles to form heterogeneous catalysts for various chemical reactions. CNF-supported catalysts have shown high activity and selectivity in reactions such as hydrogenation, oxidation, and the Fischer-Tropsch synthesis. CNFs can be used as sensing materials due to their high sensitivity and selectivity. They can be functionalized with various sensing agents, such as metal nanoparticles or organic molecules, to detect various analytes, such as gas molecules, biomolecules, and heavy metal ions. CNF-based sensors have shown high sensitivity, selectivity, and stability (Kim et al., 2008). CNFs can be used as reinforcement materials in composites to improve their mechanical and electrical properties. CNFs can be dispersed in polymers, metals, and ceramics to enhance their stiffness, strength, and electrical conductivity. CNF-reinforced composites have potential applications in the aerospace, automotive, and sports industries (Sharma et al., 2016). Overall, CNFs have shown potential applications in various fields, including energy, catalysis, sensing, and composites. As research in this field continues, more applications are expected to emerge.

1.2.2 METAL-BASED NANOPARTICLES

Metal-based nanoparticles can be synthesized using various methods, including chemical reduction, sol-gel, and thermal decomposition. The size, shape, and composition of the nanoparticles can be controlled by adjusting the reaction parameters. Metal-based nanoparticles exhibit unique optical, electrical, and magnetic properties due to their small size and high surface-to-volume ratio (Ijaz et al., 2004). Using either destructive or constructive processes, metal-based nanoparticles are created from metals to nanometric sizes. Almost all metals can be synthesized with nanoparticles (Ijaz et al., 2004; Abdelmoneim et al., 2021). For the synthesis of nanoparticles, materials like cadmium, cobalt, gold, aluminium, lead, copper, zinc, iron, and silver are frequently used (Ijaz et al., 2020; Kumar et al., 2021). Sizes between 10 and 100 nm, high surface-to-volume ratio, pore size, surface charge with density, crystalline structures, spherical shapes, colour, reactivity, and sensitivity are just a few of the distinctive characteristics of nanoparticles (Singh et al., 2021; Ranjith et al., 2019). Metal nanoparticles are produced using metal precursors. Constrained surface plasmon resonance (SPR) gives these nanoparticles their distinctive optoelectric properties (Kankala et al., 2020; Aziz et al., 2019). Cu, Ag, and Au are examples of noble metal and alkali nanoparticles that have a distinct absorbance peak in the photovoltaic electromagnetic spectrum. Shape and pattern metal NP synthesis are essential for today's cutting-edge materials.

Metal-based nanoparticles have been widely used as heterogeneous catalysts due to their high surface area and unique catalytic properties. They can catalyze various chemical reactions, such as oxidation, reduction, and hydrogenation. Metal-based nanoparticles can also be used as enzyme mimics to catalyze biorelevant reactions (Corma & Garcia, 2008). Metal-based nanoparticles have potential applications in biomedicine, such as imaging, drug delivery, and therapy. Metal-based nanoparticles can be functionalized with various biomolecules, such as antibodies and peptides, to target specific cells or tissues. They can also be used as contrast agents in imaging techniques, such as magnetic resonance imaging (MRI) and computed tomography (CT). Metal-based nanoparticles can also be used for photothermal therapy, where they absorb light and convert it into heat to selectively kill cancer cells (Dreaden et al., 2018). Metal-based nanoparticles can be used as sensing materials due to their unique optical, electrical, and magnetic properties. Metal-based nanoparticles can be functionalized with various sensing agents, such as organic dyes and enzymes, to detect various analytes, such as biomolecules and environmental pollutants. Metal-based nanoparticle-based sensors have shown high sensitivity and selectivity (Li et al., 2015). Metal-based nanoparticles have potential applications in energy storage and conversion, such as batteries, fuel cells, and solar cells. Metal-based nanoparticles can be used as catalysts in fuel cells to enhance their efficiency and durability. Metal-based nanoparticles can also be used as photoanodes in solar cells to improve their photoconversion efficiency (Zhang & Xie, 2019). Overall, metal-based nanoparticles have shown potential applications in various fields, including catalysis, biomedicine, sensing, and energy. As research in this field continues, more applications are expected to emerge.

1.2.3 METAL OXIDE NANOPARTICLES

Metal oxide nanoparticles have gained significant attention in the field of nanotechnology due to their unique properties and potential applications in various fields, such as electronics, catalysis, biomedicine, energy, and environmental remediation. According to a review article published by Wang and Chen (2013), metal oxide nanoparticles exhibit a large surface area, high surface energy, and unique electronic and optical properties, making them suitable for applications such as solar cells, gas sensors, catalysis, and drug delivery systems (Wang & Chen, 2013). Another review article stated by Liu et al. (2019) highlights the applications of metal oxide nanoparticles in energy storage and conversion technologies such as lithium-ion batteries, supercapacitors, and solar cells. The article discusses the properties of various metal oxide nanoparticles and their performance in these applications (Liu et al., 2019).

Moreover, metal oxide nanoparticles may be used in biomedical domains. Metal oxide nanoparticles, such as titanium dioxide (TiO_2), zinc oxide (ZnO), and iron oxide (Fe_2O_3), have shown immense potential in various biomedical applications due to their unique properties. These nanoparticles have high surface area-to-volume ratios, which enables them to interact effectively with biological systems. They also possess specific physical, chemical, and magnetic properties that make them useful for different applications. One advantage of using metal oxide nanoparticles in biomedical domains is their biocompatibility and low toxicity. Many metal oxide nanoparticles have been shown to be non-cytotoxic and non-immunogenic, making them suitable for use in medical devices and implants. Furthermore, metal oxide nanoparticles can be surface-functionalized with biomolecules such as proteins, peptides, and antibodies, which allows for specific targeting of cells or tissues. Another advantage of using metal oxide nanoparticles in biomedical applications is their ability to act as contrast agents for imaging. For example, iron oxide nanoparticles can be used as contrast agents in magnetic resonance imaging (MRI) due to their magnetic properties. Titanium dioxide and zinc oxide nanoparticles can also be used for optical imaging due to their ability to absorb and emit light.

Metal oxide nanoparticles have also shown potential for drug delivery applications. These nanoparticles can be functionalized with drugs and other therapeutic molecules, allowing for targeted and controlled release of the drugs. Additionally, some metal oxide nanoparticles have been shown to have intrinsic therapeutic properties, such as antibacterial and anti-inflammatory properties. The use of metal oxide nanoparticles as drug carriers for cancer treatment is proven effective. In environmental remediation, metal oxide nanoparticles have been investigated for their ability to remove pollutants from water and soil. A research article published by Kumar et al. (2017) discusses the use of metal oxide nanoparticles for the removal of organic pollutants from water. The article highlights the advantages and limitations of using metal oxide nanoparticles for this application (Kumar et al., 2017). In conclusion, metal oxide nanoparticles have a wide range of potential applications in nanotechnology, including electronics, energy storage and conversion, biomedicine, and environmental remediation. However, more research is needed to fully understand their properties and potential applications.

1.2.4 CERAMIC NANOPARTICLES

Ceramic nanoparticles are small particles made of ceramic materials, typically with a size range of 1–100 nm. They have unique properties that make them useful in various applications in nanotechnology. Throughout the last few decades, the development of goods and processes utilizing ceramic nanoparticles has resulted in unique and exciting applications of these materials. Ceramic nanoparticles' properties, such as form, particle size distribution, crystal habit, state of agglomeration, or dispersion, are defined during the synthesis process. There are numerous methods for producing ceramic nanoparticles. Li et al. (2005) discussed the various synthesis methods for ceramic nanoparticles, including sol-gel, hydrothermal, and chemical vapour deposition. The authors note that the choice of synthesis method can affect the size, shape, and properties of the resulting nanoparticles (Li et al., 2005). Ceramic nanoparticles can have unique properties due to their small size and high surface area, which can affect their mechanical, thermal, and electrical properties. Various techniques are used to characterize the structure and properties of ceramic nanoparticles, such as X-ray diffraction, transmission electron microscopy, and Fourier transform infrared spectroscopy. These techniques can provide important information about the size, shape, and chemical composition of the nanoparticles (Souza et al., 2016).

Various applications of ceramic nanoparticles include catalysis, energy storage, and biomedical engineering. The unique properties of ceramic nanoparticles make them promising candidates for these applications (Rashid et al., 2018). Ceramic nanoparticles have several advantages, such as high melting points, chemical stability, and optical properties, which make them useful in a wide range of applications. These applications include catalysis, energy storage, and biomedical engineering. However, some disadvantages of ceramic nanoparticles include their potential toxicity and difficulties in their synthesis and processing (Syafiuddin et al., 2017). The potential applications of ceramic nanoparticles are also in biomedical engineering, such as drug delivery and tissue engineering. Ceramic nanoparticles have unique properties, such as their biocompatibility and high drug-loading capacity, which make them promising candidates for these applications (Moreno-Vega et al., 2012).

In addition, are unique properties of nanoceramics, which are ceramic materials with nanoscale grain sizes. Gubicza et al. (2009) noted that nanoceramics have advantages such as increased hardness, strength, and thermal stability, which make them useful in applications such as coatings and composites (Gubicza et al., 2009). The potential applications of ceramic nanoparticles can be applied in energy storage and conversion, such as batteries and fuel cells. Ceramic nanoparticles can improve the performance and stability of these devices due to their unique properties, such as high surface area and chemical stability. However, there are challenges to using ceramic nanoparticles in these applications, such as their high cost and potential toxicity (Jayasinghe et al., 2018). Despite their various advantages, ceramic nanoparticles' potential toxicity and the factors that can affect their toxicity, such as size, shape, and surface chemistry. Zhang and Xie (2019) mentioned more research is needed to fully understand the toxicity of ceramic nanoparticles and to develop safe and effective applications for them (Zhang & Xie, 2019).

1.2.5 SEMICONDUCTOR NANOPARTICLES

Semiconductor nanoparticles, also known as quantum dots, have a wide range of applications in electronics, photonics, and biomedicine. Here are some fundamental concepts of semiconductor nanoparticles, which include quantum confinement, band gap energy, surface passivation, size-dependent optical properties, and colloidal synthesis. The fundamental concepts of semiconductor nanoparticles are closely related to their unique physical and chemical properties, which make them useful in a variety of applications. One of the most important properties of semiconductor nanoparticles is quantum confinement, which arises due to the confinement of electrons and holes in all three dimensions within the nanoparticle. This leads to the quantum size effect, which results in discrete energy levels, band gap widening, and a size-dependent emission spectrum (Klimov, 2003).

The band gap energy of a semiconductor nanoparticle is the energy difference between the valence and conduction bands. It is an important parameter that determines the optical and electronic properties of the nanoparticle, including its absorption and emission spectra. The band gap energy of a nanoparticle can be tuned by changing its size, shape, and composition (Efros & Nesbitt, 2016). Semiconductor nanoparticles have high surface-to-volume ratios, which can lead to surface defects and trap states that degrade their optical and electronic properties. Surface passivation is a process of coating the surface of the nanoparticle with a thin layer of organic or inorganic material to prevent surface defects and improve the stability and performance of the nanoparticle (Chen et al., 2013).

The size-dependent optical properties of semiconductor nanoparticles arise due to the quantum size effect, which results in a shift of the absorption and emission spectra to higher energies as the nanoparticle size decreases. This can lead to unique optical properties such as size-tunable emission, narrow emission linewidths, and quantum yield enhancement (Bawendi et al., 2003). Semiconductor nanoparticles are commonly synthesized using a colloidal approach, which involves the reaction of precursor molecules in a solution to form the nanoparticle. The size and shape of the nanoparticle can be controlled by adjusting the reaction conditions such as temperature, concentration, and reaction time (Murray et al., 1993). Semiconductor nanoparticles are widely used in biological imaging due to their bright and stable emission, tunable size and emission spectra, and low toxicity. They can be used as fluorescent probes for live-cell imaging, multiplexed detection of biomolecules, and in vivo imaging of biological processes (Algar et al., 2011). Semiconductor nanoparticles can be used as colour-converting materials in display technology to improve colour gamut, brightness, and energy efficiency. They can be integrated into LED displays, LCDs, and OLED displays to achieve high colour quality and energy efficiency (Wang & Sun, 2015).

Semiconductor nanoparticles can be used as light-harvesting materials in solar cells to improve the efficiency of solar energy conversion. They can be incorporated into the active layer of solar cells to capture a broader range of solar spectra and reduce recombination losses (Carey et al., 2015). Semiconductor nanoparticles can be used as sensing materials in chemical and biological sensors to detect and quantify analytes. They can be functionalized with specific ligands to selectively bind to target molecules and emit a detectable signal upon binding (Medintz et al., 2005). Semiconductor nanoparticles can be used as electrode materials in energy storage devices such as batteries and supercapacitors to improve their performance. They can be designed with high surface area, high conductivity, and fast charge transfer kinetics to enhance the energy and power densities of the devices. Overall, the applications of semiconductor nanoparticles are diverse and expanding, driven by their unique properties and potential for improving the performance of various devices and systems.

1.3 SYNTHESIS AND PRODUCTION METHODS OF SYNTHETIC NANOMATERIALS

Nanomaterials made from synthetic materials have many uses in various fields of industry and science. Notice that to use the applications, one will need highly dependable synthesis processes that can also create large amounts of high-purity materials. It is something that must always be kept in mind. As a result, it is essential to produce the product on a big scale and ensure its high purity. The synthesis and production of synthetic nanomaterials may be accomplished by a wide variety of methods, including ball milling, electrospinning, lithography, arc-discharge deposition, and pulsed laser ablation in liquid. This section reviews some of the most common methods for synthesizing and producing synthetic nanomaterials.

1.3.1 BALL MILLING

The process of ball milling, which is typically utilized for grinding purposes in the industrial sector, is an example of a mechanochemical technique. Instead of relying on heat as the activation energy, mechanical energy is employed in the synthesis and production processes (Zhuang et al., 2016). Compared to more traditional chemical processes, ball milling may be utilized for either dry or wet material. The operating conditions are straightforward, and the grinding occurs in an enclosed space

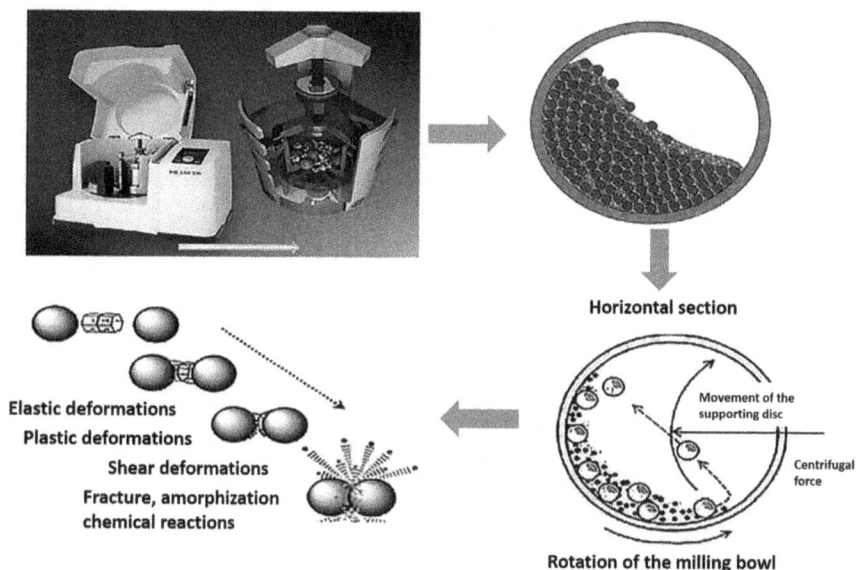

FIGURE 1.1 The working principle of the ball milling process. (Reproduced with permission from Zhuang et al., 2016.)

where no dust is blowing around. In addition, the mill may be pressurized with an inert gas rather than air, and the synthesis procedure may be carried out at room temperature (Salah et al., 2011). Ball milling is widely accepted as a practical and promising synthesis process for small-scale laboratory studies and large-scale commercial production due to its many benefits. The basic concept of the dry ball milling technique is depicted in Figure 1.1. The procedure involves using a high-energy mill with a targeted powder charge and a milling medium (Lin et al., 2017). The powder charge is subjected to the application of the kinetic energy created by the moving balls, which disrupts the chemical bonds between the molecules involved and a consequent reduction in particle size. The milling process causes a number of progressions, including the transfer of mass and energy and the development of mechanical stress, which ultimately results in the lattice structure of the materials being broken (Boldyreva, 2013).

1.3.2 ELECTROSPINNING

Although ultrafine fibres or fibrous structures of various polymers with widths down to a submicron or nanometres may be easily manufactured using this technique, the electrospinning process has garnered more attention in recent years. It is presumably due in part to a booming interest in nanotechnology. By utilizing powerful electrostatic forces, electrospinning is an efficient method for producing long polymer fibres with diameters ranging from micrometres to 100 nm or even a few nanometres. It may be accomplished via electrospinning (Greiner & Wendorff, 2008).

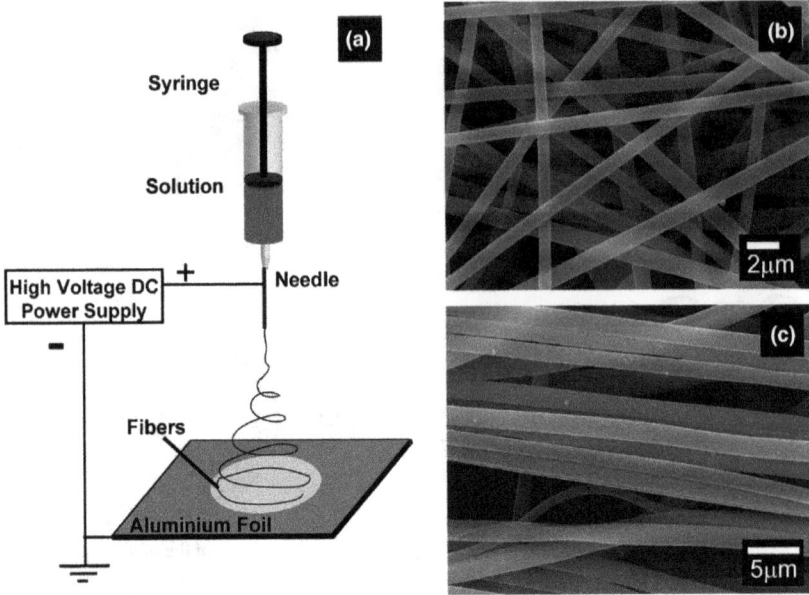

FIGURE 1.2 A graphic showing the electrospinning process used to create polymer nanofibres. (Reproduced with permission from Long et al., 2011.)

A high-voltage supply, a capillary tube fitted with a pipette or needle with a tiny diameter, and a metal collecting screen are the fundamental elements required to carry out the procedure. In the typical electrospinning method, as seen in Figure 1.2, a polymer solution is first forced out of an orifice to form a tiny droplet in the presence of an electric field, and then charged solution jets are forced out of a cone in the next step of the process.

In most cases, the fluid extension takes place first in a uniform manner. After that, the straight flow lines experience a severe whipping and splitting motion due to the fluid's instability and the instability caused by electrically induced bending. The solution jet either evaporates or becomes solidified before it reaches the collecting screen, at which point it is gathered as an interconnected web of very tiny fibres (Deitzel et al., 2001). The spinning solution or melt is put onto one electrode, while the other electrode is linked to the collector. The rest of the time, the collector is only given a grounding. At this point, the spun fibres are typically placed onto a collector as a nonwoven web. The discharged polymer solution jet goes through a process of instability and elongation, which causes the jet to stretch out and become very long and thin. While this is going on, the solvent is evaporating, which leaves a charged polymer fibre behind. In the case of the melt, the solidification process begins when the released jet is carried through the air (Huang et al., 2003). Fabrication of nanofibres that have partial or even good orientation is possible using an electrospinning apparatus that has been upgraded or changed.

1.3.3 LITHOGRAPHY

Nanofabrication can be accomplished using conventional processes already widely used in the manufacturing industry. The two techniques that are utilized the most frequently in conventional lithography are photolithography and particle beam lithography. The more established of the two is photolithography. The concept of photolithography underlies each one of its various application methods. First, an acceptable material is subjected to electromagnetic radiation, which alters the substance's solubility by causing chemical changes in its molecular structure. Next, the material is developed (Figure 1.3a). In the traditional method of lithography, the mask and the resist film are arranged to face the irradiation source in a perpendicular direction. Fabrication of inclined structures is possible using a tilting stage, which tilts the mask and resists film in relation to the beam (Figure 1.3c). UV exposures that are three or four times as slanted along multiple axes can be used to build more complicated three-dimensional objects (Figure 1.3d). As demonstrated in Figures 1.3b and e, it is possible to produce patterning by coating and exposing each successive layer, as well as tapering patterns by rotating and tilting the exposure.

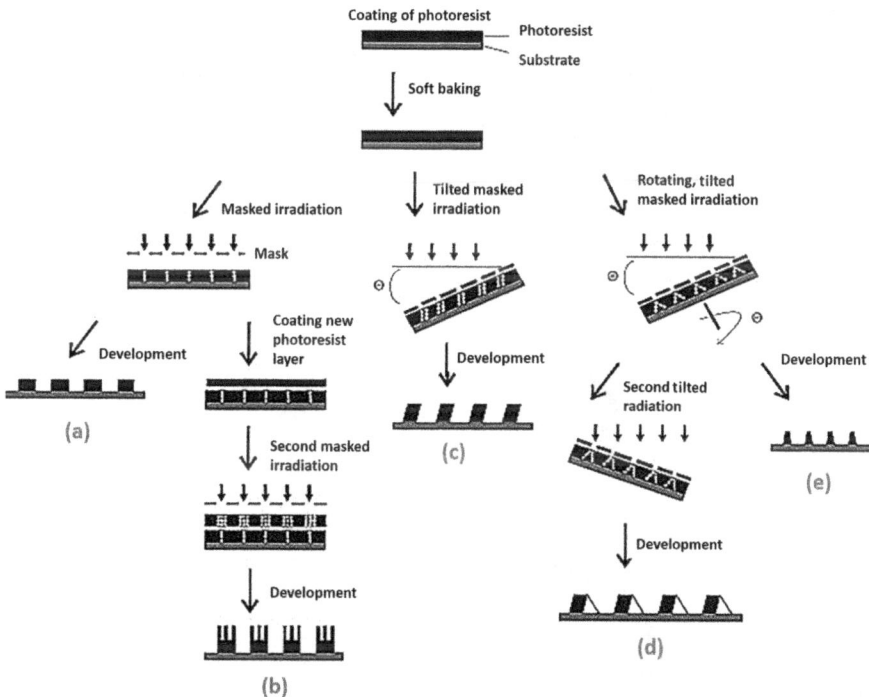

FIGURE 1.3 Methods of photolithography that involve irradiation with a mask and the use of a negative photoresist material: (a) Patterning by single exposure, (b) patterning by layer-by-layer coating and exposure, (c) tilted patterning by single inclined exposure, (d) patterning by double inclined exposure, (e) tapered patterns by rotating tilted exposure. (Reproduced with permission from del Campo & Arzt, 2008.)

By directing an electron beam towards an electron-sensitive resistor, e-beam lithography may disclose precise patterns on a surface. When the electrons strike a layer of resist, such as poly(methyl methacrylate), they produce secondary electrons with low energy, which react as free radicals and radical cations with the resist's surface (PMMA). When an electron beam interacts with a resist, the solubility of the resist changes locally; for example, in the case of polymethylmethacrylate (PMMA), the electrons will produce chain scissions locally, rendering the polymer soluble in a developer (Acikgoz et al., 2011).

1.3.4 Arc-Discharge Deposition

An arc-discharge deposition involves evaporating a graphitic rod, which functions as both a negative cathode and a positive anode when placed a few millimetres apart, with a high current to produce carbon products that deposit on the chamber walls or the cathode substrate (del Campo & Arzt, 2008). Although both alternating current (AC) and direct current (DC) arc-discharge systems are capable of synthesizing CNTs and carbon nanofibres CNFs, DC systems produce larger yields of CNTs that are deposited on the cathode (Deitzel et al., 2001).

As shown in Figure 1.4, an alternating current plasma arc is produced by placing two electrodes in a vacuum and keeping the environment inert. The carbon undergoes sublimation due to the high temperature between the electrodes, which ranges from 3,000°C to 4,000°C. The graphite that has been sublimated is deposited either on the walls of the chamber in which the procedure is carried out or on the walls of the negative electrode (Schulte, 2015). CNTs can be found in these deposits. Doping the electrodes with catalyst particles, such as Ni-Co, Co-Y, or Ni-Y, is necessary to produce single-walled carbon nanotubes (SWCNT) (Journet et al., 1997; Shi et al., 2000).

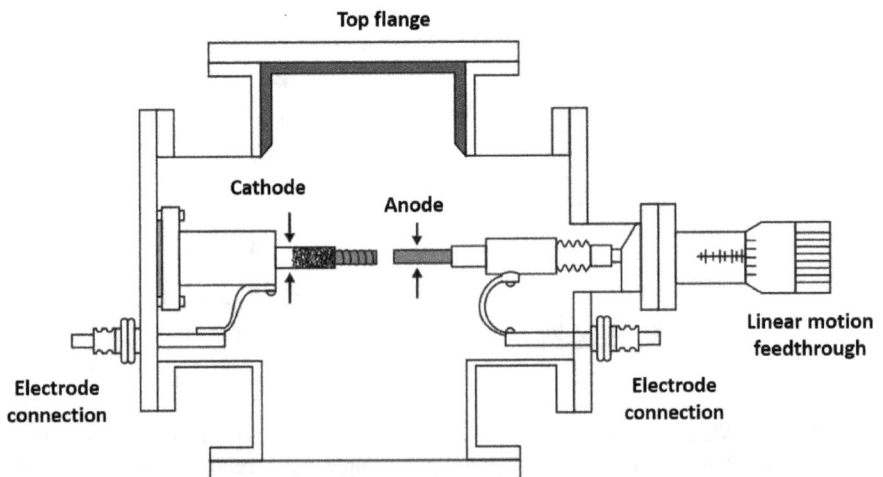

FIGURE 1.4 An abstract representation of the arc-discharge process. (Reproduced with permission from Paradise & Goswami, 2007.)

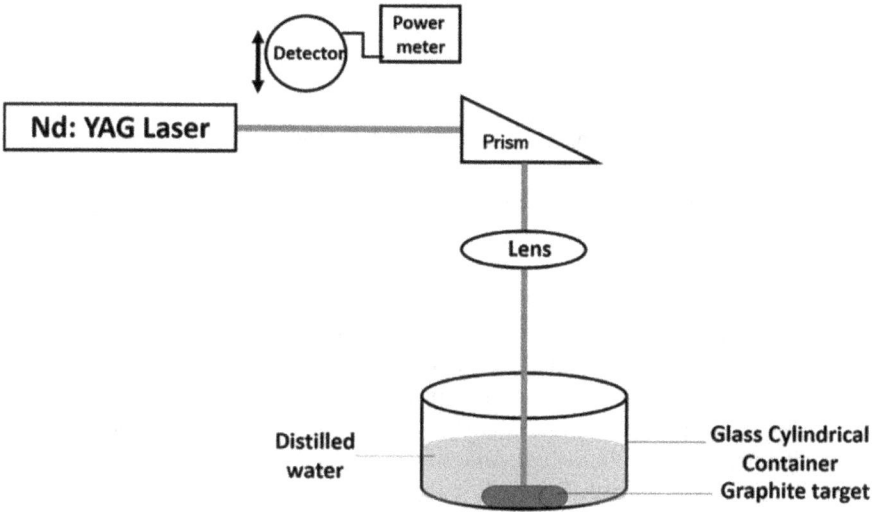

FIGURE 1.5 Experimental setup for pulsed laser ablation in distilled water. (Reproduced with permission from Ganash et al., 2020.)

1.3.5 PULSED LASER ABLATION IN LIQUID (PLAL)

The production of nanoparticles (NP) comprised a wide range of components, including gold (Kim et al., 2009), silver (Valverde-Alva et al., 2015), and copper (Moniri et al., 2017; Miranda et al., 2013) has been accomplished with the help of PLAL. In this method, a laser interacts with the studied sample immersed in a liquid. As a result of this interaction, a laser-induced breakdown occurs, as seen in Figure 1.5. After then, shock waves are produced, and the plasma that was formed expands while losing its heat. Following that, a cavitation bubble will be generated, which will then inflate until being brought to its final state of collapse by the liquid. Consequently, NP will be released into the liquid (Dell'Aglio et al., 2015).

The fact that the production of NP does not need the use of vacuum equipment is the primary benefit of employing PLAL. As a result, the method is not only inexpensive but also straightforward. In addition, once the particles have been manufactured using this technique, they may be readily collected (Thongpool et al., 2012). Nanomaterials created through laser ablation of liquids solids have properties reliant on two factors. The first part is the laser's parameters, wavelength, pulse energy, exposure time, and repetition rate. Second, material characteristics such as the bulk target, solvent, solutes, and system temperature and pressure (Amendola & Meneghetti, 2009).

1.4 CHARACTERISTICS OF SYNTHETIC NANOMATERIALS

Synthetic nanoparticles are materials that are engineered at the nanoscale, usually with precise control over their size, shape, composition, and properties. They are produced through various synthetic methods such as chemical synthesis, physical vapour deposition, or self-assembly. Synthetic nanoparticles have a wide range of applications, including in electronics, energy, medicine, and environmental remediation.

The characteristics of synthetic nanoparticles include:

- Small size: Synthetic nanoparticles have dimensions in the nanoscale, typically 1–100 nm.
- Large surface area: Due to their small size, synthetic nanoparticles have a large surface area compared to their volume.
- Unique physical and chemical properties: Synthetic nanoparticles exhibit unique properties compared to their bulk counterparts, such as increased reactivity, optical, and electronic properties.
- Tailored structure: Synthetic nanoparticles can be engineered to have specific structures, such as a particular shape, size distribution, and composition.
- High surface energy: Synthetic nanoparticles have high surface energy, making them highly reactive and susceptible to interactions with their environment

1.4.1 INORGANIC NANOPARTICLES

Nanoparticles can be divided into two categories: inorganic nanoparticles and carbon-based nanoparticles. These synthetic nanomaterials, which come in a variety of sizes and characteristics, are mass-produced on a massive scale (Zhang et al., 2022). Because of their small size (at least one dimension in the nanoscale, i.e. 1–100 nm), dynamic nature, and various features, the physicochemical characterization of nanoparticles is frequently an analytical challenge (Quevedo et al., 2021).

First inorganic nanoparticles: Inorganic nanoparticles are those without carbon atoms or those formed of metals or metal oxides (Khan et al., 2022).

Metal-based nanoparticles such as aluminium (Al), cadmium (Cd), cobalt (Co), copper (Cu), gold (Au), iron (Fe), lead (Pb), silver (Ag), and zinc (Zn) are commonly employed in nanoparticle synthesis (Khan et al., 2022). Other types of nanoparticles such as palladium (Pd) have been synthesized with high efficiency in their properties (Kurdish, 2021).

Metal oxide-based nanoparticles are positive metallic ions and negative oxygen ions combining to form ionic compounds known as metal oxides. Metal oxide-based nanoparticles are formed when positive metallic ions and negative oxygen ions combine to produce ionic compounds known as metal oxides. The most often synthesized oxides are iron oxide (Fe_2O_3), silicon dioxide (SiO_2), titanium oxide (TiO_2), zinc oxide (ZnO), and aluminium oxide (Al_2O_3). A remarkable example of the possible uses for metal oxide nanoparticles is cerium oxide (CeO_2) nanoparticles. Overall, Zn and ZnO nanoparticles were demonstrated to be more hazardous than $Al_2O_3 \cdot Fe_2O_3$, and SiO_2 nanoparticles (Kurdish, 2021).

1.4.2 CARBON-BASED NANOPARTICLES

Second carbon-based nanoparticles: Strong bonds are produced by carbon when it is joined with other materials. They have found use in varied applications due to their unique shape and variety of features. Carbon may take on a variety of allotropic forms. Examples of allotropes include buckminsterfullerene, graphite, and diamond (Khan et al., 2022).

There are five forms of carbon-based nanoparticles which are; (1) Fullerenes, (2) Graphene and Graphene Oxide (GO), (3) CNTs, (4) CNFs, and (5) Activated Carbon or Charcoal (Khan et al., 2022). Carbon-based nanoparticles are synthetic nanomaterials that can embed inside, pass through, or fuse together membranes without causing long-term harm (Atukorale et al., 2018) and have been used in a variety of research and industry applications (Ghaemi et al., 2018). Graphene and carbon nanotubes are two of the most popular nanomaterials used (Vasilescu et al., 2017).

Fullerenes: Buckminster fullerene (C60) is one of the most well-known and often used fullerenes. Its 60 carbon atoms, each with three bonds, are arranged in a cage-like pattern, giving it a soccer ball-like form. Twenty hexagons and twelve pentagons are utilized in the C60 structure (Khan et al., 2022). According to (Xiong et al., 2018), hydrophobic fullerenes have a greater potential to penetrate.

Graphene and graphene oxide: Graphene is one of the most flexible components (Xiong et al., 2018). Graphene sheets are layered to form graphite. Graphene is a new kind of carbon that consists of a single layer of atoms organized in a honeycomb pattern on a two-dimensional sheet. Due to its ease of synthesis, graphene oxide is a fine replacement for graphene (Khan et al., 2022).

Next are nanotubes: Two varieties of CNTs and CNF were previously discussed in Section 1.2.1. In-depth descriptions of the properties of both nanotubes have been provided by Ghaemi et al. (2018, 2021) and Lambert et al. (2020).

Activated carbon or charcoal: Activated carbon is a highly effective adsorbent. The term "charcoal" refers to a type of carbon that has undergone processing to reduce the volume and drastically reduce the size of the pores. It is frequently employed in the process of removing minerals from water and in the filtration of colours and gases. Most nanoporous structures are formed of carbon (Khan et al., 2022). According to (Xiong et al., 2018), active amorphous carbon particles are one of the most porous materials. Additionally, but in smaller amounts, hydrogen and oxygen exist. Depending on the precursor, production technique, and post-synthesis processing, other elements like nitrogen, sulphur, phosphorus, and inorganic components might also be present (Khan et al., 2022). Despite thorough research, the following Table 1.1 shows additional studies in the field of synthetic nanomaterials.

TABLE 1.1
Various Studies in the Field of Synthetic Nanomaterials

Authors	Descriptions	Imogolite	Laponite	Mesoporous Silica	CNF	CNTs	Structure	Sizes	Surface Areas	Properties
Cavallaro et al. (2018)	Synthetic nanomaterials (imogolite, laponite and mesoporous silica) present the advantages of well-established purity and availability with size features that are finely controlled.	x	x	x						
Cui et al. (2018)	Specific examples include PEG particles prepared through surface-initiated polymerization, mesoporous silica replication via post-infiltration, and particle assembly through metal–phenolic coordination.			x						
Lambert et al. (2020)	Use directed evolution to engineer the optoelectronic properties of DNA-wrapped single-walled carbon nanotube sensors through DNA mutation.					x				
Wang et al. (2018)	CNTs yarns are synthetic nanomaterials of interest for diverse applications. To bring fundamental understanding into the yarn formation process, authors perform mesoscopic scale distinct element method (mDEM) simulations for the stretching of CNTs networks.					x				
Ghaemi et al. (2018)	Synthetic nanoparticles such as carbon nanofibre (CNF) and CNTs with different structures, sizes, and surface areas were produced and analyzed.				x	x				
Ren et al. (2022)	A simple and efficient electrochemical sensor made of a glassy carbon electrode (GCE), modified with MoS_2 nanosheets/carboxylic multi-walled carbon nanotubes (MoS_2/MWCNT-COOH), was used to detect nitrite. The characterization of nanomaterials indicates that MoS_2/MWCNT-COOH has a big surface area (150.3 m^2/g) and abundant pores (pore volume is 0.7085 cm^3/g)					x			x	

(Continued)

TABLE 1.1 (*Continued*)
Various Studies in the Field of Synthetic Nanomaterials

Authors	Descriptions	Imogolite	Laponite	Mesoporous Silica	CNF	CNTs	Structure	Sizes	Surface Areas	Properties
Wang et al. (2021)	Characterize several typical carbon materials, such as carbon nanotubes and graphene flakes. The samples were imaged in a field of view of 32 μm × 32 μm with megapixel sampling, and the imaging time of a specific Raman band was reduced to seconds.					x				
Kim et al. (2020)	A convenient and reliable measurement protocol using dispersed carbon nanotubes and suspended graphene as test specimens is proposed to facilitate the determination of the lateral and axial resolutions of a CRM.					x				
Louro (2018)	Overview of the studies exploring the correlation between physicochemical properties of nanomaterials and their genotoxic effects in human cells, with a focus on the toxicity of two groups of NMs, titanium dioxide nanomaterials and multi-walled carbon nanotubes.					x				
Vasilescu et al. (2017)	Overview of the types of carbon nanomaterials and their composites which have been used to enhance the performance of electrochemical aptasensors. Many challenges remain related to the better characterization of nanomaterials used, clarifying the roles of specific components in multi-component nanocomposites and widening the types of food matrices and analytes tested with the aptasensors.					x				

(*Continued*)

TABLE 1.1 (*Continued*)
Various Studies in the Field of Synthetic Nanomaterials

Authors	Descriptions	Imogolite	Laponite	Mesoporous Silica	CNF	CNTs	Structure	Sizes	Surface Areas	Properties
Quevedo et al. (2021)	Standard gold (Au) colloid suspensions of different sizes (ranging from 5–100 nm) were characterized by UV-Vis at the different institutions to develop an implementable and robust protocol for NM size characterization.							x		
Peters et al. (2021)	Presented the results of a number of these ILCs for the characterization of NMs. Sunscreen lotion sample analysis by laboratories using spICP-MS and TEM/SEM identified and confirmed the TiO_2 particles as being nanoscale and compliant with the EU definition of an NM for regulatory purposes.							x		
Zhang et al. (2022)	Nanomaterials are appealing because of their incredibly small size and large surface area. Apart from the naturally occurring nanomaterials, synthetic nanomaterials are being prepared on large scales with different sizes and properties.							x		x
Lira et al. (2020)	Understanding how synthetic nanomaterials could be exploited in the allosteric regulation of enzymes. The gold nanoparticles (AuNPs) in the ultrasmall size regime could perform as allosteric effectors inducing partial inhibition of thrombin activity.							x		

(*Continued*)

TABLE 1.1 (*Continued*)
Various Studies in the Field of Synthetic Nanomaterials

Authors	Descriptions	Imogolite	Laponite	Mesoporous Silica	CNF	CNTs	Structure	Sizes	Surface Areas	Properties
Atukorale et al. (2018)	Describe structure-function relationships of highly water-soluble gold nanoparticles comprised of a $\sim$1.5–5 nm diameter metal core coated by an amphiphilic organic ligand shell, which exhibits membrane embedding and fusion activity mediated by the surface ligands.							x		
Gao et al. (2022)	Offer a microscopic molecular insight into peptide identity on AuNCs and provide a guideline in customizing nanochaperones via manipulating their nanointerfaces								x	
Xiong et al. (2018)	Discuss the most popular synthetic nanocomponents and their interfacial interactions with select biopolymers, which are currently exploited for the fabrication of novel bionanocomposites, including carbon nanomaterials, mineral nanoparticles, and metallic nanostructures.						x	x		

1.5 MODIFICATION TECHNIQUES FOR
SYNTHETIC NANOMATERIALS

Production of synthetic nanomaterials such as metal, ceramic, and carbon-based nanomaterials is in high demand due to their commercial use in industrial applications such as the automotive, marine, aerospace, and construction industries. Numerous methods have been employed for the modification of nanomaterials for surface compatibility and optimum functionalities. Techniques for surface modification of nanoparticles can be classified into physical modification and chemical modification, depending on the treatment method selected.

1.5.1 PHYSICAL MODIFICATION

Physical modification generally refers to a method of surface modification, including radiation treatment, ultrasonic treatment, and plasma treatment.

1.5.1.1 Radiation Treatment

Radiation technology is a recent method of studying, developing, and applying the laws of physics, chemistry, and biology related to the interaction of radiation with matter. Using these techniques, complex structures such as bimetallic or metal-insulator core-shell structures can be produced. In order to increase compatibility with other materials, such as polymers, active spots are generated on the surface of nanoparticles by high-energy irradiation. Havlik et al. (2018) show an easily scalable method for rapid irradiation of nanomaterials by light ions formed homogeneously in situ by a nuclear reaction. The target nanoparticles are embedded in B_2O_3 and placed in a neutron flux. The captured neutrons generate an isotropic flux of energetic α particles and Li+ ions that uniformly irradiates the surrounding nanoparticles. They produced 70 g of fluorescent nanodiamonds in an approximately 30-minute irradiation session, as well as fluorescent silicon carbide nanoparticles. This radiation technique has increased the production of ion-irradiated nanoparticles, facilitating their use in various applications (Havlik et al., 2018).

1.5.1.2 Ultrasonic Treatment

Ultrasonic treatment is an effective method for size refinement and dispersion of nanomaterials during their synthesis process. By introducing high temperature, high pressure, and microjets generated during ultrasonic cavitation, nanoparticle agglomeration can be effectively prevented, and nano-effect energy can be substantially weakened. Figure 1.6 shows the illustration of particle size with and without ultrasonic treatment.

Ultrasonic treatment shows many advantages, such as high frequency, good directionality and transmissibility, high energy concentration, strong reflectivity, and easy availability (Yang et al., 2021). So far, ultrasonic treatment has been widely used in the synthesis of different kinds of nanomaterials and their composites (Shi et al., 2000).

Yang et al. (2021) reported that ultrasonic treatment could significantly reduce the average particle size of Cu nanoparticles and the size variation. Whereas

ultrasonication

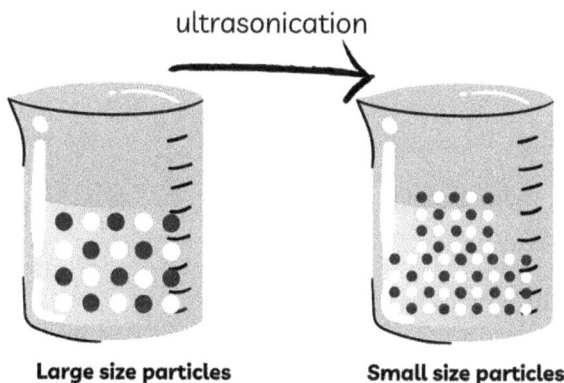

Large size particles   **Small size particles**

FIGURE 1.6   Illustration of particle size with and without ultrasonic treatment.

other studies showed that the size of the particles was maintained and could not be further reduced by increasing the ultrasound power beyond a certain value (Sabnis et al., 2020). Based on the acoustic radiation force and ultrasonic cavitation effect, the mechanism of particle refinement by ultrasound can be qualitatively explained.

1.5.1.3 Plasma Treatment

Plasma is essentially a gas that has been electrified with freely moving electrons in both the negative and positive states. It is a mixture of neutral atoms, atomic ions, electrons, molecular ions, and molecules present in excited and ground states. The positive and negative charges are balanced, thus making the charges electrically neutral. The charged particles present in plasma are responsible for its high electrical conductivity. Since plasma consists of electrons, molecules or neutral gas atoms, positive ions, UV light, and excited gas molecules and atoms, it carries a good amount of internal energy. Plasma treatment is initiated when the gas molecules, ions, and atoms come together and interact with a particular surface.

Plasma treatment has already been efficiently used for the modification of nanomaterials such as nanotubes, nanoparticles, and nanofibres. Plasma treatment is advantageous in that only the properties of a thin layer on the surface of the nanoparticle are changed, while other characteristics, like particle size and distribution, do not change significantly. Besides that, the process does not use any solvent, uses little energy and the whole process is eco-friendly. However, this treatment method needs to be performed under vacuum conditions.

Mathioudaki et al. (2018) reported on the use of low-pressure oxygen plasma treatment to remove carbon contamination from the nanoparticles' surface. Low-pressure plasma polymerization of cyclopropylamine was successfully used to functionalize ZnO, Al_2O_3, and ZrO_2 nanoparticles. Winkler et al. (2014) studied the impact of deposition and plasma cleaning on the morphology of gold (Au) nanoparticles on Si substrates. They found that after O_2 plasma treatment, significant nanoparticle coarsening (Ostwald ripening) was observed for both low and high-density Au deposition.

1.5.2 Chemical Modification

A chemical modification involves changing the structure and state of a nanoparticle by a chemical reaction between the nanoparticle surface and the treatment agent. The surface chemical modification method plays an extremely important role in the surface modification of nanoparticles. The commonly used surface modification methods are the esterification reaction method, the surfactant method, the coupling agent method, and others.

1.5.2.1 Esterification Reaction Method

Esterification is defined as a process of combining an organic acid with an alcohol to form an ester and water. Esterification reactions could be done in three ways, by the combining of acid anhydride and alcohol; acid chloride and alcohol; and carboxylic acid and alcohol. This chemical reaction also resulted in at least one ester product. The reaction of metal oxides with alcohol is an example of the esterification method. The most important aspect of surface modification of nanoparticles by esterification is to change the surface from the original hydrophilic and oleophobic to a lipophilic and hydrophobic surface.

1.5.2.2 Surfactant Modification Method

Surfactants are long-chain molecules that contain both an oleophilic group that has an affinity for oil or organic matter and a hydrophilic group that has an affinity for water or an inorganic substance. The structural characteristics of the surfactant enable it to be applied to the surface modification treatment of nanoparticles, thereby improving the affinity of the nanoparticles with organic substances, and improving compatibility and dispersibility.

Nanomaterials such as zeolites are aluminium silicates characterized by three-dimensional networks. It consists of tetrahedral units of silica and alumina, linked by the sharing of all oxygen atoms, with channels and/or interconnected voids. Zeolites show excellent affinity for metal cations, adsorption, and catalytic properties. In addition, cations of the zeolite structure can be replaced by other metal cations such as cadmium, zinc, copper, nickel, iron, and manganese. With the addition of surfactants, the zeolite surface can be modified to enhance the anion exchange capacity and to an extent the potential to remove cations, anions, and organic compounds. On the other hand, the modification of clay using cationic surfactants changes the negative basal surface charge of the clay and enhances its adsorptive capacity. For example, Jin et al. (2014) reported that HDTMA-Cl kaolinites are able to adsorb up to ten times more Cr(VI) than the natural form by significantly increasing both the specific surface area and the CEC of the clay. Surfactant-modified clays also allow for the simultaneous removal of organic contaminants and heavy metals.

1.5.2.3 Coupling Agent Method

Coupling agents are generally used for surface modification and to promote compatibility between two materials, including nanomaterials. The coupling agent is a chemical substance with an amphoteric structure that is mainly used as a compatibilizer for polymers and composite materials. One side of the molecule can react with various functional groups on the surface of the material to form a strong chemical bond, and another part of the group can undergo some chemical reaction or physical entanglement with the other material, such as an organic polymer. The coupling agent could improve the interfacial interaction between the inorganic substance and the organic substance, thereby greatly improving the performance of the composite material. These compounds may be applied directly to the substrate surface from a diluted solution or may be compounded into the matrix. Commonly used coupling agents in nanocomposites are mainly maleic anhydride and silane. Silane coupling agents play an important role in the preparation of nanocomposites from organic polymers and inorganic nanofillers such as nanoclays, carbon nanotubes, and metal oxides. The general structure of silane coupling agents is (RO)3-Si-R'-X, where X is an organofunctional group, R' is a polymer-compatible or reactable organic group, and R is either ethyl or methyl. The surface modifications of nanomaterials such as metal oxides by silane coupling agents improve the compatibility of their surfaces with a hydrophobic polymer surface. A number of studies have been performed where silane molecules are used for attachment to metal oxide surfaces. Zhao et al. (2012) modified TiO_2 NPs with 3-aminopropyltrimethoxysilane (APTMS) and 3-isocyanatopropyltrimethoxysilane (IPTMS) and investigated the effect of surface modifications on the various properties of TiO_2 NPs, such as grafting efficiency, surface composition, zeta potential, particle size, and photocatalytic activity. Prado et al. (2010) treated the alumina nanoparticles' surfaces with [2-(3,4 epoxycyclohexyl) ethyl]trimethoxysilane and found out the functionalization with this silane coupling agent imparted a hydrophobic character to the alumina samples.

1.6 RECENT PROGRESS AND TRENDS OF SYNTHETIC NANOMATERIALS

Massive quantities of nanomaterials enter our environment and begin interacting with the biota (Malakar et al., 2020). However, it is believed that synthetic NMs harm the ecosystem. The widespread use of synthetic NMs in numerous industries, such as chemistry, engineering, electronics, and medicine, makes them susceptible to release into the atmosphere, various water sources, soil, and landfill debris.

1.6.1 Lifecycle and Toxicity of Synthetic Nanomaterials

Areas such as the industrial, agricultural, and medicinal sectors use nanoparticles with unique features. However, nanomaterial toxicity is linked to some features. The "nano-paradox" raises worries about nanotechnology use and development, making nanomaterial regulation problematic for regulatory authorities. Understanding the effect of nanomaterial characteristics on nano-bio interactions is essential to ensuring optimal nanomaterial regulation (Liu et al., 2022). Recent research, as in Egbuna et al. (2021) has indicated that nanoparticles may cause severe health impacts when ingested, inhaled, or applied to the skin without precaution. The extent of toxicity depends on several factors, such as the type and size of the nanoparticle, its surface area, shape, aspect ratio, surface coating, crystallinity, dissolution, and agglomeration. The primary toxicological processes include the potential for the formation of reactive species, cytotoxicity, genotoxicity, and neurotoxicity.

In sustainable agriculture, the preservation of the environment from pollution is the most significant trade objective, and nanomaterials provide improved control and conservation of plant production inputs (Shang et al., 2019). Nanotechnology offers many benefits, yet its extensive use of nanoparticles may pose health and environmental hazards. Life cycle-based methods and risk assessments should be used to identify potential problems and introduce more environmentally friendly nanotechnologies to protect the environment and human health. Nanoparticles used in applications such as nanobioremediation (Hidangmayum et al., 2022), growth enhancers, biofertilizers (Akhtar et al., 2022) and seed quality improvers (Singh et al., 2021; Acharya et al., 2020) may also be hazardous to the environment and ecosystem. For example, bioremediation nanoparticles can induce plant stress (Rui et al., 2016). Nanomaterials may release high ion concentrations during breakdown or agglomerate afterwards, causing indirect toxicity in the food chain and environment (Maharramov et al., 2019). The promise of nanomaterials inspires a new green revolution that reduces the hazards associated with farming. Nevertheless, there are still significant gaps in our understanding of certain nanomaterials' absorption capacity, allowable limit, and ecotoxicity. Long-term effects should be studied and defined before large-scale commercial environmental uses (Shang et al., 2019).

1.6.2 Current Uses of Synthetic Nanomaterials

Synthetic nanomaterials technology has experienced remarkable growth and development in recent decades, offering a wide range of potential industrial applications. Nanomaterials are now included in numerous goods and processes, and their use is expected to increase dramatically in the coming years. In this section, we will examine the current applications of synthetic nanomaterials in a range of industries and the characteristics that make them eminently suitable for various applications. We will also look at the future prospects for this interesting topic and highlight some of the barriers and opportunities associated with the current use of synthetic nanomaterials.

1.6.2.1 In the Agricultural Sector

New technologies for agricultural food production are needed to feed the growing world population. Growing food insufficiency, dwindling land and water resources, and the lack of sustainable agriculture are significant problems limiting agricultural production. Climate change negatively impacts crop yields and is responsible for declining arable land and freshwater supplies. The agricultural sector must therefore prioritize the goal of balancing the increasing demand for food with limited resources. To prevent or mitigate the severe effects of climate change, current approaches such as the use of new materials in agriculture and the promotion of the intelligent use of natural resources must be adopted.

Nanotechnology is a promising and emerging method for improving crop productivity by administering nanoparticles through seed treatment, foliar spray on plants, nanofertilizers for balanced crop nutrition, nanoherbicides for effective weed control, nanoinsecticides for plant protection, early detection of plant diseases and nutrient deficiencies using diagnostic kits, and nanopheromones. Applications of

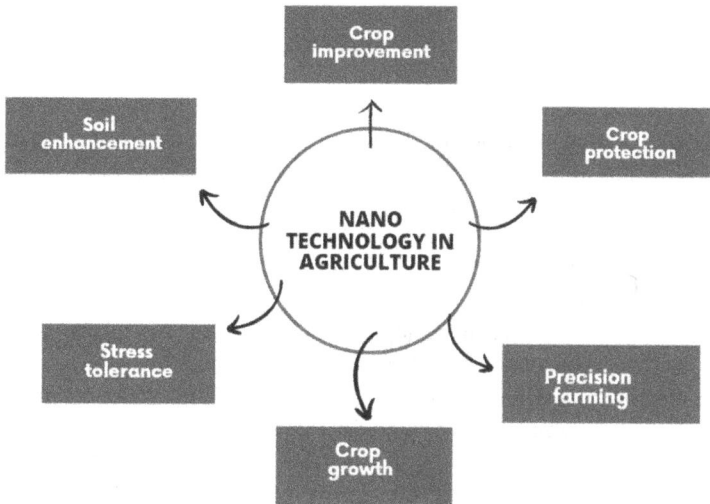

FIGURE 1.7   Applications of nanotechnology in agriculture.

nanotechnology in agriculture are shown in Figure 1.7. In addition, agriculture uses nanoparticles with unique physicochemical and biological features to promote seed germination, which boosts crop yield (Singh et al., 2021).

Numerous projects and expenditures are required to meet the rising demand for agricultural products while employing environmentally responsible strategies (Bora et al., 2022). By utilizing nanotechnology, scientists are able to create particles with unique, adaptable properties for a wide range of applications (Hochella et al., 2019).

Nanomaterials have been researched recently for various purposes, from energy generation to food production. Nanomaterials are designed to impart application-specific characteristics for these uses. For instance, research has been done to improve output and yield by examining how synthetic NMs interact with food and non-food crops. Most synthetic NMs are administered to plants through their leaves or roots (as nanofertilizers) (by foliar spraying). The physiological processes in roots and leaves are distinct. As a result, synthetic NMs are absorbed through these organs using various strategies. Exudates from the roots prevent synthetic NMs from being absorbed, whereas leaves can isolate synthetic NMs in the cell wall after being sprayed on. Nutrients given to the foliage move downward and are then dispersed to other plant components.

1.6.2.2   Seed Germination

Seed germination and seedling development are fundamental to crop output and productivity. Priming seeds is a technique used to improve the quality and tolerance of seeds to environmental conditions. The application of seed priming agent technologies to maintain seeds' seed quality and physiological parameters is an opportunity. Priming seeds with nanomaterials is an innovative technique for increasing germination rates and reducing seed ageing (Itrourwar et al., 2020). The influence of bulk zinc acetate and algal-generated zinc oxide (ZnO) synthetic NMs on the germination and growth of maize seeds has been studied. In comparison to bulk zinc acetate and hydro-primed (control) seeds, priming maize seeds with different concentrations of ZnO synthetic NMs dramatically increased early seed germination percentage. The 100 mg/L ZnO treatment yielded the best germination rate (87%) and root and shoot lengths and widths after 1 week. The influence of different concentrations of silica synthetic NMs (SiO$_2$) (100–300 mg/L) on the growth parameters of cucumber seedlings has been observed. In 10 days, 100% of the silica-primed seeds germinated at 200 and 300 mg/L. Additionally, seed priming shortened the average germination time from 4.1 (without priming) to 2.8 days (200 mg/L).

Previously, the effect of a Zinc nanoparticle (Zn NP) on growth and physiological responses in germinating rice seeds has been investigated. Rice (*Oryza sativa L.*) is soaked in zinc nanoparticle solutions (5–50 mg/L) in an incubator for 1 hour while stirring continuously. Seeds primed with Zn particles possessed longer radicals and plumules, greater radical and plumule fresh and dry weights, and greater radical and plumule relative water contents (RWC). The effect of ZnNP on the germination of growing seedlings of rice was significant. Seed priming can boost seed germination and yield. In multi-locations, low-cost, environmentally friendly, effective seed treatment for high-value speciality crops like watermelon (*Citrullus lanatus*) need to be refined and tested. Triploids, which yield seedless watermelons but have low germination rates, are especially affected. The results of the study conducted by Acharya et al. (2020) demonstrated that seed priming with AgNPs can enhance seed germination, growth, and yield while maintaining the fruit quality of watermelons (*Citrullus lanatus*) at different locations in Texas through an eco-friendly and sustainable nanotechnological approach. Turmeric oil nanoemulsions and silver nanoparticles (AgNPs) from agro-industrial wastes were employed as nanopriming agents for diploid (Riverside) and triploid (Maxima) watermelon seeds. Neutron activation, transmission electron microscopy, and gas chromatography-mass spectrometry indicated the internalization of the nanomaterial. AgNP-treated triploid seeds exhibited a significantly higher seedling emergence rate after 14 days than other treatments. AgNP-treated seeds had higher glucose and fructose contents after 96 hours.

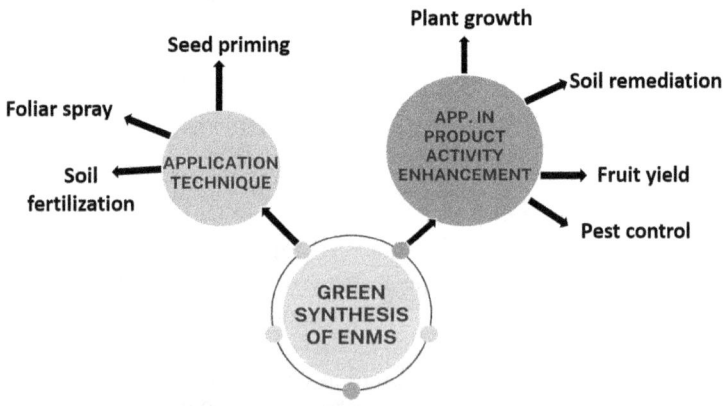

FIGURE 1.8   Application of engineered nanomaterials in agricultural productivity.

1.6.2.3 Plant Disease Control and Fruit Yield Applications

Because many nanomaterials have an inhibitory effect on pathogens that cause plant diseases, they are widely used to control them. Nanomaterials can also improve the yield of fruit crops by meeting their nutrient needs or accelerating photosynthesis (Bora et al., 2022). Nanoparticles have the ability to be directly sprayed on plant seeds, foliage, or roots to protect them from insects, bacteria, fungi, and viruses. Antibacterial and antifungal characteristics of metal nanoparticles such as silver, copper, zinc oxide, and titanium dioxide have been actively studied, as they possess antiviral effects (Worrall et al., 2018).

Tobacco (*Nicotiana benthamiana*) and Turnip mosaic virus (TuMV) were used as model systems to test synthetic NMs for crop growth and viral resistance in a study conducted by (Hao et al., 2018). A 5 mL ENM suspension of 50 or 200 mg/L of two metal-based nanoparticles (NP Fe_2O_3 or TiO_2) or two carbon-based nanomaterials (MWCNTs or C60) was foliar-sprayed onto tobacco leaves daily for 21 days. Fully developed young leaves were inoculated with GFP-tagged TuMV and cultivated for 5 days. Metal- and carbon-based NMs increased shoot biomass by 50%. Results indicated that NMs did not affect cellular integrity; NP Fe_2O_3 and TiO_2 preferentially accumulated in chloroplasts. NMs inhibited viral proliferation as measured by fluorescence intensity on newly emerged leaves. NMs may have suppressed viral infection by decreasing TuMV coat protein levels by 15%–60%. Forty percent phytohormone increases suggest that NMs stimulate plant growth and defence mechanisms.

Given the recent rapid expansion of the oomycete pathogen lineages, the need for sustainable agricultural practices has become critical. The late blight of tomatoes caused by *Phytophthora infestans* was one such devastating disease that recently caused considerable crop losses in India. Joshi et al. (2020) investigated the use of seed priming to create resistance to tomato late blight disease. Tomato plants primed with bioactive mycogenic selenium nanoparticles (SeNPs) demonstrated increased plant growth metrics compared to control plants. SeNPs-treated plants with pathogen infection exhibited a significant 72.9% defence against late blight disease. The primed plants also displayed a significant accumulation of lignin, callose, and hydrogen peroxide in comparison to the control plants. Increased levels of the enzymes lipoxygenase (LOX), phenylalanine lyase (PAL), 1,3-glucanase (GLU), and superoxide dismutase (SOD), which were also reflected in the proper transcriptome profiling of the genes encoding the enzymes, boosted the biochemical defence in primed plants further. The current work demonstrates a well-planned association between resistance and defence responses produced by SeNPs against tomato late blight disease. SeNPs can be used as a nano-biostimulant fungicide to protect tomato plants.

1.6.2.4 Synthetic Nanomaterials as Soil Improvement

Food production struggles for water, energy, nutrients, and land (Tilman et al., 2011; Bandala & Berli, 2019). Agricultural intensification, which consumes more water, energy, fertilizers, and pesticides, can harm water, biodiversity, soil, and ecosystem services. To avoid health and environmental problems, new food production technologies should be tested before they are used. According to FAO (Bandala & Berli, 2019), sustainable agriculture should promote cost-effective food production and ecosystem services. Sustainable land use and management are essential for agricultural development. However, healthy soils are often neglected, leading to soil degradation and loss of production and ecological services (Montanarella et al., 2016). Engineered nanoparticles can improve agricultural soils and promote sustainable agriculture. Synthetic NMs have improved cation exchange capacity, long-lasting nutrient release, and nutrient delivery, which makes them potentially useful for soils. Synthetic NMs have proven useful in soil remediation and the food sector (Hidangmayum et al., 2022).

Nevertheless, synthetic NMs' effects on plant-microbe ecosystems are largely unknown and the release of synthetic NMs into the environment has uncertain effects on the food chain (Gardea-torresdey et al., 2014). Recent research indicates that synthetic NMs may accumulate and/or increase concentrations of the component metal or carbon nanomaterials in the fruits and grains of agricultural crops, have detrimental or beneficial effects on the agronomic traits, yield, and productivity of plants, induce alterations in the nutritional value of food crops, and transfer within trophic levels (Montanarella et al., 2016; Batalha et al., 2012). The application of ENMs to soils can damage soil and human health and raises concerns about the uncontrolled use of ENMs in food production (Bandala & Berli, 2019).

1.6.2.5 Synthetic Nanomaterials on Plant Growth, Photosynthesis, Water Uptake, and Mineral Nutrition Applications

Nanomaterials can influence plant growth by increasing nutrient uptake, increasing capacity, increasing resistance to biotic and abiotic stressors, stimulating nitrogen metabolism, osmotic protection, and pathogen defence. In terms of nutrient availability to plants, traditional soil nutrient treatments have several drawbacks. Foliar spraying is therefore the most effective means of compensating for nutrient deficiencies while increasing plant production and quality.

Reducing the amount of fertilizer applied to the soil by using foliar spraying decreases environmental contamination and boosts nutrient utilization efficacy. Biologically engineered ZnO synthetic NMs have been employed to boost crop plant biomass at a minimal cost. In growth stimulation tests, the effect of 25–200 mg/L ZnO synthetic NMs and phosphorus on cotton seeds was investigated. The concentration of (200 mg/L) zinc oxide synthetic NMs was directly proportional to root and shoot lengths, total biomass, and the growth tolerance index. These results indicate that bioengineered ZnONPs interact with meristematic cells, triggering biochemical pathways conducive to the accumulation of biomass (Venkatachalam et al., 2016).

Cadmium (Cd) is regarded as one of the most important factors limiting wheat output due to its significant accumulation, toxicity, and non-essentiality in plants. The Cd-induced overproduction of reactive oxygen species (ROS) generated oxidative stress in wheat and significantly impacted the plant's defences. The effects of seed priming with zinc oxide (ZnO) and iron (Fe) nanoparticles (NPs) on wheat (*Triticum aestivum*) growth and cadmium (Cd) accumulation were studied by Rizwan et al. (2019). Throughout the experiment, plants were grown to maturity under natural conditions with moisture contents of 60%–70% of the total soil water holding capacity. Plant height, spike length, and dry weights of shoots, roots, spikes, and grains were all increased by NPs, especially at higher rates. The results showed that NPs had a positive effect on wheat photosynthesis when compared to the control. The NPs reduced electrolyte leakage as well as superoxide dismutase and peroxidase activities in Cd-stressed wheat leaves. The application of NPs significantly reduced Cd concentrations in roots, shoots, and grains. When the seeds were treated with higher NPs concentrations, the Cd content in the grains was lower than the threshold level of Cd (0.2 mg/kg) for cereals. ZnO NPs increased Zn concentrations, while Fe NPs increased Fe concentrations in roots, shoots, and grains. Overall, NPs play a significant role in increasing biomass, nutrients, and decreasing Cd toxicity in wheat.

Sun et al. (2020) studied the physiological and molecular processes of ZnO NP-mediated tomato plant growth. ZnO NPs (20 and 100 mg/L) foliar spraying increased chlorophyll and photosystem II activity in tomatoes. Comparative transcriptome and metabolome analyses showed that ZnO NPs increased tomato genes involved in nutrient element transport, carbon/nitrogen metabolism, and secondary metabolism. Effects of ZnO NPs on tomato plants in response to Fe deficiency after foliar spraying enhanced iron (Fe) accumulation in tomato leaves by 12.2%. Tomato foliar spraying with ZnO NPs significantly enhanced Fe deficiency tolerance. Physiological analysis showed that ZnO NPs reduced Fe deficiency-induced oxidative damage and increased tomato metal nutritional element content. Transcriptomic and metabolomic analyses showed that foliar spraying with ZnO NPs increased the expression of genes encoding antioxidative enzymes, transporters, and carbon/nitrogen metabolism and secondary metabolism enzymes or regulators, improving Fe-deficient tomato plant antioxidant, sugar, and amino acid levels.

1.6.2.6   Applications of Nanomaterials in Drilling Fluid

Recently, the oil and gas industry has shown keen interest in applying nanomaterials to drilling fluids (Hajiabadi et al., 2020). Significant success has been achieved with nanomaterials ranging in size from 1 to 100 nm, with notable studies demonstrating improved rheological performance of drilling fluids (Ma et al., 2021; Hajiabadi et al., 2020).

Due to its ability to produce tiny films, drilling fluid based on nanomaterials can greatly minimize friction between the drill pipe and the borehole (Ikram et al., 2021). Small spherical nanoparticles between the wall of boreholes and the drill pipe may further facilitate drilling by acting as lubricants in addition to generating thin lubricating layers between the wall and pipe (Ikram et al., 2021). Drill pipe movement is made easier by nanoscale ball-bearing surfaces. It is a possibility that adding nanoparticles will enhance WBDF's lubricating and rheological capabilities.

Carbon is the most adaptable substance that has ever existed. Its ability to combine with other compounds and offer variety in its structural forms has led to extraordinary progress in industrial applications (Ajdary et al., 2021). The synthesis of low-pressure diamonds, fullerene molecules, and carbon fibres has been the three most important breakthroughs in carbon research during the past 30 years (Ahmad et al., 2017; Ikram et al., 2021). Carbon-based nanoparticles have attracted the most interest among all nanomaterials for increasing Enhanced Oil Recovery (EOR) efficiency (Ikram et al., 2021; Betancur et al., 2020). The use of carbon-based nanomaterials in place of the commonly used catalyst has not been the subject of much research. The simultaneous

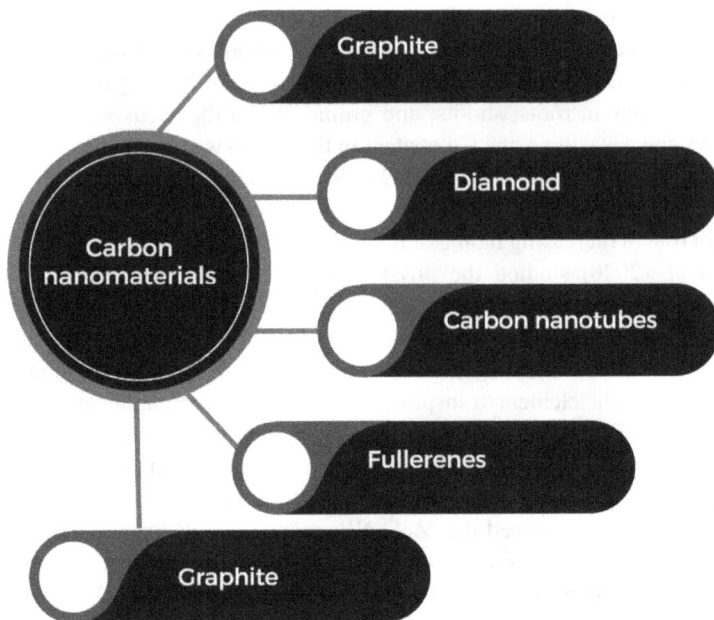

FIGURE 1.9   Type of carbon nanomaterials.

use of a surfactant blend and in situ, heavy oil upgrading has shown significant improvement in all aspects. The use of carbon nanotubes and graphene in EOR is the subject of further research, the majority of which has produced encouraging results (Betancur et al., 2020). Examples of carbon nanomaterials are shown in Figure 1.9.

Betancur et al. (2020) evaluated the effect of the simultaneous use of a surfactant mixture and magnetic iron core-carbon shell nanoparticles on oil recovery via a microfluidic study based on rock-on-a-chip technology. The impact of three injection rates corresponding to 0.1, 1, and 10 ft/day was investigated. For all injection rates, the oil recovery decreased in the following order: nanoparticle-surfactant flooding > surfactant flooding > waterflooding.

1.6.2.7   Synthetic Nanomaterials in Energy and Environmental Applications

The rapid growth of the world's population has led to a significant increase in energy consumption and environmental impact. New materials with exciting physical and chemical properties offer opportunities to address these problems. As a result of increasing energy consumption, the world's traditional energy resources are depleting at an alarming rate. Renewable energy is the best alternative to conventional energy. However, the main problem with this type of energy is that it is not constantly available. Therefore, it is important to store this form of energy when it is readily available so that it can be used when needed (Varshney et al., 2020).

In a study conducted by Xu and Wang (2019), the design and synthesis of hierarchical core/shell $MgCo_2O_4@MnO_2$ nanowall arrays on Ni-foam using a simple two-step hydrothermal process. The electrochemical measurements demonstrate that these composites containing MnO_2 provide demonstrably superior super capacitive performance than the $MgCoO_4$ electrode material. During the Faradic reaction, the nanowall structure provides additional active sites and charge transfer. Excellent electrochemical performance is exhibited by the $MgCo_2O_4$@MnO_2 nanowall (852.5 F g^{-1} at 1 A g^{-1}). The asymmetric supercapacitor is constructed of the $MgCo_2O_4$@MnO_2 nanowall and the activated carbon. The $MgCo_2O_4$@MnO_2 nanowall displays remarkable supercapacitive performance and has a huge potential for additional research and application in the asymmetric supercapacitor devices sector.

Li-ion batteries are now regarded as the most efficient energy storage option for portable devices. Nanoparticles of metal oxide have been proposed as anode materials. Li doping and usage of tin and titanium oxide mixed phases and graphene oxide composites have been studied. The SnO_2 and TiO_2 nanoparticles employed in this study were produced by distinct methods. SnO_2 is a promising anode material with a high energy density in theory. The anatase phase exhibits the best hydrolysis-obtained titanium oxide results but with lower capacity (125 mAh) values.

Transition metal oxides could have larger specific capacities than carbon-based anodes, resulting in a higher energy density compared to commercial Li-ion batteries. However, the performance of the batteries could be affected by a number of characteristics that are dependent on the processing of the electrode materials, which results in variable surface qualities, sizes, or crystalline phases. In a study conducted by Prado et al. (2020), a comparison of tin and titanium oxide nanoparticles synthesized by different methods, undoped or Li-doped, used as single components or in a mixed ratio, or alternatively forming a composite with graphene oxide, demonstrates an increase in capacity with Li doping and improved cyclability for mixed phases and composite anodes.

SnO_2, NiO, and SnO_2/NiO nanocomposites were produced at low temperatures utilizing a modified sol-gel technique and ultrasonication in a study conducted by Varshney et al. (2020). Various procedures for characterization were used to investigate the attributes of prepared samples. These exceptional electrochemical characteristics suggest that the SnO_2/NiO nanocomposite can be employed as an electrode material for supercapacitors with a high energy density. Electrochemical measurements revealed that the produced SnO_2–NiO nanocomposite electrode displayed excellent capacitive behaviour with a maximum specific capacitance of 464.67 F/g at a scan rate of 5 mV/s. These results indicate that the produced SnO_2–NiO nanocomposites are appropriate contenders and more desirable electrode materials for commercial supercapacitors applications.

Portable gadgets, electric vehicles, and grid-scale energy storage are examples of the widespread use of renewable energy storage in the developed world. Li-ion batteries, which achieve the highest massive/volumetric energy density of all existing technologies, have garnered the most interest in both academics and industry. To achieve Li-ion batteries with a higher energy density and power density, however, novel materials and architectures with a high specific capacity and high-rate performance are required.

Oil drilling, accidents involving oil vessels, procedures, and runoffs from offshore oil exploration and production can contaminate the shoreline and offshore waters. These oil spills have a negative influence on human health, vegetation, and animals. Using chemical and biological processes, oil can be extracted or destroyed on-site. These techniques consist of oil booms, dispersants, skimmers, and sorbents. Most of these approaches are expensive. In a study conducted by Shokry et al. (2020), they have developed a nano-magnetic material from the roots of water hyacinth that can absorb oil spills and help in oil spill prevention. Magnetic sorbent material derived from water hyacinth is used to clean up wastewater oil spills. The chemical and thermal treatment conditions of water hyacinth suggested that the submerged root segment has superior oil adsorption than the shoots segment. Nano-magnetic activated carbon hybrid material (NMAC) had the most ordered crystalline structure, porosity, and thermal stability of the two raw water hyacinth segments (shoots and roots). NMAC's completely saturated moment per unit mass was 17.2587 emu/g. Solution temperature and NMAC dose increase oil removal.

Raj et al. (2020) reported the biogenic synthesis of AgNPs employing Terminalia Arjuna leaf extract and its efficacy towards catalytic degradation. The effect of different pH (3–9) on the reduction of $AgNO_3$ into AgNPs was assessed by UV-visible spectroscopic analysis Fe-SEM and TEM was carried out to examine the morphology of particle surface and size of plant-based produced AgN Ps. The catalytic potential was evaluated by degrading different carcinogenic organic dyes in the presence of $NaBH_4$. This investigation has demonstrated an effective, viable, and reproducible approach for the biosynthesis of environmentally safe, inexpensive, and long-lasting AgNPs and their application as potent catalysts for the breakdown of hazardous dyes.

Previously, Mustapha et al. (2020), comparison between kaolin and kaolin/ZnO nanocomposites were conducted in terms of the adsorption of Cr(VI), Fe(III), COD, BOD and chloride in tannery effluents. Sol-gel and wet impregnation processes were used to prepare ZnO nanoparticles and kaolin/ZnO nanocomposites. Due to their larger surface area, the kaolin/ZnO nanocomposites adsorb better than kaolin. The Jovanovic isothermal model showed the highest correlation (R2 > 0.99) for both nanoadsorbents, indicating adsorption on monolayers and heterogeneous surfaces. The Weber Morris intraparticle diffusion model and Boyd diagram indicated that the metal ions in the tannery wastewater adsorbed on the nanoadsorbents by intraparticle and film diffusion. Enthalpy modification showed that the adsorption of metal ions and other parameters was possible, spontaneous and endothermic. Even after six applications, the ZnO/clay nanocomposites were recyclable and reusable. The results demonstrate the practical usability of kaolin/ZnO nanocomposites for the adsorption of contaminants from tannery wastewater, making the adsorbent a suitable choice for water and wastewater treatment.

1.6.2.8 Other Applications of Synthetic Nanomaterials

Synthetic nanomaterials also have the potential to revolutionize many industries and contribute to the Fourth Industrial Revolution (Industry 4.0). Table 1.2 displays numerous various other applications of synthetic nanomaterials towards approaching Industry 4.0.

Overall, synthetic nanomaterials have great potential to contribute to Industry 4.0 in several areas, including advanced manufacturing, energy storage, sensors and IoT, biomedical applications, and environmental applications. As research continues in this area, more applications are expected to be added. In addition, synthetic nanomaterials have the potential to contribute to several Sustainable Development Goals (SDGs) UN, including SDGs 2, 3, 7, and 9. SDG 2: Zero Hunger shows how synthetic nanomaterials can be used in agriculture to increase crop yields and food production. For example, nanofertilizers can improve the nutrient uptake and water retention

TABLE 1.2

Numerous Applications of Synthetic Nanomaterials towards Approaching Industry 4.0

Industry	Applications	References
Advanced manufacturing	The ability of additive manufacturing to produce complex patterns has generated interest and the potential to revolutionize the market. Because of their desirable properties, nanomaterials are being used in more and more products. Additive manufacturing can be used to create a variety of materials, including sensors, electronics, solar cells and scaffolds. The inclusion of nanoparticles in additive manufacturing is expected to increase the availability and affordability of cutting-edge technologies and materials for rural areas, potentially transforming the healthcare infrastructure. Nanomaterials can be used in advanced manufacturing processes, such as additive manufacturing (3D printing), to improve product quality, reduce costs, and increase production efficiency. For example, nanomaterials can be used as additives in polymer-based 3D printing to enhance the mechanical properties and printability of the final product	Challagulla et al. (2020) and Khan et al. (2020)
Sensors and the Internet of Things (IoT)	The Internet of Things (IoT) is a system of interconnected physical objects equipped with sensors and other technologies that enable the exchange of relevant data over the Internet. The IoT network connects not only people but also objects in their environment. The number of IoT devices is expected to grow to 75 billion by 2025 and 125 billion by 2030, enabling industries and governments to improve manufacturing processes and transform communities into smart cities. Key components of the IoT include sensors and devices, network connectivity, data storage and processing, user interfaces, and security, which can be enhanced by nanotechnologies. Nanomaterials can be used to create smaller, more sensitive sensors that can sense a wide range of parameters. Nanomaterials improve sensor and IoT device sensitivity, selectivity, and reliability. Nanomaterials can detect contaminants in gas sensors with excellent sensitivity and selectivity. IoT devices can harvest energy, store data, and communicate using nanomaterials.	Michael Berger (2023) and Moinudeen et al. (2017)
Biomedical applications	Nanomaterials can be used in biomedical applications, such as drug delivery, tissue engineering, and biosensing. For example, nanomaterials can be used as drug carriers to improve drug efficacy and reduce side effects. Nanomaterials can also be used as scaffolds in tissue engineering to promote tissue regeneration. In recent years, the use of nanomaterials in biomedical applications has garnered considerable interest because of their unique features and prospective uses in a variety of sectors. Using nanoparticles for targeted drug delivery is one of the most promising fields of study. It is possible to create nanoparticles to carry therapeutic compounds and selectively distribute them to specific places in the body, thereby minimizing the risk of adverse effects and enhancing medicinal efficacy.	McNamara and Tofail (2017) and Smith and Gambhir (2017)

capacity of plants, leading to higher yields and more sustainable agriculture (Usman et al., 2020). In addition, synthetic nanomaterials can be used in medicine to improve drug delivery and develop new therapeutics, which is in line with SDG 3: Good health and well-being. For example, nanoparticles can be tailored to specific cells or tissues in the body, leading to more effective and targeted treatment of diseases such as cancer (Oroojalian et al., 2020).

Moreover, synthetic nanomaterials can be employed to increase energy storage and conversion under SDG 7: Affordable and Clean Energy. Nanomaterials, for example, can be utilized in batteries and supercapacitors to boost energy density and longevity, leading to more effective and sustainable energy storage (Ahmad et al., 2022). Also, synthetic nanomaterials have the potential to revolutionize many industries, from electronics and telecommunications to transportation and construction in accordance with SDG 9: Industry, Innovation and Infrastructure. Nanomaterials, for example, can be utilized to make stronger and more durable materials, enhance energy efficiency, and generate new products and uses.

In general, the use of synthetic nanomaterials could enable more sustainable and efficient solutions to global problems that could contribute to many of the SDGs. Although nanotechnology has great potential, it must be developed and used carefully to avoid negative impacts on human and environmental health.

1.7 CONCLUSION

In summary, the development of synthetic nanomaterials represents a critical step forward for sustainable materials and global sustainability efforts. Given the scarcity of non-renewable raw materials and the environmental impact of waste disposal, society has an urgent need for more environmentally friendly materials that respect nature. Nanotechnology, with its enormous potential and versatility, has emerged as a promising solution to this challenge. Synthetic nanoparticles, with their unique properties and multiple applications, offer significant benefits in various industries, including protective coatings, computer systems, textiles, cosmetics, packaging, and pharmaceuticals. By using synthetic nanoparticles in these products, companies can improve their function and quality while reducing environmental impact. Although challenges remain in areas such as toxicity and scalability, synthetic nanomaterials remain a critical aspect of nanotechnology that will continue to permeate all industrial sectors in the medium term. Overall, continued research and development of synthetic nanomaterials is essential to promote sustainable materials and create a more sustainable future for all.

REFERENCES

Abdelmoneim, H. E. M., Wassel, M. A., Elfeky, A. S., Bendary, S. H., Awad, M. A., Salem, S. S., & Mahmoud, S. A. (2021). Multiple applications of CdS/TiO2 nanocomposites synthesized via microwave-assisted sol-gel. *Journal of Cluster Science*, *4*(33), 1119–1128. DOI: 10.1007/s10876-021-02041-4.

Acharya, P., Jayaprakasha, G. K., Crosby, K. M., Jifon, J. L., & Patil, B. S. (2020). Nanoparticle-mediated seed priming improves germination, growth, yield, and quality of watermelons (*Citrullus lanatus*) at multi-locations in texas. *Scientific Reports*, *10*(5037), 1–16. DOI: 10.1038/s41598-020-61696-7.

Acikgoz, C., Hempenius, M. A., Huskens, J., & Vancso, G. J. (2011). Polymers in conventional and alternative lithography for the fabrication of nanostructures. *European Polymer Journal, 47*(11), 2033–2052. DOI: 10.1016/j.eurpolymj.2011.07.025.

Ahmad, I., Ahmad, W., Qadir, S., & Ahmad, T. (2017). Synthesis and characterization of molecular imprinted nanomaterials for the removal of heavy metals from water. *Journal of Materials Research and Technology, 7*(3), 270–282. DOI: 10.1016/j.jmrt.2017.04.010.

Ahmad, T., Zhu, H., Zhang, D., Tariq, R., Bassam, A., Ullah, F., AlGhamdi, Ah., & Alshamrani, S. S. (2022). Energetics systems and artificial intelligence: Applications of industry 4.0. *Energy Reports, 4*(8), 334–361. DOI: 10.1016/j.egyr.2021.11.256

Ajdary, R., Tardy, B. L., Mattos, B. D., Bai, L., & Rojas, O. J. (2021). Plant nanomaterials and inspiration from nature: Water interactions and hierarchically structured hydrogels. *Advanced Materials*, 2001085, 33, 1–31. DOI: 10.1002/adma.202001085.

Akhtar, N., Ilyas, N., Meraj, T. A., Pour-aboughadareh, A., Sayyed, R. Z., Mashwani, Z., & Poczai, P. (2022). Improvement of plant responses by nanobiofertilizer: A step towards sustainable agriculture. *Nanomaterials, 12*(6), 965. DOI: 10.3390/nano12060965.

Algar, W. R., Susumu, K., Delehanty, J. B., Medintz, I. L., & Gates, R. (2011). Semiconductor quantum dots in bioanalysis: Crossing the valley. *Analytical Chemistry, 83*, 8826–8837.

Amendola, V., & Meneghetti, M. (2009). Laser ablation synthesis in solution and size manipulation of noble metal nanoparticles. *Physical Chemistry Chemical Physics, 11*(20), 3805–3821. DOI: 10.1039/b900654k.

Atukorale, P. U., Guven, Z. P., Bekdemir, A., Carney, R. P., Lehn, R. C. Van, Yun, D. S., Silva, P. H. J., Demurtas, D., Yang, Y., & Alexander-katz, A. (2018). Structure – property relationships of amphiphilic nanoparticles that penetrate or fuse lipid membranes. *Bioconjugate Chemistry, 29*(4), 1131–1140. DOI: 10.1021/acs.bioconjchem.7b00777.

Avouris, P., Chen, Z., & Perebeinos, V. (2007). Carbon based electronics. *Nature Nanotechnology, 2*, 605–615.

Aziz, N., Faraz, M., Sherwani, M. A., Fatma, T., & Prasad, R. (2019). Illuminating the anticancerous efficacy of a new fungal chassis for silver nanoparticle synthesis. *Frontiers in Chemistry, 7*(February), 1–11. DOI: 10.3389/fchem.2019.00065.

Bandala, E. R., & Berli, M. (2019). Engineered nanomaterials (ENMs) and their role at the nexus of food, energy, and water. *Materials Science for Energy Technologies, 2*(1), 29–40. DOI: 10.1016/j.mset.2018.09.004.

Batalha, L. A. R., Colodette, J. L., Gomide, J. L., Barbosa, L. C. A., Maltha, C. R. A., & Gomes, F. J. B. (2012). Dissolving pulp production from bamboo. *Bioresources, 7*, 640–651.

Bawendi, M., Steigerwald, M., & Brus, L. (2003). The quantum mechanics of larger semiconductor clusters ("quantum dots"). *Annual Reviews Physical Chemistry*, November, *41*(1), 477–496. DOI: 10.1146/annurev.pc.41.100190.002401.

Berger, M. (2023). *Nanotechnology and the Internet of Things: Boosting Efficiency and Capability*. https://www.nanowerk.com/spotlight/spotid=62112.php.

Betancur, S., Olmos, C. M., Pérez, M., Lerner, B., Franco, C. A., Riazi, M., Gallego, J., Carrasco-Marín, F., & Cortés, F. B. (2020). A microfluidic study to investigate the effect of magnetic iron core-carbon shell nanoparticles on displacement mechanisms of crude oil for chemical enhanced oil recovery. *Journal of Petroleum Science and Engineering, 184*, 106589. DOI: 10.1016/j.petrol.2019.106589.

Boldyreva, E. (2013). Mechanochemistry of inorganic and organic systems: What is similar, what is different? *Chemical Society Reviews, 42*(18), 7719–7738. DOI: 10.1039/c3cs60052a.

Bora, K. A., Hashmi, S., Zul, F., & Siddique, K. H. M. (2022). Recent progress in bio-mediated synthesis and applications of engineered nanomaterials for sustainable agriculture. *Frontiers in Plant Science*, October, 1–21. DOI: 10.3389/fpls.2022.999505.

Carey, G. H., Abdelhady, A. L., Ning, Z., Thon, S. M., Bakr, O. M., & Sargent, E. H. (2015). Colloidal quantum dot solar cells. *Chemical Reviews, 115*(23), 12732–12763. DOI: 10.1021/acs.chemrev.5b00063.

Cavallaro, G., Lazzara, G., & Fakhrullin, R. (2018). Mesoporous inorganic nanoscale particles for drug adsorption and controlled release. *Therapeutic Delivery*, *9*(4), 287–301. DOI:10.4155/tde-2017-0120.

Challagulla, N. V., Rohatgi, V., Sharma, D., & Kumar, R. (2020). Recent developments of nanomaterial applications in additive manufacturing: A brief review. *Current Opinion in Chemical Engineering*, *28*, 75–82. DOI: 10.1016/j.coche.2020.03.003.

Chen, O., Zhao, J., Chauhan, V. P., Cui, J., Wong, C., & Harris, D. K. (2013). Compact high-quality CdSe/CdS core/shell nanocrystals with narrow emission linewidths and suppressed blinking. In *MIT Libraries*, *12*(5), 445–451. DOI: 10.1038/nmat3539

Coleman, J., Khan, U., Blau, W. J., & Gun'ko, Y. K. (2006). Small but strong: A review of the mechanical properties of carbon nanotube-polymer composites small but strong: A review of the mechanical properties of carbon nanotube – Polymer composites. *Carbon*, December 2016, *44*(9), 1624–1652. DOI: 10.1016/j.carbon.2006.02.038.

Corma, A., & Garcia, H. (2008). Supported gold nanoparticles as catalysts for organic reactions. *Chemical Society Reviews*, *37*(9), 2096–2126. DOI: 10.1039/b707314n.

Cui, J., Ju, Y., & Caruso, F. (2018). Nanoengineering of poly(ethylene glycol) particles for stealth and targeting. *Langmuir*, *34*(37), 10817–10827. DOI: 10.1021/acs.langmuir.8b02117.

Deitzel, J. M., Kleinmeyer, J. D., Hirvonen, J. K., & Beck Tan, N. C. (2001). Controlled deposition of electrospun poly(ethylene oxide) fibers. *Polymer*, *42*(19), 8163–8170. DOI: 10.1016/S0032-3861(01)00336-6.

del Campo, A., & Arzt, E. (2008). Fabrication approaches for generating complex micro- and nanopatterns on polymeric surfaces. *Chemical Reviews*, *108*(3), 911–945. DOI: 10.1021/cr050018y.

Dell'Aglio, M., Gaudiuso, R., De Pascale, O., & De Giacomo, A. (2015). Mechanisms and processes of pulsed laser ablation in liquids during nanoparticle production. *Applied Surface Science*, *348*, 4–9. DOI: 10.1016/j.apsusc.2015.01.082.

Dreaden, E. C., Alkilany, A. M., Huang, X., Murphy, C. J., & El-sayed, M. A. (2018). The golden age: Gold nanoparticles for biomedicine. *HHS Public Access*, *41*(7), 2740–2779. DOI: 10.1039/c1cs15237h.

Dresselhaus, M. S., Dresselhaus, G., & Avouris, P. (2001). *Carbon Nanotubes Synthesis, Structure, Properties, and Applications.pdf*, Springer. Berlin. ISBN:9783662307892

Efros, A. L., & Nesbitt, D. J. (2016). Origin and control of blinking in quantum dots. *Nature Nanotechnology*, *11*(8), 661–671. DOI: 10.1038/nnano.2016.140.

Egbuna, C., Parmar, V. K., Jeevanandam, J., Ezzat, S. M., Patrick-iwuanyanwu, K. C., Adetunji, C. O., Khan, J., Onyeike, E. N., Uche, C. Z., Akram, M., Ibrahim, M. S., Mahdy, N. M. El, Awuchi, C. G., Saravanan, K., Tijjani, H., Odoh, U. E., Messaoudi, M., Ifemeje, J. C., Olisah, M. C., … Ibeabuchi, C. G. (2021). Toxicity of nanoparticles in biomedical application: Nanotoxicology. *Journal of Toxicology*, *2021*, 29, 1–21. DOI: 10.1155/2021/9954443

Foong, L. K., Foroughi, M., Mirhosseini, A. F., Safaei, M., Jahani, S., Mostafavi, M., Ebrahimpoor, N., Sharifi, M., Varma, R. S., & Khatami, M. (2020). Applications of nano-materials in diverse dentistry. *Royal Society of Chemistry*, *10*, 15430–15460. DOI: 10.1039/d0ra00762e.

Ganash, E. A., Al-Jabarti, G. A., & Altuwirqi, R. M. (2020). The synthesis of carbon-based nanomaterials by pulsed laser ablation in water. *Materials Research Express*, *7*(1), 1–11. DOI: 10.1088/2053-1591/ab572b.

Gao, G., Liu, X., Gu, Z., Mu, Q., Zhu, G., Zhang, T., & Zhang, C. (2022). Engineering nanointerfaces of Au 25 clusters for chaperone-mediated peptide amyloidosis. *Nano Letters*, *22*(7), 2964–2970. DOI: 10.1021/acs.nanolett.2c00149.

Gardea-torresdey, J. L., Rico, C. M., & White, J. C. (2014). Trophic transfer, transformation, and impact of engineered nanomaterials in terrestrial environments. *Environmental Science & Technology*, *48*(5), 2526–2540. DOI: 10.1021/es4050665

Ghaemi, F., Abdullah, L. C., Kargarzadeh, H., & Abdi, M. M. (2018). Comparative study of the electrochemical, biomedical, and thermal properties of natural and synthetic nanomaterials. *Nanoscale Research Letters*, *13*, 1–8. DOI: 10.1186/s11671-018-2508-3

Ghaemi, F., Amiri, A., Bajuri, M. Y., Yuhana, N. Y., & Ferrara, M. (2021). Role of different types of nanomaterials against diagnosis, prevention and therapy of COVID-19. *Sustainable Cities and Society*, *72*, 103046, 1–22. DOI: 10.1016/j.scs.2021.103046

Greiner, A., & Wendorff, J. H. (2008). Functional self-assembled nanofibers by electrospinning. *Advances in Polymer Science*, *219*(1), 107–171. DOI: 10.1007/12_2008_146.

Gubicza, J., Chinh, N. Q., Labar, J. L., Tichy, G., Hegedus, Z., Xu, C., & Langdon, T. G. (2009). Stability of microstructure in silver processed by severe plastic deformation. *International Journal of Materials Research*, *100*(6), 884–887. DOI: 10.3139/146.110113

Hajiabadi, S. H., Aghaei, H., Ghabdian, M., Kalateh-aghamohammadi, M., Esmaeilnezhad, E., & Choi, H. J. (2020). On the attributes of invert-emulsion drilling fluids modified with graphene oxide/inorganic complexes. *Journal of Industrial and Engineering Chemistry*, *93*, 290–301. DOI: 10.1016/j.jiec.2020.10.005.

Hao, Y., Yuan, W., Ma, C., White, J. C., Zhang, Z., Adeel, M., Zhou, T., Rui, Y., & Xing, B. (2018). Engineered nanomaterials suppress turnip mosaic virus infection in tobacco (*Nicotiana benthamiana*). *Environmental Science: Nano*, *5*(7), 1685–1693. DOI: 10.1039/C8EN00014J.

Havlik, J., Petrakova, V., Kucka, J., Raabova, H., Panek, D., Stepan, V., Cilova, Z. Z., Reineck, P., Stursa, J., Kucera, J., Hruby, M., & Cigler, P. (2018). Extremely rapid isotropic irradiation of nanoparticles with ions generated in situ by a nuclear reaction. *Nature Communications*, *2018*, 6–15. DOI: 10.1038/s41467-018-06789-8.

Hidangmayum, A., Debnath, A., Guru, A., Singh, B. N., Upadhyay, S. K., & Dwivedi, P. (2022). Mechanistic and recent updates in nano - bioremediation for developing green technology to alleviate agricultural contaminants. *International Journal of Environmental Science and Technology*, 0123456789, 1–26. DOI: 10.1007/s13762-022-04560-7.

Hochella, M. F., Mogk, D. W., Ranville, J., Allen, I. C., Luther, G. W., Marr, L. C., Mcgrail, B. P., Murayama, M., Qafoku, N. P., Rosso, K. M., Sahai, N., Schroeder, P. A., Vikesland, P., Westerhoff, P., & Yang, Y. (2019). Natural, incidental, and engineered nanomaterials and their impacts on the Earth system. *Science (New York, N.Y.)*, *363*(6434), 1–12. DOI: 10.1126/science.aau8299.

Huang, Z. M., Zhang, Y. Z., Kotaki, M., & Ramakrishna, S. (2003). A review on polymer nanofibers by electrospinning and their applications in nanocomposites. *Composites Science and Technology*, *63*(15), 2223–2253. DOI: 10.1016/S0266-3538(03)00178-7.

Ijaz, I., Gilani, E., Nazir, A., & Bukhari, A. (2004). Gold nanoparticles: Assembly, supramolecular chemistry, quantum-size-related properties, and applications toward biology, catalysis, and nanotechnology. *Chemical Reviews*, *104*(1), 293–346. DOI: 10.1021/cr030698+

Ijaz, I., Gilani, E., Nazir, A., & Bukhari, A. (2020). Detail review on chemical, physical and green synthesis, classification, characterizations and applications of nanoparticles. *Green Chemistry Letters and Reviews*, *13*(3), 223–245. DOI: 10.1080/17518253.2020.1802517.

Ikram, R., Jan, B. M., & Vejpravova, J. (2021). Towards recent tendencies in drilling fluids: Application of carbon-based nanomaterials. *Journal of Materials Research and Technology*, *15*, 3733–3758. DOI: 10.1016/j.jmrt.2021.09.114.

Itrourwar, P., Kasivelu, G., Raguraman, V., Malaichamy, K., & Kizhaeral, S. (2020). Effects of biogenic zinc oxide nanoparticles on seed germination and seedling vigor of maize (Zea mays). *Biocatalysis and Agricultural Biotechnology*, *29*, 101778, 1–5. https://doi.org/10.1016/j.bcab.2020.101778

Jana, S., Bandyopadhyay, A., Datta, S., Bhattacharya, D., & Jana, D. (2021). Emerging properties of carbon based 2D material beyond graphene. *Journal of Physics: Condensed Matter*, *34*(5), 053001.

Jayasinghe, A. S., Payne, M. K., Unruh, D. K., Johns, A., Leddy, J., & Forbes, T. Z. (2018). Diffusion and selectivity of water confined within metal-organic nanotubes. *Journal of Materials Chemistry A*, 1531–1539. DOI: 10.1039/C7TA06741K.

Jin, X., Jiang, M., Du, J., & Chen, Z. (2014). Journal of industrial and engineering chemistry removal of Cr (VI) from aqueous solution by surfactant-modified kaolinite. *Journal of Industrial and Engineering Chemistry*, 20(5), 3025–3032. DOI: 10.1016/j.jiec.2013.11.038.

Joshi, S. M., De Britto, S., & Jogaiah, S. (2020). Myco-engineered selenium nanoparticles elicit resistance against tomato late blight disease by regulating differential expression of cellular, biochemical and defense responsive genes. *Journal of Biotechnology*, 325, 196–206. DOI: 10.1016/j.jbiotec.2020.10.023.

Journet, C., Maser, W. K., Bernier, P., & Loiseau, A. (1997). Large-scale production of single-walled carbon nanotubes by the electric-arc technique. *Letters to Nature*, 388(August), 20–22.

Kaempgen, M., Chan, C. K., Ma, J., Cui, Y., & Gruner, G. (2009). Printable thin film supercapacitors using single-walled carbon nanotubes. *Nano Letters*, 9(5), 1872–1876.

Kam, N. S. W., & Dai, H. (2005). Carbon nanotubes as intracellular protein transporters: Generality and biological functionality. *Journal of the American Chemical Society*, 127(16), 6021–6026. DOI: 10.1021/ja050062v

Kankala, R. K., Han, Y., Na, J., Lee, C., & Sun, Z. (2020). Nanoarchitectured structure and surface bio-functionality of mesoporous silica nanoparticles. *Advanced Materials*, 32(23), 1–74. DOI: 10.1002/adma.201907035.

Khan, M. Z. R., Srivastava, S. K., & Gupta, M. K. (2020). A state-of-the-art review on particulate wood polymer composites: Processing, properties and applications. *Polymer Testing*, 89, 1–47, 106721. DOI: 10.1016/j.polymertesting.2020.106721.

Khan, Y., Sadia, H., Zeeshan, S., Shah, A., Khan, M. N., Shah, A. A., Ullah, N., Ullah, M. F., Bibi, H., Bafakeeh, O. T., Khedher, N. Ben, Eldin, S. M., Fadhl, B. M., & Khan, M. I. (2022). Nanoparticles, and their applications in various fields of nanotechnology: A review. *Catalysts, 12*(11), 1386.

Kim, H. J., Bang, I. C., & Onoe, J. (2009). Characteristic stability of bare Au-water nanofluids fabricated by pulsed laser ablation in liquids. *Optics and Lasers in Engineering*, 47(5), 532–538. DOI: 10.1016/j.optlaseng.2008.10.011.

Kim, T. H., Lee, H. J., & Park, H. G. (2008). Carbon nanofiber-based biosensor for the detection of circulating tumor cells. *Nano Letters*, 8, 2878.

Kim, Y., Ji, E., Roy, S., Sharbirin, A. S., Ranz, L., Dieing, T., & Kim, J. (2020). Measurement of lateral and axial resolution of confocal Raman microscope using dispersed carbon nanotubes and suspended graphene. *Current Applied Physics*, 20(1), 71–77. DOI: 10.1016/j.cap.2019.10.012.

Klimov, V. I. (2003). Nanocrystal quantum dots. *Los Alamos Science*, 28, 214–220.

Kumar, A., Pandey, A. K., Singh, S. S., & Shanker, R. (2017). Metal oxide nanoparticles as efficient pollutant sorbents: Challenges and opportunities. *Environmental Science and Pollution Research*, 24, 2017.

Kumar, J. A., Krithiga, T., Manigandan, S., Sathish, S., Renita, A. A., Prakash, P., … Crispin, S. (2021). A focus to green synthesis of metal/metal based oxide nanoparticles: Various mechanisms and applications towards ecological approach. *Journal of Cleaner Production*, 324, 129198, DOI: 10.1016/j.jclepro.2021.129198.

Kurdish, I. K. (2021). Natural and synthetic nanomaterials in microbial biotechnologies for crop production. *Мікробіол. журн.*, 81–91. Doi:10.15407/microbiolj83.03.081

Lambert, B. P., Gillen, A. J., & Boghossian, A. A. (2020). Synthetic biology: A solution for tackling nanomaterial challenges. *The Journal of Physical Chemistry Letters, 11*(12), 4791–4802. DOI: 10.1021/acs.jpclett.0c00929.

Li, J., Li, W., & Li, Q. (2015). Metal nanoparticles for diagnosis and therapy of bacterial infection. *Advanced Healthcare Materials*, *7*(13), 1701392. DOI:10.1002/adhm.201701392

Li, Z., Hou, B., Xu, Y., Wu, D., Sun, Y., Hu, W., & Deng, F. (2005). Comparative study of sol-gel-hydrothermal and sol-gel synthesis of titania-silica composite nanoparticles. *Journal of Solid State Chemistry*, *178*(5), 1395–1405. DOI: 10.1016/j.jssc.2004.12.034.

Lin, X., Liang, Y., Lu, Z., Lou, H., Zhang, X., Liu, S., Zheng, B., Liu, R., Fu, R., & Wu, D. (2017). Mechanochemistry: A green, activation-free and top-down strategy to high-surface-area carbon materials. *ACS Sustainable Chemistry and Engineering*, *5*(10), 8535–8540. DOI: 10.1021/acssuschemeng.7b02462.

Lira, A. L., Ferreira, R. S., Oliva, M. L. V, & Sousa, A. A. (2020). Regulation of thrombin activity with ultrasmall nanoparticles: Effects of surface chemistry. *Langmuir*, *36*(27), 7991–8001. DOI: 10.1021/acs.langmuir.0c01352.

Liu, J., Li, Y., & Liang, J. (2019). Metal oxide nanoparticles and their applications in energy storage and conversion: From li-ion batteries to supercapacitors and solar cells. *Advanced Materials*, *267*, 26–46. DOI:https://doi.org/10.1016/j.cis.2019.03.001

Liu, Y., Zhu, S., Gu, Z., Chen, C., & Zhao, Y. (2022). Toxicity of manufactured nanomaterials. *Particuology*, *69*, 31–48. DOI: 10.1016/j.partic.2021.11.007

Long, Y. Z., Li, M. M., Gu, C., Wan, M., Duvail, J. L., Liu, Z., & Fan, Z. (2011). Recent advances in synthesis, physical properties and applications of conducting polymer nanotubes and nanofibers. *Progress in Polymer Science (Oxford)*, *36*(10), 1415–1442. DOI: 10.1016/j.progpolymsci.2011.04.001.

Louro, H. (2018). Relevance of physicochemical characterization of nanomaterials for understanding nano-cellular interactions. *Advances in Experimental Medicine and Biology*, 123–142. DOI: 10.1007/978-3-319-72041-8_8.

Luo, J., Xia, Y., Kim, J. et al. (2013). Carbon nanofiber-supported SNO2 nanoparticles as high-capacity anodes for lithium-ion batteries. *ACS Applied Materials & Interfaces*, *5*, 9596.

Ma, J., Xu, J., Pang, S., Zhou, W., Xia, B., & An, Y. (2021). Novel environmentally friendly lubricants for drilling fluids applied in shale formation. *Energy Fuels*. DOI: 10.1021/acs.energyfuels.1c00495.

Maharramov, A. M., Hasanova, U. A., Suleymanova, I. A., Osmanova, G. E., & Hajiyeva, N. E. (2019). The engineered nanoparticles in food chain: Potential toxicity and effects. *SN Applied Sciences*, *1*(11), 1–25. DOI: 10.1007/s42452-019-1412-5.

Malakar, A., Kanel, S. R., Ray, C., Snow, D. D., & Nadagouda, M. N. (2020). Nanomaterials in the environment, human exposure pathway, and health effects: A review. *Science of the Total Environment*, *1*(937), 1–76. https://doi.org/10.1016/j.scitotenv.2020.143470.

Mathioudaki, S., Barthelémy, B., Detriche, S., Vandenabeele, C., Delhalle, J., Mekhalif, Z., & Lucas, S. (2018). Plasma treatment of metal oxide nanoparticles: Development of core – shell structures for a better and similar dispersibility, *1*(7), 3464–3473. DOI: 10.1021/acsanm.8b00645.

McNamara, K., & Tofail, S. A. M. (2017). Nanoparticles in biomedical applications. *Advances in Physics: X*, *2*(1), 54–88. DOI: 10.1080/23746149.2016.1254570.

Medintz, I., Tetsuo, U., Goldman, E. R., & Mattoussi, H. (2005). Quantum dot bioconjugates for imaging, labelling and sensing. *Nature Materials*, *4*, 435–446.

Miranda, M., Gellini, C., Simonella, A., Tiberi, M., Giammanco, F., & Giorgetti, E. (2013). Characterization of copper nanoparticles obtained by laser ablation in liquids. *Applied Physics A: Materials Science and Processing*, *110*, 829–833. DOI: 10.1007/s00339-012-7160-7.

Moinudeen, G. K., Ahmad, F., Kumar, D., Al-douri, Y., & Ahmad, S. (2017). IoT applications in future foreseen guided by engineered nanomaterials and printed intelligence technologies a technology review. *International Journal of Internet of Things*, *August*, *6*(3), 106–148. DOI: 10.5923/j.ijit.20170603.03.

Moniri, S., Ghoranneviss, M., Hantehzadeh, M. R., & Asadabad, M. A. (2017). Synthesis and optical characterization of copper nanoparticles prepared by laser ablation. *Bulletin of Materials Science*, *40*(1), 37–43. DOI: 10.1007/s12034-016-1348-y.

Montanarella, L., Commission, E., Mckenzie, N. J., Scientific, T. C., & Baptista, I. (2016). World's soils are under threat. *Soil, February*, *2*(1), 79–82. DOI: 10.5194/soild-2-1263-2015.

Moreno-Vega, A. I., Gómez-Quintero, T., Nuñez-Anita, R. E., Acosta-Torres, L. S., & Castaño, V. (2012). Polymeric and ceramic nanoparticles in biomedical applications. *Journal of Nanotechnology*, 1–11. DOI: 10.1155/2012/936041.

Murray, C. B., Noms, D. J., & Bawendi, M. G. (1993). Synthesis and characterization of nearly monodisperse cde (E = S, Se, Te) semiconductor nanocrystallites. *Journal of American Chemical Society*, *4*, 8706–8715.

Mustapha, S., Tijani, J. O., Ndamitso, M. M., Abdulkareem, S. A., Shuaib, D. T., Mohammed, A. K., & Sumaila, A. (2020). The role of kaolin and kaolin/ZnO nanoadsorbents in adsorption studies for tannery wastewater treatment. *Scientific Reports*, *10*(1), 1–22. DOI: 10.1038/s41598-020-69808-z.

Oroojalian, F., Charbgoo, F., Hashemi, M., & Amani, A. (2020). Recent advances in nanotechnology-based drug delivery systems for the kidney. *Journal of Controlled Release*, *321*(February), 442–462. DOI: 10.1016/j.jconrel.2020.02.027.

Paradise, M., & Goswami, T. (2007). Carbon nanotubes - production and industrial applications. *Materials and Design*, *28*(5), 1477–1489. DOI: 10.1016/j.matdes.2006.03.008.

Peters, R., Elbers, I., Undas, A., Sijtsma, E., Briffa, S., Carnell-morris, P., Siupa, A., Yoon, T., Burr, L., Schmid, D., Tentschert, J., Hachenberger, Y., Jungnickel, H., Luch, A., Meier, F., Kocic, J., Kim, J., & Park, B. C. (2021). Benchmarking the acenano toolbox for characterisation of nanoparticle size and concentration by interlaboratory comparisons. *Molecules*, *26*(17), 5315. DOI: 10.3390/molecules26175315

Prado, F., Andersen, H. F., Taeño, M., Mæhlen, J. P., Ramírez-castellanos, J., Maestre, D., Karazhanov, S., & Cremades, A. (2020). Comparative study of the implementation of tin and titanium oxide nanoparticles as electrodes materials in li-ion batteries. *Scientific Reports*, *2027*, 1–8. DOI: 10.1038/s41598-020-62505-x.

Prado, L. A., Sriyai, M., Ghislandi, M., Barros-Timmons, A., & Schulte, K. (2010). Surface modification of alumina nanoparticles with silane coupling agents. *Journal of the Brazilian Chemical Society*, *21*, 12, 2238–2245. DOI: 10.1590/S0103-50532010001200010

Quevedo, A. C., Guggenheim, E., Briffa, S. M., Adams, J., Lofts, S., Kwak, M., Geol, T., Johnston, C., Wagner, S., Holbrook, T. R., Hachenberger, Y. U., Tentschert, J., & Valsami-jones, E. (2021). UV-Vis spectroscopic characterization of nanomaterials in aqueous media. *Journal of Visualized Experiments*, *October*, 1–11. DOI: 10.3791/61764.

Raj, S., Singh, H., Trivedi, R., & Soni, V. (2020). Biogenic synthesis of agnps employing *Terminalia arjuna* leaf extract and its efficacy towards catalytic degradation of organic dyes. *Scientific Reports*, *10*(1), 9616. DOI: 10.1038/s41598-020-66851-8.

Ranjith, K. S., Kwak, C. H., Hwang, J. U., Ghoreishian, S. M., Seeta, G., Raju, R., Huh, Y. S., Im, J. S., & Han, Y. (2019). High-performance all-solid-state hybrid supercapacitors based on surface-embedded bimetallic oxide nanograins loaded onto carbon nanofiber and activated carbon. *Electrochimica Acta*, *332*, 135494. DOI: 10.1016/j.electacta.2019.135494.

Rashid, F., Hadi, A., Humood Al-Garah, N., & Hashim, A. (2018). Novel phase change materials, MgO nanoparticles, and water based nanofluids for thermal energy storage and biomedical applications. *International Journal of Pharmaceutical and Phytopharmacological Research*, *8*(1), 46–56. www.eijppr.com.

Ren, S., Zhang, Y., Qin, R., Xu, H., Ye, M., & Nie, P. (2022). MoS$_2$/MWCNT-COOH-modified glassy carbon electrode for nitrite detection in water environment. *Chemosensors*, *10*, 419.

Rizwan, M., Ali, S., Ali, B., Adrees, M., Arshad, M., Hussain, A., Zia ur Rehman, M., & Waris, A. A. (2019). Zinc and iron oxide nanoparticles improved the plant growth and reduced the oxidative stress and cadmium concentration in wheat. *Chemosphere*, *214*, 269–277. DOI: 10.1016/j.chemosphere.2018.09.120.

Rui, M., Ma, C., Hao, Y., Guo, J., Rui, Y., Tang, X., & Sperotto, R. A. (2016). Iron oxide nanoparticles as a potential iron fertilizer for peanut (*Arachis hypogaea*). *Frontiers in Plant Science*, *7*(June), 1–10. DOI: 10.3389/fpls.2016.00815.

Sabnis, S. S., Raikar, R., & Gogate, P. R. (2020). Ultrasonics – sonochemistry evaluation of different cavitational reactors for size reduction of dadps. *Ultrasonics – Sonochemistry*, *69*(July), 105276. DOI: 10.1016/j.ultsonch.2020.105276.

Salah, N., Habib, S. S., Khan, Z. H., Memic, A., Azam, A., Alarfaj, E., Zahed, N., & Al-Hamedi, S. (2011). High-energy ball milling technique for zno nanoparticles as antibacterial material. *International Journal of Nanomedicine*, *6*, 863–869. DOI: 10.2147/ijn.s18267.

Schulte, K. (2015). Production of CNTs and risks to health. *Carbon Nanotube Reinforced Composites: CNR Polymer Science and Technology, Cvd*, 103–123. DOI: 10.1016/B978-1-4557-3195-4.00004-7.

Serp, P., & Figueiredo, J. L. (2009). *Carbon Materials for Catalysis*. John Wiley & Sons Inc., Publication, Hoboken, New Jersey, 579.

Shang, Y., Hasan, K., Ahammed, G. J., Li, M., & Yin, H. (2019). Applications of nanotechnology in plant growth and crop protection: A review. *Molecules*, *24*(14), 2558. DOI: 10.3390/molecules24142558

Sharma, S., Chandra, R., & Kumar, P. (2016). Mechanical properties of carbon nanofiber reinforced polymer composites-molecular dynamics approach. *JoM*, *68*, 1717–1727. DOI: 10.1007/s11837-016-1933-y.

Shi, Z., Lian, Y., Liao, F. H., Zhou, X., Gu, Z., Zhang, Y., Iijima, S., Li, H., Yue, K. T., & Zhang, S. L. (2000). Large scale synthesis of single-wall carbon nanotubes by arc-discharge method. *Journal of Physics and Chemistry of Solids*, *61*(7), 1031–1036. DOI: 10.1016/S0022-3697(99)00358-3.

Shokry, H., Elkady, M., & Salama, E. (2020). Eco-friendly magnetic activated carbon nano-hybrid for facile oil spills separation. *Scientific Reports*, 1–17. DOI: 10.1038/s41598-020-67231-y.

Singh, N., Bhuker, A., & Jeevanadam, J. (2021). Effects of metal nanoparticle-mediated treatment on seed quality parameters of different crops effects of metal nanoparticle-mediated treatment on seed quality parameters of different crops. *Naunyn-Schmiedeberg's Arch Pharmaco*, *March*, *394*, 1067–1089. DOI: 10.1007/s00210-021-02057-7.

Smith, B. R., & Gambhir, S. S. (2017). Nanomaterials for in vivo imaging. *Chemical Reviews*, *117*(3), 901–986. DOI: 10.1021/acs.chemrev.6b00073.

Souza, T. G. F., Ciminelli, V. S. T., & Mohallem, N. D. S. (2016). A comparison of TEM and DLS methods to characterize size distribution of ceramic nanoparticles. *Journal of Physics: Conference Series*, *733*(1), 012039. DOI: 10.1088/1742-6596/733/1/012039.

Sun, L., Wang, Y., Wang, R., Wang, R., Zhang, P., Ju, Q., & Xu, J. (2020). Physiological, transcriptomic, and metabolomic analyses reveal zinc oxide nanoparticles modulate plant growth in tomato. *Environmental Science: Nano*, *7*(11), 3587–3604. DOI: 10.1039/d0en00723d.

Syafiuddin, A., Salmiati, Salim, M. R., Beng Hong Kueh, A., Hadibarata, T., & Nur, H. (2017). A review of silver nanoparticles: Research trends, global consumption, synthesis, properties, and future challenges. *Journal of the Chinese Chemical Society*, *64*(7), 732–756. DOI: 10.1002/jccs.201700067.

Tehrani, M., & Khanbolouki, P. (2018). *Carbon Nanotubes: Synthesis, Characterization, and Applications*, 3–35. Springer, Pennsylvania. DOI: 10.1007/978-3-319-64717-3.

Thongpool, V., Asanithi, P., & Limsuwan, P. (2012). Synthesis of carbon particles using laser ablation in ethanol. *Procedia Engineering*, *32*, 1054–1060. DOI: 10.1016/j.proeng.2012.02.054.

Tilman, D., Balzer, C., Hill, J., & Befort, B. L. (2011). Global food demand and the sustainable intensification of agriculture. *Proceedings of the National Academy of Sciences of the United States of America*, *108*(50), 20260–20264. DOI: 10.1073/pnas.1116437108.

Usman, M., Farooq, M., Wakeel, A., Nawaz, A., Alam, S., Ashraf, I., & Sanaullah, M. (2020). Nanotechnology in agriculture: Current status, challenges and future opportunities. *Science of the Total Environment*, *721*, 137778. DOI: 10.1016/j.scitotenv.2020.137778.

Valverde-Alva, M. A., García-Fernández, T., Villagrán-Muniz, M., Sánchez-Aké, C., Castañeda-Guzmán, R., Esparza-Alegría, E., Sánchez-Valdés, C. F., Llamazares, J. L. S., & Herrera, C. E. M. (2015). Synthesis of silver nanoparticles by laser ablation in ethanol: A pulsed photoacoustic study. *Applied Surface Science*, *355*, 341–349. DOI: 10.1016/j.apsusc.2015.07.133.

Vamvakaki, V., Fouskaki, M., Chaniotakis, N., Vamvakaki, V., Fouskaki, M., & Electrochemical, N. C. (2021). Electrochemical biosensing systems based on carbon nanotubes and carbon electrochemical biosensing systems based on carbon nanotubes and carbon nanofibers. *Analytical Letters, March*, *40*(12), 2271–2287. DOI: 10.1080/00032710701575520.

Varshney, B., Siddiqui, M. J., Anwer, A. H., Khan, M. Z., Ahmed, F., Aljaafari, A., Hammud, H. H., & Azam, A. (2020). Synthesis of mesoporous – Nanocomposite using modified sol - gel method and its electrochemical performance as electrode material for supercapacitors. *Scientific Reports*, 1–13. DOI: 10.1038/s41598-020-67990-8.

Vasilescu, A., Hayat, A., Gaspar, S., & Marty, J.-L. (2017). Advantages of carbon nanomaterials in electrochemical aptasensors for food analysis. *Electroanalysis*, 1–19. DOI: 10.1002/elan.201700578.

Venkatachalam, P., Priyanka, N., Manikandan, K., Ganeshbabu, I., Indiraarulselvi, N., Geetha, N., Muralikrishna, K., Bhattacharya, R. C., Tiwari, M., Sharma, N., & Sahi, S. V. (2016). Enhanced plant growth promoting role of phycomolecules coated zinc oxide nanoparticles with P supplementation in cotton (*Gossypium hirsutum* L.). *Plant Physiology and Biochemistry*, *110*, 118–127. DOI: 10.1016/j.plaphy.2016.09.004.

Wang, C., & Chen, D. (2013). Synthesis and applications of metal oxide nanoparticles in electronic devices. *Chemical Society Reviews*, *42*, 2959.

Wang, J., & Sun, X. W. (2015). Semiconducting quantum dots in electronic and optoelectronic devices. *Journal of Materials Chemistry C*, *3*, 5947.

Wang, M., Zhang, C., Yan, S., Chen, T., Fang, H., & Yuan, X. (2021). Wide-field super-resolved Raman imaging of carbon materials. *ACS Photonics*. DOI: 10.1021/acsphotonics.1c00392.

Wang, Y., Xu, H., Drozdov, G., & Dumitrică, T. (2018). Mesoscopic friction and network morphology control the mechanics and processing of carbon nanotube yarns. *Carbon*, *139*, 94–104. DOI: 10.1016/j.carbon.2018.06.043.

Winkler, K., Wojciechowski, T., Liszewska, M., Górecka, E., & Fiałkowski, M. (2014). Morphological changes of gold nanoparticles due to adsorption onto silicon substrate and oxygen plasma treatment. *Royal Society of Chemistry*, 12729–12736. DOI: 10.1039/c4ra00507d.

Worrall, E. A., Hamid, A., Mody, K. T., Mitter, N., & Pappu, H. R. (2018). Nanotechnology for plant disease management. *Agronomy*, 1–24. DOI: 10.3390/agronomy8120285.

Xiong, R., Grant, A. M., Ma, R., Zhang, S., & Tsukruk, V. V. (2018). Naturally-derived biopolymer nanocomposites: Interfacial design, properties and emerging applications. *Materials Science & Engineering R*, *125*, 1–41. DOI: 10.1016/j.mser.2018.01.002.

Xu, J., & Wang, L. (2019). Fabrication of hierarchical core/arrays on ni-foam as high – rate electrodes for asymmetric supercapacitors. *Scientific Reports, August*, 1–11. DOI: 10.1038/s41598-019-48931-6.

Yang, G., Lin, W., Lai, H., Tong, J., Lei, J., Yuan, M., Zhang, Y., & Cui, C. (2021). Understanding the relationship between particle size and ultrasonic treatment during the synthesis of metal nanoparticles. *Ultrasonics Sonochemistry*, *73*, 105497. DOI: 10.1016/j.ultsonch.2021.105497.

Zhang, Q., & Xie, J. (2019). Metal-based nanomaterials for energy applications. *Advanced Energy Materials*, *9*, 1803078.

Zhang, S., Malik, S., Ali, N., & Khan, A. (2022). Covalent and non-covalent functionalized nanomaterials. In *Topics in Current Chemistry*. Springer International Publishing, Verlag. DOI: 10.1007/s41061-022-00397-3.

Zhao, J., Milanova, M., Warmoeskerken, M. M. C. G., & Dutschk, V. (2012). Colloids and surfaces A: Physicochemical and engineering aspects surface modification of TiO_2 nanoparticles with silane coupling agents. *Colloids and Surfaces A: Physicochemical and Engineering Aspects*, *413*, 273–279. DOI: 10.1016/j.colsurfa.2011.11.033.

Zhuang, S., Lee, E. S., Lei, L., Nunna, B. B., Kuang, L., & Zhang, W. (2016). Synthesis of nitrogen-doped graphene catalyst by high- energy wet ball milling for electrochemical systems. *International Journal of Energy Research*, *40*(15), 2136–2149. DOI: 10.1002/er.3595.

2 The Characterization Techniques of Nanomaterials

Mohd Ridhwan Adam, Muhammad Hakimin Shafie, and Mohd Saiful Shamsudin
Universiti Sains Malaysia

Siti Khadijah Hubadillah
Universiti Utara Malaysia

Mohd Riduan Jamalludin
Universiti Malaysia Perlis (UniMAP)

Mohd Haiqal Abd Aziz
Universiti Tun Hussein Onn Malaysia

Atikah Mohd Nasir
Universiti Kebangsaan Malaysia

2.1 INTRODUCTION

Nanotechnology encompasses the development of many nanomaterials (NM), such as nano-objects and nanoparticles (NP). Nanomaterials had at least a dimension of about 100 nm, while nano-objects such as carbon nanotubes have two dimensions of around 100 nm. On the other hand, nanoparticles are classified as particles having three dimensions below 100 nm. Given the speedy evolution of nanotechnology along with the enormous diversity of nanomaterials within fabrication and research, it is crucial to mitigate the possible effects of nanomaterials on environmental and human health. Owing to their diminutive size, NP has relatively higher surface area than comparable conventional materials. Furthermore, the sizes frequently lead to enhanced reactivity and modified surface characteristics that can be utilized and manipulated in a wide range of customer goods, including paints, personal care products, medicinal products, food, and also systems that instantaneously emit NPs into the surroundings, namely the treatment of contaminated environments [1]. Hence, any possible adverse consequences must be weighed in order to comprehend

DOI: 10.1201/9781003400998-2

consequences on the environment and prospective detrimental impacts on human health. This will necessitate correlating the physicochemical properties of NP to their biological properties.

The identification and measurement of NMs in analytes such as nutrition remains an unresolved problem that must be resolved in order to protect people and categorize or govern commodities. The greatest obstacle for analytical sciences is that NMs are new variants of substance that require chemical (composition, volume, and concentration) and physical analysis (dimension, form, and accumulation). The presence of NMs in natural and environmental samples must be determined. As a consequence, accurate analytical strategies are needed to quantify NM quantities in a diverse variety of matrices. Product diversity may result in a change in the sources of pollution emission levels [2]. Thin films, surfaces, and coatings are examples of one-dimensional NMs while nanotubes, nanowires, and nanofibers are examples of two-dimensional NMs. On the other hand, fullerenes, quantum dots, and dendrimers are examples of three-dimensional NMs. The multiple types of NMs thus enhanced the chance that NMs are being introduced into the environment, either intentionally or accidentally.

Understanding the various types of nanomaterials and associated possible exposure pathways is critical for characterizing the possible risks of nanomaterials and their potential demise in the ecosystem. The type of nanomaterial might well be unknown due to its manufacturing technique (vapour deposition gas phase, attrition, colloidal, and many more), which could also lead to human interaction via respiratory, skin penetration, or consumption [3]. According to research, man-made nanomaterials induce unforeseen toxicities in living creatures as well as negative human health consequences. The modified characteristics of nanomaterials such as aggressiveness, melting point, mechanical characteristics, fluorescence, and electrical conductivity at the nanoscale render predicting whether nanomaterials will interfere with biological beings and the ecosystem considerably complex. Nevertheless, certain natural nanoparticles can be harmful and originate from a wide range of biological and geological mechanisms. The researchers have found that both humans and other organisms have developed, changed, and multiplied as a consequence of potentially hazardous exposures to natural nanomaterials.

The characteristics of NMs can indeed be altered through a thorough analysis of the fabricated NMs. Unless NMs are assessed immediately after such a production process modification, it is feasible to detect if that particular construction process changes impaired or enhanced the attributes of the NMs [4]. Furthermore, a proper and exact definition of NMs in relation to the proposed function is critical. In the decades that followed, NMs characterization approaches improved dramatically and continued to evolve. As a result, the challenge is to identify the most appropriate characterization methodologies for researching the properties of NMs. The investigation of techniques capable of identifying the scale, structure, and content of complex and heterogeneous substances, interfaces, surface layers, and inhomogeneities on the nanometre scale persists. Characterization methods could not maintain up in terms of transverse and depth precision to the nanometre scale in the previous era. Therefore, this chapter will discuss physicochemical evaluation approaches with respect to NMs traits or qualities.

2.2 CHARACTERIZATION OF PHYSICOCHEMICAL PROPERTIES OF NANOMATERIALS

Many approaches have been invented to assess the physicochemical features of NMs. Assessment of NPs encompasses not only dimension and shape but also polymorphism, electrical properties, crystalline structure, purity, surface morphology, and resilience, all of which are important in the physical and chemical engagement with the surroundings [5]. These approaches comprise optical spectroscopic, electron microscopy, light scattering, and nuclear magnetic resonance spectroscopy, along with thermogravimetric analysis, electrophoresis, chromatography, and other techniques. This section divides the existing characterization methodologies for characterizing NMs into the physical and chemical characteristics of the materials. Table 2.1 summarizes some of the most frequent techniques used for the characterization of nanomaterials in previous studies. The table also describes briefly the theories, techniques, advantages, and constraints of several methods typically employed to examine the physicochemical properties of nanomaterials.

2.2.1 Physical Properties Determination of Nanomaterials

Analytical techniques are still incapable of maintaining up in terms of transverse and specific resolutions to the nanoscale range in the past. Currently, a multitude of characterization methods have been developed, which can provide crucial insights into the structure and behavior of NMs down to the particle level, especially in relation to the imaging process. The invention of new characterization instruments and the progression of nanotechnology, including electron microscopy (EM), dynamic light scattering (DLS), and atomic force microscopy (AFM), are employed in visualizing materials to ascertain their dimensions, form, proportions, and morphological characteristics.

2.2.1.1 Scanning Electron Microscopy

Converse to typical microscopy techniques that use light bases and glass lens system to clarify samples and generate high-resolution images, electron microscopy (EM) employs rays of electron and electromagnetic or electrostatic lenses to create images with a significantly greater resolution, owing to the shorter wavelength range of electrons compared to visible light photons. SEM is a technique for surface topography that involves the emitted electron beam sweeping across the surface of the specimen and interacting with the specimen to produce signals that represent the elemental structure and topographical information of the sample exterior [23]. Inbound electrons trigger several phenomena, including the emission of backscattered electrons through elastic scattering, the generation of low-energy secondary electrons through inelastic scattering, and the occurrence of cathodoluminescence from molecules on the surface of the samples or from materials in close proximity. From these releases, the most prevalent method in SEM is the identification of secondary electrons, which may attain a precision of less than 1 nm.

SEM can precisely assess the size, size distribution, and appearance of nanomaterials; nonetheless, drying and contrasting samples can induce specimen shrinking

TABLE 2.1

Commonly Used Characterization Techniques of Nanomaterials

Measurement	Techniques	Characteristics Analyzed	Advantages	Disadvantages	References
Physical	Scanning electron microscopy (SEM)	• Surface morphology. • Shape. • Particle size and its distribution. • Dispersion. • Agglomeration.	• High resolution. • Direct measurement. • Biomolecule images from natural samples.	• Requires conductive material coating. • Requires dry samples. • Expensive measurement. • Requires cryogenic techniques for sampling.	[6,7]
Physical/chemical	Transmission electron microscopy (TEM)	• Shape heterogeneity. • Particle size and its distribution. • Dispersion. • Agglomeration.	• High spatial resolution. • Direct measurement. • Capable of determination of electrical and chemical properties.	• Requires sample in nonphysiological form. • Destructive technique. • Expensive measurement. • Requires ultrathin samples.	[8,9]
Chemical	Zeta potential	• Surface charge. • Stability.	• Simultaneous measurement of multiple samples.	• Less precise. • Affected by electro-osmotic effect. • Repeatable measurement.	[10,11]
Physical	Dynamic light scattering (DLS)	• Hydrodynamic particle size distribution.	• Modest cost of measurement. • Rapid and reproducible measurement. • Non-destructive.	• Limited in polydisperse sample measurement. • Induced by large particles. • Assumption of spherical sample-shaped measurement.	[12]
Chemical	Mass spectroscopy (MS)	• Composition and structure. • Molecular weight. • Surface properties.	• High precision and accuracy. • High detection sensitivity.	• Limited database information for sample identification. • Expensive measurement.	[13,14]

(Continued)

TABLE 2.1 (*Continued*)
Commonly Used Characterization Techniques of Nanomaterials

Measurement	Techniques	Characteristics Analyzed	Advantages	Disadvantages	References
Chemical	Infrared spectroscopy (IR)	• Composition and structure. • Surface properties.	• Rapid measurement. • Inexpensive measurement. • Small sample requirement. • Not affected by sample thickness.	• Highly affected by H₂O presence. • Relatively low sensitivity.	[15]
Chemical	Raman spectroscopy (RS)	• Chemical, electrical, and structural properties. • Hydrodynamic particle size distribution.	• Acquire topological properties of nanomaterials. • Complementing IR data. • Simple and small sample requirements.	• High fluorescence interference • Irreplicable measurement. • Low spatial resolution.	[16,17]
Physical	Atomic force microscopy (AFM)	• Surface morphology. • Shape. • Particle size and its distribution. • Dispersion. • Agglomeration.	• High topographical resolution. • Simultaneous measurement in any condition of samples (dry, wet, or ambient conditions)	• Time-consuming and poor sampling. • Restricted to the exterior properties of the nanomaterials. • Overvaluation of adjacent dimension.	[18]
Chemical	X-ray diffraction (XRD)	• Shape, size, and structural properties. • Crystallinity properties.	• High spatial resolution. • Well-known method.	• Restricted to crystalline nanomaterials. • Accessible for a single binding state of sample. • Low intensity.	[19,20]
Chemical	Nuclear magnetic resonance (NMR)	• Composition, size, and structural properties. • Conformational change. • Purity.	• Non-destructive technique. • Simple sample preparation.	• Time-consuming. • Less sensitivity. • Requires relatively substantial amount of sample. • Restricted to particular nuclei NMR-active materials.	[21,22]

and modify the nanomaterials' properties. Furthermore, when being examined by a beam of electrons, most non-conducting biomolecule samples develop a charge and inadequately divert the electron beam, resulting in imaging errors or abnormalities [24]. Usually, it is necessary to apply an ultrathin coating of an electrical conductor substance on the biomolecules during this process of sample preparation. Since the cryogenic freezing approach is frequently employed in EM to scan surface groups associated with NPs, nanomaterial size cannot be examined under physiological settings. The exception is environmental scanning electron microscopy (ESEM), which may capture materials in their natural form without alteration or pre-treatment. Although this specimen compartment of ESEM operates in a low-pressure vapour phase of 10–50 Torr and relatively prominent moisture, charging objects may be removed, and it is not necessary to coat the samples with a conducting substance. However, a significant drawback of many electron microscopy methods, including SEM, is that they often require sample preparation that can result in damage, making the samples unsuitable for examination using other techniques. Owing to the little quantity of specimen particles in the scanning zone, skewed statics of the size distribution of heterogeneous samples are also inevitable with SEM.

2.2.1.2 Transmission Electron Microscopy

Known as the highly common method for analyzing nanomaterials in EM, transmission electron microscopy (TEM) delivers immediate descriptions and chemical details of nanomaterials to atomic-level spatial resolution (less than 1 nm). In the standard TEM procedure, an incident electron ray is passed across an extremely thin foil sample, where the incident electrons contact the sample and are converted into either unscattered electrons, elastic scattering electrons, or inelastically scattered electrons. The proportion of the distance between the objective lens and the sample, and the length between the objective lens and its imaging plane largely determines the magnification of TEM. A set of electromagnetic lenses focusses unscattered or scattered electrons, which are then reflected onto a display to form an electron scattering, fullness-contrast projection, phase-contrast pattern, or a shadow appearance of variable shade based on the denseness of unscattered electrons.

Intriguingly, wet TEM may be conducted to identify the size of the particles, dispersion, dynamic migration, and agglomeration of nanomaterials in a fluid setting [25]. Additional to adjusting the feature of ESEM for monitoring measurements beneath partial water vapour pressure in the microscope sample holder, a newly invented wet scanning transmission electron microscopy (STEM) image processing procedure permits transmission assessment of organisms entirely covered in a fluid state, as opposed to the challenges of deprived contrast and possible drifting of particles that occur in the image data of the upper fluid surface while using ESEM. Therefore, the wet mode STEM enables viewing with nanoscale precision and excellent contrast across several micrometres of water without the requirement for contrast agents and stains.

2.2.1.3 Dynamic Light Scattering

Dynamic light scattering (DLS) (sometimes called as photon correlation spectroscopy or quasielastic light scattering) is a valuable and commonly used method for determining the in-situ sizes of metal nanoparticles in a fluid between a few nanometres and a few micrometres. The approach permits the measurement of nanoparticle dispersion stability; dimensions; size distribution; and, in certain conditions, structures. DLS is a non-invasive and rapid approach that provides strong statistically significant results despite some misinterpreted and unevaluated factors, including the thickness of the colloid, the dispersion angle, and the structural anisotropy of nanomaterials. In particular, the approach offers several benefits over a conventional sizing method (microscopy) which is costly, laborious, and needs sample groundwork in a dry condition, which could result in nanomaterial aggregation. Biological specimens, emulsions, polymers, micelles, nanoparticles, proteins, and colloids are some of the examples of liquid-phase operations. The fundamental principle of the DLS method is that nanoparticles submerged in a liquid move according to "Brownian Motion" [26]. After a laser ray strikes a liquid enclosing randomly moving nanoparticles, the light is dispersed at a frequency that is directly proportional to the nanoparticles' size. The typical diagram of DLS is seen in Figure 2.1, which contains the principal constituents namely the laser beam which lights the specimen that scatters the light, and thus, fluctuations at a specified angle are observed [27]. The rate of decay of this exponential function is proportionate to the diffusion rate, which the Stokes–Einstein equation indicates as the radius of the particles.

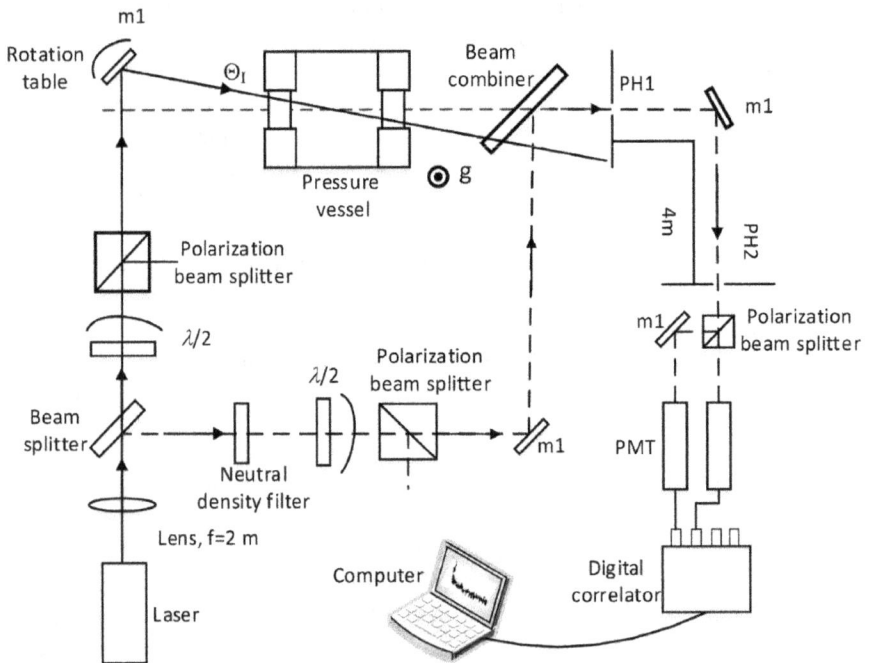

FIGURE 2.1   Schematic diagram dynamic light scattering spectroscopy using transmission grafting. Reproduced from [27] with permission (Licence number #5550251238234).

2.2.1.4 Atomic Force Microscopy

The atomic force microscope (AFM) is a vital instrument in the advancement of nanotechnology and is commonly utilized in nanoscale examination that provides details about the topography of any material. The topographical images are generated by scanning a probe with a sharp tip positioned at the open edge of a cantilever across the surface of any specimen, and this procedure is dependent on the interparticle interactions between the specimen and the tip [28]. The energies involved are on the nano-newton scale or lower, acting over distances of nanometers. In the case of longer distances (beyond 0.5 nm), the interaction is attractive, following van der Waals forces, whereas at closer ranges, it transitions to repulsive interactions. A key component of the AFM is the cantilever, which comprises a sharp edge for measuring the tiny forces between the tip and the sample's particles. The little force causes a minor distortion of the cantilever, which is recorded by the displacement of a laser beam, which is subsequently converted into electronic signals by a photodetector.

2.2.2 CHEMICAL PROPERTIES VERIFICATION OF NANOMATERIALS

Unlike the physical testing methods, there are some analyses utilized for the chemical properties' determination of nanomaterials. These analyses are used to determine the composition, structural properties, crystallinity, electrical properties, purity, molecular weight, surface charge, and many others. The analyses covered in this chapter are zeta potential, mass spectroscopy (MS), infrared spectroscopy (IR), Raman spectroscopy (RS), X-ray diffraction (XRD), and nuclear magnetic resonance (NMR) spectroscopy.

2.2.2.1 Zeta Potential

The approach of zeta potential assessment is used to determine the dispersibility of NPs in solution. The electric potential at the interface of the double layer on the surface of the particle is referred to as the particle's Zeta potential. The intensity of electrostatic interactions between two charged particles is related to the square of the zeta potential values. The approach of electrophoretic light scattering (ELS), which might concurrently detect the paces of numerous charged atoms in fluids, is the very frequently employed method for determining zeta potential. Zeta potential studies are frequently employed to enhance dispersion, emulsion, and suspension composition. Particles' Zeta potentials normally vary from −100 to 100 mV. Zeta potential values for well-scattered NPs are larger than or even less than 30 mV. NPs with zeta potential values ranging from −10 mV to +10 mV is considered to be neutral. The Zeta potential is extremely susceptible to revolutions in pH and ionic force that occur during dilution [29]. As a consequence, measurements in diluted solutions could accurately depict the true value of its Zeta potential.

2.2.2.2 Mass Spectroscopy

MS is among the most important analytical methods for determining a particle's or molecule's weight, constituent makeup, and structural properties. MS works on the basis of mass-to-charge relations to identify charged particles of various compositions. MS has a significant level of precision and efficiency for determining molecular mass, in addition to detection accuracy and sensitivity, requiring just a few moles of a sample. Various physicochemical properties of nanoparticles, such as weight, content, and structure, may be portrayed using different MS approaches, which differ in terms of separation methods, ion sources, and detection techniques. Matrix-assisted laser desorption/ionization (MALDI) and electrospray ionization (ESI) are two ionization procedures often employed with MS instruments to ionize and solubilize thermally labile biomolecular constituents without causing substantial fragmentation or disintegration of the compounds. Ionization by inductively coupled plasma (ICP) is used mostly in the study of metal-comprising nanoparticles. Time of flight (TOF)-MS is used to evaluate the dimension and distribution of nanoparticles; MALDI-TOF-MS is used to quantify the molecular weights of polymers, macromolecules, and dendrimers other than to depict proteins bonding to NPs. On the other hand, ICP-MS was employed to confirm the coupling interaction of a synthesized NP with an altered contrast agent, as well as secondary ion detection. MS was also utilized to investigate the chemical and physical characteristics of the upper layer of NPs, in addition to the exterior properties of biomaterials under physiological settings. Although specific MS methods have been utilized to analyze the physicochemical properties of numerous biomolecules, the absence of comprehensive MS spectrum databases makes it difficult to accurately identify molecular species. This challenge is particularly evident in studies involving outcomes from MALDI-TOF-MS analysis. Furthermore, presently, the implementations of MS methods for nanomaterials have been limited in nanomaterial bioconjugate description, owing to the high charge of equipment, sample degradation, and prerequisite tools that are often offered for other studies. Figure 2.2 depicts the ESI and MALDI ionization sources and mass spectrometry instruments for nanomaterials with increasing mass and dispersion [30]. High-resolution mass spectrometry (HRMS) is typically utilized by atomically precise nanostructures, and multiplicative

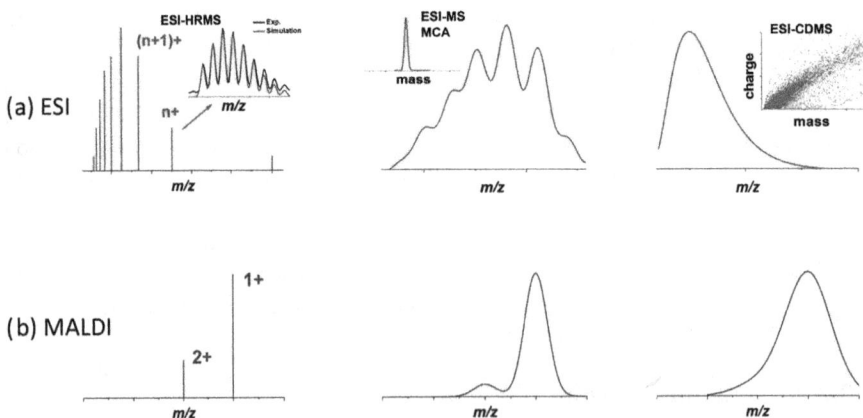

FIGURE 2.2 Illustration of the (a) ESI and (b) MALDI ionization sources and mass spectrometry instruments for nanomaterials with increasing mass and dispersion. Reproduced from [30] with permission from the Royal Society of Chemistry.

correlation algorithm (MCA) in combined application with ESI-MS could be beneficial for mass identification of polydisperse nanostructures (in the mass range 10–100 kDa), and charge detection mass spectrometry (CDMS) can be employed to determine the mass distribution of nanomaterials with masses surpassing one megadalton.

2.2.2.3 Infrared Spectroscopy

Fourier-transform infrared (FTIR) spectroscopy is a method used to assess the occurrence (and quantity) of IR-active functional groups or bonding in crystalline, microcrystalline, or amorphous organic or inorganic materials. FTIR is often used in nano research to discover distinctive spectral information that shows proteins, surfactants, or even other functional compounds attached to NP surfaces [10]. IR radiation is transmitted through a substance in infrared spectroscopy. The material absorbs a specified range of wavelengths of infrared light, which excites oscillations (e.g. stretching or bending vibrations) and rotations in bound molecules and atoms. The acquired infrared spectrum is a structural identity of the material. Generally, sample preparation is pounding the solid sample into a small particle and scattering it in semi-transparent materials such as KBr, NaCl, or CaF_2. To produce a mull, the powdered material can also be disseminated in a liquid (mineral oil). The samples can be put in the spectrometer in any form for transmission examination. Attenuated total reflection (ATR)-FTIR spectroscopy addresses one of the most difficult elements of infrared analysis that analyzes both liquid and solid specimens while minimizing the limitations of specimen preparation in transmission. Total inner reflectance arises in the internal reflection element (IRE) crystal in contact with the specimen in an ATR-FTIR system, creating evanescent vibrations that absorb into the material from the IRE crystal-sample interface, leading in a penetration depth of 0.5–5 μm. The material absorbs some of the evanescent currents, and the echo is then recorded on the detector.

2.2.2.4 Raman Spectroscopy

RS is a frequently applied technique for the structural description of nanomaterials and nanoparticles, providing sub-micron spatial resolution for light-transparent materials exclusive of the necessity for specimen groundwork, rendering it suitable for in-situ investigations. The concept of RS is to detect the non-elastic scattering of photons of various frequencies from the incident beam once it contacts with the particle's electric dipoles. The RS technique reveals differences in frequency between the photons that are scattered inelastically and those incident photons. These differences are linked to vibrational energy state characteristics. In the Raman spectrum, the photons resulting from non-elastic scattering, which have lower energy than the incident photons, are labeled as Stokes lines. Conversely, if the non-elastic scattered photons have higher frequencies than the incident photons, they are termed anti-Stokes lines. In the case of moderately symmetrical particles, Raman spectroscopy is often seen as analogous to infrared (IR) spectroscopy. This means that vibrational modes considered Raman active should be inactive in IR, and vice versa. This principle is based on the fact that Raman shifts result from nuclear movements, influencing the polarizability of particles, rather than arising from changes in the particles' dipole moment. Tip-enhanced Raman spectroscopy (TERS) uses an apertureless metallic tip as opposed to an optical fibre to obtain surface amplification of Raman data by applying the notion of restricting the light field in Raman nearfield scanning optical microscopy to surpass diffraction-limited resolution (the SERS effect) [31]. SERS and TERS, in contrast to standard RS, give topological information on nanomaterials in relation to their structure, chemical, and electrical characteristics. Nevertheless, the paucity of detection consistency in SERS due to shape and size change, along with unfavourable NP agglomeration, is an impediment to in vitro or in vivo imaging modalities.

2.2.2.5 X-Ray Diffraction

The polymorph, crystal structures, crystalline nature, and crystallite size of NMs have previously been determined using X-ray diffraction (XRD). The X-ray wavelength is equivalent to the distance of atoms in a scattering investigation. When a crystalline substance is exposed to X-rays, it forms a diffraction pattern with many sharp areas known as Bragg diffraction peaks. The crystalline structure of the substance being investigated may be identified by examining the locations and concentrations of these peaks. XRD is frequently utilized to identify atomic configurations in rigid crystal structures; however, it has limitations in obtaining data from single crystal formation. Unidentified crystals are recognized by comparing their peaks, locations, and related concentrations to a known pattern of a standard sample. Attributing a diffraction peak to a sample comprising numerous phases may be challenging; a computerized match is frequently utilized. The molecular pair distribution function (PDF) method, a non-Bragg-type approach, has been employed for weakly crystalline materials with extremely scattered XRD patterns [32]. Particle PDF data analysis appears to be distinct from standard XRD.

While XRD is a well-known method and has been widely utilized to identify the molecular structure of various materials, the difficulties in forming crystalline and the inability to obtain findings from more than one structural or binding state of the material restrict its usefulness. In comparison to electron diffractions, the additional drawback of XRD is the barely noticeablediffracted X-rays, especially for substances with lower atomic masses. A latest X-ray diffraction study indicated a fresh strategy utilizing femtosecond signals from a hard-X-ray free-electron laser for structural characterization, which could be advantageous for understanding the structure of macromolecules that do not produce appropriate crystal size for customary radioactive sources or do not attune to radiation damage.

2.2.2.6 Nuclear Magnetic Resonance

Unlike images and diffraction approaches, which provide information about the structure at long-range quest or the crystallographic feature, NMR is specific to the local surroundings and can reveal the geometries of polymers, amorphous materials, and biomolecules that possess less variety sequence. Besides assessing the configurations and contents of the materials, NMR spectroscopy includes facilities for investigating complex interactions of the species under different environments, such as relaxation, chemical composition, and molecular adaptability, which can be assessed using specially intended rf and/or gradient pulse patterns [33]. NMR spectroscopy has been employed to identify numerous physicochemical features of nanomaterials, namely structural homogeneity, and features in dendrimers, polymers, and fullerene constituents, along with structural changes that take place during ligand-nanomaterial engagements. Pulsed field gradient NMR has been utilized to assess the dispersibility of nanomaterials, allowing the dimensions and interconnections of the species beneath research to be computed.

NMR is a non-detrimental and non-aggressive approach that necessitates simple specimen processing. Contrary to visual approaches, NMR's poor detection capability necessitates a relatively larger sample volume for analysis. Spectral analysis might also be laborious because a specific degree of the signal-to-noise ratio is required. High-resolution magic angle spinning (HR-MAS) NMR has gained significant recognition in the fields of biology and biomedicine over the past few decades due to its capacity to generate spectra comparable to those obtained through high-resolution NMR. This technique is particularly valuable for studying diverse biological tissues and cells, thanks to its ability to provide detailed insights. The benefits of HR-MAS NMR for precise analysis of surface-enclosed ligands and customized exteriors have been employed for examining each synthesized stage of the cyclo-peptide debilitated on the top of poly(vinylidene fluoride)-based NPs and thermolytically produced thiol-constituents silver groups. Figure 2.3 displays the one-dimensional 1H NMR spectra acquired to determine the quality of the synthesized

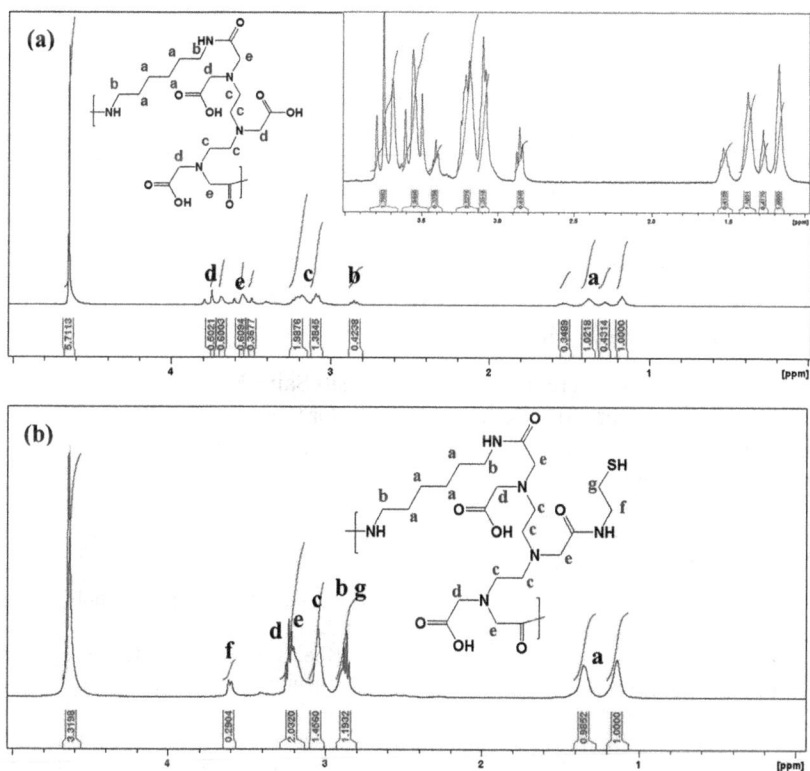

FIGURE 2.3 One-dimensional 1H NMR spectra of (a) synthesized HMDA–DTPA and (b) thiol group-terminated HMDA–DTPA polymer. Reproduced from [34] with permission from the Royal Society of Chemistry.

gold nanoparticles and to gather chemical shift details on the ligands that attach to the gold nanoparticles {hexamethylenediamine-diethylenetriaminepentaacetic dianhydride (HMDA-DTPA)} [34].

2.3 CONCLUDING REMARKS

Considering the distinctiveness of physicochemical properties at the nanoscale level, nanomaterials have the ability to influence physiological engagement from the molecule to the structural level, rendering the in vivo application of nanomaterials an intriguing area of investigation. Presently, there is no dearth of established methods or an appropriate approach for characterizing NMs. The fast research and manufacturing of nanomaterials for the purpose of multiple applications demonstrate the need for and prudence of nanomaterials regulation. For regulatory standards to ensure the security of nanomaterials in overall, robust methodologies for the characterization of nanomaterials are essential. This study discusses the fundamental physicochemical features of nanomaterials, comprehended by a description of the various approaches used to characterize nanomaterials. Hence, it is required to describe the nanomaterial intended for numerous applications in its original state upon insertion into a physiological environment. The quick summary of each approach, along with its advantages and disadvantages, presents us with a framework for choosing the proper methods for characterizing a prospective nanomaterial. There is no procedure that can be utilized for all NM varieties. On the account of the qualities of NMs, analytical methodologies and operating parameters should be selected in a case-by-case scenario. As there are currently no empirically appropriate procedures for all classes of NMs, it is vital not only to provide input values but also to provide comprehensive information on the sample processing, applicable technique, and any changes that were required to execute the analysis.

ACKNOWLEDGEMENTS

The authors express their gratitude to the Universiti Sains Malaysia Short-term grant scheme (Project number: 304/PKIMIA/6315731) for their awards. The authors would also like to thank the School of Chemical Sciences, Universiti Sains Malaysia, for their technical assistance.

REFERENCES

[1] W.J. Stark, P.R. Stoessel, W. Wohlleben, A. Hafner, Industrial applications of nanoparticles, *Chemical Society Reviews*, 44 (2015) 5793–5805.

[2] M. Bundschuh, J. Filser, S. Lüderwald, M.S. McKee, G. Metreveli, G.E. Schaumann, R. Schulz, S. Wagner, Nanoparticles in the environment: where do we come from, where do we go to?, *Environmental Sciences Europe*, 30 (2018) 1–17.

[3] S.A. Mazari, E. Ali, R. Abro, F.S.A. Khan, I. Ahmed, M. Ahmed, S. Nizamuddin, T.H. Siddiqui, N. Hossain, N.M. Mubarak, Nanomaterials: applications, waste-handling, environmental toxicities, and future challenges-A review, *Journal of Environmental Chemical Engineering*, 9 (2021) 105028.

[4] P. Laux, J. Tentschert, C. Riebeling, A. Braeuning, O. Creutzenberg, A. Epp, V. Fessard, K.-H. Haas, A. Haase, K. Hund-Rinke, Nanomaterials: certain aspects of application, risk assessment and risk communication, *Archives of Toxicology*, 92 (2018) 121–141.

[5] C.J. Chirayil, J. Abraham, R.K. Mishra, S.C. George, S. Thomas, Instrumental techniques for the characterization of nanoparticles, In Sabu Thomas, Raju Thomas, Ajesh K. Zachariah, Raghvendra Kumar Mishra (Eds.), *Thermal and rheological measurement techniques for nanomaterials characterization*, Elsevier, Amsterdam, 2017, pp. 1–36.

[6] M.-C. Bernier, M. Besse, M. Vayssade, S. Morandat, K. El Kirat, Titanium dioxide nanoparticles disturb the fibronectin-mediated adhesion and spreading of pre-osteoblastic cells, *Langmuir*, 28 (2012) 13660–13667.

[7] H. Jin, N. Wang, L. Xu, S. Hou, Synthesis and conductivity of cerium oxide nanoparticles, *Materials Letters*, 64 (2010) 1254–1256.

[8] S. Dominguez-Medina, S. McDonough, P. Swanglap, C.F. Landes, S. Link, In situ measurement of bovine serum albumin interaction with gold nanospheres, *Langmuir*, 28 (2012) 9131–9139.

[9] Z. Khatun, M. Nurunnabi, K.J. Cho, Y.-k. Lee, Oral delivery of near-infrared quantum dot loaded micelles for noninvasive biomedical imaging, *ACS Applied Materials & Interfaces*, 4 (2012) 3880–3887.

[10] K.E. Sapsford, K.M. Tyner, B.J. Dair, J.R. Deschamps, I.L. Medintz, Analyzing nanomaterial bioconjugates: a review of current and emerging purification and characterization techniques, *Analytical Chemistry*, 83 (2011) 4453–4488.

[11] J. Choi, V. Reipa, V.M. Hitchins, P.L. Goering, R.A. Malinauskas, Physicochemical characterization and in vitro hemolysis evaluation of silver nanoparticles, *Toxicological Sciences*, 123 (2011) 133–143.

[12] T. Zhao, K. Chen, H. Gu, Investigations on the interactions of proteins with polyampholyte-coated magnetite nanoparticles, *The Journal of Physical Chemistry B*, 117 (2013) 14129–14135.

[13] J.-P. Lavigne, P. Espinal, C. Dunyach-Remy, N. Messad, A. Pantel, A. Sotto, Mass spectrometry: a revolution in clinical microbiology?, *Clinical Chemistry and Laboratory Medicine (CCLM)*, 51 (2013) 257–270.

[14] I. Gmoshinski, S.A.e. Khotimchenko, V.O. Popov, B.B. Dzantiev, A. Zherdev, V. Demin, Y.P. Buzulukov, Nanomaterials and nanotechnologies: methods of analysis and control, *Russian Chemical Reviews*, 82 (2013) 48.

[15] A.K. Zak, W.H.A. Majid, M. Darroudi, R. Yousefi, Synthesis and characterization of ZnO nanoparticles prepared in gelatin media, *Materials Letters*, 65 (2011) 70–73.

[16] C.S. Kumar, Raman spectroscopy for nanomaterials characterization, In Challa S.S.R. Kumar (Ed.), Springer Science & Business Media, Verlag Berlin Heidelberg, 2012.

[17] F. Sinjab, B. Lekprasert, R.A.J. Woolley, C.J. Roberts, S.J.B. Tendler, I. Notingher, Near-field Raman spectroscopy of biological nanomaterials by in situ laser-induced synthesis of tip-enhanced Raman spectroscopy tips, *Optics Letters*, 37 (2012) 2256–2258.

[18] F. Schacher, E. Betthausen, A. Walther, H. Schmalz, D.V. Pergushov, A.H.E. Müller, Interpolyelectrolyte complexes of dynamic multicompartment micelles, *ACS Nano*, 3 (2009) 2095–2102.

[19] V.M. Gun'ko, J.P. Blitz, V.I. Zarko, V.V. Turov, E.M. Pakhlov, O.I. Oranska, E.V. Goncharuk, Y.I. Gornikov, V.S. Sergeev, T.V. Kulik, B.B. Palyanytsya, R.K. Samala, Structural and adsorption characteristics and catalytic activity of titania and titania-containing nanomaterials, *Journal of Colloid and Interface Science*, 330 (2009) 125–137.

[20] C. Zhou, Z. Liu, X. Du, D.R.G. Mitchell, Y.-W. Mai, Y. Yan, S. Ringer, Hollow nitrogen-containing core/shell fibrous carbon nanomaterials as support to platinum nanocatalysts and their TEM tomography study, *Nanoscale Research Letters*, 7 (2012) 165.

[21] A. Marchetti, J. Chen, Z. Pang, S. Li, D. Ling, F. Deng, X. Kong, Understanding surface and interfacial chemistry in functional nanomaterials via solid-state NMR, *Advanced Materials*, 29 (2017) 1605895.

[22] T. Iline-Vul, N. Adiram-Filiba, I. Matlahov, Y. Geiger, M. Abayev, K. Keinan-Adamsky, U. Akbey, H. Oschkinat, G. Goobes, Understanding the roles of functional peptides in designing apatite and silica nanomaterials biomimetically using NMR techniques, *Current Opinion in Colloid & Interface Science*, 33 (2018) 44–52.

[23] M. Abd Mutalib, M. Rahman, M. Othman, A. Ismail, J. Jaafar, Scanning electron microscopy (SEM) and energy-dispersive X-ray (EDX) spectroscopy, In Nidal Hilal, Ahmad Fauzi Ismail, Takeshi Matsuura, Darren Oatley-Radcliffe (Eds.), *Membrane characterization*, Elsevier, 2017, pp. 161–179.

[24] E. Ortiz Ortega, H. Hosseinian, M.J. Rosales López, A. Rodríguez Vera, S. Hosseini, Characterization techniques for morphology analysis, In E. Ortiz Ortega, H. Hosseinian, M.J. Rosales López, A. Rodríguez Vera, S. Hosseini (Eds.), Material characterization techniques and applications, Springer, Singapore, 2022, pp. 1–45.

[25] N. De Jonge, F.M. Ross, Electron microscopy of specimens in liquid, *Nature Nanotechnology*, 6 (2011) 695–704.

[26] S. Angayarkanni, J. Philip, Review on thermal properties of nanofluids: recent developments, *Advances in Colloid and Interface Science*, 225 (2015) 146–176.

[27] J. Cui, S. Yan, J. Wu, S. Bi, Determination of thermal and mutual diffusivity of n-heptane with dissolved carbon dioxide by dynamic light scattering, *Fluid Phase Equilibria*, 526 (2020) 112804.

[28] K. Bian, C. Gerber, A.J. Heinrich, D.J. Müller, S. Scheuring, Y. Jiang, Scanning probe microscopy, *Nature Reviews Methods Primers*, 1 (2021) 36.

[29] C.N. Lunardi, A.J. Gomes, F.S. Rocha, J. De Tommaso, G.S. Patience, Experimental methods in chemical engineering: zeta potential, *The Canadian Journal of Chemical Engineering*, 99 (2021) 627–639.

[30] C. Comby-Zerbino, X. Dagany, F. Chirot, P. Dugourd, R. Antoine, The emergence of mass spectrometry for characterizing nanomaterials. Atomically precise nanoclusters and beyond, *Materials Advances*, 2 (2021) 4896–4913.

[31] H.-L. Wang, E.-M. You, R. Panneerselvam, S.-Y. Ding, Z.-Q. Tian, Advances of surface-enhanced Raman and IR spectroscopies: from nano/microstructures to macro-optical design, *Light: Science & Applications*, 10 (2021) 161.

[32] V. Petkov, Nanostructure by high-energy X-ray diffraction, *Materials Today*, 11 (2008) 28–38.

[33] T.D. Claridge, High-resolution NMR techniques in organic chemistry, In T.D. Claridge (Ed.), Elsevier, Amsterdam, 2016.

[34] P.F. Hsiao, R. Anbazhagan, C. Hsiao-Ying, A. Vadivelmurugan, H.-C. Tsai, Thermoresponsive polyamic acid-conjugated gold nanocarrier for enhanced light-triggered 5-fluorouracil release, *RSC Advances*, 7 (2017) 8357–8365.

3 Nanosafety
Exposure, Detection, and Toxicology

Nusrat Tara
Jamia Millia Islamia

Mukul Pratap Singh
GL Bajaj Institute of Technology and Management

3.1 INTRODUCTION OF NANO-SAFETY

The fast improvement and life style changes have made life more attractive, comfortable, and expensive. The new generation is moving toward smaller, safer, cheaper, and faster working materials since these materials lessen the work consignment and support to convey easier at a plentiful pace with minimal effort. The term nanotechnology has been coined from 'nanometer'. It is a unit used to measure one billionth of a meter of length. The prefix 'nano' in Greek means 'dwarf', and it signifies the one-billionth part (10^{-9}m) of hair of human being [1–3]. Nanotechnology is relatively a novel term, although the fundamental technology dates back to the the period when "submicro" was used to nmanufacture enormously small constituent parts of polymers and copolymers. Now, the technology compact materials with the engineering and science at the magnitude of length 1–100 nm are baptized nanotechnology (NT) [4]. The word nanotechnology describes the manipulation, length scale, internal and external properties of the atom or bulk substances that can be developed through improved materials like devices, structure and well-developed systems for new generation [5,6].

3.2 EXPOSURE TO NANOMATERIALS (METAL NANOPARTICLES, FULLERENES, AND CARBON NANOTUBES)

There have been various gadgets developed for everyday life that are much smaller, such as microsized chip, nanosized tablets, nanocarbon tubes, small memory chips, and pen drives, which can make transportation and storage capacity much easier and consistency in less time. Nanomaterials with various dimensions, such as 0D, 1D, 2D, and 3D, are extensively applied for several applications. Various metal nanoparticles like: iron oxide (Fe_3O_4) is applied for remediation of heavy metals and color stuff from ground water; titania and titanium oxide nanoparticles are used for paints and cosmetic products; fullerenes are used for tennis rackets and LED video screens;

DOI: 10.1201/9781003400998-3

silica is used for the manufacturing of electronics products; zinc its oxides used for industrial purpose like coating to electronic screens from protect the ultraviolet rays; Ag nanoparticles are used as a antimicrobial agents; and carbon- and graphene-based nanoparticles are used as electrodes in fuel cells [7].

3.2.1 NANOPARTICLES

Nanoparticles are the model of archetypical nanomaterials and are considered extremely affective strategy associated with the nanotoxicity. Nanotoxicity is generally based on the interpretations and the treatment of nanoparticles (NPs). A nanoparticle also gives emphasis to the information about the dimension of nanoscale, which assists unreactive constituents and turns out to be reactive catalytic positions. Gold (Au) nanoparticle, a noble metal, is unreactive in bulky amounts [8]; however, the size of particle >10 nm (less than) is extremely reactive material and can be formed into improved geometries like nanorods of gold, which are biologically non-hazardous [9]. Gold (Au) nanoparticles have high congeniality for sulfur, most significantly conjugated with biological macromolecules containing unique group like thiols [10]. The cytotoxicity degree of the material is interrelated to the modification or coating of the used area. For example, gold nanoparticles coated with cetyl-trimethyl-ammonium bromide or poly-di-allyl-dimethyl ammonium chloride are both affected by cells. On the other hand, the coating of diallyl products is nontoxic, while the coating of cetyl products is toxic. The mechanism of toxicology through nanoparticles (NPs) is given in Figure 3.1. Once nanoparticles arrive into the body, they may pass over numerous cellular barricades and enter the highly sensitive body organs such as the kidney, liver,

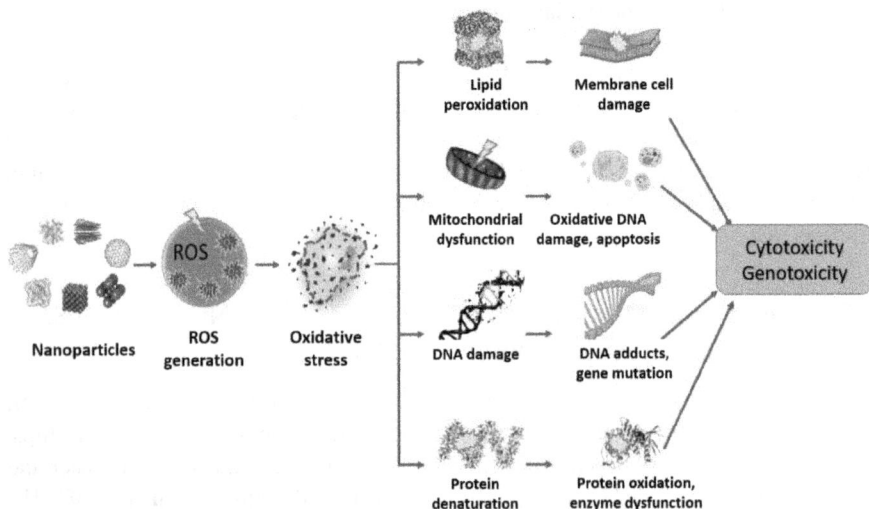

FIGURE 3.1 Toxicity mechanism of nanoparticles mediated by ROS generation. Intracellular ROS can be engendered from the mitochondria cell and later cause lipid peroxidation, damage of DNA, and protein denaturation. Reproduced with permission from Ref. [20], Elsevier license number 501760542.

and lungs, causing damage to mitochondria, DNA structure, and ultimately cell apoptosis or death [11–14]. The reactive oxygen species (ROS) production, which could be the main reason for oxidative stress, inflammation, protein damage, DNA, and membrane of cells, is the largest production mechanism leading to toxicity [15–19].

3.2.2 FULLERENES

A fullerene is a carbon allotrope whose materials contain carbon atoms linked through bonds like single bond or double bonds so as to form a closed or partially closed web, with fused rings of carbon atoms (five to seven). The fullerene molecule may be a tube, a hollow sphere, ellipsoid or available in many more shape and size. The carboxylated negatively charged fullerenes show the prospective selectivity of migrating across membranes, including −ve charged cerebral micro-vessel endothelial (cells), where the indigenous negativity is decreased in the membrane by the stress of oxidative response, allowing the derivative fullerenes to transfer across the membrane. Hence, the nano-system can defend against hydrogen peroxide (H_2O_2)-induced stress and F-actin de-polymerization at the entry point and facilitate cells into a viable state in discriminating against injured cells that suffer programmed death cells [21]. In another study, the hydroxylated derivative attached to the gadolinium is the supreme optimal [22]. The insertion of a metal component in center of fullerene and the manufacturing of the metallo-fullerene form allow the nanosytem to become more hydrophilic, while specific clusters aggregate and others are comparable to the carbon parent [23]. In general, there are various types of metallo-fullerenes available in the environment, and comparison between various metallo-fullerenes depends on the size, shape, and surface modification of any nanomaterials. In the hepatoma animal model, the gadolinium-hydroxylated metallo-fullerene intra-peritoneal injection in mice shows various toxic effects on the mice's body parts. After injection, the intra-peritoneal injection confirmed the circulation of the whole kidney (more copious), stomach, liver, and then less copious in the spleen, with the insignificant concentration of intra-peritoneal injection in the brain or serum. However, the derivative of carboxylated generally accumulates in the kidney, with minimal accumulation in the liver. Gadolinium-based hydroxylated metallo-fullerene is not cytotoxic in human being and animal models in vitro cell-based analyses [24]. These types of metallo-fullerene show excellent antioxidant properties through assets of high electron affinity and deficiency of toxicity and allow the anti-proliferation of neoplastic cells with metallo-fullerenes of properties to cyclophosphamide [25].

3.2.3 CARBON NANOTUBES

Carbon nanotubes (CNTs) are a type of fullerene with the specific cylindrical or tube-like shape and are very similar to the structure of the fullerenes in that they are hydrophobic in nature, altered fullerenes. According to their hydrophobic nature, CNTs are identified as being stored in the lungs and liver; they are affecting inflammation and pulmonary injury because of the induction of mesothelioma. However, CNTs may be bioconverted through Fenton-like reactions, in contrast to more polar hydroxylase CNTs that are biodegraded through the peroxidases reaction [26–32]. In asbestos-connected fibrosis, additionally, pulmonary responses are described by the fiber length; subsequently, lengthier fibers are lesser clean continue with the longer duration as comparison to short fibers; however, there is no clear clarification between the length and harshness of pathological reaction [33]. The CNTs are available in two different lengths, such as single-walled carbon nanotubes (SWCNTs) and multi-walled carbon nanotubes (MWCNTs), while asbestos fibers have a very different structure. A colorimetric tetrazolium response in vitro cell evaluation confirmed the SWCNTs are highly cytotoxic as compared to MWCNTs in the alveolar macrophage stage [34]. Unlike asbestos fibers, CNTs are easily conjugated, which can be easily bound with the help of weak interactions (like van der Waals) to other systems, including pep stacking, which means facilitating the adsorption of protein and CNTs cellular cytotoxicity to turn the "tunes" [35–37]. There are also various methods available for the evaluation of tune solubility such as surface modification [38], insertion of metal [39], and addition of covalent [40]. For instance, dual-end CNTs structure modified with hydroxyl groups (–OH) can demonstrate physical characteristics parallel to dimyristoyl-phosphatidyl-choline established in membranes. The CNTs sidewall surface is simply modified with sulfo– or –COOH groups, resulting in a smaller amount of cytotoxicity in cultivated human dermal fibroblasts than individuals stabilized with matter like detergent, due to the selective membrane of SWCNTs [41]. In other studies, CNTs coated with polyethylene glycol (PEG), polyethyleneimine, and doped of nitrogen-based data indicate a negligible immune response unless directly injected into the trachea of mice [42–44].

3.2.4 METAL/METAL OXIDE AND QUANTUM DOTS

Metal nanoparticles are also known as the modern era of nanomaterials in that these nanosystems facilitat various novel characteristics such as magnetic, optical, and catalytic functionalities. Variation type of metal oxide sizes depends upon the toxicity, where nano-sized metal oxides are more toxic as compared to the milli or micro-sized metal oxide, due to increased catalytic efficiency, and organs are affected by the high amount of toxicity. For instance, if iron oxide nanoparticles can be directly inhaled, they affect the central nervous and olfactory nerve systems [45]. Another material zinc oxide is generally known as an antibiotic characteristic of the action of metal oxide. This incidence was also observed in zebrafish embryos. If they applied a higher concentration of metal oxide in zebrafish, the larvae died, but if the amount of concentration was less, then the embryos showed only deformation, and the results confirmed the metal oxide was more toxic as compared to zinc ion. This phenomenon proposes that the toxicity of the metal oxide is greater than M-ion (metal ion), clusters of bulky metal oxide can be combined and show less toxicity, whereas middle- and small-size metal oxide shows the slow development of the zebra fish embryos [40]. Titania was one the earliest metal oxides that was displayed excellent properties like photocatalytic oxidant activity in the ultraviolet region and was highly applied for the water purification [46–48].

Various sizes of titania nanoparticles inject directly into the stomach, which can show toxicity for varying parts of the body such as the liver, heart, kidney, etc.;

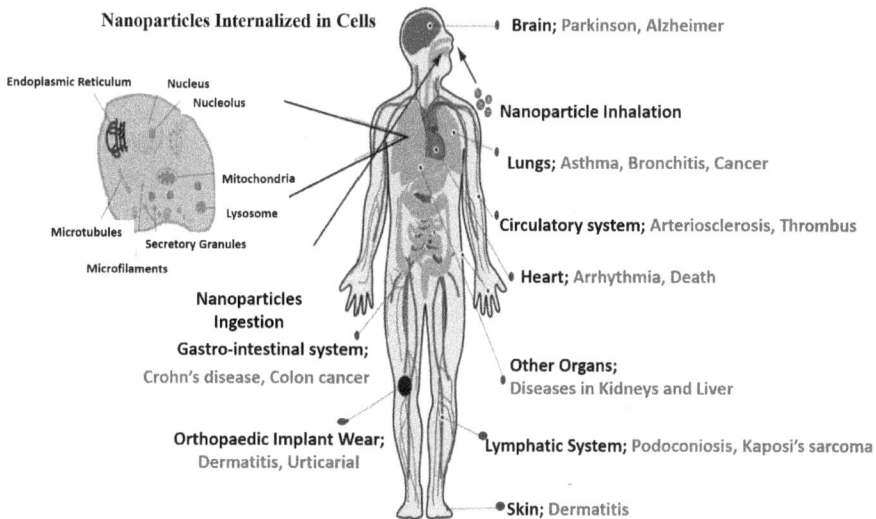

FIGURE 3.2   Diseases associated with nanoparticles. Reproduced with permission from Ref. [20], Elsevier license number 501760542.

however, no morphological injury to the lungs or spleen with the increased amount of titania nanoparticles in the lungs, liver, kidney, and spleen (Figure 3.2) [49,50]. Another nanomaterial ceria has also catalytic activity due to its storage capacity for oxygen [51,52]. The observed study of ceria in rats proposed that nanoparticles used as an inhaled process can be cumulative in the broncho-alveolar and redistributed through the rotation to other organs [53]. Other metals are also in core-shelled systems commonly recognized as quantum dots and widely used for imaging [54]. While there are various metals such as calcium, iron, lithium, manganese, magnesium, chromium, cobalt, copper, molybdenum, tin, vanadium, nickel, selenium, and zinc, and halogens such as iodine fluorine, chloride, and non-metal silicon, all are maintaining the cellular homeostasis performance. Additional metals such as arsenic, lead, cadmium, beryllium, mercury, hexavalent chromium, and beryllium are toxic in trace quantities [55–62]. The core-shelled quantum dots have interesting features that include the use of essential and toxic or available inert metals in their chemical composition. For instance, cadmium metal is enormously toxic and well known to cause various infections such as proteinuria, renal dysfunction, and other allied health problems, while selenium (Se) is an indispensible metal and is present in very low amounts. In cadmium-selenide-based quantum dots, there exists a fascinating arrangement where the discharge of cadmium ions highly display the toxicity. Zinc sulfide or proteins conjugates composite coating the core shell would reduce cytotoxicity, and the cytotoxicity is interrelated to the presence of cadmium or tellurium quantum dots [63].

Coating process leads to inhibiting the cadmium (Cd) suspension and may produce a health possibility, although various polar groups such as (poly) alcohols, zinc sulfide, and acids are combined with the core-shell and show nontoxic results. The coating of tri-noctyl-phosphine oxide is very toxic in human cells (B lymphoblast)

[64,65]. Fluronic 68 and sodium dodecyl sulfate quantum dots materials modified the surface in cell-cultured bioassay, and results suggested the material was non-toxic; however, cetyl-trimethyl-ammonium bromide coated with the quantum dots was toxic [66]. The L- or D-forms of glutathione used with protein coating demonstrated distinction toxicity, with the L-form coating being more toxic, then easily treated with L-form enzymes to process, and nanotoxicity is normally connected to size; minor size endorses the degradation of intracellular and apoptosis [67–69]. Assessment in animal primates exhausting phospholipid micelle-encapsulated core shell quantum dots of cadmium-selenide indicates no longer-term toxicity and may reflect various physiologies in binary types of animal representations [70].

3.3 DETECTION AND ITS PROCESS

The detection of toxicology divided into three major categories: cellular evaluation in cultured cells, investigation of the tissue/homogenate for organs in animal models, and morphologically changes. Morphological changes are of various types, like cardiovascular, nephrological, psychological, neurological, circulation, and olfactory analysis of animals to gauge variations in specific organisms such as the heart, lungs, variation in skin, perception of pain, function in the kidney, cognitive abilities, circulation, and behavior [58,70–79]. Discuss only one part in this section, like cellular evaluation to analyze the cellular homeostasis through ultrafine nanoparticles [80]. The cytotoxicity experiment and some reagents applied to the evaluation of cytotoxicity are shown in Figure 3.3.

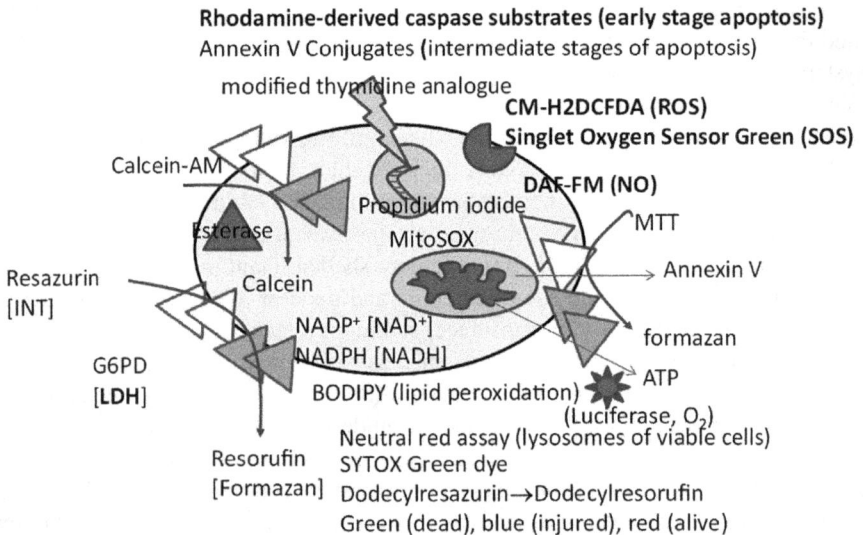

FIGURE 3.3 Experiments of cytotoxicity and some reagents applying to evaluation of nanomaterial cytotoxicity, as well as test for geno-toxicity. Reproduced with permission from Ref. [58], Elsevier license number 5384381158847.

Various numbers of cell culture have been established to estimate cytotoxicity and cell viability. The trypan blue technique is mostly used for the detection of cytotoxicity, which may include holding through the unviable cells or as well as counting of the colonies after the distinct period of incubation of each viable cell to gauge cytotoxicity [81]. The earlier stage may not investigate the cell viability but also show the metabolically inactive properties, while the final stage is a time-consuming stage and not cooperative to multiplexing. At this time, the common techniques applied for the enzymes within the cells, such as cell viability indicators, were the detection of each consumption of substrate or manufacture of the product using the absorbance's variations, luminescence, or fluorescence [82–84]. A common technique involves lactate dehydrogenase (LDH) or glucose-6-phosphate dehydrogenase (G6PD) as a viability function through measure formazan (LDH) or resorufin. The declining power tendency of a cell can be utilized to assess its health by converting the decreasing MTT to its insoluble formazan (LDH) through mitochondrial reductase activity [85]. Calcein AM is a non-fluorescent cell dye that can easily hydrolyze through the intracellular esterases to a green-colored fluorescent calcein in cells [86]. Determination of the fluorescence can be applied to the evaluation of cell proliferation. The calcein AM can applied in tandem using propidium-iodide, which is a color stuff (dye) for use on the chromosomes of dead cells and stains nuclear; it is cooperatively used to evaluate live or dead cells, which stain color is available in green or red [87]. Instead, two-color availability may be performed through the non-fluorescent dodecylresazurin, which is decreased by the metabolically active cells converted into the dodecylresorufin (they emit fluorescence like dark red) attached with the SYTOX fluorescence like green (a stain of nucleic acid) [88]. The membrane of plasma is injured, enabling the color of dye to accumulate. Imaging through green color fluorescence in the opposite injured cell is metabolically active and combines specific SYTOX green color florescence and dodecyl-resorufin, which emit green and red colors and may also appear blue color florescence in dead cells [89].

Finally, the unambiguous indicators for oxidative stress (OA) can be studied through selecting the effective binding dyes that undergo fluorescence upon binding agents such as 70-difluoro-fluorescein diacetate and 4-amino-5-methylamino-20, which certainly convert into the benzo-triazole, binding with NO, which is produced greatly fluorescent, and acetyl ester oxidation by ROS (like OH·radical), fluorescently produced adduct that is surrounded inside the whole cell. In the last calculation, the Singlet Oxygen Sensor (SOS) gives the green color, as a singlet oxygen ($1O_2$) detailed investigation become binding upon the fluorescent [90–92].

3.4 PROCEDURES

3.4.1 CELL CULTURES PROCEDURE

The h-TERT-RPE cell cultures in humans were continued in a buffer solution involving indispensable medium with a solution of Earle's salts increased within heat-inactivated fetal bovine serum (10%), penicillin (amount 100 units/mL), L-glutamine (1.4 mM), streptomycin (100 mg/mL), etc., constant at low temperature (35°C) in an incubator given other atmospheres like air (95%), CO_2 (5%), and humidity (75%). These types of cells were applied for the growth of the log-phase. The morphology of cells was determined through cellular integrity and microscopic techniques. A part of a cell is randomly selected for cellular integrity by using the lactate dehydrogenase test (evaluation of the integrity of the membrane) [93].

3.4.2 Monitoring of Nitric Oxide, Singlet Oxygen Sensor, Reactive Oxygen Species, MMP, and LDH Level(s)

In this section, their is a brief discussion about monitoring the various parts such as nitric oxide (NO) level, protein content, detection of singlet oxygen sensor (SOS) through the singlet oxygen species, (ROS), LDH level, MMP quantity, Caspase level, etc.

A brief discussion of the evaluation the protein content: here the cells were harvested from the culture plate and suspended in the buffer lysis solution by a simple procedure [94]. The cells were centrifuged for 10 minutes at a very low temperature of 4°C. After that, the pellet was re-suspended in a pH solution (~7.5) of phosphate buffered saline (PBS) to a limited volume. The result of the final volume obtained a high yield in the ranges of 100,000 and 150,000 cells per measurement [95]. The determination of protein content was observed at 1:8 ratios of cells after the addition of Bradford reagent (v/v) in dark medium and also adding BSA calibration (100–2,000 mg/mL) in a cuvette [96]. The prepared samples were incubated at a constant temperature of 37°C for a half-hour in a dark medium and then measured the absorption performance. For the activity of bioassay, a 1:1 v/v ratio was applied for the measurements of sample, reagent, and fluorescence [97]. The maximum excitation and emission are detailed at each level in Table 3.1.

According to the comparison of inhibitors, the entire cell was re-suspended in PBS accompanied with pyruvate (8 mM) and added to the specific additive or inhibitor between 1 and 900 μM. All compounds were soluble in water and properly mixed in deionized water or PBS; otherwise, they were assorted with dimethyl-sulfoxide (DMSO) solution. The stock solution of the concentration was highly abundant, such that only 0.5–25 mL of the prepared sample was added to the entire cell. It was always observed that the obtained volume was lesser than 10%, which removed the probability of DMSO toxicity on the entire cells [98,99].

TABLE 3.1

Monitoring the Nitric Oxide, Singlet Oxygen Sensor, Reactive Oxygen Species, MMP, and LDH Level(s)

S. No	Monitoring Agents	Excitation λ	Emission λ
1.	NO	495	515
2.	ROS	492	517
3.	LDH	490	680
4.	SOS	504	525
5.	MNP	549	575

3.5 TOXICOLOGY

There are various research studies that represent the influence of nanomaterials in forming various diseases such as interstitial lung disease, rhinitis, pleural, asthma, lung infections, respiratory embolism, dermatitis, tuberculosis, bosom distortion, immune system illnesses, growth of lung, and so on [20,50,55–58]. Therefore, the crucial understanding of toxicology through nanomaterials (nano-toxicology) is very impressive in this current situation. The systematic figure can easily understand that nano-toxicology can help scientists choose nanomaterials that are environmentally kind and prioritize further research and minor health effects on human health and the environment. A brief discussion of the various types of cell death and their toxicities also includes the factors affecting the physicochemical characteristics of nanomaterials and their influence on the cell through physiological activity.

3.5.1 EVALUATION OF CYTOTOXICITY

The cytotoxicity bio-evaluation was based on the yield of the total fluorescence, which was the ratio of the entire cell in PBS solution (pH 7.4). This reaction did not obligate any explanation to stop to permit the kinetic investigation for long-term. The limit of evaporating the cells after a long time duration (17 hours) in the 96-well plate or cuvette and as probably the fluorescence rose as a time function. These two evaluations have investigated the fluorescence induced through the LDH leakage, which is directly evaluating the integrity of cell membrane and indirectly necrosis; after that, caspase 3/7 also measures apoptosis [100]. These results show the entire cell in PBS controls various functions in "Ag-series" of MOFs such as different readings for initial and final products, time function, ratio, and bust level [100]. The other experiments were compared and partly discussed with further elaboration in Section 3.5.4.

3.5.2 EVALUATION OF LDH

According to morphological studies using SEM and TEM analysis (scanning electron microscopy and transmission electron microscopy), the nanoparticles arrived at the nuclei and vesicles and affected the nucleus, although nanowires only covered the upper layer of the cells. When the upper layer area (surface) was increased, they promoted the level of increasing toxicity of the cell. The ceria nanorods and nanowires increased the ratio aspect of the toxicity by exposing the monomyelocytic leukemia cell lines (THP-1) for 1 day in humans and also studying the evaluation of LDH. The outcomes exhibited that the tiny size of ceria nanorods were not at all toxic; however, the ratio of nanorods of intermediary did not source the cell death; it only induced the production of IL-1β. The nanorods attached with the maximum aspect ratio demonstrated the highest toxicity and removed the IL-1β [7,101].

In another study, CTAB modified with the gold nanoparticles unpleasantly affected the microtubules, nuclear lamina, and filamentous actin [102]. In Figure 3.3, the measurement of the fluorescence by the initiation of INT (iodophenyl, nitrophenyl, and tetrazolium) to formazan is represented. The bioassay principle is placed on the LDH leakage into the extracellular medium, which converted the lactate to simply pyruvate in presence of catalyzes in an oxidation state with decreased the oxidizing agents nicotinamide adenine dinucleotide (NAD⁺), and correspondingly converted the INT to formazan in a couple oxidation process measurement. Two absorbance values were given different data, which evaluated the cytotoxicity effort-lessly. These results showed that the minimum quantity controlled the fluorescence, and after that addition, the 7-ethyl-10-hydroxycamptothecin, an anti-topoisomerase, or H_2O_2 gave higher fluorescence while nitrite provided practically no response [103–105] (Figure 3.3). The 7-ethyl-10-hydroxycamptothecin is generally known as code SN-38, and that is usually anticancer drug. The SN-38 is known for its DNA topoisomerase I competency in addition to inhibition of synthesis of DNA, resulting in in greater interruptions in only single-strand DNA, probable to create a maximum oxidative atmosphere and loss of the plasma membrane. H_2O_2 is commonly known as a highly oxidizing mediator that can easily lesion the plasma membrane. Adding the H_2O_2 gives two separate profiles on the different sides of toxicity: at a high amount of dosage, the oxidizing degree causes the lesion of the plasma membrane and ion leakage, while a low amount of dosage of H_2O_2 has protective properties [106–108].

3.5.3 EVALUATION OF CASPASE-3/7

To further investigate cytotoxicity through the MOFs, we used the caspase 3/7 bioas-say method. To provide an overview, the association between cytotoxicity and bio-assay was described in Figure 3.4. Caspase 3/7 is a copious cysteine protease that cleaves the substrate portion at the C-terminus of the aspartate filtrate residue; there-fore, introducing the new member of peptide derivatives and cleaving through cas-pase, the product is fluorescence obtained and can be evaluated certainly [109]. In a

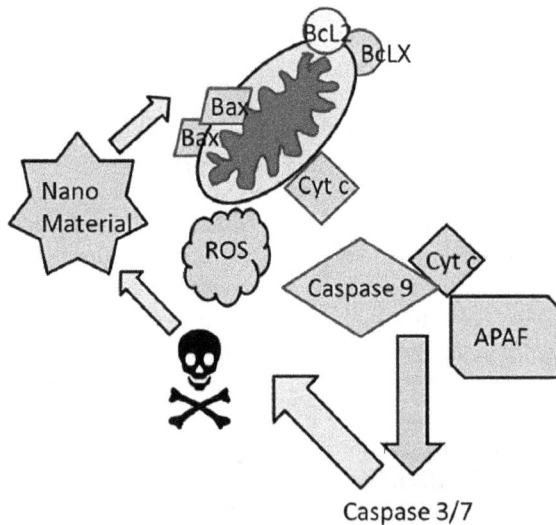

FIGURE 3.4 Correlation between cytotoxicity and bioassay. The flow shows the work as Mitochondria/ROS/Cyt C/Caspase-9 (inactive stage)/Caspase-3/7 (active stage) and Apoptosis. Reproduced with permission from Ref. [58], Elsevier license number 5384381158847.)

fixed amount of peptide concentration, the maximum availability of the fluorescence obtains a maximum number of apoptosis and caspases enzymes. Figure 3.4 shows that the "Ag" MOF (nanomaterial) series does not encourage the levels of caspase to be superior to the resistor. The entire cells are treated with numerous MOFs (metal-organic framework) series, such as "Ag", and "Fe". The Ag has maximum level and was measured while Fe has the minimum level and was measured relative to resistor, representing the previous outcomes by the evaluation of LDH and portentous role of the MOFs, which show all are not cytotoxic [58].

3.5.4 CHANGES IN INTRACELLULAR LEVELS OF NITRIC OXIDE (NO)

The outcomes with the combination of nitric indicated that further biochemical, such as NO generation, possibly would be activated. The NO is a neutral molecule with one unpaired e− and has miscellaneous activities shown in mammalian cells. NO has various functions. One of the major functions is a vasodilator and the other is used as neurotransmitters. NO may be created by the nitrite reduction through the reductase enzyme or oxidases originating in the mitochondria inner portion or in a single cell. Systematically, the proteins help the respiratory system in mitochondria that can enable the redox interactions, for instance, with iron nanoparticles [110]. Iron (Fe) can source two e− in nitrite reduction and convert them into the oxidized state (Eqs. 3.1 and 3.2).

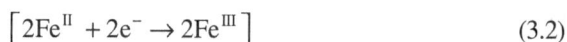

$$\left[2H^+ + 2e + NO_2^- \rightarrow H_2O + NO \right] \tag{3.1}$$

$$\left[2Fe^{II} + 2e^- \rightarrow 2Fe^{III} \right] \tag{3.2}$$

The created NO is surrounded through dye DAF-FM fluorescence [93], which can estimate the environment of local NO. The reactions are systematically summarized in Figure 3.5. Each possible mechanism of the reaction is shown here, such as centers of iron, potential sites for inhibition area with rotenone, carbon monoxide or cyanide,

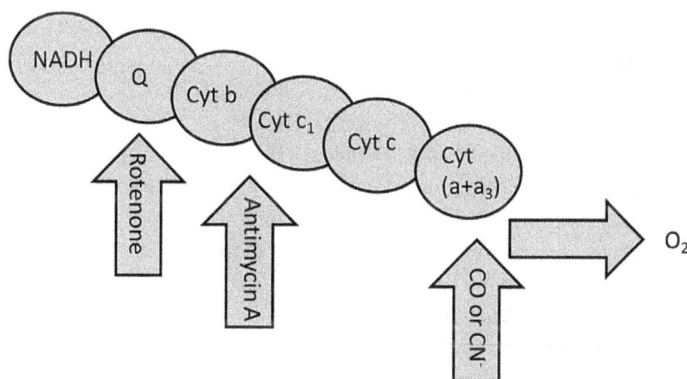

FIGURE 3.5 Illustration show the relation of inhibitor and inhibition place within the e− transport chain. Reproduced with permission from Ref. [58], Elsevier license number 5384381158847.

and Antimycin A [111–115]. Another well-known incident is the ability of NO to resist the poisoning of the cyanide, probably by changing the position of the cyanide (CN) from the centers of iron of oxidase in cytochrome c [116,117].

Finally, the CN was located in the perspective of new inhibitors attached to cytochrome c oxidase [118]. The last electron acceptors attach the rotenone or antimycin A from the transport chain of electrons as opposed to both enzymes, which bind earlier, as represented in Figure 3.5. Rotenone generally inhibits the electron transfer from centers of Fe-S in complex I to ubiquinone [119]. However, antimycin A attached to the Q spot of Cyt-C reductase, ubiquinol oxidation was inhibited, and CN & CO attached to the Cyt-C oxidase and electrons transportation was inhibited from the Cyt-C to oxygen, bringing a stop to the production and respiration systems of ATP [120]. NO is well known to inverse the effect of CN [121]. The level of NO increased with the support of time; as greater NO comes into the surface of the dye, the interaction generated greater fluorescence. Still continuously increasing the levels of NO with the addition of nitrite, this is a simple method for producing NO. This reaction may be completed by the heme active sites of cytochrome c oxidase. The addition of sodium nitroprusside, potassium ferrocyanide, and NaCN has decreased the levels of NO. The previous data indicate that NaCN is a very strong inhibitor of the cytochrome c oxidase and inhibits the cytochrome c oxidase active functions, which are intricate in the reduction of NO. Therefore, increasing the level of CN will prevent the enzyme growth from the nitrite converted to NO through the reduction process and modify the oxygen demand of the mitochondria, causing the main factor for the production of oxidizing species like H_2O_2 [122].

3.5.5 CHANGES IN MITOCHONDRIAL MEMBRANE POTENTIAL (MMP)

In the MMP assay, Fluorescence Resonance Energy Transfer (FRET) peptide is used as an indicator to find out the generic MMP activity. For the whole FRET peptide, one part shows active fluorescence while the first one quenches another one. After the cleavage, they were divided into two different fragments through the MMP and then recovered the fluorescence. The previous section clearly indicates the NO levels widely fluctuated through the "Ag" MOFs, which simply defined the silver outcomes that the silver show harmful effects to the cell and/or enzymes, also known as nanosilver, and silver MOFs trapped in the NO. Therefore, the MMPs procedure was applicable for mitochondrial stress due to its silver MOFs. A general fluorescent dye, TMRE (tetra-methyl-rhodamine ethyl ester), aggregates in mitochondria owing to their relative −ve charge on the membrane, the membrane was damaged mitochondrial cell will discharge the dye molecule to the cytoplasm, let down the measuring the fluorescence [123]. The fluorescence is openly interconnected to quantity of the dye and also interconnected to mitochondria due to its membrane potential. MMPs indicator gave similar measurements of TMRE fluorescence for every cell, but in the case of sodium nitroprusside and sodium nitrite gives different measurements, which suggested the production and oxidation of NO both are associated, ensuing in the lowering of the oxidation and membrane potential.

3.5.6 CHANGES IN ROS

The ROS was investigated through the changes in the ROS process. In simple words, a dye molecule binds the oxidizing species and fluorescence; therefore, this result is directly proportional to the oxidizing environment (Figure 3.3), as shown in the labeling where ROS is specifically denoted to the hydroxyl or superoxide anions). Herein, a control investigation that was carried out previously verified that the fluorescence linearly increased with the addition of a huge dosage of H_2O_2, which is a strong oxidizing species, while ROS dye does not interact with reactive nitrogen species, like peroxy-nitrite [124]. These obtained results suggested that specific constituents are not cytotoxic through the ROS procedure of cellular inactivation.

3.5.7 DISCUSSION ON MOFs TOXICOLOGY THROUGH THE CELL CULTURE

The synthesis of structured combining labile metal ions is accomplished by consuming various methods such as solvothermal and organic ligands methods. The MOFs are extremely porous, crystalline from long-term series, and have the opportunity to contain hard patterns for the structure like 3D form (three dimensional), and soft stress-free linkers are available for contact with the environment. This takes a different process (like vapor phase) with a distillation of the solution of metals such as Au and Pt (noble metals) [125,126]. Various approaches were applied for the redox reaction chemistry to decrease the Au nanoparticles; however, outcomes indicate instability with prepared structures that are present in 3D forms. Recently, few scientists have focused on sheet-like structures with two-dimensional forms. At this point, the metal clusters in this area are some nanometers bigger than the pore size due to the mobility of ions [127]. Three common structural categories can be used as a beginning point. For instance, Cu-coordinate with benzene-tri-carboxylic acid develops a simple cubic structure with a minimum pore size (>2 nm). One more structure is formed in strong penetrating cubes, with a highly flexible sheet linked at the vertices, enlargement of outer 1D form channels with one ligand like Zn and benzene-dicarboxylic acid and another one is bi-pyridine as a soft linker. After that, kagome structure can be made through the indium and terephthalic acid used as metals and ligands. One-dimensional is formed with two sizes in kagome structure due to the two linkers or ligands used [128].

The consistency of the formed MOFs was studied through the IR spectroscopy analysis. For instance, the M–O bond or C=O stretches can be investigated by the IR spectroscopy. The zinc (Zn) MOFs demonstrated the ZnO bond in structure at ~1,640 cm^{-1} and C=O stretching appeared at ~1,390 cm^{-1}, respectively. The Cu ion can coordinate with the –OH functional group of the ligand and confirm the MOF structure in the compound through the FT-IR spectroscopy. The Cu-OH stretching appeared at 1,570 cm^{-1} and conformation the metal ligands bonds formed correctly. Also, generated the iron and silver series based MOFs using the above pattern [129]. The metal application is an extremely old technique in biology as the human race, with awareness of the characteristics of the metals, and could be used for various treatments like disorders and catalysis process [130,131]. Metals are fundamental to the inhalation system due to the incidence of iron and copper metals in proteins respiratory system, represented in Figure 3.6.

The final terminal acceptor electron is cytochrome c oxidase in the respiration system, however cytochrome c oxidase is always intricate with the generation of nitric oxide [132]. It has been clearly revealed that the nitric oxide scavenging decreased the nitrite-assisted liberations of poisoned cells with CN, also signifying that nitric oxide can compete with CN at the site of cytochrome c oxidase (Figure 3.7) [133].

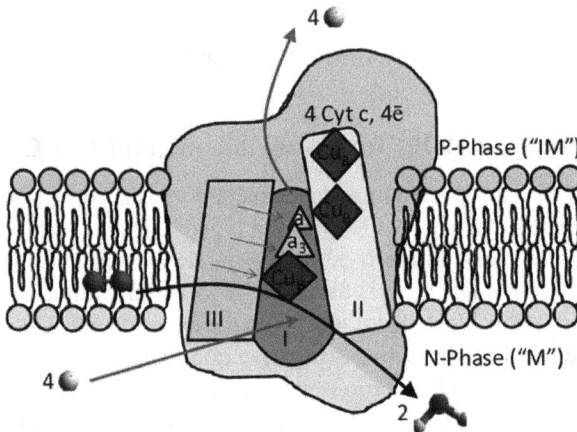

FIGURE 3.6 A systematically image of cytochrome c oxidase, with the represent of main subunits and route of electron movement. Cytochrome c oxidase is the fourth part of the complex contained by the transportation of electron chain. Reproduced with permission from Ref. [58], Elsevier license number 5384381158847.

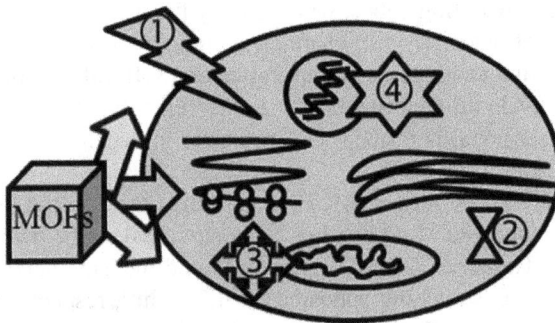

FIGURE 3.7 MOFs, atoms, or ions are released ("dissolution," denoted by ⇨), plasma membrane (1), ROS (4), membrane integrity (⑤); denaturation of proteins by the covalent bond, lipids, prominent to folding, show poor solubility or improved fragmentation etc. The recycling process of macromolecule (⑥); and one transferred into the cytoplasm, through the cellular commitment of additional oxidation of acute proteins (2) and present nucleic acids (⑦), which leads to degradation of protein and the other hand the nucleic acid expanding; and directly bind to respiratory proteins (RP) hindering the transportation the electron (⑧). Reproduced with permission from Ref. [58], Elsevier license number 5384381158847.

When CN is bound to the cytochrome c oxidase, it will interrupt the oxidative phosphorylation, decrease the demand for oxygen, and increase ATP [133,134]. The ligands having extraordinary binding capacity to CN as possible treatments, one excellent compound is $NaNO_2$. Other sources are the transformation of the CN through nitrile using thiosulfate, which is used as a substrate for enzymes to oxidize the meta-hemoglobin to the hemoglobin phase, which the reaction treats as scavenging for CN [135]. Naturally, NO converted the arginine and then produced the NOS; however, under the minimum concentration of oxygen, the activity of NOS decreases, and the main source of internal nitrite is generated by NO, which enlarges in the acidic environment by means of reactions as described by Eqs. 3.3–3.8 [136]:

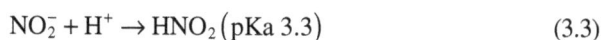

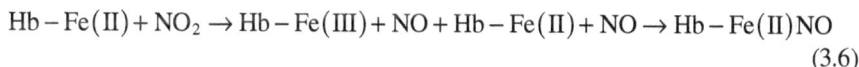

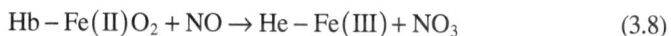

$$NO_2^- + H^+ \rightarrow HNO_2 \,(pKa\ 3.3) \tag{3.3}$$

$$2HNO_2^- \rightarrow N_2O_3 + H_2O \tag{3.4}$$

$$N_2O_3^- \rightarrow NO + NO_2 \tag{3.5}$$

$$Hb - Fe(II) + NO_2 \rightarrow Hb - Fe(III) + NO + Hb - Fe(II) + NO \rightarrow Hb - Fe(II)NO \tag{3.6}$$

$$Hb(beta93 - cys) + NO \rightarrow Hb(beta - cys - NO) \tag{3.7}$$

$$Hb - Fe(II)O_2 + NO \rightarrow He - Fe(III) + NO_3 \tag{3.8}$$

Hence, NO is easily produced from another nitrite (at a very lower concentration of oxygen) or other arginine (at a very higher concentration of oxygen) (Eq. 3.9) by using the xanthine oxidase substance, which is denoted by the XO or the NOS by superoxide radical from the oxygen atom in Eqs. 3.10 and 3.11 [137], representing the chemical reaction as follows:

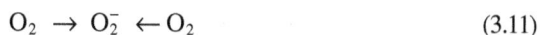

$$Nitrite \rightarrow NO \rightarrow arginine \tag{3.9}$$

$$XO \rightarrow ONOO^- \leftarrow NOS \tag{3.10}$$

$$O_2 \rightarrow O_2^- \leftarrow O_2 \tag{3.11}$$

3.5.8 CONCLUSION

Nanomaterials contain various forms, such as metal-based, carbon-based, organic-based, and combination-based nanoparticles. The manufacturing rate of nanomaterials is high, and available results show the maximum exposure to human beings and the environment, therefore increasing the toxicity issues. Here we outline the nanomaterial's characteristics, applications, and nanoparticle toxicity. Nanomaterials are accumulated in sensitive body parts like the liver, spleen, heart, kidney, brain, skin, ingestion, and so on. In vivo and in-vitro, both studies point out that exposure through the nanomaterials could persuade ROS fabrication, whose most popular mechanism is leading to toxicity. ROS extreme production causes various factors such as inflammation, oxidative stress, protein damage, DNA, and cell membrane. In vitro analysis is widely applicable for the toxicological investigation based on ROS oxidative stress, cellular metabolic activity (MTT assay), and membrane integrity (LDH assay). In this chapter, the MOF-type structures suggest that the cellular probes such as $K_4 Fe(CN)_6$ (potassium ferricyanide) also work on therapeutically healing the killing factors of wounded or carcinomatous cells through ROS and SOS generation.

CONFLICT OF INTEREST

The authors declare no conflict of interest, financial or otherwise.

ACKNOWLEDGEMENTS

Declared none.

REFERENCES

1. K.P. Chong, Nanoscience and engineering in mechanics and materials. *J. Phys. Chem. Solids* 65 (2004) 1501.
2. Q.Q. Zhao, A. Boxman, U. Chowdhry, Nanotechnology in the chemical industry opportunities and challenges. *J. Nanoparticle Res.* 5 (2013) 567.
3. M. Joshi, A. Bhattacharyya, Nanotechnology – A new route to high-performance functional textiles. *Textile Prog.* 43 (2011) 155.
4. R. Saini, S. Saini, S. Sharma, Nanotechnology: The future medicine. *J. Cutan. Aesthet. Surg.* 3 (1) (2010) 32–33.
5. F.L. Zhou, R.H. Gong, Manufacturing technologies of polymeric nanofibres and nanofibre yarns. *Poly. Int.* 57 (6) (2008) 837–845.
6. R. Feynman, Lecture at the California Institute of Technology. December 29, 1959.
7. P. Ganguly, A. Breen, S.C. Pillai, Toxicity of nanomaterials: Exposure, pathways, assessment, and recent advances. *ACS Biomater. Sci. Eng.* 4 (2018) 2237–2275.
8. M. Valden, X. Lai, D.W. Goodman, Onset of catalytic activity of gold clusters on titania with the appearance of nonmetallic properties, *Science* 281 (5383) (1998) 1647–1650.
9. H. Häkkinen, The gold-sulfur interface at the nanoscale, *Nat. Chem.* 4 (6) (2012) 443–455.
10. Z. Zhang, L. Wang, J. Wang, X. Jiang, X. Li, Z. Hu, C. Chen, Mesoporous silica-coated gold nanorods as a light-mediated multifunctional theranostic platform for cancer treatment, *Adv. Mater.* 24 (11) (2012) 1418–1423.
11. M. Ahamed, M. Siddiqui, M. Akhtar, I. Ahmad, A. Pant, H. Alhadlaq, Genotoxic potential of copper oxide nanoparticles in human lung epithelial cells. *Biochem. Biophys. Res. Commun.* 396 (2) (2010) 578–583.
12. H. Bahadar, F. Maqbool, K. Niaz, M. Abdollahi, Toxicity of nanoparticles and an overview of current experimental models. *Iran Biomed. J.* (2016) DOI: 10.7508/ibj.2016.01.001.

13. S.W. Shin, I.H. Song, S.H. Um, Role of physicochemical properties in nanoparticle toxicity. *Nanomaterials* 5 (3) (2015) 1351–1365.

14. K.X. Tan, A. Barhoum, S. Pan, M.K. Danquah, Risks and toxicity of nanoparticles and nanostructured materials. In: Barhoum A, Makhlouf ASH (eds) Emerging Applications of Nanoparticles and Architecture Nanostructures. Elsevier, Amsterdam, 121–139, 2018.

15. J. Fard, S. Jafari, M. Eghbal, A review of molecular mechanisms involved in toxicity of nanoparticles. *Adv. Pharm. Bull.* 5 (4) (2015) 447–454.

16. P. Fu, Q. Xia, H. Hwang, P. Ray, H. Yu, Mechanisms of nanotoxicity: Generation of reactive oxygen species. *J. Food Drug Anal.* 22 (1) (2014) 64–75.

17. X. He, W.G. Aker, P.P. Fu, H-M. Hwang, Toxicity of engineered metal oxide nanomaterials mediated by nano-bio-eco-interactions: A review and perspective. *Environ. Sci. Nano* 2 (6) (2015) 564–582.

18. G. Liu, J. Gao, H. Ai, X. Chen, Applications and potential toxicity of magnetic iron oxide nanoparticles. *Small* 9 (9–10) (2013) 1533–1545.

19. A. Manke, L. Wang, Y. Rojanasakul, Mechanisms of nanoparticle-induced oxidative stress and toxicity. *Biomed. Res. Int.* 2013 (2013) 942916.

20. A.B. Sengul, and E. Asmatulu, Toxicity of metal and metal oxide nanoparticles: A review. *Environ. Chem. Lett.* (2020) DOI: 10.1007/s10311-020-01033-6.

21. F. Lao, L. Chen, W. Li, C. Ge, Y. Qu, Q. Sun, C. Chen, Fullerene nanoparticles selectively enter oxidation-damaged cerebral micro vessel endothelial cells and inhibit JNK-related apoptosis. *ACS Nano* 3 (11) (2009) 3358–3368.

22. J.J. Yin, F. Lao, P.P. Fu, W.G. Warner, Y. Zhao, P.C. Wang, C. Chen, The scavenging of reactive oxygen species and the potential for cell protection by functionalized fullerene materials, *Biomaterials* 30 (4) (2009) 611–621.

23. S. Laus, B. Sitharaman, E. Toth, R.D. Bolskar, L. Helm, S. Asokan, A.E. Merbach, Destroying gadofullerene aggregates by salt addition in aqueous solution of Gd@C60(OH)x and Gd@C60[C(COOH2)]10, *J. Am. Chem. Soc.* 127 (26) (2005) 9368–9369.

24. J. Wang, C. Chen, B. Li, H. Yu, Y. Zhao, J. Sun, L. Wan, Antioxidative function and biodistribution of [Gd@C82(OH)22]n nanoparticles in tumor-bearing mice, *Biochem. Pharmacol.* 71 (6) (2006) 872–881.

25. J.T. Chen, T. Kuhlmann, G.H. Jansen, D.L. Collins, H.L. Atkins, M.S. Freedman, Canadian MS/BMT Study Group, Voxel-based analysis of the evolution of magnetization transfer ratio to quantify remyelination and demyelination with histopathological validation in a multiple sclerosis lesion, *Neuroimage* 36 (4) (2007) 1152–1158.

26. J. Palomaiki, E. Valimaiki, J. Sund, M. Vippola, P.A. Clausen, K.A. Jensen, H. Alenius, Long, needlelike carbon nanotubes and asbestos activate the NLRP3 inflammasome through a similar mechanism, *ACS Nano* 5 (9) (2011) 6861–6870.

27. C.W. Lam, J.T. James, R. McCluskey, R.L. Hunter, Pulmonary toxicity of single-wall carbon nanotubes in mice 7 and 90 days after intratracheal instillation, *Toxicol. Sci.* 77 (1) (2004) 126–134.

28. G. Jia, H. Wang, L. Yan, X. Wang, R. Pei, T. Yan, X. Guo, Cytotoxicity of carbon nanomaterials: Single-wall nanotube, multi-wall nanotube, and fullerene, *Environ. Sci. Technol.* 39 (5) (2005) 1378–1383.

29. A.R. Murray, E. Kisin, S.S. Leonard, S.H. Young, C. Kommineni, V.E. Kagan, A.A. Shvedova, Oxidative stress and inflammatory response in dermal toxicity of single-walled carbon nanotubes, *Toxicology* 257 (3) (2009) 161–171.

30. J.P. Ryman-Rasmussen, M.F. Cesta, A.R. Brody, J.K. Shipley-Phillips, J.I. Everitt, E.W. Tewksbury, J.C. Bonner, Inhaled carbon nanotubes reach the subpleural tissue in mice, *Nat. Nanotechnol.* 4 (11) (2009) 747–751.

31. C.A. Poland, R. Duffin, I. Kinloch, A. Maynard, W.A. Wallace, A. Seaton, K. Donaldson, Carbon nanotubes introduced into the abdominal cavity of mice show asbestos-like pathogenicity in a pilot study, *Nat. Nanotechnol.* 3 (7) (2008) 423–428.

32. A. Bianco, K. Kostarelos, M. Prato, Making carbon nanotubes biocompatible and biodegradable, *Chem. Commun.* 47 (37) (2011) 10182–10188.
33. R.F. Dodson, M.A. Atkinson, J.L. Levin, Asbestos fiber length as related to potential pathogenicity: A critical review, *Am. J. Ind. Med.* 44 (3) (2003) 291–297.
34. C. Ge, J. Du, L. Zhao, L. Wang, Y. Liu, D. Li, C. Chen, Binding of blood proteins to carbon nanotubes reduces cytotoxicity, *Proc. Natl. Acad. Sci. U.S.A.* 108 (41) (2011) 16968–16973.
35. Y. Wang, Y. Bu, Noncovalent interactions between cytosine and SWCNT: Curvature dependence of complexes via π-π stacking and cooperative CH π/NH π, *J. Phy Chem. B* 111 (23) (2007) 6520–6526.
36. M. Horie, K. Nishio, K. Fujita, S. Endoh, A. Miyauchi, Y. Saito, Y. Yoshida, Protein adsorption of ultrafine metal oxide and its influence on cytotoxicity toward cultured cells, *Chem. Res. Toxicol.* 22 (3) (2009) 543–553.
37. W. Zhao, C. Song, P.E. Pehrsson, Water-soluble and optically pH-sensitive single-walled carbon nanotubes from surface modification, *J. Am. Chem. Soc.* 124 (42) (2002) 12418–12419.
38. A.S. Kumar, P. Barathi, K.C. Pillai, In situ precipitation of nickel-hexacyanoferrate within multi-walled carbon nanotube modified electrode and its selective hydrazine electrocatalysis in physiological pH, *J. Electroanal. Chem.* 654 (1) (2011) 85–95.
39. M.J. Park, J.K. Lee, B.S. Lee, Y.W. Lee, I.S. Choi, S.G. Lee, Covalent modification of multiwalled carbon nanotubes with imidazolium-based ionic liquids: Effect of anions on solubility, *Chem. Mater.* 18 (6) (2006) 1546–1551.
40. B. Liu, X. Li, B. Li, B. Xu, Y. Zhao, Carbon nanotube based artificial water channel protein: Membrane perturbation and water transportation, *Nano Lett.* 9 (4) (2009) 1386–1394.
41. D. Zhang, X. Deng, Z. Ji, X. Shen, L. Dong, M. Wu, Y. Liu, Long-term hepatotoxicity of polyethyleneglycol functionalized multi-walled carbon nanotubes in mice, *Nanotechnology* 21 (17) (2010) 175101.
42. M. Shen, S.H. Wang, X. Shi, X. Chen, Q. Huang, E.J. Petersen, W.J. Weber Jr., Polyethyleneimine mediated functionalization of multiwalled carbon nanotubes: Synthesis, characterization, and in vitro toxicity assay, *J. Phys. Chem. C* 113 (8) (2009) 3150–3156.
43. J.C. Carrero-Sanchez, A.L. Elias, R. Mancilla, G. Arrellin, H. Terrones, J.P. Laclette, M. Terrones, Biocompatibility and toxicological studies of carbon nanotubes doped with nitrogen, *Nano Lett.* 6 (8) (2006) 1609–1616.
44. A.K. Patlolla, S.M. Hussain, J.J. Schlager, S. Patlolla, P.B. Tchounwou, Comparative study of the clastogenicity of functionalized and nonfunctionalized multiwalled carbon nanotubes in bone marrow cells of Swiss-Webster mice, *Environ. Toxicol.* 25 (6) (2010) 608–621.
45. I. Raynal, P. Prigent, S. Peyramaure, A. Najid, C. Rebuzzi, C. Corot, Macrophage endocytosis of superparamagnetic iron oxide nanoparticles: Mechanisms and comparison of ferumoxides and ferumoxtran-10, *Invest. Radiol.* 39 (1) (2004) 56–63.
46. N. Tara, S.I. Siddiqui, G. Rathi, S.A. Chaudhry, Inamuddin, A.M. Asiri, Nano-engineered adsorbent for the removal of dyes from water: A review. *Curr. Anal. Chem.* 15 (2019) 1.
47. N. Tara, S.I. Siddiqui, R.K. Nirala, N.K. Abdulla, S.A. Chaudhry, Synthesis of anti-bacterial, antioxidant and magnetic *Nigella sativa*-graphene oxide based nano- composite BC-GO@ Fe3O4 for water treatment, *Colloid Interfac. Sci.* 37 (2020) 100281.
48. N. Tara, S.I. Siddiqui, Q.V. Bach, S.A. Chaudhry, Reduce graphene oxide-manganese oxide-black cumin based hybrid composite (rGO-MnO2/BC): A novel material for water remediation, *Mater. Today Commun.* 25 (2020) 101560.
49. J. Wang, G. Zhou, C. Chen, H. Yu, T. Wang, Y. Ma, Z. Chai, Acute toxicity and biodistribution of different sized titanium dioxide particles in mice after oral administration, *Toxicol. Lett.* 168 (2) (2007) 176–185.

50. H.C. Yao, Y.Y. Yao, Ceria in automotive exhaust catalysts: I. Oxygen storage, *J. Catal.* 86 (2) (1984) 254–265.

51. Q. Fu, H. Saltsburg, M. Flytzani-Stephanopoulos, Active nonmetallic Au and Pt species on ceria-based water-gas shift catalysts, *Science* 301 (5635) (2003) 935–938.

52. X. He, H. Zhang, Y. Ma, W. Bai, Z. Zhang, K. Lu, Z. Chai, Lung deposition and extrapulmonary translocation of nano-ceria after intratracheal instillation, *Nanotechnology* 21 (28) (2010) 285103.

53. R. Hardman, A toxicologic review of quantum dots: Toxicity depends on physicochemical and environmental factors, *Environ. Health Perspect.* 114 (2) (2006) 165–172.

54. H. Ozmen, S. Akarsu, F. Polat, A. Cukurovali, The levels of calcium and magnesium, and of selected trace elements, in whole blood and scalp hair of children with growth retardation, *Iran J. Pediatr.* 23 (2) (2013) 125.

55. N.K. Abdulla, S.I. Siddiqui, N. Tara, A.A. Hashmi, S.A. Chaudhry, *Psidium guajava* leave-based magnetic nanocomposite γ-Fe2O3@ GL: A green technology for methylene blue removal from water, *J. Environ. Chem. Eng.* 7 (2019) 103423.

56. S.I. Siddiqui, P.N. Singh, N. Tara, S. Pal, S.A. Chaudhry, I. Sinha, Arsenic removal from water by starch functionalized maghemite nano-adsorbents: Thermodynamics and kinetics investigations, *Colloid Interface Sci.* 36 (2020) 100263.

57. N.K. Abdulla, S.I. Siddiqui, B. Fatima, R. Sultana, N. Tara, A.A. Hashmi, R. Ahmad, Md. Mohsin, R.K. Nirala, N.T. Linh, Q.V. Vach, S.A. Chaudhry, Silver based hybrid nanocomposite for dye removal: A novel antibacterial material for water cleansing, *J. Clean. Prod.* (2020) Article 124746.

58. S. Bashir, J. Liu, Nanosafety: Exposure, Measurement, and Toxicology, Advanced Nanomaterials and Their Applications in Renewable Energy, Elsevier, Amsterdam, 367–421, 2015. DOI: 10.1016/B978-0-12-801528-5.00008-7.

59. N. Tara, S.A. Chaudhry, Polysaccharide nanocomposite materials for the removal of Methylene blue (MB) dye from water. Innovation in Nano-Polysaccharides for Eco-Sustainability, Elsevier, Amsterdam, 1, 277–295, 2022.

60. N. Tara, S.A. Chaudhry, Nanopolysaccharide-based composite materials for photocatalysis applications. Innovation in Nano-Polysaccharides for Eco-Sustainability, Elsevier, Amsterdam, 255–275, 2022.

61. N. Tara, A. Sharma, A. Choudhry, N.K. Abdulla, G. Rathi, A.M Khan, S.A. Chaudhry, Graphene, graphene oxide, and reduced graphene oxide-based materials: A comparative adsorption performance. *Contamin. Water* 495–507, 2021.

62. S.I. Siddiqui, R. Ravi, G. Rathi, N. Tara, S.Ul Islam, S.A. Chaudhry, Decolorization of textile wastewater using composite materials. In: Islam SU, Butola BS (eds.) Nanomaterials in the Wet Processing of Textiles, John Wiley & Sons, Inc., 187–218, ISBN: 1119459915.

63. A. Hoshino, S. Hanada, K. Yamamoto, Toxicity of nanocrystal quantum dots: The relevance of surface modifications, *Arch. Toxicol.* 85 (7) (2011) 707–720.

64. J.L. Schwartz, R. Jordan, H. Liber, J.P. Murnane, H.H. Evans, TP53-dependent chromosome instability is associated with transient reductions in telomere length in immortal telomerase-positive cell lines, *Gene Chromsome Cancer* 30 (3) (2001) 236–244.

65. G. Guo, W. Liu, J. Liang, Z. He, H. Xu, X. Yang, Probing the cytotoxicity of CdSe quantum dots with surface modification, *Mater. Lett.* 61 (8) (2007) 1641–1644.

66. Y. Qu, W. Li, Y. Zhou, X. Liu, L. Zhang, L. Wang, C. Chen, Full assessment of fate and physiological behavior of quantum dots utilizing *Caenorhabditis elegans* as a model organism, *Nano Lett.* 11 (8) (2011) 3174–3183.

67. N.F. Ho, T.G. Geary, T.J. Raub, C.L. Barsuhn, D.P. Thompson, Biophysical transport properties of the cuticle of *Ascaris suum*, *Mol. Biochem. Parasitol.* 41 (2) (1990) 153–165.

68. Y. Li, Y. Zhou, H.Y. Wang, S. Perrett, Y. Zhao, Z. Tang, G. Nie, Chirality of glutathione surface coating affects the cytotoxicity of quantum dots, *Angew. Chem. Int. Ed. Engl.* 50 (26) (2011) 5860–5864.

69. L. Ye, K.T. Yong, L. Liu, I. Roy, R. Hu, J. Zhu, P.N. Prasad, A pilot study in non-human primates shows no adverse response to intravenous injection of quantum dots, *Nat. Nanotechnol.* 7 (7) (2012) 453–458.

70. A.K. Cheetham, G. Ferey, T. Loiseau, Open-framework inorganic materials, *Angew. Chem. Int. Ed. Engl.* 38 (22) (1999) 3268–3292.

71. A.S. Daar, J.W. Fabre, Organ-specific IgM autoantibodies to liver, heart and brain in man: Generalized occurrence and possible functional significance in normal individuals, and studies in patients with multiple sclerosis, *Clin. Exp. Immunol.* 45 (1) (1981) 37–38.

72. M. Eydner, D. Schaudien, O. Creutzenberg, H. Ernst, T. Hansen, W. Baumga"rtner, S. Rittinghausen, Impacts after inhalation of nano-and fine-sized titanium dioxide particles: Morphological changes, translocation within the rat lung, and evaluation of particle deposition using the relative deposition index, *Inhal Toxicol.* 24 (9) (2012) 557–569.

73. N. Khlebtsov, L. Dykman, Biodistribution and toxicity of engineered gold nanoparticles: A review of in vitro and in vivo studies, *Chem. Soc. Rev.* 40 (3) (2011) 1647–1671.

74. Z. Yang, Z.W. Liu, R.P. Allaker, P. Reip, J. Oxford, Z. Ahmad, G. Ren, A review of nanoparticle functionality and toxicity on the central nervous system, *J. R. Soc. Interface* 7 (Suppl. 4) (2010) S411–S422.

75. L. Ma-Hock, S. Burkhardt, V. Strauss, A.O. Gamer, K. Wiench, B.van Ravenzwaay,R. Landsiedel, Development of a short-term inhalation test in the rat using nano-titanium dioxide as a model substance, *Inhal. Toxicol.* 21 (2) (2009) 102–118.

76. S.L. Harper, J.L. Carriere, J.M. Miller, J.E. Hutchison, B.L. Maddux, R.L. Tanguay, Systematic evaluation of nanomaterial toxicity: Utility of standardized materials and rapid assays, *ACS Nano* 5 (6) (2011) 4688–4697.

77. K.T. Fitzgerald, C.A. Holladay, C. McCarthy, K.A. Power, A. Pandit, W.M. Gallagher, Standardization of models and methods used to assess nanoparticles in cardiovascular applications, *Small* 7 (6) (2011) 705–717.

78. R.G. Ellis-Behnke, Y.X. Liang, S.W. You, D.K. Tay, S. Zhang, K.F. So, G.E. Schneider, Nano neuro knitting: Peptide nanofiber scaffold for brain repair and axon regeneration with functional return of vision, *Proc. Natl. Acad. Sci. U.S.A.* 103 (13) (2006) 5054–5059.

79. R.B. Schlesinger, The health impact of common inorganic components of fine particulate matter (PM2.5) in ambient air: A critical review, *Inhal. Toxicol.* 19 (10) (2007) 811–832.

80. G. Fotakis, J.A. Timbrell, In vitro cytotoxicity assays: Comparison of LDH, neutral red, MTT and protein assay in hepatoma cell lines following exposure to cadmium chloride, *Toxicol. Lett.* 160 (2) (2006) 171–177.

81. E. Borenfreund, J.A. Puerner, Toxicity determined in vitro by morphological alterations and neutral red absorption, *Toxicol. Lett.* 24 (2) (1985) 119–124.

82. B. Page, M. Page, C. Noel, A new fluorometric assay for cytotoxicity measurements in-vitro, *Int. J. Oncol.* 3 (3) (1993) 473–476.

83. A.A. Bulich, K.K. Tung, G. Scheibner, The luminescent bacteria toxicity test: Its potential as an in vitro alternative, *J. Biolumin. Chemilumin.* 5 (2) (1990) 71–77.

84. S. Ganesan, N.D. Chaurasiya, R. Sahu, L.A. Walker, B.L. Tekwani, Understanding the mechanisms for metabolism-linked hemolytic toxicity of primaquine against glucose 6-phosphate dehydrogenase deficient human erythrocytes: Evaluation of eryptotic pathway, *Toxicology* 294 (1) (2012) 54–60.

85. R. Lichtenfels, W.E. Biddison, H. Schulz, A.B. Vogt, R. Martin, CARE-LASS (calcein-release-assay), an improved fluorescence-based test system to measure cytotoxic T lymphocyte activity, *J. Immunol. Methods* 172 (2) (1994) 227–239.

86. S. Neri, E. Mariani, A. Meneghetti, L. Cattini, A. Facchini, Calcein-acetyoxymethyl cytotoxicity assay: Standardization of a method allowing additional analyses on recovered effector cells and supernatants, *Clin. Diagn. Lab. Immunol.* 8 (6) (2001) 1131–1135.

87. J.A. Fortune, B.I. Wu, A.M. Klibanov, Radio frequency radiation causes no nonthermal damage in enzymes and living cells, *Biotechnol. Prog.* 26 (6) (2010) 1772–1776.

88. J.R. Ferreira, R. Padilla, G. Urkasemsin, K. Yoon, K. Goeckner, W.S. Hu, C.C. Ko, Titanium-enriched hydroxyapatiteegelatin scaffolds with osteogenically differentiated progenitor cell aggregates for calvaria bone regeneration, *Tissue Eng. Part A* 19 (15–16) (2013) 1803–1816.

89. D.L. Farkas, M.D. Wei, P. Febbroriello, J.H. Carson, L.M. Loew, Simultaneous imaging of cell and mitochondrial membrane potentials, *Biophys. J.* 56 (6) (1989) 1053.

90. R.M. Touyz, E.L. Schiffrin, Ang II stimulated superoxide production is mediated via phospholipase D in human vascular smooth muscle cells, *Hypertension* 34 (4) (1999) 976–982.

91. C. Flors, M.J. Fryer, J. Waring, B. Reeder, U. Bechtold, P.M. Mullineaux, N.R. Baker, Imaging the production of singlet oxygen in vivo using a new fluorescent sensor, Singlet Oxygen Sensor Green, *J. Exp. Bot.* 57 (8) (2006) 1725–1734.

92. K.A. Kristiansen, P.E. Jensen, I.M. Møller, A. Schulz, Monitoring reactive oxygen species formation and localisation in living cells by use of the fluorescent probe CM-H2DCFDA and confocal laser microscopy, *Physiol. Plant.* 136 (4) (2009) 369–383.

93. Q. Zhang, M. Raoof, Y. Chen, Y. Sumi, T. Sursal, W. Junger, C.J. Hauser, Circulating mitochondrial DAMPs cause inflammatory responses to injury, *Nature* 464 (7285) (2010) 104–107.

94. P. Rieck, D. Peters, C. Hartmann, Y. Courtois, A new, rapid colorimetric assay for quantitative determination of cellular proliferation, growth inhibition, and viability, *J. Tissue Cult. Methods* 15 (1) (1993) 37–41.

95. T. Zor, Z. Selinger, Linearization of the bradford protein assay increases its sensitivity: Theoretical and experimental studies, *Anal. Biochem.* 236 (2) (1996) 302–308.

96. M. Koresawa, T. Okabe, High-throughput screening with quantitation of ATP consumption: A universal non-radioisotope, homogeneous assay for protein kinase, *Assay Drug Dev. Technol.* 2 (2) (2004) 153–160.

97. D. Yang, S.G. Elner, Z.M. Bian, G.O. Till, H.R. Petty, V.M. Elner, Pro-inflammatory cytokines increase reactive oxygen species through mitochondria and NADPH oxidase in cultured RPE cells, *Exp. Eye Res.* 85 (4) (2007) 462–472.

98. E.S. Öz, E. Aydemir, K. Fiskin, DMSO exhibits similar cytotoxicity effects to thalidomide in mouse breast cancer cells, *Oncol. Lett.* 3 (4) (2012) 927.

99. S. Cao, G.B. Walker, X. Wang, J.Z. Cui, J.A. Matsubara, Altered cytokine profiles of human retinal pigment epithelium: Oxidant injury and replicative senescence, *Mol. Vis.* 19 (2013) 718.

100. G. Fotakis, J.A. Timbrell, In vitro cytotoxicity assays: Comparison of LDH, neutral red, MTT and protein assay in hepatoma cell lines following exposure to cadmium chloride, *Toxicol. Lett.* 160 (2) (2006) 171–177.

101. A. Roch, R.N. Muller, P. Gillis, Theory of proton relaxation induced by superparamagnetic particles, *J. Chem. Phys.* 110 (11) (1999) 5403–5411.

102. Z.Q. Li, L.G. Qiu, W. Wang, T. Xu, Y. Wu, X. Jiang, Fabrication of nanosheets of a fluorescent metal-organic framework [Zn(BDC)(H2O)]n (BDC 1/4 1,4-benzenedicarboxylate): Ultrasonic synthesis and sensing of ethylamine, *Inorg. Chem. Comm.* 11 (11) (2008) 1375–1377.

103. Y. Kawato, M. Aonuma, Y. Hirota, H. Kuga, K. Sato, Intracellular roles of SN-38, a metabolite of the camptothecin derivative CPT-11, in the antitumor effect of CPT-11, *Cancer Res.* 51 (16) (1991) 4187–4191.

104. R. Dringen, B. Hamprecht, Involvement of glutathione peroxidase and catalase in the disposal of exogenous hydrogen peroxide by cultured astroglial cells, *Brain Res.* 759 (1) (1997) 67–75.

105. L. Iyer, C.D. King, P.F. Whitington, M.D. Green, S.K. Roy, T.R. Tephly, M.J. Ratain, Genetic predisposition to the metabolism of irinotecan (CPT-11). Role of uridine diphosphate glucuronosyltransferase isoform 1A1 in the glucuronidation of its active metabolite (SN-38) in human liver microsomes, *J. Clin. Invest.* 101 (4) (1998) 847.

106. G.F. Jin, J.S. Hurst, B.F. Godley, Hydrogen peroxide stimulates apoptosis in cultured human retinal pigment epithelial cells, *Curr. Eye Res.* 22 (3) (2001) 165–173.

107. S. Qin, A.P. McLaughlin, G.W. De Vries, Protection of RPE cells from oxidative injury by 15-deoxy-n-12,14-prostaglandin-J2 by augmenting GSH and activating MAPK, *Invest. Ophthalmol. Vis. Sci.* 47 (11) (2006) 5098–5105.

108. G. Hristov, T. Marttila, C. Durand, B. Niesler, G.A. Rappold, A. Marchini, SHOX triggers the lysosomal pathway of apoptosis via oxidative stress, *Hum. Mol. Genet.* 23 (6) (2014) 1619–1630.

109. J.O.N. Lundberg, T. Farkas-Szallasi, E. Weitzberg, J. Rinder, J. Lidholm, A. Anggaard, K. Alving, High nitric oxide production in human paranasal sinuses, *Nat. Med.* 1 (4) (1995) 370–373.

110. M. Feelisch, B.O. Fernandez, N.S. Bryan, M.F. Garcia-Saura, S. Bauer, D.R. Whitlock, H. Ashrafian, Tissue processing of nitrite in hypoxia an intricate interplay of nitric oxide-generating and -scavenging systems, *J. Biol. Chem.* 283 (49) (2008) 33927–33934.

111. W.S. Choi, S.E. Kruse, R.D. Palmiter, Z. Xia, Mitochondrial complex I inhibition is not required for dopaminergic neuron death induced by rotenone, MPP+, or paraquat, *Proc. Natl. Acad. Sci. U.S.A.* 105 (39) (2008) 15136–15141.

112. E.C. Slater, The mechanism of action of the respiratory inhibitor, antimycin, *BBA Bioenerg.* 301 (2) (1973) 129–154.

113. R.F. Furchgott, D. Jothianandan, Endothelium-dependent and -independent vasodilation involving cyclic GMP: relaxation induced by nitric oxide, carbon monoxide and light, *J. Vasc. Res.* 28 (1–3) (1991) 52–61.

114. B. Commoner, Cyanide inhibition as a means of elucidating the mechanisms of cellular respiration, *Biol. Rev.* 15 (2) (1940) 168–201.

115. E.Å. Jansson, L. Huang, R. Malkey, M. Govoni, C. Nihlen, A. Olsson, J.O. Lundberg, A mammalian functional nitrate reductase that regulates nitrite and nitric oxide homeostasis, *Nat. Chem. Biol.* 4 (7) (2008) 411–417.

116. M.T. Gladwin, A.N. Schechter, D.B. Kim-Shapiro, R.P. Patel, N. Hogg, S. Shiva, J.O. Lundberg, The emerging biology of the nitrite anion, *Nat. Chem. Biol.* 1 (6) (2005) 308–314.

117. J.P. Collman, A. Dey, R.A. Decreau, Y. Yang, A. Hosseini, E.I. Solomon, T.A. Eberspacher, Interaction of nitric oxide with a functional model of cytochrome c oxidase, *Proc. Natl. Acad. Sci. U.S.A.* 105 (29) (2008) 9892–9896.

118. T. Pan, P. Rawal, Y. Wu, W. Xie, J. Jankovic, W. Le, Rapamycin protects against rotenone-induced apoptosis through autophagy induction, *NeuroScience* 164 (2) (2009) 541–551.

119. V.R. Potter, A.E. Reif, Inhibition of an electron transport component by antimycin A, *J. Biol. Chem.* 194 (1) (1952) 287–297.

120. J. Vásquez-Vivar, B. Kalyanaraman, P. Martásek, N. Hogg, B.S.S. Masters, H. Karoui, K.A. Pritchard, Superoxide generation by endothelial nitric oxide synthase: The influence of cofactors, *Proc. Natl. Acad. Sci. U.S.A.* 95 (16) (1998) 9220–9225.

121. J.L. Way, D. Sylvester, R.L. Morgan, G.E. Isom, G.E. Burrows, C.B. Tamulinas, J.L. Way, Recent perspectives on the toxicodynamic basis of cyanide antagonism, *Fund Appl. Toxicol.* 4 (2) (1984) S231–S239.

122. J.N. Bates, M.T. Baker, R. Guerra Jr., D.G. Harrison, Nitric oxide generation from nitroprusside by vascular tissue: Evidence that reduction of the nitroprusside anion and cyanide loss are required, *Biochem. Pharmacol.* 42 (1991) S157–S165.

123. M. Skrzypski, M. Sassek, S. Abdelmessih, S. Mergler, C. Gro¨tzinger, D. Metzke, M.Z. Strowski, Capsaicin induces cytotoxicity in pancreatic neuroendocrine tumor cells via mitochondrial action, *Cell. Signal.* 26 (1) (2014) 41–48.

124. Q. Chen, E.J. Vazquez, S. Moghaddas, C.L. Hoppel, E.J. Lesnefsky, Production of reactive oxygen species by mitochondria central role of complex III, *J. Biol. Chem.* 278 (38) (2003) 36027–36031.

125. S. Hermes, F. Schröder, S. Amirjalayer, R. Schmid, R.A. Fischer, Loading of porous metaleorganic open frameworks with organometallic CVD precursors: Inclusion compounds of the type [LnM]a@MOF-5, *J. Mater. Chem.* 16 (25) (2006) 2464–2472.

126. M.P. Suh, H.R. Moon, E.Y. Lee, S.Y. Jang, A redox-active two-dimensional coordination polymer: Preparation of silver and gold nanoparticles and crystal dynamics on guest removal, *J. Am. Chem. Soc.* 128 (14) (2006) 47104718.

127. S. Turner, O.I. Lebedev, F. Schröder, D. Esken, R.A. Fischer, G.V. Tendeloo, Direct imaging of loaded metaleorganic framework materials (Metal@MOF-5), *Chem. Mater.* 20 (17) (2008) 5622–5627.

128. M. Sabo, A. Henschel, H. Fröde, E. Klemm, S. Kaskel, Solution infiltration of palladium into MOF-5: Synthesis, physisorption and catalytic properties, *J. Mater. Chem.* 17 (36) (2007) 3827–3832.

129. C.S. Liu, Z. Chang, J.J. Wang, L.F. Yan, X.H. Bu, S.R. Batten, A photoluminescent 3D silver (I) coordination polymer with mixed ligands anthracene-9, 10-dicarboxylate and hexamethylenetetramine, showing binodal 4-connected (4363)2(42•62•82)3 topology, *Inorg. Chem. Comm.* 11 (8) (2008) 889–892.

130. I. Fischer-Hjalmars, A. Henriksson-Enflo, Metals in biology: An attempt at classification, *Adv. Quantum Chem.* 16 (1982) 1–41.

131. M. Brunori, G. Antonini, F. Malatesta, P. Sarti, M.T. Wilson, Cytochrome-c oxidase, *Eur. J. Biochem.* 169 (1) (1987) 1–8.

132. S. Yoshikawa, Cytochrome-c oxidase, In Edward J. Massaro (Ed.), *Handbook of Copper Pharmacology and Toxicology*, Humana Press, Berlin, Germany, 131–152, 2002.

133. F.J. Baud, S.W. Borron, B. Me´garbane, H. Trout, F. Lapostolle, E. Vicaut, C. Bismuth, Value of lactic acidosis in the assessment of the severity of acute cyanide poisoning, *Crit. Care Med.* 30 (9) (2002) 2044–2050.

134. R. Gracia, G. Shepherd, Cyanide poisoning and its treatment, *Pharmacotheraphy* 24 (10) (2004) 1358–1365.

135. N.A. Paitian, K.A. Markossian, R.M. Nalbandyan, The effect of nitrite on cytochrome oxidase, *Biochem. Biophys. Res. Commun.* 133 (3) (1985) 1104–1111.

136. A. Webb, R. Bond, P. McLean, R. Uppal, N. Benjamin, A. Ahluwalia, Reduction of nitrite to nitric oxide during ischemia protects against myocardial ischemiaereperfusion damage, *Proc. Natl. Acad. Sci. U.S.A.* 101 (37) (2004) 13683–13688.

137. J.L. Way, Cyanide intoxication and its mechanism of antagonism, *Ann. Rev. Pharmacol. Toxicol.* 24 (1) (1984) 451–481.

4 Natural Nanofillers
Preparation and Properties

Syed Umar Faruq Syed Najmuddin
Universiti Malaysia Sabah

Mohd Nor Faiz Norrrahim
Universiti Pertahanan Nasional Malaysia

Mohd Nurazzi Norizan
Universiti Sains Malaysia

Tengku Arisyah Tengku Yasim-Anuar
Nextgreen Pulp & Paper Sdn. Bhd

Naveen Jesuarockiam
Vellore Institute of Technology

Mohammed Abdillah Ahmad Farid and Yoshito Andou
Kyushu Institute of Technology

Mohd Idham Hakimi
Universiti Putra Malaysia

4.1 INTRODUCTION

Nanofiller is a set of substances or particles (with regards to its distinct state: solid, liquid, or gaseous) in a nanoscale material where at least one dimension is less than 100nm (Shankar & Rhim, 2018). Consider the fact that nanoplates, nanofibers, and nanoparticles are all part of the three-dimensional nanorange (Figure 4.1). When two or more materials with unique traits are combined to form a composite, the ingenious mechanical, physical, optical, and conductivity properties of the composite are often greatly strengthened by the addition of nanofiller (Kapoor et al., 2020). For example, nanofillers can improve the composite's mechanical properties by lowering the interfacial tension and interfacial slip upon filler-polymer interaction. However, if the fillers and polymer chains fall short to adhere to one another, the composite will fail at form interface interaction, hence suffering from poor mechanical properties

DOI: 10.1201/9781003400998-4

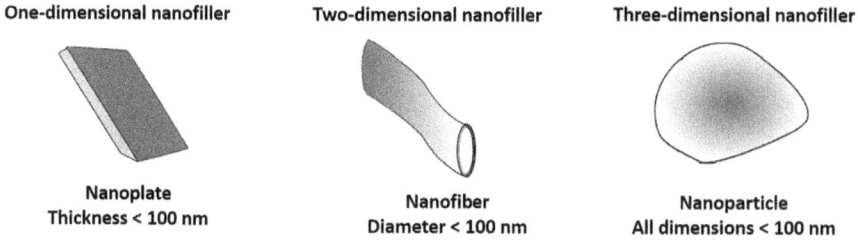

One-dimensional nanofiller   **Two-dimensional nanofiller**   **Three-dimensional nanofiller**

Nanoplate   **Nanofiber**   **Nanoparticle**
Thickness < 100 nm   **Diameter < 100 nm**   **All dimensions < 100 nm**

FIGURE 4.1   Nano objects used for nanocomposites. Reproduced from Marquis et al., 2011.

(Šupová et al., 2011). Apart from that, the addition of nanofillers could also reduce the consumption of a more expensive binder (i.e., economic benefits) and controlling the processing procedure (Šupová et al., 2011). Considerations among other strength-to-mass ratio, versatility, stiffness, thermal behavior, procurement of materials, and environmental footprint should always be made prior to actually deciding the preferred or relevant character traits of nanofiller reinforcement to be embedded within the polymer matrix, such as morphology, hardness, relative density, absorbency, and chemical reactivity or inertness (Kapoor et al., 2020). Other than that, it is important to note that the properties of composite materials can be significantly impacted by the mixture ratio between the organic matrix and the nanofillers, the dispersion of the filler, and its interaction with the matrix (Kapoor et al., 2020).

A few decades ago, synthetic fillers like calcium carbonate, titanium dioxide, and aluminum oxide have been used in improving the properties of composites for various uses, such as in the production of paints, coatings, sealants, adhesives, concrete, rubbers, and plastic products, but environmental concerns and depleting natural resources coupled with advancements in technologies have shifted its vision toward greener alternatives fillers to the conventional polymeric material ones. For example, the inclusion of nanofiller in nanocomposite has been known to improve heat distortion temperature with loading as well as to decrease gas permeability and flammability, which are advantageous properties for protective barrier packaging (Sorrentino et al., 2007). There are multiple types of green natural nanofillers which comes from plant- and animal-based sources as well as other natural sources, including organic and inorganic materials. Some common instances of animal-based sources are typically industrial byproducts or leftover that manufacture animal products, such as eggshells, seashells, and silkworms (Kapoor et al., 2020).

The main components that function as nanofillers are calcium carbonate crystals, chitin fibers, and silk fibers extracted from the aforementioned eggshells, seashells, and silkworms, respectively (Kapoor et al., 2020). Variations of nanofillers derived from plants include lignin, lignocellulose, nanocellulose, and palm oil wax, as well as plant fibers, which later be mechanically tenderized and/or chemically treated to optimize their characteristics for the purpose of reinforced composite architectures (Kapoor et al., 2020). On the other hand, organic and inorganic materials like carbon/graphite (Blokhin et al., 2019), natural clays (e.g., smectite, illite, sepiolite, chlorite, and kaolinite) (Rajeshkumar et al., 2021), and nano-oil palm fibers (Norrrahim et al., 2022) have also been extensively studied and utilized as nanofillers in polymer composites for various purposes.

4.2 NANOCELLULOSE

In line with the growing concern about environmental sustainability, especially the biodegradability of materials, toxicity issues, and depletion of non-renewable materials, particular attention has been given to nanofillers made from biological resources, and one of them is nanocellulose (Khalid et al., 2021). Nanocellulose, short for nanostructured cellulose, is cellulose that has undergone one or more varying nanofibrillation techniques, providing the basis for many variants (Ariffin et al., 2021). The three primary types of nanocellulose are cellulose nanofiber (CNF), cellulose nanocrystals (CNC), and bacterial nanocellulose (BNC). It is generally agreed that CNC and CNF are examples of nanocellulose derived from plants, whereas BNC is derived from microorganisms. Fabrication methods for CNF and CNC included mechanical or chemical disintegration of plant cellulose, whereas fabrication methods for BNC included bacterial bioformation of cellulose (Trache et al., 2020). Although all of them are nanoscale, each of them has different physical and chemical properties, e.g., shape and size, due to different production methods and sources (Thomas et al., 2018). The differences between these three classes of nanocellulose are tabulated in Table 4.1. Nonetheless, they all have superior properties, such as a large specific area, high porosity with good pore interconnectivity, low weight, high biodegradability, and high durability (Bitounis et al., 2019; Sharma et al., 2019; Phanthong et al., 2018; Gamelas et al., 2015; Rantanen et al., 2015; and Norrrahim et al., 2021). Due to its superior properties, nanocellulose is in high demand to be used mainly as a filler for a variety of applications such as biocomposites (Yasim-Anuar et al., 2020; Khattab et al., 2017), filtration (Boujemaoui, 2016), paper (Chun et al., 2011), and the biomedicine industry (Metreveli et al., 2014; Gumrah Dumanli, 2017).

Nanocellulose, a material generated from a wide range of biological resources, has been the subject of intensive basic study over the last decade. According to the information provided by lens.org, by typing 'nanocellulose' as a keyword, there have been approximately 5,553 patents for nanocellulose-related works and 7,567 scientific articles published worldwide in March 2022. This data indirectly shows that nanocellulose is definitely destined for a major part in a commercial society, as seen by the rising study into the material.

4.3 NANOCLAYS

Because they are so common in the environment, nanoclays have found widespread use in the food packaging sector. In 2014, they accounted for a staggering USD 343 million share in the food packaging industry, and they are forecasted to expand rapidly through 2022 onward. (Bumbudsanpharoke & Ko, 2019). Several reports have shown that adding nanoclays to polymers in small amounts (10 wt%) improves the material's resistance to moisture, heat, and deterioration (Thakur & Kessler, 2015; Guo et al., 2018). Alkali metals, alkaline earth metals, metal oxides, calcium, and other organic elements are found in abundance in clays, which are layered silicates (sheet-like structures of hydrous aluminum phyllosilicates) (Valapa et al., 2017). The number of SiO_2 layers and the orientation of the AlO_6 layers determine which of two

TABLE 4.1
Nanocellulose Classifications Based on Morphological and Methodological Differences

	CNF	CNC	BNC
Morphological images	Source: Burhani et al. (2021)	Source: Burhani et al. (2021)	Source: Budaeva et al. (2019)
Physical appearance	CNF consists of interconnected, long, flexible fibers with a diameter of 20–100 nm and a length of several micrometers (Tibolla et al., 2014).	CNC are characterized by a diameter of 5–70 nm and a length of 100 nm, making them resembling rods or whiskers (Kaboorani & Riedl, 2015).	BNC has a thick lateral surface and a gelatinous layer on the other side, and its diameter ranges from 20 to 80 nm (Dima et al., 2017; Jozala et al., 2016; Shabanpour et al., 2018).
Nanofibrillation method	CNF can be produced by subjecting any lignocellulosic materials to mechanical methods such as sonication (Yasim-Anuar et al., 2018), electrospinning (Härdelin et al., 2013), high pressure homogenization (Wang et al., 2015), milling (Norrahim et al., 2019), grinding (Berglund et al., 2016), and cryocrushing (Alemdar & Sain, 2008).	CNC can solely be produced by acid hydrolysis (Börjesson & Westman, 2015; Zhang et al., 2018).	BNC can be obtained via bacterial synthesis process and is produced mostly by the bacterium *Gluconacetobacter xylinus* (*G. xylinus*) (Mondal, 2017).

categories each material belongs to: (1) 2:1 clay (e.g., smectite, talc-pyrophyllite, micas, and vermiculite) and (2) 1:1 clay (e.g., kaolinite, halloysite, chrysotile, and rectorite) (Abulyazied & Ene, 2021). In contrast to the 1:1 clay, montmorillonite (MMT) has unique intercalation-exfoliation capabilities, making it especially important and prevailing as reinforcing fillers for polymers and therefore one of the most well-known smectites used in polymer nanocomposites (Abulyazied & Ene, 2021). It is important to note that the MMT structure consists of an octahedral sheet of alumina bonded between two tetrahedral sheets of silica by the sharing of oxygen atoms due to the presence of sodium and calcium ions inside the framework (Kiliaris & Papaspyrides, 2010). Therefore, because of the higher charge density, hydration of interlayer positive ions or penetration of polymer chains into the layers is very challenging, if not impossible (Murray, 1991).

The challenges involving the use of clay particles as nanofiller can be resolved via modification prior to be incorporated into any polymer matrices. This is because (1) the hydrophilic nature of clays prevents them from reacting and dispersing with most polymer matrices, (2) the electrostatic forces tightly keep the stacks of clay layers together, and (3) the existence of strong covalent bonds between the interlayers of clay layers (Abulyazied & Ene, 2021). Modification on clay particles can be carried out by either physical or chemical modification techniques, which consequently increase the distance between the sheets of clay-layered silicates as a result of surfactant intercalation (Kakuta et al., 2021). Hydrophobic ingredients/additives could also be introduced into clay minerals to allow the dispersion of very fine structures of modified clay in polymer matrices (Abulyazied & Ene, 2021). One such prime example is MMT, which is obtained via sedimentation from bentonite. In order to increase the distance between the layers of monomers and polymers of varying polarities, sodium ions are used to activate the MMT clay, and the addition of organic cations like silanes, ammonium bromides, or ammonium chlorides (e.g., vinylbenzyloctadecyl dimethylammonium chloride (VOAC) and vinylbenzyldodecyl ammonium chloride (VDAC)) reduces the surface energy (Marquis et al., 2011). Although the clay particles themselves are not altered by the physical modification process, the characteristics of the resulting polymer composites are somewhat enhanced owing to the adsorption of modifying chemicals on the clay surface via weak van der Waals interactions (aided by cationic or anionic functional groups in the ion exchange process) (Wang et al., 2021).

Either of the following synthesis techniques may be used to create a nanoclay/polymer composite, a homogeneous dispersion of nanoclay in the polymer matrix may be achieved using one of three methods: (1) solution-blending, (2) melt-blending, or (3) in situ polymerization (Jawaid et al., 2016). The solution-blending approach is preferred for dispersing clay layers in polymer matrices because of its low viscosity and high agitation power, whereas the melt-blending method is seen as industrially feasible and has significant economic potential (Valapa et al., 2017). On the other hand, in situ polymerization allows for modest customization by varying polymerization settings and homogeneous dispersion (Vo et al., 2016). Polymer and layered clay particles are dissolved in a specific solvent, such as water, chloroform, or toluene, in a separate container, prior the clay-solvent mixture is put into the polymer solution and homogenized for some time (Beyer, 2002). When polymer and layered clay are combined, the polymer chains displace the solvent inside the interlayer of the clay, and the polymer chains gradually intercalate with the clay (Gurses, 2015). The intercalated sheets reassemble into a polymer/nanoclay composite (composite film casting) once the solvent is removed, either by evaporation or precipitation (Guo et al., 2018). The melt-blending technique, on the other hand, combines the polymers with the necessary amount of intercalated nanoclay particles at a temperature above the melting point of the polymers while under the influence of an inert gas (Guo et al., 2018).

One approach to melt blending is known as static melt annealing, in which melting occurs in a melt mixer at a temperature around 50°C above the transition temperature without any mixing (Hesami & Jalali-Arani, 2018). When utilizing this technique for the synthesis of the composite, it is important to keep in mind that a number of melting conditions or parameters, including dosage, melting temperature profile, rotor/screw rapidity, homogenization period, oxidizing atmosphere, pressure die, materials quality, and the chemical nature of the nanoclay filler and polymers, must be taken into account (Dennis et al., 2001).

The in situ polymerization procedure involves the first steps of swelled organo-modified clay suspended in a liquid monomer, followed by the dispersion of monomer into the clay sphere, and finally the production of reinforced polymer molecules (Ray, 2013). After the polymerization procedure is complete, a nanocomposite is generated with polymer molecules attached to nanoclay (Gao, 2004). The in situ polymerization technique provides an efficient way for synthesizing various polymer/nanoclay composites with a wider variety of properties and permits the design of the interface between the nanoclays and the polymers through the tunable composition and structure of the matrix.

The incorporation of nanoclay into polymer matrices has been lauded in previous research for its ability to significantly improve characteristics at a very modest volume dosage while still retaining the polymer's original desirable qualities (Pavlidou & Papaspyrides, 2008). However, the resultant qualities of a polymer-clay nanocomposite depend on a number of factors, including the type and properties of the constituents, the processing technique, and the conditions under which the nanocomposite is synthesized (Pavlidou & Papaspyrides, 2008). Previous research has demonstrated that improving the mechanical characteristics of polymer matrices by introducing a certain quantity (wt%) of nanoclay improves the Young's modulus of the resulting nanocomposite. Kaushik et al. (2009) combined 5% nanoclay and polyester to create a nanocomposite with a modulus of 3.79 GPa, which is much greater than that of pure polyester (2.87 GPa). Nylon 6/clay nanocomposite modulus was found to be 2.43 GPa, which is an increase over the pure nylon 6 polymer modulus of 1.2 GPa, according to research utilizing 5% nanoclay (Shelley et al., 2001). When the volume proportion of organoclay is greater than a certain value, completely exfoliated structures may transform into partly intercalated ones, accompanied by a reduction in Young's modulus (Abulyazied & Ene, 2021). Researchers also consider other mechanical qualities, such as tensile strength (MPa) and elongation at break point value (%).

The use of nanoclay often results in a significant improvement in the aforementioned characteristics. When added to other polymers, nanoclay may increase their thermal stability (Vyazovkin et al., 2004). In comparison to neat polystyrene, the degradation temperature of the polystyrene nanoclay composite was elevated by 30°C–40°C (Vyazovkin et al., 2004). Furthermore, owing to the lateral length of the layered silicates and the degree of intercalation or exfoliation, layered nanoclay may improve the gas barrier characteristics of polymer by orchestrating a labyrinth or convoluted route that increases the distance traversed by diffusive molecules (Tortora et al., 2002; Koh et al., 2008). The water permeability of a PLA-clay nanocomposite was reported to be reduced by 95% with a 15 wt% addition of MMT to the polylactic acid (Żenkiewicz et al., 2010). Nanoclays are fascinating since they diminish the combustibility of rubbers and other polymers once employed as reinforcement (Kiliaris & Papaspyrides, 2010). Other than that, a mixture between MMT and animal skins can produce a flame-retardant nanocomposite called hide powder (Hiujian et al., 2012).

4.4   CHITOSAN

Natural biopolymer chitosan is produced by the alkaline deacetylation of chitin, a structural component of the exoskeletons of crustaceans. The degree of deacetylation determines the percentage of positively charged primary amines (the ratio between N-acetyl glucosamine and D-glucosamine units) and, by extension, the charge density of chitosan (Thomas et al., 2019). Also known as chitin with a deacetylation degree of 75% or above, chitosan is a linear polysaccharide comprised of D-glucosamine and N-acetyl-glucosamine units connected by (1–4) glycosidic linkages (Figure 4.2) (Thomas et al., 2019). The chitosan structure may be altered at either the amine or hydroxyl groups (Pillai & Ray, 2012). Crystallinity, solubility, and degradation are among the physicochemical characteristics of chitosan that may be affected by its

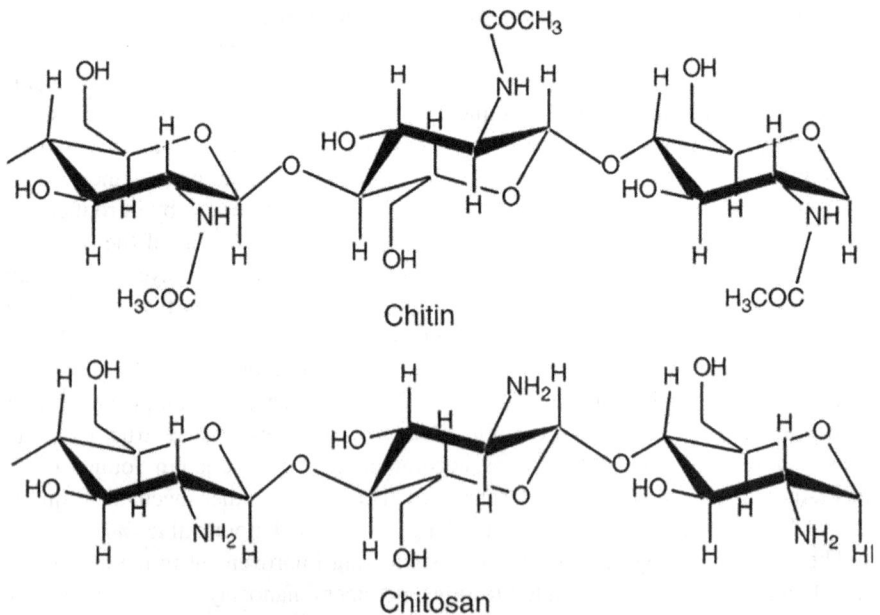

FIGURE 4.2   Chemical structure of chitin and chitosan.

wide range of molecular weights and degrees of deacetylation (Levengood & Zhang, 2014). As a semicrystalline polymer in its solid state, the molecular weight of industrially produced chitosan can range from 5 to 1,000 kg/mol, and its ability to exist in a variety of allomorphs is dependent on the degree of acetylation, the distribution of acetyl groups along the carbohydrate chain, and the chitosan preparation method (Chivrac et al., 2009). Chitosan's unique biological features, including biocompatibility, biodegradability, anti-bacterial activity, and release mechanism, make it especially appropriate as a carrier in a variety of industries, including biomedicine, biotechnology, pharmaceuticals, packaging, wastewater treatment, and cosmetics (Peniche et al., 2008). Chitosan, for example, has been used to make enzymatic biosensors for the analysis of metals, proteins, and lipids; its hemostatic property makes it suitable for blood anticoagulants and anti-thrombogenic purposes; and it can be shaped into a wide variety of forms, including gels, membranes, beads, microparticles, nano-fibers, scaffolds, and sponges (Nagahama et al., 2008; Madhumathi et al., 2009; Peter et al., 2010). It was also used by the textile industry to preserve and strengthen textile fibers and by the food packaging industry to coat paper and paperboard (Liu et al., 2001).

Cross-linking with reagents like glutaraldehyde, tripolyphosphate salts, genipin, epichlorohydrin, ethylene glycol, or diglycidylether can stabilize chitosan in acid solutions and increase its mechanical properties, and they can also improve its adsorption performance for metals (Fan et al., 2011; Monier et al., 2012). According to Kyzas et al. (2014), chitosan (CS), and magnetic chitosan (mCS) serve as nanofillers of graphite oxide (GO) for enhanced removal of hazardous mercury ions, Hg(II). Analyses evidenced that the oxygen moieties of GO interacted with amino groups of chitosan had created emerging binding sites for Hg(II) adsorption (Kyzas et al., 2014). Interestingly, the research also demonstrated that, in contrast to GO (i.e., unmodified), the adsorption behavior of the GO/CS and GO/mCS was highly impacted by pH settings (Kyzas et al., 2014). The combination of chitosan with magnetic nanoparticles, including Fe_3O_4 and $CoFe_2O_4$ has found usage in a variety of bio-applications, such as the immobilization of proteins, peptides, and enzymes; bioaffinity adsorbents; and drug delivery (Ngwuluka et al., 2016). The exceptional photocatalytic efficiency and stability of composites combining chitosan and metal oxide nanoparticles (e.g., TiO_2 or ZnO) have been shown in both acidic and basic conditions (Zhu et al., 2009; Zainal et al., 2009).

Table 4.2 summarizes the several ways in which chitosan nanoparticles may be prepared: through emulsification, precipitation, ionic gelation, covalent cross-linking, reversed micelles (micro-emulsion), and other processes. Aqueous chitosan solution is emulsified with the oil phase, utilizing Span 80 as a stabilizer and toluene and glutaraldehyde as cross-linker (Ohya et al., 1994; Yanat & Schroën, 2021). After intensive mixing, chitosan nanoparticles are produced, which are then separated from the emulsion through centrifugation, purified with petroleum ether, acetone, sodium metabisulfite, and water, followed by vacuum- or freeze-drying (Yanat & Schroën, 2021). Glutaraldehyde was formerly employed in this method, but it was phased out due to safety concerns and issues with maintaining the integrity of the drugs. Producing chitosan nanoparticles requires combining chitosan and glutaraldehyde in the inorganic phase with a lipophilic surfactant (e.g., cetyltrimethylammonium

TABLE 4.2

Overview of Chitosan Nanofiller Preparation Methods

Technique	Fundamental	Benefits	Drawbacks	References
Emulsification and cross-linking	Covalent cross-linking	Easy procedure	Exposure to harmful chemicals	Gan et al. (2005)
Reversed micelles	Covalent cross-linking	Produce fine nanofillers <100 nm	1. Laborious operation 2. Application complexity 3. Exposure to harmful chemicals	Kafshgari et al. (2012)
Phase inversion precipitation	Precipitation	High encapsulation capacity for specific compounds	1. Exposure to harmful chemicals	Ana Grenha (2012)
Emulsion-droplet coalescence	Precipitation		2. Demand exceptional shear force	Tokumitsu et al. (1999)
Ionic gelation-based	Ionic cross-linking	1. Easy procedure 2. Use of non-hazourdous chemicals 3. Tunable nanofillers size	1. Laborious operation 2. Application complexity	Fan et al. (2012)

bromide (CTAB) or sodium bis (2-ethylhexyl) sulfosuccinate (AOT)) and an organic solvent (e.g., n-hexane) (Kafshgari et al., 2012). For many uses wherein a particular surface area is deemed vital (for example, loading capacity and sustained release), the fact that this technology made the production of ultrafine nanoparticles (size less than 100 nm) feasible is noteworthy (Grenha, 2012). Micelles containing chitosan at their center serve as nanoreactors, producing cross-linked chitosan nanoparticles that are then separated by three consecutive steps: (1) surfactant precipitation with $CaCl_2$, (2) dialysis to remove unreacted components, and (3) freeze-drying (Yanat & Schroën, 2021). It's also worth noting that alternatives to glutaraldehyde have been developed, including non-harmful solvents and cross-linkers, such as glutaric acid (Grenha, 2012).

Alternatively, nanometer-sized, uniformly distributed emulsion chitosan droplets can be obtained via phase inversion precipitation, which involved high-pressure homogenization of the organic phase (e.g., dichloromethane and acetone) and an aqueous solution of chitosan stabilized by polyoxamer (El-Shabouri, 2002). During the evaporation process at room temperature and low pressure, methylene chloride will be removed from the emulsion, resulting in acetone diffusing out of the droplets and the simultaneous precipitation of chitosan nanoparticles (Yanat & Schroën, 2021). This technique, however, is less favored since it requires organic solvents and a high intensity homogenization stage to form chitosan nanoparticles of 600–800 nm in size (El-Shabouri, 2002). In spite of this, hydrophobic pharmaceuticals like cyclosporin A have benefitted greatly from the method's high encapsulation efficiency, hence, it is still widely used (El-Shabouri, 2002). The process of ionic cross-linking, also known as ionic gelation, begins with the addition of chitosan to an aqueous solution of an acid, such as acetic acid. Subsequently, an aqueous solution of sodium tripolyphosphate (TPP) is added to the mixture while it is vigorously stirred and completed after the mixture reached the desired consistency (Fan et al., 2012). Spherical nanoparticles are formed when negatively charged anionic TPP molecules diffuse into a solution containing positively charged amino groups of chitosan molecules, causing cross-linking between the two molecules (Yanat & Schroën, 2021). It is worth noting that this procedure does not require the use of a hazardous cross-linker or solvent and may be performed at ambient temperatures (Liu & Gao, 2009). The effectiveness of drug encapsulation and delivery is also influenced by the final nanoparticle size, which may be altered by modifying the chitosan/TPP ratio (Hejjaji et al., 2018).

4.5 CARBON NANOTUBES

Long-distance electricity transmission results in energy losses of 8%–15% as heat. Conductive nanotubes are able to conduct electrical currents with such efficiency that they barely lose any heat. All of this is possible as a direct consequence of their one-of-a-kind topologically tuned electrical behavior, which also enables them to serve as the functional semiconductor in nanodevices. CNTs (carbon nanotubes) had previously been the discipline for researchers who were intrigued mostly in the distinct morphological qualities they had; nonetheless, well over course of the last few years, emphasis has began to center on the surface chemistry of CNTs. Because of their low cost, high aspect ratio, and exceptional physical features, most notably their conductivity, CNTs are widely recognized as excellent reinforcement. However, CNTs aggregation limits their applicability. Improving CNTs' performance as a nanofiller in nanocomposites products is feasible by either non-covalent or covalent alterations with regards to reducing their inherently erratic clustering and bolstering their dispersibility. This is essential to devising nanoarchitectures by modulating their interplay with polymeric materials or even biological species like proteins and DNA for next generation CNTs medicate composite materials (Dubey et al., 2021).

CNTs, or tubular fullerenes, are sp^2-bonded carbon sheets rolled into a cylinder. Multi-walled carbon nanotubes (MWCNTs) and single-walled carbon nanotubes (SWCNTs) are both allotropes of CNTs formed by rolling a flat 2D-graphene sheet on top of itself (SWCNTs). Depending on the number of layers of graphene, the diameter of a single MWCNT may be anywhere from 7 to 100 nm (Marina, 2022). The average spacing between adjacent layers in these tubes is around 0.34 nm. Contrarily, SWCNTs are composed of a single graphene sheet wrapped onto itself and have a diameter of 1–2 nm. Depending on the preparation techniques, the length might vary (Ibrahim, 2013). Both the archways (terminals) and the sidewalls of CNTs may be functionalized, where the terminals being considerably more reactive than the sidewalls, and they are easily removed during chemical treatments, perhaps as a result of the extent contours (Rathinavel et al., 2021; Dubey et al., 2021). An overview of CNTs and their derivatives is illustrated in Figure 4.3.

The mechanical strength, surface area, and aspect ratio of CNTs are said to be exceptional. CNTs have a Young's modulus of 0.3–1.8 TPa, which is up to five times more than steel and twice higher than carbon fibers, and a tensile strength of 11–63 GPa, which is nearly 100 times stronger than that of steel while being six times lighter

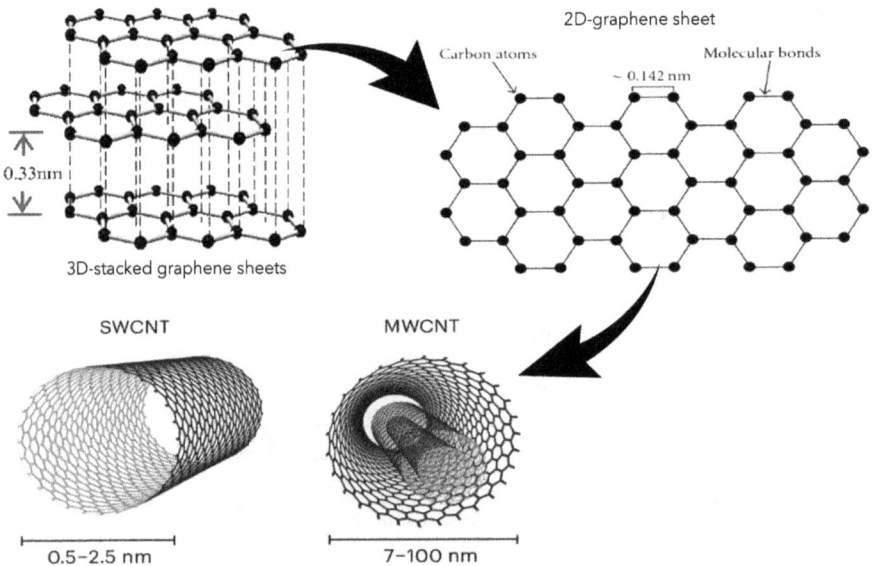

FIGURE 4.3 Graphene sheets are building blocks for CNTs, MWCNT, and SWCNT.

(Hassanzadeh-Aghdam et al., 2019). Because of their distinctive qualities, CNTs may be used as additives in a wide selection of polymers and ceramics to produce high-quality consumer goods, especially those that need superior mechanical resilience.

According to the existing findings, MWCNTs are pliable and may be bent at acute angles without suffering structural fractures. Also, the CNTs are notoriously soft in the radial direction (Ruoff et al., 1993) while exhibiting tremendous axial strength (Yu et al., 2000a). It was formerly anticipated that SWCNTs may reach a Young's modulus as high as diamond's 1.22 TPa (Sinnott et al., 1998). To this end, CNTs' radial direction elasticity is crucial, particularly in the context of CNT nanocomposites and their mechanical characteristics, where embedded tubes experience substantial deformation in the transverse direction when a load is applied to a composite structure (Ibrahim, 2013). There is a strong theoretical consensus that the diameter of SWCNTs plays a crucial role in determining their mechanical characteristics. Theoretical calculations by Gao et al. (1998) suggest that nanotubes with a diameter of >1 nm have a Young's modulus of 0.6–0.7 TPa. According to Hernandez et al. (1998), the Young's moduli of the tubes approach those of planar graphite (45.4 GPa) after their diameters reach a particular threshold value, suggesting that the increase in diameter is related to the strengthening of the mechanical characteristics. (Yu et al., 2000b) determined the breaking strength of SWCNTs on the perimeter of each rope to be anywhere from 13 to 52 GPa by measuring the Young's modulus of individual SWCNTs and reported results between 320 and 1,470 GPa.

CNTs also exhibit desirable electrical characteristics, which are variant-dependent: chiral, armchair, or zigzag. Resistivity measurements for SWCNTs fall between 0.34 and 1.0×10^{-4} $\Omega \cdot$cm (Ebbesen et al., 1996). Each carbon atom in CNTs forms covalent bonds with its three nearest neighbors using sp^2 molecular orbitals. Therefore, all of the valence electrons in each unit remain unpaired and free, and these free electrons are dispersed throughout all of the atoms, giving CNTs their electrical properties (Ibrahim, 2013).

The low temperature specific heat and thermal conductivity of CNTs provide a clear proof of their 1-D quantized phonon subbands, demonstrating the significance of quantum effects despite their tiny size (Hone et al., 2000). With an aspect ratio equal to 10^2, CNTs have a longitudinal thermal conductivity of 2,800–6,000 W/m K for a single nanotube at ambient temperature, making them similar to diamond and greater than graphite and carbon fibers (Han & Fina, 2011). Nanotube composite materials may be valuable for thermal management applications in industries since incorporating pristine and functionalized nanotubes into diverse materials may double the thermal conductivity with a loading of ~1%. It was discovered by Kim et al. (2001) that at room temperature, the thermal conductivity of single MWCNTs is 3,000 W/m K, which is much greater than graphite. The researchers also found that this value is two orders of magnitude larger than the ones obtained for MWCNTs in bulk. For SWCNTs, a comparable investigation yielded a value more than 200 W/m K. Many parameters, including the number of phonon-active modes, the phonon free route length, and boundary surface scattering, all play a role in determining the thermal characteristics (Yu et al., 2005). To a lesser extent, the mimetic patterns, tube dimensions, structural flaws, and impurities all have a role in the manifestation of these features in CNTs (Maeda & Horie, 1999).

4.6 CONCLUSION

As discussed in this article, nanocellulose, nanoclays, chitosan, and carbon nanotubes could contribute as natural nanofillers and served as alternatives to the conventional ones. On top of their abundance in nature, making them reliable and renewable sources, their physical and chemical properties matched and sometimes even surpassed those of conventional fillers. Each natural nanofiller discussed has a significant contribution in the enhancement of the physical and mechanical performance of the polymer matrices. In conclusion, the utilization of the aforementioned natural nanofillers has the huge ability to partially substitute or replace the current synthetic nanofiller, which would be beneficial in many applications such as medical, packaging, industrial, structural engineering, and military fields.

ACKNOWLEDGEMENTS

Authors would like to acknowledge the Universiti Pertahanan Nasional Malaysia toward its support in the preparation of this manuscript.

REFERENCES

Abulyazied, D. E., & Ene, A. (2021). An investigative study on the progress of nanoclay-reinforced polymers: Preparation, properties, and applications: A review. *Polymers*, 13(24), 4401. https://doi.org/10.3390/polym13244401.

Alemdar, A., & Sain, M. (2008). Isolation and characterization of nanofibers from agricultural residues-Wheat straw and soy hulls. *Bioresource Technology*, 99(6), 1664–1671. https://doi.org/10.1016/j.biortech.2007.04.029.

Ariffin, H., Tengku Yasim-Anuar, T. A., Norrrahim, M. N. F., & Hassan, M. A. (2021). Synthesis of cellulose nanofiber from oil palm biomass by high pressure homogenization and wet disk milling. In Nanocellulose: Synthesis, Structure, Properties and Applications (pp. 51–64). https://doi.org/10.1142/9781786349477_0002.

Berglund, L., Noël, M., Aitomäki, Y., Öman, T., & Oksman, K. (2016). Production potential of cellulose nanofibers from industrial residues: Efficiency and nanofiber characteristics. *Industrial Crops and Products*, 92, 84–92. https://doi.org/10.1016/j.indcrop.2016.08.003.

Beyer, G. (2002). Nanocomposites: A new class of flame retardants for polymers. *Plastics, Additives and Compounding*, 4(10), 22–28. https://doi.org/10.1016/S1464-391X(02)80151-9.

Bitounis, D., Pyrgiotakis, G., Bousfield, D., & Demokritou, P. (2019). Dispersion preparation, characterization, and dosimetric analysis of cellulose nano-fibrils and nano-crystals: Implications for cellular toxicological studies. *NanoImpact*, 15, 100171. https://doi.org/10.1016/j.impact.2019.100171.

Blokhin, A., Sukhorukov, A., Stolyarov, R., Zaytsev, I., Yashchishin, N., & Yagubov, V. (2019). Carbon nanofillers used in epoxy polymeric composites: A brief review. In *IOP Conference Series: Materials Science and Engineering (693,1,012015)*. IOP Publishing. https://doi.org/10.1088/1757-899X/693/1/012015.

Börjesson, M., & Westman, G. (2015). Crystalline nanocellulose-preparation, modification, and properties. *Cellulose-Fundamental Aspects and Current Trends*, 7.

Boujemaoui, A. (2016). *Surface Modification of Nanocellulose towards Composite Applications* (Doctoral dissertation, KTH Royal Institute of Technology).

Budaeva, V. V., Gismatulina, Y. A., Mironova, G. F., Skiba, E. A., Gladysheva, E. K., Kashcheyeva, E. I., Baibakova, O. V., Korchagina, A. A., Shavyrkina, N. A., Golubev, D. S. and Bychin, N. V. (2019). Bacterial nanocellulose nitrates. *Nanomaterials*, 9(12), 1694. https://doi.org/0.3390/nano9121694.

Bumbudsanpharoke, N., & Ko, S. (2019). Nanoclays in food and beverage packaging. *Journal of Nanomaterials*, 2019. https://doi.org/10.1155/2019/8927167.

Burhani, D., Septevani, A. A., Setiawan, R., Djannah, L. M., Putra, M. A., Kusumah, S. S., & Sondari, D. (2021). Self-assembled behavior of ultralightweight aerogel from a mixture of CNC/CNF from oil palm empty fruit bunches. *Polymers*, 13(16), 2649. https://doi.org/10.3390/polym13162649.

Chivrac, F., Pollet, E., & Averous, L. (2009). Progress in nano-biocomposites based on polysaccharides and nanoclays. *Materials Science and Engineering: R: Reports*, 67(1), 1–17. https://doi.org/10.1016/j.mser.2009.09.002.

Chun, S. J., Lee, S. Y., Doh, G. H., Lee, S., & Kim, J. H. (2011). Preparation of ultrastrength nanopapers using cellulose nanofibrils. *Journal of Industrial and Engineering Chemistry*, 17(3), 521–526. https://doi.org/10.1016/j.jiec.2010.10.022.

Dennis, H., Hunter, D. L., Chang, D., Kim, S., White, J. L., Cho, J. W., & Paul, D. R. (2001). Effect of melt processing conditions on the extent of exfoliation in organoclay-based nanocomposites. *Polymer*, 42(23), 9513–9522. https://doi.org/10.1016/S0032-3861(01)00473-6.

Dima, S. O., Panaitescu, D. M., Orban, C., Ghiurea, M., Doncea, S. M., Fierascu, R. C., ... Oancea, F. (2017). Bacterial nanocellulose from side-streams of kombucha beverages production: Preparation and physical-chemical properties. *Polymers*, 9(8), 374. https://doi.org/10.3390/polym9080374.

Dubey, R., Dutta, D., Sarkar, A., & Chattopadhyay, P. (2021). Functionalized carbon nanotubes: Synthesis, properties and applications in water purification, drug delivery, and material and biomedical sciences. *Nanoscale Advances*, 3(20), 5722–5744. https://doi.org/10.1039/D1NA00293G.

Ebbesen, T. W., Lezec, H. J., Hiura, H., Bennett, J. W., Ghaemi, H. F., & Thio, T. (1996). Electrical conductivity of individual carbon nanotubes. *Nature*, 382(6586), 54–56. https://doi.org/10.1038/382054a0.

El-Shabouri, M. H. (2002). Positively charged nanoparticles for improving the oral bioavailability of cyclosporin-A. *International Journal of Pharmaceutics*, 249(1–2), 101–108. https://doi.org/10.1016/S0378-5173(02)00461-1.

Fan, L., Luo, C., Lv, Z., Lu, F., & Qiu, H. (2011). Preparation of magnetic modified chitosan and adsorption of Zn2+ from aqueous solutions. *Colloids and surfaces B: Biointerfaces*, 88(2), 574–581. https://doi.org/10.1016/j.colsurfb.2011.07.038.

Fan, W., Yan, W., Xu, Z., & Ni, H. (2012). Formation mechanism of monodisperse, low molecular weight chitosan nanoparticles by ionic gelation technique. *Colloids and surfaces B: Biointerfaces*, 90, 21–27. https://doi.org/10.1016/j.colsurfb.2011.09.042.

Gamelas, J. A., Pedrosa, J., Lourenço, A. F., & Ferreira, P. J. (2015). Surface properties of distinct nanofibrillated celluloses assessed by inverse gas chromatography. *Colloids and Surfaces A: Physicochemical and Engineering Aspects*, 469, 36–41. https://doi.org/10.1016/j.colsurfa.2014.12.058.

Gan, Q., Wang, T., Cochrane, C., & McCarron, P. (2005). Modulation of surface charge, particle size and morphological properties of chitosan-TPP nanoparticles intended for gene delivery. *Colloids and Surfaces B: Biointerfaces*, 44(2–3), 65–73. https://doi.org/10.1016/j.colsurfb.2005.06.001.

Gao, F. (2004). Clay/polymer composites: The story. *Materials Today*, 7(11), 50–55. https://doi.org/10.1016/S1369-7021(04)00509-7.

Gao, G., Cagin, T., & Goddard III, W. A. (1998). Energetics, structure, mechanical and vibrational properties of single-walled carbon nanotubes. *Nanotechnology*, 9(3), 184. https://doi.org/10.1088/0957-4484/9/3/007.

Grenha, A. (2012). Chitosan nanoparticles: A survey of preparation methods. *Journal of Drug Targeting*, 20(4), 291–300. https://doi.org/10.3109/1061186X.2011.654121.

Gumrah Dumanli, A. (2017). Nanocellulose and its composites for biomedical applications. *Current Medicinal Chemistry*, 24(5), 512–528. https://doi.org/10.2174/0929867323666161014124008.

Guo, F., Aryana, S., Han, Y., & Jiao, Y. (2018). A review of the synthesis and applications of polymer- nanoclay composites. *Applied Sciences*, 8(9), 1696. https://doi.org/10.3390/app8091696.

Gurses, A. (2015). *Introduction to Polymer-Clay Nanocomposites*. CRC Press.

Han, Z., & Fina, A. (2011). Thermal conductivity of carbon nanotubes and their polymer nanocomposites: A review. *Progress in Polymer Science*, 36(7), 914–944. https://doi.org/10.1016/j.progpolymsci.2010.11.004.

Härdelin, L., Perzon, E., Hagström, B., Walkenström, P., & Gatenholm, P. (2013). Influence of molecular weight and rheological behavior on electrospinning cellulose nanofibers from ionic liquids. *Journal of Applied Polymer Science*, 130(4), 2303–2310. https://doi.org/10.1002/app.39449.

Hassanzadeh-Aghdam, M. K., Mahmoodi, M. J., & Safi, M. (2019). Effect of adding carbon nanotubes on the thermal conductivity of steel fiber-reinforced concrete. *Composites Part B: Engineering*, 174, 106972. https://doi.org/10.1016/j.compositesb.2019.106972.

Hejjaji, E. M., Smith, A. M., & Morris, G. A. (2018). Evaluation of the mucoadhesive properties of chitosan nanoparticles prepared using different chitosan to tripolyphosphate (CS: TPP) ratios. *International Journal of Biological Macromolecules*, 120, 1610–1617. https://doi.org/10.1016/j.ijbiomac.2018.09.185.

Hernandez, E., Goze, C., Bernier, P., & Rubio, A. (1998). Elastic properties of C and $B_xC_yN_z$ composite nanotubes. *Physical Review Letters*, 80(20), 4502. https://doi.org/10.1103/PhysRevLett.80.4502.

Hesami, M., & Jalali-Arani, A. (2018). Morphology development via static crosslinking of (polylactic acid/acrylic rubber) as an immiscible polymer blend. *Macromolecular Materials and Engineering*, 303(3), 1700446. https://doi.org/10.1002/mame.201700446.

Hiujian, L., Jinwei, Y., & Ling, X. (2012). The synthesis and application of a high performance amino resin nanocomposite as a leather flame retardant. *Journal of the Society of Leather Technologists and Chemists*, 96(1), 5–10.

Hone, J., Batlogg, B., Benes, Z., Johnson, A. T., & Fischer, J. E. (2000). Quantized phonon spectrum of single-wall carbon nanotubes. *Science*, 289(5485), 1730–1733. https://doi.org/10.1126/science.289.5485.1730.

Ibrahim, K. S. (2013). Carbon nanotubes-properties and applications: A review. *Carbon Letters*, 14(3), 131–144. https://doi.org/10.5714/cl.2013.14.3.131.

Jawaid, M., Qaiss, A., & Bouhfid, R. (2016). *Nanoclay Reinforced Polymer Composites*. Springer.

Jozala, A. F., de Lencastre-Novaes, L. C., Lopes, A. M., de Carvalho Santos-Ebinuma, V., Mazzola, P. G., Pessoa-Jr, A., Grotto, D., Gerenutti, M. & Chaud, M. V. (2016). Bacterial nanocellulose production and application: A 10-year overview. *Applied Microbiology and Biotechnology*, 100, 2063–2072. https://doi.org/10.1007/s00253-015-7243-4.

Kaboorani, A., & Riedl, B. (2015). Surface modification of cellulose nanocrystals (CNC) by a cationic surfactant. *Industrial Crops and Products*, 65, 45–55. https://doi.org/10.1016/j.indcrop.2014.11.027.

Kafshgari, M. H., Khorram, M., Mansouri, M., Samimi, A., & Osfouri, S. (2012). Preparation of alginate and chitosan nanoparticles using a new reverse micellar system. *Iranian Polymer Journal*, 21, 99–107. https://doi.org/10.1007/s13726-011-0010-1

Kakuta, T., Baba, Y., Yamagishi, T. A., & Ogoshi, T. (2021). Supramolecular exfoliation of layer silicate clay by novel cationic pillar [5] arene intercalants. *Scientific Reports*, 11(1), 10637. https://doi.org/10.1038/s41598-021-90122-9.

Kapoor, A., Shankar, P., & Ali, W. (2020). Green synthesis of metal nanoparticles for electronic textiles. *Green Nanomaterials: Processing, Properties, and Applications*, 81–97. https://doi.org/10.1007/978-981-15-3560-4_4.

Kaushik, A. K., Podsiadlo, P., Qin, M., Shaw, C. M., Waas, A. M., Kotov, N. A., & Arruda, E. M. (2009). The role of nanoparticle layer separation in the finite deformation response of layered polyurethane-clay nanocomposites. *Macromolecules*, 42(17), 6588–6595. https://doi.org/10.1021/ma901048g.

Khalid, M. Y., Al Rashid, A., Arif, Z. U., Ahmed, W., & Arshad, H. (2021). Recent advances in nanocellulose-based different biomaterials: Types, properties, and emerging applications. *Journal of Materials Research and Technology*, 14, 2601–2623. https://doi.org/10.1016/j.jmrt.2021.07.128.

Khattab, M. M., Abdel-Hady, N. A., & Dahman, Y. (2017). Cellulose nanocomposites: Opportunities, challenges, and applications. *Cellulose-Reinforced Nanofibre Composites*, 483–516. https://doi.org/10.1016/B978-0-08-100957-4.00021-8.

Kiliaris, P., & Papaspyrides, C. D. (2010). Polymer/layered silicate (clay) nanocomposites: An overview of flame retardancy. *Progress in Polymer Science*, 35(7), 902–958. https://doi.org/10.1016/j.progpolymsci.2010.03.001.

Kim, P., Shi, L., Majumdar, A., & McEuen, P. L. (2001). Thermal transport measurements of individual multiwalled nanotubes. *Physical Review Letters*, 87(21), 215502. https://doi.org/10.1103/PhysRevLett.87.215502.

Koh, H. C., Park, J. S., Jeong, M. A., Hwang, H. Y., Hong, Y. T., Ha, S. Y., & Nam, S. Y. (2008). Preparation and gas permeation properties of biodegradable polymer/layered silicate nanocomposite membranes. *Desalination*, 233(1–3), 201–209. https://doi.org/10.1016/j.desal.2007.09.043.

Kyzas, G. Z., Travlou, N. A., & Deliyanni, E. A. (2014). The role of chitosan as nanofiller of graphite oxide for the removal of toxic mercury ions. *Colloids and Surfaces B: Biointerfaces*, 113, 467–476. https://doi.org/10.1016/j.colsurfb.2013.07.05.

Levengood, S. K. L., & Zhang, M. (2014). Chitosan-based scaffolds for bone tissue engineering. *Journal of Materials Chemistry B*, 2(21), 3161–3184. https://doi.org/10.1039/C4TB00027G.

Liu, H., & Gao, C. (2009). Preparation and properties of ionically cross-linked chitosan nanoparticles. *Polymers for Advanced Technologies*, 20(7), 613–619. https://doi.org/10.1002/pat.1306.

Liu, X. D., Nishi, N., Tokura, S., & Sakairi, N. (2001). Chitosan coated cotton fiber:preparation and physical properties. *Carbohydrate Polymers*, 44(3), 233–238. https://doi.org/10.1016/S0144-8617(00)00206-X.

Madhumathi, K., Kumar, P. S., Kavya, K. C., Furuike, T., Tamura, H., Nair, S. V., & Jayakumar, R. (2009). Novel chitin/nanosilica composite scaffolds for bone tissue engineering applications. *International Journal of Biological Macromolecules*, 45(3), 289–292. https://doi.org/10.1016/j.ijbiomac.2009.06.009.

Maeda, T., & Horie, C. (1999). Phonon modes in single-wall nanotubes with a small diameter. *Physica B: Condensed Matter*, 263, 479–481. https://doi.org/10.1016/S0921-4526(98)01415-X.

Marina, F. (2022). *Multi-Walled Carbon Nanotubes Production, Properties and Applications*, OCSiAl, 2022. https://tuball.com/articles/multi-walled-carbon-nanotubes.

Marquis, D. M., Guillaume, E., & Chivas-Joly, C. (2011). Properties of nanollers in polymer. *Nanocomposites and Polymers with Analytical Methods*, 261. InTechOpen.

Metreveli, G., Wågberg, L., Emmoth, E., Belák, S., Strømme, M., & Mihranyan, A. (2014). A size-exclusion nanocellulose filter paper for virus removal. *Advanced Healthcare Materials*, 3(10), 1546–1550. https://doi.org/10.1002/adhm.201300641.

Mondal, S. (2017). Preparation, properties and applications of nanocellulosic materials. *Carbohydrate Polymers*, 163, 301–316. https://doi.org/10.1016/j.carbpol.2016.12.050.

Monier, M., Ayad, D. M., & Abdel-Latif, D. A. (2012). Adsorption of Cu (II), Cd (II) and Ni (II) ions by cross-linked magnetic chitosan-2-aminopyridine glyoxal Schiff's base. *Colloids and Surfaces B: Biointerfaces*, 94, 250–258. https://doi.org/10.1016/j.colsurfb.2012.01.051.

Murray, H. H. (1991). Overview-clay mineral applications. *Applied Clay Science*, 5(5–6), 379–395. https://doi.org/10.1016/0169-1317(91)90014-Z.

Nagahama, H., Nwe, N., Jayakumar, R., Koiwa, S., Furuike, T., & Tamura, H. (2008). Novel biodegradable chitin membranes for tissue engineering applications. *Carbohydrate Polymers*, 73(2), 295–302. https://doi.org/10.1016/j.carbpol.2007.11.034.

Ngwuluka, N. C., Ochekpe, N. A., & Aruoma, O. I. (2016). Functions of bioactive and intelligent natural polymers in the optimization of drug delivery. *Industrial Applications for Intelligent Polymers and Coatings*, 165–184. https://doi.org/10.1007/978-3-319-26893-4_8.

Norrrahim, M. N. F., Ariffin, H., Hassan, M. A., Ibrahim, N. A., Yunus, W. M. Z. W., & Nishida, H. (2019). Utilisation of superheated steam in oil palm biomass pretreatment process for reduced chemical use and enhanced cellulose nanofibre production. *International Journal of Nanotechnology*, 16(11–12), 668–679. https://doi.org/10.1504/IJNT.2019.107360.

Norrrahim, M. N. F., Ariffin, H., Yasim-Anuar, T. A. T., Hassan, M. A., Ibrahim, N. A., Yunus, W. M. Z. W., & Nishida, H. (2021). Performance evaluation of cellulose nanofiber with residual hemicellulose as a nanofiller in polypropylene-based nanocomposite. *Polymers*, 13(7), 1064. https://doi.org/10.3390/polym13071064.

Norrrahim, M. N. F., Farid, M. A. A., Lawal, A. A., Yasim-Anuar, T. A. T., Samsudin, M. H., & Zulkifli, A. A. (2022). Emerging technologies for value-added use of oil palm biomass. *Environmental Science: Advances*, 1(3), 259–275. https://doi.org/10.1039/d2va00029f.

Ohya, Y., Shiratani, M., Kobayashi, H., & Ouchi, T. (1994). Release behavior of 5-fluorouracil from chitosan-gel nanospheres immobilizing 5-fluorouracil coated with polysaccharides and their cell specific cytotoxicity. *Journal of Macromolecular Science-Pure and Applied Chemistry*, 31(5), 629–642. https://doi.org/10.1080/10601329409349743.

Pavlidou, S., & Papaspyrides, C. D. (2008). A review on polymer-layered silicate nanocomposites. *Progress in Polymer Science*, 33(12), 1119–1198. https://doi.org/10.1016/j.progpolymsci.2008.07.008.

Peniche, C., Argüelles-Monal, W., & Goycoolea, F. M. (2008). Chitin and chitosan: Major sources, properties and applications. In *Monomers, Polymers and Composites From Renewable Resources* (pp. 517–542). Elsevier.

Peter, M., Ganesh, N., Selvamurugan, N., Nair, S. V., Furuike, T., Tamura, H., & Jayakumar, R. (2010). Preparation and characterization of chitosan-gelatin/nanohydroxyapatite composite scaffolds for tissue engineering applications. *Carbohydrate Polymers*, 80(3), 687–694. https://doi.org/10.1016/j.carbpol.2009.11.050.

Phanthong, P., Reubroycharoen, P., Hao, X., Xu, G., Abudula, A., & Guan, G. (2018). Nanocellulose: Extraction and application. *Carbon Resources Conversion*, 1(1), 32–43. https://doi.org/10.1016/j.crcon.2018.05.004.

Pillai, S. K., & Ray, S. S. (2012). Chitosan-based nanocomposites. *Natural Polymers*, 2, 33–68.

Rajeshkumar, G., Seshadri, S. A., Ramakrishnan, S., Sanjay, M. R., Siengchin, S., & Nagaraja, K. C. (2021). A comprehensive review on natural fiber/nano-clay reinforced hybrid polymeric composites: Materials and technologies. *Polymer Composites*, 42(8), 3687–3701. https://doi.org/10.1002/pc.26110.

Rantanen, J., Dimic-Misic, K., Kuusisto, J., & Maloney, T. C. (2015). The effect of micro and nanofibrillated cellulose water uptake on high filler content composite paper properties and furnish dewatering. *Cellulose*, 22, 4003–4015. https://doi.org/10.1007/s10570-015-0777-x.

Rathinavel, S., Priyadharshini, K., & Panda, D. (2021). A review on carbon nanotube: An overview of synthesis, properties, functionalization, characterization, and the application. *Materials Science and Engineering: B*, 268, 115095. https://doi.org/10.1016/j.mseb.2021.115095.

Ray, S. S. (2013). *Clay-Containing Polymer Nanocomposites: From Fundamentals to Real Applications*. Newnes.

Ruoff, R. S., Tersoff, J., Lorents, D. C., Subramoney, S., & Chan, B. (1993). Radial deformation of carbon nanotubes by van der Waals forces. *Nature*, 364, 514–516.

Shabanpour, B., Kazemi, M., Ojagh, S. M., & Pourashouri, P. (2018). Bacterial cellulose nanofibers as reinforce in edible fish myofibrillar protein nanocomposite films. *International Journal of Biological Macromolecules*, 117, 742–751. https://doi.org/10.1016/j.ijbiomac.2018.05.038.

Shankar, S., & Rhim, J. W. (2018). Bionanocomposite films for food packaging applications. *Reference Module in Food Science*, 1, 1–10.

Sharma, A., Thakur, M., Bhattacharya, M., Mandal, T., & Goswami, S. (2019). Commercial application of cellulose nano-composites–A review. *Biotechnology Reports*, 21, e00316. https://doi.org/10.1016/j.btre.2019.e00316.

Shelley, J. S., Mather, P. T., & DeVries, K. L. (2001). Reinforcement and environmental degradation of nylon-6/clay nanocomposites. *Polymer*, 42(13), 5849–5858. https://doi.org/10.1016/S0032-3861(00)00900-9.

Sinnott, S. B., Shenderova, O. A., White, C. T., & Brenner, D. W. (1998). Mechanical properties of nanotubule fibers and composites determined from theoretical calculations and simulations. *Carbon*, 36(1–2), 1–9. https://doi.org/10.1016/S0008-6223(97)00144-9.

Sorrentino, A., Gorrasi, G., & Vittoria, V. (2007). Potential perspectives of bio-nanocomposites for food packaging applications. *Trends in Food Science & Technology*, 18(2), 84–95. https://doi.org/10.1016/j.tifs.2006.09.004.

Šupová, M., Martynková, G. S., & Barabaszová, K. (2011). Effect of nanofillers dispersion in polymer matrices: A review. *Science of Advanced Materials*, 3(1), 1–25. https://doi.org/10.1166/sam.2011.1136.

Thakur, V. K., & Kessler, M. R. (2015). Self-healing polymer nanocomposite materials: A review. *Polymer*, 69, 369–383. https://doi.org/10.1016/j.polymer.2015.04.086.

Thomas, B., Raj, M. C., Joy, J., Moores, A., Drisko, G. L., & Sanchez, C. (2018). Nanocellulose, a versatile green platform: From biosources to materials and their applications. *Chemical Reviews*, 118(24), 11575–11625. https://doi.org/10.1021/acs.chemrev.7b00627.

Thomas, M. S., Koshy, R. R., Mary, S. K., Thomas, S., & Pothan, L. A. (2019). *Starch, Chitin and Chitosan Based Composites and Nanocomposites* (pp. 19–42). Cham: Springer International Publishing.

Tibolla, H., Pelissari, F. M., & Menegalli, F. C. (2014). Cellulose nanofibers produced from banana peel by chemical and enzymatic treatment. *LWT-Food Science and Technology*, 59(2), 1311–1318. https://doi.org/10.1016/j.lwt.2014.04.011.

Tokumitsu, H., Ichikawa, H., & Fukumori, Y. (1999). Chitosan-gadopentetic acid complex nanoparticles for gadolinium neutron-capture therapy of cancer: Preparation by novel emulsion-droplet coalescence technique and characterization. *Pharmaceutical Research*, 16, 1830–1835. https://doi.org/10.1023/A:1018995124527.

Tortora, M., Vittoria, V., Galli, G., Ritrovati, S., & Chiellini, E. (2002). Transport properties of modified montmorillonite-poly (ε-caprolactone) nanocomposites. *Macromolecular Materials and Engineering*, 287(4), 243–249. https://doi.org/10.1002/1439-205.

Trache, D., Tarchoun, A. F., Derradji, M., Hamidon, T. S., Masruchin, N., Brosse, N., & Hussin, M. H. (2020). Nanocellulose: From fundamentals to advanced applications. *Frontiers in Chemistry*, 8, 392. https://doi.org/10.3389/fchem.2020.00392.

Valapa, R. B., Loganathan, S., Pugazhenthi, G., Thomas, S., & Varghese, T. O. (2017). An overview of polymer-clay nanocomposites. *Clay-Polymer Nanocomposites*, 29–81. https://doi.org/10.1016/B978-0-323-46153-5.00002-1.

Vo, V. S., Mahouche-Chergui, S., Babinot, J., Nguyen, V. H., Naili, S., & Carbonnier, B. (2016). Photo-induced SI-ATRP for the synthesis of photoclickable intercalated clay nanofillers. *RSC Advances*, 6(92), 89322–89327. https://doi.org/10.1039/C6RA14724K.

Vyazovkin, S., Dranca, I., Fan, X., & Advincula, R. (2004). Kinetics of the thermal and thermo-oxidative degradation of a polystyrene-clay nanocomposite. *Macromolecular Rapid Communications*, 25(3), 498–503. https://doi.org/10.1002/marc.200300214.

Wang, H., Zhang, X., Jiang, Z., Li, W., & Yu, Y. (2015). A comparison study on the preparation of nanocellulose fibrils from fibers and parenchymal cells in bamboo (*Phyllostachys pubescens*). *Industrial Crops and Products*, 71, 80–88. https://doi.org/10.1016/j.indcrop.2015.03.086.

Wang, R., Li, H., Ge, G., Dai, N., Rao, J., Ran, H., & Zhang, Y. (2021). Montmorillonite-based two-dimensional nanocomposites: Preparation and applications. *Molecules*, *26*(9), 2521. https://doi.org/10.3390/molecules26092521.

Yanat, M., & Schroën, K. (2021). Preparation methods and applications of chitosan nanoparticles; with an outlook toward reinforcement of biodegradable packaging. *Reactive and Functional Polymers*, 161, 104849. https://doi.org/10.1016/j.reactfunctpolym.2021.104849.

Yasim-Anuar, T. A. T., Ariffin, H., & Hassan, M. A. (2018). Characterization of cellulose nanofiber from oil palm mesocarp fiber produced by ultrasonication. In *IOP Conference Series: Materials Science and Engineering* (Vol. 368, No. 1, p. 012033). IOP Publishing. https://doi.org/10.1088/1757-899X/368/1/012033.

Yasim-Anuar, T. A. T., Ariffin, H., Norrrahim, M. N. F., Hassan, M. A., Andou, Y., Tsukegi, T., & Nishida, H. (2020). Well-dispersed cellulose nanofiber in low density polyethylene nanocomposite by liquid-assisted extrusion. *Polymers*, *12*(4), 927. https://doi.org/10.3390/polym12040927.

Yu, M. F., Kowalewski, T., & Ruoff, R. S. (2000b). Investigation of the radial deformability of individual carbon nanotubes under controlled indentation force. *Physical Review Letters*, *85*(7), 1456. doi.org/10.1103/PhysRevLett.85.1456.

Yu, M. F., Lourie, O., Dyer, M. J., Moloni, K., Kelly, T. F., & Ruoff, R. S. (2000a). Strength and breaking mechanism of multiwalled carbon nanotubes under tensile load. *Science*, *287*(5453), 637–640. https://doi.org/10.1126/science.287.5453.637.

Yu, C., Shi, L., Yao, Z., Li, D., & Majumdar, A. (2005). Thermal conductance and thermopower of an individual single-wall carbon nanotube. *Nano Letters*, *5*(9), 1842–1846. https://doi.org/10.1021/nl051044e.

Zainal, Z., Hui, L. K., Hussein, M. Z., & Abdullah, A. H. (2009). Characterization of TiO_2-chitosan/glass photocatalyst for the removal of a monoazo dye via photodegradation-adsorption process. *Journal of Hazardous Materials*, *164*(1), 138–145. https://doi.org/10.1016/j.jhazmat.2008.07.154.

Żenkiewicz, M., Richert, J., & Różański, A. (2010). Effect of blow moulding ratio on barrier properties of polylactide nanocomposite films. *Polymer Testing*, *29*(2), 251–257. https://doi.org/10.1016/j.polymertesting.2009.11.008.

Zhang, T., Zhang, Y., Wang, X., Liu, S., & Yao, Y. (2018). Characterization of the nano-cellulose aerogel from mixing CNF and CNC with different ratio. *Materials Letters*, 229, 103–106. https://doi.org/10.1016/j.matlet.2018.06.101.

Zhu, Y., Cao, H., Tang, L., Yang, X., & Li, C. (2009). Immobilization of horseradish peroxidase in three-dimensional macroporous TiO_2 matrices for biosensor applications. *Electrochimica Acta*, *54*(10), 2823–2827. https://doi.org/10.1016/j.electacta.2008.11.025.

5 Compatibility Study of Nanofillers-Based Polymer Composites

Ismail M. Fareez
Universiti Teknologi MARA

Ramli M. Zaki
Universiti Kuala Lumpur Royal College of Medicine Perak

Ainil Hawa Jasni
International Islamic University of Malaysia

Azirah Akbar Ali
Universiti Sains Malaysia

Suresh Sagadevan
University of Malaya

Zaharah Wahid
International Islamic University of Malaysia

5.1 INTRODUCTION

Polymer nanocomposites (PNCs) are a blend of two or more materials with a polymer matrix and a dispersed phase of at least one dimension less than 100 nm (Müller et al., 2017). In recent decades, it has been discovered that incorporating small amounts of these nanofillers into polymers can increase their mechanical, thermal, barrier, and flammability properties without compromising their processability (de Oliveira & Beatrice, 2018). Individual nanoparticles (NPs) need to be homogeneously scattered in a matrix polymer to make up an ideal nanocomposite. In fact, the dispersion state of NPs remains a key indicator to unlocking the real potential of PNCs (Müller et al., 2017). The homogeneous dispersion of nanofillers can result in a large interfacial area between the constituents of the nanocomposites (de Oliveira & Beatrice, 2018). The enhancement of the nanocomposite by the filler as reinforcing materials is associated with a number of variables, including the nature and type of the nanofiller, its concentration, particle aspect ratio, particle size, particle orientation, particle distribution, and polymeric properties of the matrix (de Oliveira & Beatrice, 2018; Guchait et al., 2022; Wongvasana et al., 2022). Various types of NPs have been used as a filler to obtain nanocomposites with different polymers that include clays (Naidu & John, 2020), graphene (Alkhouzaam et al., 2021), carbon nanotubes (CNTs) (Cho et al., 2018), and nanocellulose (NC) (Mamat Razali et al., 2021) among others.

The mechanical and thermal properties are strongly related to the morphologies formed, therefore, evaluating the nanofiller dispersion in the polymer matrix is crucial. Three phases of nanocomposite morphologies are available depending on the degree of separation of the NPs: conventional composites (or microcomposites), intercalated nanocomposites, and exfoliated nanocomposites (Alexandre & Dubois, 2000). When the polymer fails to intercalate between the silicate layers, a composite of different phases is formed, with properties similar to those of conventional composites. A well-ordered multi-layered structure with intercalated layers of polymer and clay is formed by an intercalated structure between the layers of the silicate, either a single or multiple extended polymer chains. An exfoliated structure is obtained when the silicate layers are completely and uniformly dispersed in a continuous polymer matrix. The large surface area of contact between the matrix and NPs provides maximum reinforcing effect on the exfoliated composites (de Oliveira & Beatrice, 2018). This is one of the key differences between nanocomposites and traditional composites.

The aim of this chapter is to review the interactions of polymer matrix interfaces with supplementing organic nanofillers [i.e., (NC) and poly-(lactic acid) (PLA)], inorganic nanofillers [i.e., nanoclays, calcium carbonate, silver nanoparticles, and ferrous (III) oxide (Fe_2O_3), and carbon-based nanofillers (CNTs and graphene)]. Recent studies that evaluate the electrostatic interactions among the constituent materials and their role in creating a distinct functionalized composite that is attributable to its characters will be detailed. Also, some properties achieved in recently developed natural nanofiller-added composites are highlighted and their potential applications are envisioned. The characteristics of nanofiller will be initially discussed to provide further understanding of their unique attributes.

5.2 CLASSIFICATIONS AND CHARACTERIZATION OF NANOFILLERS

Nanofillers are mainly categorized based on their dimensions (Figure 5.1), which are classified as one-dimensional nanofiller or nanoplatelet, two-dimensional nanofillers or nanofibers, and three-dimensional nanofillers or nanoparticulates.

Generally, the dimensions for 1D nanofillers are less than 100nm, and they are in the form of sheets that are one to a few nm thick to 100 and 1,000nm long (Akpan et al., 2019). They are also in the shape of nanosheets, nanowalls, nanodisks, nanoprism, nanoplates, and branced structures, for example, montmorillonite (MMT) clay (Sharma & Kaith, 2021) and nanographene platelets (Jamróz et al., 2019). Some of the well-known 1D nanofillers are zinc oxide (ZnO) nanosheets, carbon nanowalls, Fe_3O_4 nanodiscs, and graphite nanoplatelets (GNP) (Mutiso & Winey, 2015; Russo et al., 2017; Zhang et al., 2018). Normally, 1D nanofiller show unique forms that make them suitable and fix into the main components of nanodevices. Tremendous applications by 1D nanofillers such as biomedical (Yang et al., 2015), biosensors (Muñoz et al., 2016), coating (Shafiq et al., 2020), and microelectronics (Mutiso & Winey, 2015; Russo et al., 2017) because of their exceptional magnetic, electric, and optic properties.

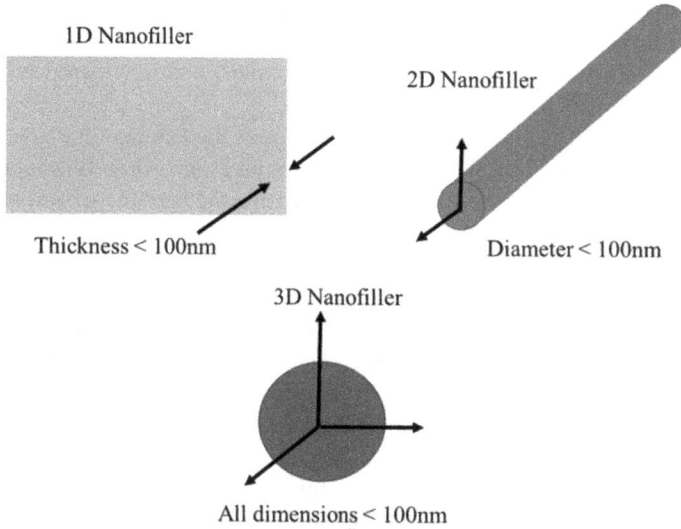

FIGURE 5.1   Classification of nanofillers. Adapted from Sharma and Kaith (2021).

Two-dimensional (2D) nanofillers, on the other hand, have a measurement of less than 100 nm and are usually in the form of tubes or fibers (Bhattacharya, 2016). Examples of 2D nanofillers include gold (AuNTs) or silver nanotubes (AgNTs), clay nanotubes, CNTs, cellulose whiskers, and black phosphorus. Most 2D nanofillers are in nature of PNC such as ZnO, titanium dioxide, silica, carbon fibers, and copper oxide (Sun et al., 2018; Xie et al., 2018). Their purpose is mostly in catalysis, sensors, energy, nonareactors, and electronics. 2D nanofillers also give a superior character in flame-retardant properties (Muñoz et al., 2016; Shen et al., 2020).

Three-dimensional (3D) nanofillers exist in a nanometer scale with relatively equal dimensions in all directions (Gomathi et al., 2019). Cubical and spherical shapes are the most common forms of 3D nanofillers. Generally, they have been known as zero-dimensional or isodimensional NPs. 3D nanofiller are also named nanocrystal, nanogranules, and nanospheres. An instance of 3D nanofillers is nanosilica, nanotitanium oxide, carbon black, silicon carbide, and nanoalumina (Panse et al., 2016; Xu et al., 2016). They are essential in the development of PNC as they exhibit excellent stability, hydrophilicity, ultraviolet resistance, a high refractive index, nontoxicity, and low cost (Paszkiewicz et al., 2020). These fantastic features make them useful in applications such as coatings, separation and purification, and biomedicine (Panse et al., 2016; Paszkiewicz et al., 2020; Vashist et al., 2018).

5.3   PROCESSING METHODS OF NANOFILLER-REINFORCED POLYMER COMPOSITES

Poor hydrophilicity, mechanical properties, and low electrical conductivity are the disadvantages of bio-based materials during processing (Väisänen et al., 2017). Nanofillers are the key to rectifying the above disadvantages and enhancing the properties of composite materials. Therefore, nanofillers are used to improve biodegradable materials, with many advancing methods for processing and preparing the nanofiller biodegradable composite materials. This chapter mainly discusses the processing of NC-based biocomposites, nanoclay-based biocomposites, and polymer-CNT-based biocomposites.

Currently, solvent casting and melt processing are the leading treatment methods for using NC reinforced into biocomposites (Abhijit et al., 2020; Bharimalla et al., 2017). Both treatments are the solutions to overcome the hard dispersion of NC into a non-polar medium. The polarity of a solvent has become the main problem that affect the dispersion of NC due to its hydrophilicity nature. Solvents with a high polarity, such as water, will generally result in better dispersion of NC than solvents with low polarity, such as hydrocarbons (Dufresne, 2013). Additionally, the type of NC (such as CNC or CNF) can also affect the dispersion in different polar solvents. So, the best solvent to disperse NC will depend on the specific type of NC being used and the desired properties of the final product (Zhang et al., 2022). For solvent casting, three types of polymers are used frequently: first is a water-soluble polymer; the second is a polymer emulsion; and the third is a water-insoluble polymer. Melt extrusion is the most common method used in the industry. It refers to the process of adding the plasticized material to the extruder for forming. Nevertheless, this method facing the biggest problems, the use of dry NC is the issue. The NC particles easily form hydrogen bonds in the amorphous state during the extrusion process. Due to the strong adsorption force of these bonds, the material is likely to aggregate when it is dry, which will cause problems to evenly disperse in the polymer (Kedzior et al., 2020). Currently, the feed process has been studied to overcome this matter. Better dispersibility has been obtained by using the pumping suspension method and creating nanocomposite-enhanced PLA biodegradable composites. Another technique is wet extrusion. Compared to melt extrusion, wet extrusion has a lower temperature and is suitable for applying in biomedical applications (Peterson et al., 2019).

Nowadays, there are three methods to mix nanoclay into biocomposites: (1) polymer solution embedding; (2) in situ polymerization; and (3) melt embedding method. The polymer solution embedding method works by macromolecule clay intercalation solvent in the polymer. By this mechanism, the nanoclay can be embedded in the polymer and will not deteriorate the internal structure of the polymer. Nevertheless, this process needs a large amount of solvent, thus resulting in a large amount of waste liquid, which caused pollution on the environment (Taghaddosi et al., 2017). In situ polymerization uses the polymerization of monomers in phyllosilicates (Bumbudsanpharoke et al., 2017; Eckert et al., 2020). The nanoclays expand in liquid monomer during this method. They are effectively embedded in the polymer, a process requiring catalyst initiation of polymerization. Melt embedding is currently widely used method in industrial production. The nanoclay is annealed, and once the temperature reaches the melting point, the polymer chains can penetrate the silicate interlayer and form a sandwich structure (Herrero et al., 2019). This method does not produce waste liquid and is more environmentally friendly (Torres-Giner et al., 2016).

CNTs are used to enhance biodegradable composites, their enhancement depends mainly on their molecular orientation and degree of dispersion. Currently, single-walled CNTs (SWCNTs) and multi-walled CNTs (MWCNTs) are mostly dispersed in polymer composites (Anzar et al., 2020). The van der Waals actions are applied to allow the polymer matrix to condense CNTs. These forces are relatively weak interactions between molecules, but they can be significant when the molecules are in close proximity (Alosfur et al., 2021). The van der Waals forces can help hold the CNTs in place within the polymer matrix, which can improve the mechanical properties of the final composite material (Jiang et al., 2006). Additionally, other types of interactions such as covalent or chemical interactions can also be applied to enhance the dispersion and stability of CNTs in the polymer matrix. The ability of dispersion depends on the network structure, while uniformity depends on the molecular orientation of the matrix and its compatibility. Moreover, the differences in size of the SWCNTs and the larger surface energy caused polymerization of the MWCNTs. Hence, it is difficult to uniformly distribute the surface of polymers (Anzar et al., 2020). The existing methods of uniformly dispersing CNTs, such as in situ polymerization, coating of CNTs, chemical modification, addition of surfactants, melt processing, and crystallization. Uniform dispersion of CNTs can expressively improve the strength and electrical conductivity of composites. It is an important measure for degradable biocomposites (Morimune-Moriya, 2022).

5.4   COMPATIBILIZATION OF NANOFILLER-REINFORCED POLYMER COMPOSITES

Nanofilling technology involves adding many fillers in relatively small quantities (<10%) of specially processed nano-size clay particles (Rothon, 2017). It can dramatically improve polymer performance, including heat strength, barrier strength, rigidity or dimensional stability, and fire retardance (Wong et al., 2022). All of these performance benefits are offered without increasing density or reducing the base polymer's light transmission properties. Traditional microscale fillers are being replaced by nanofillers. Fillers are important (Anzar et al., 2020) components that contribute to the formation of particles, fragments, fiber, sheets, and whisker forms of NPs. Nanofillers are basically understood to be additives in solid form that improve in mechanical or physical properties. The activity of active fillers involves filling of a certain volume, disruption of the conformational position of a polymer matrix, immobilization of adjacent molecule groups, and possible orientation of the polymer material are the key contributors of its performance. Compatibilization is the addition of a material to a mixture of polymers that will improve its stability. The purpose of this section is to review the common types of nanofillers used in PNC, explain the current production and characterization of nanocomposites, and make known their applications. Figure 5.2 listed different type of nanofillers.

Polymer blends can be compatibilized in a variety of ways. The industrial suitability of compatibilization techniques is determined by a number of factors, including cost, final efficiency, recyclability, and potential biodegradability (Ajitha et al., 2020). The following Figure 5.3 depicts some compatibilization strategies.

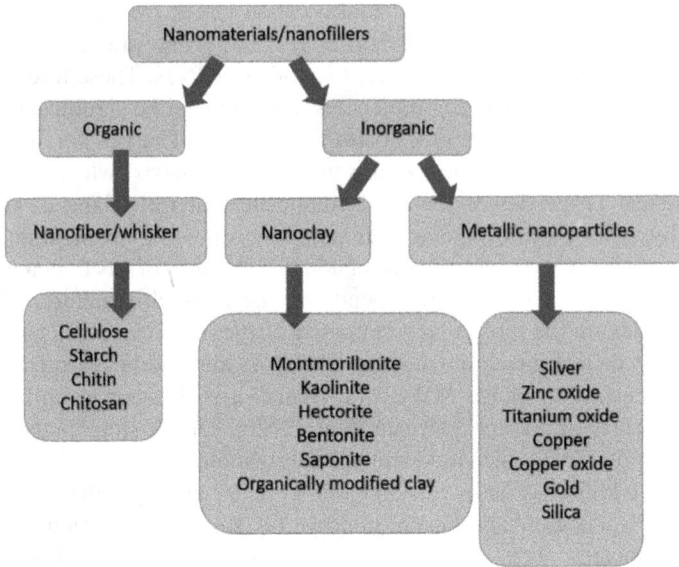

FIGURE 5.2   Types of nanofillers based on Chan et al. (2021) and El-Makaty et al. (2021).

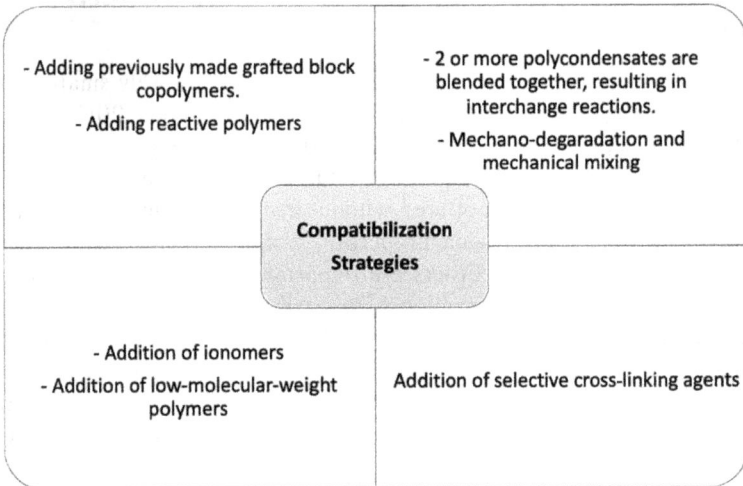

FIGURE 5.3   The compatibilzation strategies of nanofillers.

5.4.1   ORGANIC NANOFILLERS

In both the academic and industrial worlds, biopolymers prepared from renewable organic materials have drawn considerable interest in recent decades. Scientists and engineers have worked together in recent years, and in conjunction with natural green polymers, the intrinsic strength and performance of the fibers and NPs was used to create a new class of bio-based composites. For this sort of bio-composites, the unique problems are their eco-friendly qualities, which make them environmentally friendly, fully degradable, and renewable.

5.4.1.1 Cellulosic Nanofiller

NC is a product of natural cellulose consisting of nanoscale basic materials, and it can be divided into three groups: first is the cellulose nanocrystals (CNC), second is the nanofibrillated cellulose (CNF), and third is the bacterial cellulose (BC) (Blanco et al., 2018). They originated from flour, algae, beets, potato tubers, algae wood, and other plants. For example, BC can be reproduced rapidly by changing size under acclimatization conditions. It is secreted by microorganisms and has been proved as an artificial blood tube in tissue engineering (Tang et al., 2017). Good mechanical properties and biocompatibility, ultrafine fiber, and high porosity make BC an excellent nanofillers. The process of BC production, which is naturally synthesized by bacteria in a pure form (de Oliveira Barud et al., 2016). CNF can be extracted by using thermochemical treatments (alkali treatment) and mechanical processes, such as grinding, to defibrillate the cellulose into nanofibrils (Fareez et al., 2018). The optimization of the grinding process can greatly save time and energy. Meanwhile, CNC undergo further chemical processes (called acid hydrolysis) to extract them from natural cellulose in order to maintain their crystalline structure (Dufresne, 2018). The diversity of properties in NC makes them better suited for specific applications.

NC-based materials have a low carbon footprint, are affordable, renewable, recyclable, and nontoxic, and have the potential to be green nanofillers with a wide range of useful and unforeseen properties. For instance, rice husk and rice husk ash from agricultural waste were prepared as organic nanofillers for the production of PNC by Arifin et al. (2018), in which XRF analysis showed that rice husk ash has the highest silica (SiO_2) content of 89. 835%, while rice husk has SiO_2 contents of 82.540%. Many other previous findings were tabulated in Table 5.1. The effect of integrating arecanut nanofiller on the tribological behavior of a coir-reinforced epoxy-based polymer matrix composite has been investigated by Vishwas et al. (2020). The findings showed that the inclusion of areca nut nanofiller improved the microhardness of the composite. The tensile strength increased up to 5% with the inclusion of arecanut filler, but then decreased. Flexural intensity increases dramatically as the filler percentage increases from 0% to 5%, but the difference from 5% to 10% is negligible. The presence of filler has no impact on the interlaminar shear strength of composites. The composite's impact strength and wear resistance were increased.

5.4.1.2 PLA Nanofiller

PLA is a group of crystalline biodegradable thermoplastics known for their relatively high melting points and outstanding mechanical properties. Lately, the PLA nanofiller has been reported because of its convenience from renewable resources. High-advanced technologies have been used to develop the high-molecular weight and pure PLA, which have a long enough lifetime to maintain mechanical properties without fast hydrolysis. The PLA is widely being studied as a biomaterial in medicine. Due to growing environmental issues, interest in biodegradable polymers and natural fiber-reinforced polymers has increased lately. Natural fiber reinforcements could substantially lower the level of the price of bio-based composites,

TABLE 5.1

Different Class of Nanofillers and the Inclusion Study with Different Type of Polymers

Reinforcement	Improvements	References
	Organic Nanofiller	
Rice husk and rice husk ash	Increased crystallinity and strength	Arifin et al. (2018)
Coconut coir	Improved the microhardness of the composite.	Vishwas et al. (2020)
	Inorganic Nanofiller	
$CaCO_3$	Precipitation filling to conserve the binding of structural materials	Ortega-Villamagua et al. (2020)
AgNPs	Lower percolation threshold in electrical conduction	Fang and Lafdi (2021)
	Clay Nanofillers	
MMT	Better mechanical property	Mat Yazik et al. (2021)
	Carbon-Based Nanofillers	
GO	Increased mechanical attributes	Daniel and Liu (2021)
CNTs	Decreased domain coalescence	Guo et al. (2017)
	Hybrid Nanofiller	
$PLA + AgNPs + TiO_2 + ZnO$	Activation values of PLA decreased	Tarani et al. (2021)
$Fe_2O_3 + PCL$	Improves its complex permittivity properties	Mensah et al. (2020)

but their broader use is now the greater hurdle. Since they come from recycled materials, they may be environmentally sustainable alternatives to synthetic fiber reinforcement. Good basic mechanical properties, reduced tool wear, improved energy recovery, and biodegradability are further benefits of natural fibers over synthetic fibers. Besides this, the mechanical properties of bio-matrices can also be affected by natural fibers such as hybrid PLA nanofiller, which have been reported to be more superior and advanced in terms of their performance (Amjad et al., 2021; Mofokeng & Luyt, 2015; Norrrahim et al., 2022). Tarani et al. (2021) demonstrated that incorporation of 1 wt.% of ZnO nanofillers catalytically affects the thermal degradation and crystallization of PLA samples when compared to TiO_2 and AgNPs, while all these nanofillers cause a decrease in the activation energy values of PLA.

5.4.2 Inorganic Fillers

Mineral or metallic fillers are reviewed as inorganic fillers. Clay, nanoclay MMT, AgNPs, and calcium carbonate ($CaCO_3$) are the most common inorganic reinforcement material-based composites. As an inorganic filler, clay is frequently used as a filler in an area of the composite field because of the prehistoric civilization due to its high natural abundance.

5.4.2.1 Calcium Carbonate

The other standard inorganic filler widely used in industry and labs is $CaCO_3$. $CaCO_3$ is acquired from carbonatite-lava, stalactites, stalagmites, skeletons, and the shells of some animals. It is also procured from excavating carbonate-containing rock, causing sound, air, and water pollution (Abdul Khalil et al., 2019). Recent studies have indicated an alternative way to enhance compatibility features and decrease the environmental affect by developing $CaCO_3$ with biological agents or bacteria such as *Bacillus sphaericus* and *B. subtilis*. This process is termed biomineralization, the formation of minerals by a living organism. Biologically induced mineralization is the organism's mechanism to synthesize minerals; the metabolic activity of the organism adjusts the local environment, hence giving rise to mineral precipitations (Ortega-Villamagua et al., 2020). The foremost biological pathway of bacterial activities in carbonate precipitation is dependent on the hydrolysis of urea, catalyzed by the enzyme Urease.

5.4.2.2 Silver Nanoparticles

AgNPs are also a popular metallic filler due to their physicochemical properties, including optical, electrical, catalytic, antimicrobial, and therapeutic properties. It can be synthesized either chemically or biologically. Conventional physical and chemical methods are more costly and toxic than biological methods that can produce NPs in high yields, with good solubility and stability (Zhang et al., 2016). The biological method is most preferred and is very simple, cost-effective, fast, environmentally friendly, and non-toxic. The green approach is preferred for synthesizing AgNPs because it uses bacteria, plant extracts, fungi, vitamins, and amino acids far from chemical methods.

5.4.2.3 Ferric Oxide (Fe_2O_3) Nanofillers

Fe_2O_3 also known as iron oxide or hematite, is a common inorganic compound consisting of an inorganic magnetic element in particles, lamellae, or fiber form, with 1-D in nanometer range incorporated in an organic polymer. Fe_2O_3 has a high surface area, which can increase the overall surface area of the composite and improve its mechanical properties (Lopez Maldonado et al., 2014). These magnetic materials are applied in several technological applications, such as microwave absorbers (Wang et al., 2019), sensors (Koli et al., 2022), reflectors (Balachandran et al., 2020), optical filters (Balachandran et al., 2020), and polarizers (Tripathi et al., 2023). They are very familiar because of their potential characteristics, including magnetic, electrical, thermal, dielectric, mechanical, electronic, and optical, the viability of processing, and nonetheless, their structural flexibility (Lopez Maldonado et al., 2014). Fe_2O_3 is biocompatible and has been used in various biomedical applications such as bone repair, drug delivery, and cancer therapy (Lunin et al., 2019).

5.4.3 CLAY NANOFILLERS

Originating in the late 1980s at Toyota Central Research Laboratories, Japan. Toyota developed the first nanoclay-based nanofiller (Arifin et al., 2018). Clay has an atom's thickness relative to or below 1 nm. It has a 1000:1 aspect ratio and is hydrophilic.

It is not compatible with most ammonia cation applications. MMT is already a polymer that is being treated and chemically modified. These are phyllosilicate minerals produced from the chemical weathering of other silicate minerals on earth. It has a right intercalation property, and it can swell upon absorption of water (Mekhemer et al., 2000). MMT is a member of the smectite group used widely in composites because of its abundance, eco-friendliness, and well-documented literature. The MMT has become a promising candidate because of its large surface area and aspect ratio as a reinforcement material in polysaccharide-based matrices (Daniel & Liu, 2021). MMT, consisting of two tetrahedral silica sheets that are sandwiched by weak dipolar forces between the central aluminum octahedral sheet, gives rise to interlayer galleries. Graphene, on the other hand, is a single layer of hybridized carbon atoms arranged in a 2D matrix. It can be manufactured by the peeling off of graphite nanosheets (Ferrari et al., 2015). In general, the surface area of graphene sheets is between 2,600 and 3,000 m^2, and the aspect ratio is more than 2,000 (Sun et al., 2018). With this unique structure, graphene makes good nanofillers in terms of mechanical and thermal properties. However, poor solubility and van der Waals bonding are their main weaknesses in most cases due to their large surface area (Seo et al., 2019). Therefore, graphene has good optical activity that can be monitored by simple optical methods. Different numbers of atomic layers of graphene can be notable using transmission optical microscopy (Gass et al., 2008).

Clay is synthesized in the laboratory by tuning controlled conditions such as pH, temperature, composition, and starting materials. It is differentiated into many groups consisting of kaolinite, MMT chlorite, and fibrous silicate (e.g., sepiolite and palygorskite), which is termed attapulgite (Pozo & Calvo, 2018). Clay consists of hydrated alumina silicate with neutral or negatively charged layers and positive counterions in the interlamellar space organized and connected by weak electrostatic forces in the parallel form (Luckham & Rossi, 1999; Pozo & Calvo, 2018). MMT/smectite is the most commonly used clay group in recent years because of its stabilizing and rheological properties in water. Smectite particles have thousands of platelets attached like a sandwich.

Nonetheless, the separation of the sandwich structure takes place in aqueous dispersion, resulting in the weakly positively charged platelet edges connecting with the negatively charged platelet, hence forming a three-dimensional colloidal structure that provides its rheological properties (Bailey et al., 2015). Sepiolite is also among the clay group, has acquired attention, especially in preparing polysaccharide-based nanocomposites materials, due to its vast range of industrial applications ranging from its supportive, rheological, and catalytic properties. It is a hydrous magnesium silicate having the chemical formula $Mg_4Si_6O_{15}$ (OH) $6H_2O$ (Mekhemer et al., 2000). It has a needle-like morphology with a high aspect ratio and has been reported earlier to enhance mechanical properties and increase resistance in aqueous media. The field of nanotechnology has shepherded the development of nanoclays, most chosen in industries.

5.4.4   Carbon-Based Nanofillers

5.4.4.1   Graphene Oxide

Apart from mineral fillers, a carbon-based material, especially graphene oxide (GO), has also gained attention as a reinforcing material in the polysaccharide-based matrix.

Graphene can be viewed as a single flat layer of carbon atoms bonded covalently with each other and packed densely in order of 2D honeycomb structures. The GO applications in the industries are still in their deficiency, yet they have been widely used for research purposes as reinforcing filler in polymer matrix composites/nanocomposites due to their significant enhancement in mechanical, thermal, and electrical properties (Del Castillo, 2013).

Graphene is a single layer of hybridized carbon atoms arranged in a 2D matrix. It can be manufactured by peeling off of graphite nanosheets (Ferrari et al., 2015). In general, the surface area of graphene sheets is between 2,600 and 3,000 m², and the aspect ratio is more than 2,000 (Sun et al., 2018). With this unique structure, graphene makes good nanofillers in terms of mechanical and thermal properties. However, poor solubility and van der Waals bonding are their main weaknesses in most cases due to their large surface area (Seo et al., 2019). GO also has unique optical properties due to its unique electronic structure, which includes a very high electron mobility and a very high optical absorption coefficient. The optical absorption coefficient of graphene is about 2.3% per layer, which is about 200 times higher than that of silicon which can be observed using transmission optical microscopy (Gass et al., 2008). This makes it an ideal material for use in optical applications such as sensors, photodetectors, and optoelectronics with atomic layers of graphene.

The developments in the thermal conductivity of composites based on GO must be checked and summarized. More researchers are attempting to use GO as a filler to prepare high composites with thermal conductivity due to the excellent thermal transfer properties of graphene (Akbari et al., 2020; Liang et al., 2019). According to the thermal conductive mechanism discussions, graphene dispersion in the polymer matrix is crucial to achieving high thermal conductivity in graphene/polymer composites. Popular techniques include powder blending, melt blending, and solution blending, and many other techniques are commonly used to produce graphene-based polymer composites with high thermal conductivity (Terzopoulou et al., 2015). Many polymer materials are well known to have the benefits of low density, fast handling, and low cost (Chen et al., 2018). Due to their varying intrinsic properties, various polymer matrices are used in composites that are thermally conductive. The GO interface bonding effect, however, is not optimal. As well as outstanding electrical, electronic, and chemical properties. In order to produce products with excellent properties, graphene is functionalized and then compounded with a polymer matrix. It demonstrated comparatively excellent properties and choices for the production of modern materials used under extreme conditions such as high speeds, fluctuations in temperature, and efficient protection of the environment (Daniel & Liu, 2021).

Maintaining the consistent qualities of the GO-based PNC in large-scale production remains a challenge. This is because the characteristics of the polymer composites depend solely on the distribution in the polymer matrices. Since GO is hydrophilic and tends to aggregate in aqueous solutions, making it challenging to disperse in a polymer matrix (Zhao et al., 2020). Furthermore, GO is a highly anisotropic material, and it is difficult to achieve homogeneous dispersion in the polymer matrix using conventional processing techniques (Zhao et al., 2020). GO is also relatively expensive, and scaling up the production of GO nanocomposites can be costly.

Another challenge in the production of the GO nanocomposites is related to their technical barriers, which include the control of structures, graphene matrix dispersion, graphene and composite matrix interfacial interaction, and individual graphene contact (Zhao et al., 2020). Since GO can have a wide range of properties depending on the synthesis method, quality control of GO is a critical aspect in the production of GO nanocomposites.

5.4.4.2 CNTs

There are many ways to synthesize nanotubes, for instance, arc discharge, laser ablation, and chemical vapor deposition (Das et al., 2016; Hou et al., 2022; Manawi et al., 2018; Salah et al., 2022). CNTs are a perfect material for super-strength polymers because of their tremendous abilities in mechanical, electrical, and magnetic aspect. Nevertheless, CNTs are stable structure and are really hard to disperse in the polymer due to van der Waals bonds. (Ghavaminezhad et al., 2020). This is the major concern with the manufacturer to integrates CNTs and polymers, which is their capability to disperse, arrange, and control the CNTs in the matrix. There are solutions to overcome this problem, such as solution mixing and melt mixing (Faraguna et al., 2017). Both processes have developed the construction of CNTs, which has stimulated the widespread industrial application of this process.

CNTs have similar mechanical to graphene properties. The quality of the polymer matrix nanofiller dispersal is directly linked to its performance in enhancing mechanical, electrical, thermal, waterproofing, and other properties. CNT is a heterophasic polymer-based material where the dispersed phase, i.e., nanofiller(s), has at least one of its three dimensions in the order of a few nanometers. The formation of CNT networks decreases the rate of domain coalescence (Ajitha et al., 2020). Research done by Guo et al. (2017) prepared dyestuff based through a dye-printing methodology. The polyester yarn was treated with MWCNT-based dyestuff (40°C) via a dye-bath. The CNT-dyed polyester yarn was treated at 170°C, resulting in black dye-stuff (Fugetsu et al., 2009). Polyester has been extensively used for biomedical applications due to its greater physical characteristics. Carbonaceous nanofillers are valuable polyester nanofillers in biomedical engineering due to their cytocompatibility, biocompatibility, non-toxicity, and increased resistance to erosion. In order to improve interactions with cells and proteins in vivo, graphene and GO NPs have been chemically modified, thereby increasing cell adhesion and proliferation (Kausar, 2019). Nanofiller composites have also generally been used in many fields and have led to many advanced developments in electronics, including electromagnetic applications, solar cells, and diodes. Most of the electronic devices cause environmental pollution, and the occurrence of biodegradable bio-composites successfully relieves the environmental pressure (Lee et al., 2011).

5.5 EMERGING APPLICATIONS

Nowadays, an emerging hybrid nanofiller application is Electromagnetic Interference (EMI) shielding (Jin et al., 2021; Lee et al., 2022; Sun et al., 2018). Several carbon-based nanomaterials, including graphene, graphite, carbon black, carbon nano-fiber and CNTs, have been used to improve the electromagnetic interference shielding performance (EMISE) of PNC (Avella et al., 2009). Furthermore, in nanocomposite membranes, GO and sulfonated GO applied to the Nafion membranes enhance the bond water content and proton conduction. The mechanical efficiency and proton conductivity have also been improved by GO, with reduced methanol permeability for polyimide membranes. Modified GO self-assembly was produced through layer-by-layer technology on polyester fiber mats. Interactions with the polyester were formed by the updated GO. With growing assembly layers, the nanocomposite proton conductivity and methanol resistance grew. The mechanical properties were also enhanced by the layered GO structure (Daniel & Liu, 2021). Flexible solid-state supercapacitors with fine charge-discharge capability, stability, and improved cyclic performance can be provided by polymer/nanocomposites (Ashraf et al., 2014). To achieve high performance, low power consumption, rechargeability, and energy storage, these devices may be configured (Sun et al., 2016). Form memory goods are intelligent materials that can memorize a temporary shape with an external stimulus such as heat, light, pressure, pH, and solvents and restore their permanent form (Kausar, 2019). The latest research focuses on high recoverable strain and low density above the temperature of transition (T_m or T_g). Still, shape memory materials have convinced shortcomings for several advanced applications. The PNC filled with carbonaceous nanofillers has been investigated to fix its shortcomings (Yang & Mai, 2014). As studied by Song et al. (2014), vapor-grown carbon nanofiber (VGCF-G)-prepared graphene is coated and hardened into bio-based polyester. Moreover, polyester and carbonaceous nanofiller-based materials also used in the textile industry.

5.6   CONCLUSION AND FUTURE CONCERNS

PNC provides a tremendous opportunity to discover additional functionality that is not available in traditional materials. PNC has indeed been one of the most promising and emerging study fields because of its unique qualities, such as light weight, ease of production, and flexibility. A defining criteria of PNC is that the small amount of nanofillers leads to a significant increase in interfacial area as compared to traditional composites. The interfacial area produces a significant volume fraction of interfacial polymer with properties different from the bulk polymer, even at low loadings of the nanofiller. There are several ways to mitigate issues related to the incompatibility of the nanofiller in a nanocomposite, which include the use of compatibilizers, surface modification of the nanofiller, the right selection of processing techniques, and the filler materials. Furthermore, it is also possible to improve the compatibility between the nanofiller and the matrix by controlling the size and distribution of the nanofiller in the composite. Nevertheless, more research is still needed to explore the potential risks of inorganic metallic fillers (e.g., AgNPs) toward ecological, human, and animal activities. Inorganic fillers are not biodegradable, and it varies with different synthesis approaches and methods for the quality. Hence, clay and calcium carbonate are the most common inorganic fillers used in research since they are usually environmentally friendly and inexpensive.

REFERENCES

Abdul Khalil, H., Chong, E., Owolabi, F., Asniza, M., Tye, Y., Rizal, S., Nurul Fazita, M., Mohamad Haafiz, M., Nurmiati, Z., & Paridah, M. (2019). Enhancement of basic properties of polysaccharide-based composites with organic and inorganic fillers: a review. Journal of Applied Polymer Science, *136*(12), 47251.

Abhijit, V., Johannes, T., Sahlin-Sjövold, K., Mikael, R., & Boldizar, A. (2020). Melt processing of ethylene-acrylic acid copolymer composites reinforced with nanocellulose. *Polymer Engineering & Science*, *60*(5), 956–967.

Ajitha, A., Mathew, L. P., & Thomas, S. (2020). Compatibilization of polymer blends by micro and nanofillers. In *Compatibilization of Polymer Blends* (pp. 179–203). Elsevier.

Akbari, A., Cunning, B. V., Joshi, S. R., Wang, C., Camacho-Mojica, D. C., Chatterjee, S., Modepalli, V., Cahoon, C., Bielawski, C. W., & Bakharev, P. (2020). Highly ordered and dense thermally conductive graphitic films from a graphene oxide/reduced graphene oxide mixture. *Matter*, *2*(5), 1198–1206.

Akpan, E., Shen, X., Wetzel, B., & Friedrich, K. (2019). Design and synthesis of polymer nanocomposites. In *Polymer Composites with Functionalized Nanoparticles* (pp. 47–83). Elsevier.

Alexandre, M., & Dubois, P. (2000). Polymer-layered silicate nanocomposites: preparation, properties and uses of a new class of materials. *Materials Science and Engineering: R: Reports*, *28*(1–2), 1–63.

Alkhouzaam, A., Qiblawey, H., & Khraisheh, M. (2021). Polydopamine functionalized graphene oxide as membrane nanofiller: spectral and structural studies. *Membranes*, *11*(2), 86.

Alosfur, F. K. M., Ridha, N. J., Haji Jumali, M. H., Radiman, S., Tahir, K. J., Hadi, H. T., Madlol, R. A., & Al-Dahan, N. (2021). *Functionalization of Multi-Walled Carbon Nanotubes Using Microwave Method*. Materials Science Forum.

Amjad, A., Abidin, M., Alshahrani, H., & Rahman, A. A. A. (2021). Effect of fibre surface treatment and nanofiller addition on the mechanical properties of flax/PLA fibre reinforced epoxy hybrid nanocomposite. *Polymers*, *13*(21), 3842.

Anzar, N., Hasan, R., Tyagi, M., Yadav, N., & Narang, J. (2020). Carbon nanotube-A review on synthesis, properties and plethora of applications in the field of biomedical science. *Sensors International*, *1*, 100003.

Arifin, B., Aprilia, S., Alam, P. N., Mulana, F., Amin, A., Anaska, D. M., & Putri, D. E. (2018). *Characterization Nanofillers from Agriculture Waste for Polymer Nanocomposites Reinforcement*. MATEC Web of Conferences.

Ashraf, A., Tariq, M., Naveed, K., Kausar, A., Iqbal, Z., Khan, Z. M., & Khan, L. A. (2014). Design of carbon/glass/epoxy-based radar absorbing composites: microwaves attenuation properties. *Polymer Engineering & Science*, *54*(11), 2508–2514.

Avella, M., Buzarovska, A., Errico, M. E., Gentile, G., & Grozdanov, A. (2009). Eco-challenges of bio-based polymer composites. *Materials*, *2*(3), 911–925.

Bailey, L., Lekkerkerker, H. N., & Maitland, G. C. (2015). Smectite clay-inorganic nanoparticle mixed suspensions: phase behaviour and rheology. *Soft Matter*, *11*(2), 222–236.

Balachandran, G. B., David, P. W., Mariappan, R. K., Kabeel, A. E., Athikesavan, M. M., & Sathyamurthy, R. (2020). Improvising the efficiency of single-sloped solar still using thermally conductive nano-ferric oxide. *Environmental Science and Pollution Research*, *27*(26), 32191–32204.

Bharimalla, A., Deshmukh, S., Vigneshwaran, N., Patil, P., & Prasad, V. (2017). Nanocellulose-polymer composites for applications in food packaging: current status, future prospects and challenges. *Polymer-Plastics Technology and Engineering*, *56*(8), 805–823.

Bhattacharya, M. (2016). Polymer nanocomposites-a comparison between carbon nanotubes, graphene, and clay as nanofillers. *Materials*, *9*(4), 262.

Blanco, A., Monte, M. C., Campano, C., Balea, A., Merayo, N., & Negro, C. (2018). Nanocellulose for industrial use: cellulose nanofibers (CNF), cellulose nanocrystals (CNC), and bacterial cellulose (BC). In *Handbook of Nanomaterials for Industrial Applications* (pp. 74–126). Elsevier, Amsterdam.

Bumbudsanpharoke, N., Lee, W., Choi, J. C., Park, S.-J., Kim, M., & Ko, S. (2017). Influence of montmorillonite nanoclay content on the optical, thermal, mechanical, and barrier properties of low-density polyethylene. *Clays and Clay Minerals, 65*(6), 387–397.

Chan, J. X., Wong, J. F., Petrů, M., Hassan, A., Nirmal, U., Othman, N., & Ilyas, R. A. (2021). Effect of nanofillers on tribological properties of polymer nanocomposites: A review on recent development. Polymers, 13(17), 2867.

Chen, Y.-F., Tan, Y.-J., Li, J., Hao, Y.-B., Shi, Y.-D., & Wang, M. (2018). Graphene oxide-assisted dispersion of multi-walled carbon nanotubes in biodegradable Poly (ε-caprolactone) for mechanical and electrically conductive enhancement. Polymer Testing, 65, 387-397.

Cho, B.-G., Lee, S., Hwang, S.-H., Han, J. H., Chae, H. G., & Park, Y.-B. (2018). Influence of hybrid graphene oxide-carbon nanotube as a nano-filler on the interfacial interaction in nylon composites prepared by in situ interfacial polymerization. *Carbon, 140*, 324–337.

Daniel, N., & Liu, H. (2021). Graphene polymer composites: art of review on fabrication method, properties, and future perspectives. *Advances in Science and Technology Research Journal, 15*(1), 37–49.

Das, R., Shahnavaz, Z., Ali, M. E., Islam, M. M., & Abd Hamid, S. B. (2016). Can we optimize arc discharge and laser ablation for well-controlled carbon nanotube synthesis? *Nanoscale Research Letters, 11*(1), 1–23.

de Oliveira, A. D., & Beatrice, C. A. G. (2018). Polymer nanocomposites with different types of nanofiller. In *Nanocomposites-Recent Evolutions* (pp. 103–104), IntechOpen, London.

de Oliveira Barud, H. G., da Silva, R. R., da Silva Barud, H., Tercjak, A., Gutierrez, J., Lustri, W. R., de Oliveira Junior, O. B., & Ribeiro, S. J. (2016). A multipurpose natural and renewable polymer in medical applications: bacterial cellulose. *Carbohydrate Polymers, 153*, 406–420.

Del Castillo, P. D. (2013). The use of special clays in the rheological modification of paints & coatings. *Abrafati, 2013*(177), 1–4.

Dufresne, A. (2013). Nanocellulose: a new ageless bionanomaterial. *Materials Today, 16*(6), 220–227.

Dufresne, A. (2018). Cellulose nanomaterials as green nanoreinforcements for polymer nanocomposites. *Philosophical Transactions of the Royal Society A: Mathematical, Physical and Engineering Sciences, 376*(2112), 20170040.

Eckert, A., Abbasi, M., Mang, T., Saalwächter, K., & Walther, A. (2020). Structure, mechanical properties, and dynamics of polyethylenoxide/nanoclay nacre-mimetic nanocomposites. *Macromolecules, 53*(5), 1716–1725.

El-Makaty, F. M., Ahmed, H. K., & Youssef, K. M. (2021). The effect of different nanofiller materials on the thermoelectric behavior of bismuth telluride. *Materials & Design, 209*, 109974.

Fang, Q., & Lafdi, K. (2021). Effect of nanofiller morphology on the electrical conductivity of polymer nanocomposites. *Nano Express, 2*(1), 010019.

Faraguna, F., Pötschke, P., & Pionteck, J. (2017). Preparation of polystyrene nanocomposites with functionalized carbon nanotubes by melt and solution mixing: investigation of dispersion, melt rheology, electrical and thermal properties. *Polymer, 132*, 325–341.

Fareez, I. M., Ibrahim, N. A., Yaacob, W. M. H. W., Razali, N. A. M., Jasni, A. H., & Aziz, F. A. (2018). Characteristics of cellulose extracted from Josapine pineapple leaf fibre after alkali treatment followed by extensive bleaching. *Cellulose, 25*(8), 4407–4421.

Ferrari, A. C., Bonaccorso, F., Fal'Ko, V., Novoselov, K. S., Roche, S., Bøggild, P., Borini, S., Koppens, F. H., Palermo, V., & Pugno, N. (2015). Science and technology roadmap for graphene, related two-dimensional crystals, and hybrid systems. *Nanoscale, 7*(11), 4598–4810.

Fugetsu, B., Akiba, E., Hachiya, M., & Endo, M. (2009). The production of soft, durable, and electrically conductive polyester multifilament yarns by dye-printing them with carbon nanotubes. *Carbon*, *47*(2), 527–530.

Gass, M. H., Bangert, U., Bleloch, A. L., Wang, P., Nair, R. R., & Geim, A. (2008). Free-standing graphene at atomic resolution. *Nature Nanotechnology*, *3*(11), 676–681.

Ghavaminezhad, E., Mahnama, M., & Zolfaghari, N. (2020). The effects of van der Waals interactions on the vibrational behavior of single-walled carbon nanotubes using the hammer impact test: a molecular dynamics study. *Physical Chemistry Chemical Physics*, *22*(22), 12613–12623.

Gomathi, T., Rajeshwari, K., Kanchana, V., Sudha, P., & Parthasarathy, K. (2019). Impact of nanoparticle shape, size, and properties of the sustainable nanocomposites. In *Sustainable Polymer Composites and Nanocomposites* (pp. 313–336). Springer, Cham.

Guchait, A., Saxena, A., Chattopadhyay, S., & Mondal, T. (2022). Influence of nanofillers on adhesion properties of polymeric composites. *ACS Omega*, *7*(5), 3844–3859.

Guo, H., Li, X., Li, B., Wang, J., & Wang, S. (2017). Thermal conductivity of graphene/poly (vinylidene fluoride) nanocomposite membrane. *Materials & Design*, *114*, 355–363.

Herrero, M., Asensio, M., Núñez, K., Merino, J. C., & Pastor, J. M. (2019). Morphological, thermal, and mechanical behavior of polyamide11/sepiolite bio-nanocomposites prepared by melt compounding and in situ polymerization. *Polymer Composites*, *40*(S1), E704–E713.

Hou, P. X., Zhang, F., Zhang, L., Liu, C., & Cheng, H. M. (2022). Synthesis of carbon nanotubes by floating catalyst chemical vapor deposition and their applications. *Advanced Functional Materials*, *32*(11), 2108541.

Jamróz, E., Kulawik, P., & Kopel, P. (2019). The effect of nanofillers on the functional properties of biopolymer-based films: a review. *Polymers*, *11*(4), 675.

Jiang, L. Y., Huang, Y., Jiang, H., Ravichandran, G., Gao, H., Hwang, K., & Liu, B. (2006). A cohesive law for carbon nanotube/polymer interfaces based on the van der Waals force. *Journal of the Mechanics and Physics of Solids*, *54*(11), 2436–2452.

Jin, B., Meng, F., Ma, H., Zhang, B., Gong, P., Park, C. B., & Li, G. (2021). Synergistic manipulation of zero-dimension and one-dimension hybrid nanofillers in multi-layer two-dimension thin films to construct light weight electromagnetic interference material. *Polymers*, *13*(19), 3278.

Kausar, A. (2019). Review of fundamentals and applications of polyester nanocomposites filled with carbonaceous nanofillers. *Journal of Plastic Film & Sheeting*, *35*(1), 22–44.

Kedzior, S. A., Gabriel, V. A., Dubé, M. A., & Cranston, E. D. (2020). Nanocellulose in emulsions and heterogeneous water-based polymer systems: a review. *Advanced Materials*, *33*(28), 2002404.

Koli, P. B., Birari, M. D., Ahire, S. A., Shinde, S. G., Ingale, R. S., & Patil, I. J. (2022). Ferroso-ferric oxide (Fe3O4) embedded g-C3N4 nanocomposite sensor fabricated by photolithographic technique for environmental pollutant gas sensing and relative humidity characteristics. *Inorganic Chemistry Communications*, *146*, 110083.

Lee, J.-H., Kim, Y.-S., Ru, H.-J., Lee, S.-Y., & Park, S.-J. (2022). Highly flexible fabrics/epoxy composites with hybrid carbon nanofillers for absorption-dominated electromagnetic interference shielding. *Nano-Micro Letters*, *14*(1), 1–17.

Lee, W. C., Lim, C. H. Y., Shi, H., Tang, L. A., Wang, Y., Lim, C. T., & Loh, K. P. (2011). Origin of enhanced stem cell growth and differentiation on graphene and graphene oxide. *ACS Nano*, *5*(9), 7334–7341.

Liang, C., Qiu, H., Han, Y., Gu, H., Song, P., Wang, L., Kong, J., Cao, D., & Gu, J. (2019). Superior electromagnetic interference shielding 3D graphene nanoplatelets/reduced graphene oxide foam/epoxy nanocomposites with high thermal conductivity. *Journal of Materials Chemistry C*, *7*(9), 2725–2733.

Lopez Maldonado, K., De La Presa, P., De La Rubia, M., Crespo, P., De Frutos, J., Hernando, A., Matutes Aquino, J., & Elizalde Galindo, J. (2014). Effects of grain boundary width and crystallite size on conductivity and magnetic properties of magnetite nanoparticles. *Journal of Nanoparticle Research*, *16*(7), 1–12.

Luckham, P. F., & Rossi, S. (1999). The colloidal and rheological properties of bentonite suspensions. *Advances in Colloid and Interface Science*, *82*(1–3), 43–92.

Lunin, A. V., Kolychev, E. L., Mochalova, E. N., Cherkasov, V. R., & Nikitin, M. P. (2019). Synthesis of highly-specific stable nanocrystalline goethite-like hydrous ferric oxide nanoparticles for biomedical applications by simple precipitation method. *Journal of Colloid and Interface Science*, *541*, 143–149.

Mamat Razali, N. A., Ismail, M. F., & Abdul Aziz, F. (2021). Characterization of nanocellulose from Indica rice straw as reinforcing agent in epoxy-based nanocomposites. *Polymer Engineering & Science*, *61*(5), 1594–1606.

Manawi, Y. M., Samara, A., Al-Ansari, T., & Atieh, M. A. (2018). A review of carbon nanomaterials' synthesis via the chemical vapor deposition (CVD) method. *Materials*, *11*(5), 822.

Mat Yazik, M. H., Sultan, M. T. H., Jawaid, M., Abu Talib, A. R., Mazlan, N., Md Shah, A. U., & Safri, S. N. A. (2021). Effect of nanofiller content on dynamic mechanical and thermal properties of multi-walled carbon nanotube and montmorillonite nanoclay filler hybrid shape memory epoxy composites. *Polymers*, *13*(5), 700.

Mekhemer, W. K., Abou El-Ala, A. A., & El-Rafey, E. (2000). Clay as a filler in the thermoplastic compounding. *Molecular Crystals and Liquid Crystals Science and Technology. Section A. Molecular Crystals and Liquid Crystals*, *354*(1), 13–21.

Mensah, E. E., Abbas, Z., Ibrahim, N. A., Khamis, A. M., & Abdalhadi, D. M. (2020). Complex permittivity and power loss characteristics of α-Fe2O3/polycaprolactone (PCL) nanocomposites: effect of recycled α-Fe2O3 nanofiller. *Heliyon*, *6*(12), e05595.

Mofokeng, J. P., & Luyt, A. S. (2015). Dynamic mechanical properties of PLA/PHBV, PLA/PCL, PHBV/PCL blends and their nanocomposites with TiO2 as nanofiller. *Thermochimica Acta*, *613*, 41–53.

Morimune-Moriya, S. (2022). Polymer/nanocarbon nanocomposites with enhanced properties. *Polymer Journal*, 1–8.

Müller, K., Bugnicourt, E., Latorre, M., Jorda, M., Echegoyen Sanz, Y., Lagaron, J. M., Miesbauer, O., Bianchin, A., Hankin, S., & Bölz, U. (2017). Review on the processing and properties of polymer nanocomposites and nanocoatings and their applications in the packaging, automotive and solar energy fields. *Nanomaterials*, *7*(4), 74.

Muñoz, J., Riba-Moliner, M., Brennan, L. J., Gun'ko, Y. K., Céspedes, F., González-Campo, A., & Baeza, M. (2016). Amperometric thyroxine sensor using a nanocomposite based on graphene modified with gold nanoparticles carrying a thiolated β-cyclodextrin. *Microchimica Acta*, *183*(5), 1579–1589.

Mutiso, R. M., & Winey, K. I. (2015). Electrical properties of polymer nanocomposites containing rod-like nanofillers. *Progress in Polymer Science*, *40*, 63–84.

Naidu, D. S., & John, M. J. (2020). Effect of clay nanofillers on the mechanical and water vapor permeability properties of xylan-alginate films. *Polymers*, *12*(10), 2279.

Norrrahim, M., Nurazzi, N., Shazleen, S., Najmuddin, S., Yasim-Anuar, T., Naveen, J., & Ilyas, R. (2022). Biocompatibility, biodegradability, and environmental safety of PLA/cellulose composites. In *Polylactic Acid-Based Nanocellulose and Cellulose Composites* (pp. 251–264). CRC Press, London.

Ortega-Villamagua, E., Gudiño-Gomezjurado, M., & Palma-Cando, A. (2020). Microbiologically induced carbonate precipitation in the restoration and conservation of cultural heritage materials. *Molecules*, *25*(23), 5499.

Panse, P., Anand, A., Murkute, V., Ecka, A., Harshe, R., & Joshi, M. (2016). Mechanical properties of hybrid structural composites reinforced with nanosilica. *Polymer Composites*, *37*(4), 1216–1222.

Paszkiewicz, S., Pypeć, K., Irska, I., & Piesowicz, E. (2020). Functional polymer hybrid nanocomposites based on polyolefins: a review. *Processes*, *8*(11), 1475.

Peterson, A., Ostergren, I., Lotsari, A., Venkatesh, A., Thunberg, J., Strom, A., Rojas, R., Andersson, M., Berglund, L. A., & Boldizar, A. (2019). Dynamic nanocellulose networks for thermoset-like yet recyclable plastics with a high melt stiffness and creep resistance. *Biomacromolecules*, *20*(10), 3924–3932.

Pozo, M., & Calvo, J. P. (2018). An overview of authigenic magnesian clays. *Minerals*, *8*(11), 520.

Rothon, R. (2017). *Fillers for Polymer Applications* (Vol. 489). Springer, Cham.

Russo, P., Xiao, M., & Zhou, N. Y. (2017). Carbon nanowalls: a new material for resistive switching memory devices. *Carbon*, *120*, 54–62.

Salah, N., Alfawzan, A. M., Allafi, W., Alshahrie, A., & Al-Shawafi, W. M. (2022). Synthesis of carbon nanotubes using pre-sintered oil fly ash via a reproducible process with large-scale potential. *Methods*, *199*, 37–53.

Seo, J., Lee, J., Jeong, G., & Park, H. (2019). Site-selective and van der waals epitaxial growth of rhenium disulfide on graphene. *Small*, *15*(2), 1804133.

Shafiq, M., Anjum, S., Hano, C., Anjum, I., & Abbasi, B. H. (2020). An overview of the applications of nanomaterials and nanodevices in the food industry. *Foods*, *9*(2), 148.

Sharma, A. K., & Kaith, B. S. (2021). Polymer Nanocomposite matrices: classification, synthesis methods, and applications. In *Handbook of Polymer and Ceramic Nanotechnology* (pp. 403–428). Springer, Cham.

Shen, X., Zheng, Q., & Kim, J.-K. (2020). Rational design of two-dimensional nanofillers for polymer nanocomposites toward multifunctional applications. *Progress in Materials Science*, 115, 100708.

Song, W.-L., Cao, M.-S., Hou, Z.-L., Lu, M.-M., Wang, C.-Y., Yuan, J., & Fan, L.-Z. (2014). Beta-manganese dioxide nanorods for sufficient high-temperature electromagnetic interference shielding in X-band. *Applied Physics A*, *116*(4), 1779–1783.

Sun, J., Shen, J., Chen, S., Cooper, M. A., Fu, H., Wu, D., & Yang, Z. (2018). Nanofiller reinforced biodegradable PLA/PHA composites: current status and future trends. *Polymers*, *10*(5), 505.

Sun, Z. G., Qiao, X. J., Wan, X., Ren, Q. G., Li, W. C., Zhang, S. Z., & Guo, X. D. (2016). The synthesis and microwave absorbing properties of MWCNTs and MWCNTs/ferromagnet composites. *Applied Physics A*, *122*(2), 1–13.

Taghaddosi, S., Akbari, A., & Yegani, R. (2017). Preparation, characterization and anti-fouling properties of nanoclays embedded polypropylene mixed matrix membranes. *Chemical Engineering Research and Design*, *125*, 35–45.

Tang, J., Li, X., Bao, L., Chen, L., & Hong, F. F. (2017). Comparison of two types of bioreactors for synthesis of bacterial nanocellulose tubes as potential medical prostheses including artificial blood vessels. *Journal of Chemical Technology & Biotechnology*, *92*(6), 1218–1228.

Tarani, E., Pušnik Črešnar, K., Zemljič, L. F., Chrissafis, K., Papageorgiou, G. Z., Lambropoulou, D., Zamboulis, A., Bikiaris, D., & Terzopoulou, Z. (2021). Cold crystallization kinetics and thermal degradation of PLA composites with metal oxide nanofillers. *Applied Sciences*, *11*(7), 3004.

Terzopoulou, Z., Kyzas, G. Z., & Bikiaris, D. N. (2015). Recent advances in nanocomposite materials of graphene derivatives with polysaccharides. *Materials*, *8*(2), 652–683.

Torres-Giner, S., Montanes, N., Boronat, T., Quiles-Carrillo, L., & Balart, R. (2016). Melt grafting of sepiolite nanoclay onto poly (3-hydroxybutyrate-co-4-hydroxybutyrate) by reactive extrusion with multi-functional epoxy-based styrene-acrylic oligomer. *European Polymer Journal*, *84*, 693–707.

Tripathi, P., Singh, D., Yadav, T., Singh, V., Srivastava, A., & Negi, Y. (2023). Enhancement of birefringence for liquid crystal with the doping of ferric oxide nanoparticles. *Optical Materials*, *135*, 113298.

Väisänen, T., Das, O., & Tomppo, L. (2017). A review on new bio-based constituents for natural fiber-polymer composites. *Journal of Cleaner Production*, *149*, 582–596.

Vashist, A., Kaushik, A., Ghosal, A., Bala, J., Nikkhah-Moshaie, R., A Wani, W., Manickam, P., & Nair, M. (2018). Nanocomposite hydrogels: advances in nanofillers used for nanomedicine. *Gels*, *4*(3), 75.

Vishwas, M., Vinyas, M., & Puneeth, K. (2020). Influence of areca nut nanofiller on mechanical and tribological properties of coir fibre reinforced epoxy based polymer composite. *Scientia Iranica*, *27*(4), 1972–1981.

Wang, D.-J., Zhang, J.-Y., He, P., & Hou, Z.-L. (2019). Size-modulated electromagnetic properties and highly efficient microwave absorption of magnetic iron oxide ceramic opened-hollow microspheres. *Ceramics International*, *45*(17), 23043–23049.

Wong, J. F., Chan, J. X., Hassan, A., Mohamad, Z., Hashim, S., Abd Razak, J., Ching, Y. C., Yunos, Z., & Yahaya, R. (2022). Use of synthetic wollastonite nanofibers in enhancing mechanical, thermal, and flammability properties of polyoxymethylene nanocomposites. *Polymer Composites*, *43*(11), 7845–7858.

Wongvasana, B., Masa, A., Saito, H., Sakai, T., & Lopattananon, N. (2022). Influence of nanofiller types on morphology and mechanical properties of natural rubber nanocomposites. *IOP Conference Series: Materials Science and Engineering*, *1234*, 012007.

Xie, H., Shao, J., Ma, Y., Wang, J., Huang, H., Yang, N., Wang, H., Ruan, C., Luo, Y., & Wang, Q.-Q. (2018). Biodegradable near-infrared-photoresponsive shape memory implants based on black phosphorus nanofillers. *Biomaterials*, *164*, 11–21.

Xu, H., Adolfsson, K. H., Xie, L., Hassanzadeh, S., Pettersson, T. r., & Hakkarainen, M. (2016). Zero-dimensional and highly oxygenated graphene oxide for multifunctional poly (lactic acid) bionanocomposites. *ACS Sustainable Chemistry & Engineering*, *4*(10), 5618–5631.

Yang, P., & Mai, W. (2014). Flexible solid-state electrochemical supercapacitors. *Nano Energy*, *8*, 274–290.

Yang, Y., Liu, X., Lv, Y., Herng, T. S., Xu, X., Xia, W., Zhang, T., Fang, J., Xiao, W., & Ding, J. (2015). Orientation mediated enhancement on magnetic hyperthermia of Fe3O4 nanodisc. *Advanced Functional Materials*, *25*(5), 812–820.

Zhang, H., Zhang, G., Tang, M., Zhou, L., Li, J., Fan, X., Shi, X., & Qin, J. (2018). Synergistic effect of carbon nanotube and graphene nanoplates on the mechanical, electrical and electromagnetic interference shielding properties of polymer composites and polymer composite foams. *Chemical Engineering Journal*, *353*, 381–393.

Zhang, H.-C., Yu, C.-N., Li, X.-Z., Wang, L.-F., Huang, J., Tong, J., Lin, Y., Min, Y., & Liang, Y. (2022). Recent developments of nanocellulose and its applications in polymeric composites. *ES Food & Agroforestry*, *9*, 1–14.

Zhang, X.-F., Liu, Z.-G., Shen, W., & Gurunathan, S. (2016). Silver nanoparticles: synthesis, characterization, properties, applications, and therapeutic approaches. *International Journal of Molecular Sciences*, *17*(9), 1534.

Zhao, L., Zhu, S., Wu, H., Zhang, X., Tao, Q., Song, L., Song, Y., & Guo, X. (2020). Deep research about the mechanisms of graphene oxide (GO) aggregation in alkaline cement pore solution. *Construction and Building Materials*, *247*, 118446.

6 Inclusion of Nano-Fillers in Natural Fiber-Reinforced Polymer Composites
Overviews and Applications

M. R. M. Asyraf, S. A. Hassan, and R. A. Ilyas
Universiti Teknologi Malaysia

K. Z. Hazrati
Taman Universiti

A. Syamsir, M. A. F. M. Zaki, and L. Y. Tee
Universiti Tenaga Nasional

D. D. C. Vui Sheng
Universiti Teknologi Malaysia

N. M. Nurazzi
Universiti Sains Malaysia

M. N. F. Norrrahim
Universiti Pertahanan Nasional Malaysia

M. Rafidah
Universiti Putra Malaysia

Ahmad Rashedi
Nanyang Technological University

S. Sharma
Chandigarh University

DOI: 10.1201/9781003400998-6

6.1 INTRODUCTION

The need to create novel composite goods with superior mechanical and structural performance for structural applications was recognized by customers in the modern era. Several problems, including high impact, creep, and fatigue loads from external environments, could compromise the structural integrity and shorten the serviceable lifespan of power transmission, marine, automotive, and aerospace structures [1–5]. These impacts can result in significant delamination, leading to a decline in the structural performance of the composite [6–8]. This is in addition to the possibility that the composite will be penetrated. The product's primary goal is to have a high ratio of strength-to-weight structure for efficacious high-performance applications, including extinguishing purposes, aircraft, medical applications, manufacturing tools and armored vehicles [9–12]. Composite materials often have excellent strength characteristics but limited damping capacities. Since the polymer matrix exhibits outstanding viscoelastic properties, it can enable high strength and good damping capabilities for polymeric composites such as thermoplastic composites [13–15]. This might result in valuable qualities that could be used for structural purposes.

The expanding market demand for and popularity of lignocellulosic fiber-based composites can currently be seen as a result of its benefits of affordability, lightweight, and flexibility [16,17]. Nevertheless, when reinforced in polymer composites, lignocellulosic fiber, such as kenaf, sugar palm, coir, and others, display inferior mechanical characteristics due to their hydrophilic properties and high moistness. [18–20]. The insufficient surface adhesion between lignocellulosic fiber and polymeric resin may cause the biocomposite laminate to have limited capabilities. [21–23].

Modifiable fiber loading, orientations, stacking order, treatment, and nanofillers additives can all be used to overcome the shortcomings mentioned above of lignocellulosic fiber composites [24–27]. There have been reports of several different ways of treating fibers, some of which are physical, such as steam explosion [16], plasma or corona treatment, and others which are chemicals, such as alkaline treatment and acetylation procedures [28–30]. The addition of synthetic fibers like E-glass significantly reduces the water absorption of jute-reinforced polyester composites [31]. According to a study, one important way to reduce the moisture absorption of composite materials is to include impermeable fillers and lignocellulosic fiber. Filler content, distribution, size, and shape are filler factors that may affect a composite product's mechanical, thermal, and water barrier properties. Novel platy-structured, nanofiller-filled materials known as nanocomposites have lately seen an upsurge in attention and advancement mainly because of their high aspect ratio [32].

Due to their capacity to significantly alter the general properties of polymer composites, nanoparticle fillers are currently experiencing an increase in demand on a global scale [33,34]. The hydrophilicity of most nano-filler, such as nano clay and MMT, led to difficulty in homogenously distributing in polymer matrix despite the numerous promising advantages of nano-filler in the composite. As a result, the electrical field distribution within the material would be uneven, and nano composites' electrical performance would be substandard. Therefore, appropriate adjustments to the nanoclays must be made to encourage the improvement of the nanoparticles inside lignocellulosic fiber composites.

Since it greatly improves composites' mechanical performance and strong electrical resistance performance, nano-filler can be utilized as an add-on for the transmission tower's cross-arm structure, as described in the previous paragraph [35–37]. The properties of lignocellulosic composites must be simpler to modify and more adaptable to change in order to match the desired performance of varied applications [38–42]. This article focused on the overview and applications of hybrid nano-filler/lignocellulosic fiber-reinforced polymer composites. This study reviews recent developments in lignocellulosic fiber-reinforced polymer composites that use nano-fillers for enhanced applications.

6.2   LIGNOCELLULOSIC FIBER

Natural fibers are obtainable from many sources in the world and can be classified according to their origin. The vegetable or plant fiber class, animal fiber class, and mineral fiber class. Figure 6.1 shows the classification of natural fiber.

Plant extractions yield lignocellulosic fiberes, which can be further broken down into smaller groupings according to where in the plant they originated. Seed and fruit fibers produce cotton and coir, respectively, whereas bast fibers consist of single, long and narrow cells such as jute, hemp, flax, and ramie. All plant fibers are mainly made of cellulose and other substances like hemicellulose, lignin, pectin, and wax that must be removed or reduced during the process. Plant fibers are much easier to access than other natural fibers due to their wide availability worldwide [44]. The plant fibers' chemical characteristics, including cellulose, hemicellulose, and pectin, are shown in Table 6.1, whereas Table 6.2 displays the mechanical characteristics of plant fibers.

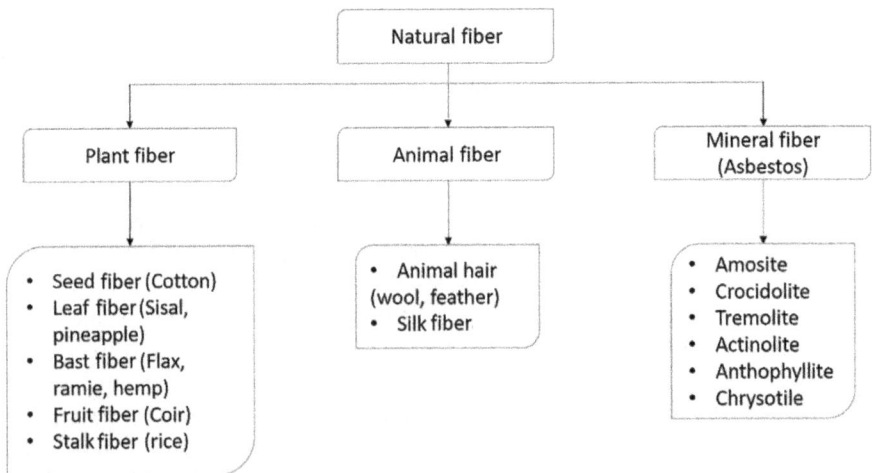

FIGURE 6.1   Classifications of natural fibers. (Reproduced from ref [43].)

TABLE 6.1

Chemical Properties of Lignocellulosic Fibers from Different Parts of Plants [43,45,46]

Fiber Type	Cellulose	Hemicellulose	Lignin	Pectin	Waxes	Moisture
Bast Fiber (Surrounds the Central Core of the Plant)						
Flax	64.1	16.7	2	1.8	1.5	10
Banana	31.26–63.9	12.04	13.88	-	-	12.43
Hemp	55–80.2	12–22.4	2.6–13	0.9–3.0	0.2	6.5
Ramie	68.2	13.1	0.6	1.9	0.3	10
Kenaf	37–49	18.24	15–21	8.9	0.5	-
Jute	64.4	12	0.2	11.8	0.5	10
Leaf Fibers (Extracted from the Leaf of the Plant)						
Sisal	65.8	12	9.9	0.8	0.3	10
Cabuya	68–77	4–8	13	-	2	-
Abacca	56–63	15–17	7–10	-	3	-
Fruit Fibers (Collected from the Fruits)						
Coir	20–36.7	12–15.4	32.7–53	4.7-7	-	0.2–0.5
Banana	48–60	10.2–16	144.4–21.6	2.1–4	3–5	2–3
Betelnut	35–64.8	29–33.1	13–26	92–15.4	0.5–0.7	-
Stalk Fibers (Collected from the Stalk of the Plant)						
Rice	28–48	23–28	14	-	20	-
Wheat	29–51	26–32	16–21	-	7	-
Oat	31–48	27–38	16–19	-	7.5	-
Grass Fibers (These Fibers Are Cells Which Occur in Different Parts of Plants)						
Sea Grass	57	38	5	10	-	-
Cane Fibers (Cane Fiber Is a Natural Fiber Obtained from Fiber-Rich Parts of Plants)						
Bagasse	28.4–56	20–36.4	21.4–24	-	0.9	-
Bamboo	48.1–73.8	12.5–73	10.2–21.4	0.37	-	11.7

TABLE 6.2

Mechanical Properties of Lignocellulosic Fiber [47–49]

Fiber		Density	Tensile Modulus	Specific Tensile Modulus	Tensile Strength	Specific Tensile Strength	Failure Strain
Bast	Flax	1.45–1.55	28–100	19–65	343–1,035	237–668	2.7–3.2
	Banana	0.75–0.95	27–32	20–24	529–914	392–677	1–3
	Hemp	1.45 1.55	32–60	22–39	310–900	214–581	1.3–21
	Jute	1.35–1.45	25–55	19–38	393–773	291–533	1.4-3.1
	Sisal	1.4–1.45	9–28	6–19	347–700	248–483	2.0–2.9

(Continued)

TABLE 6.2 (*Continued*)
Mechanical Properties of Lignocellulosic Fiber [47–49]

Fiber		Density	Tensile Modulus	Specific Tensile Modulus	Tensile Strength	Specific Tensile Strength	Failure Strain
Leaf	Pineapple	1.44–1.56	6–42	4–27	170–727	118–466	0.8–1.6
	Banana	1.3–1.35	8–32	6–24	503–790	387–585	3.0–10.0
Seed	Cotton	1.5–1.60	5–13	3–8	287–597	191–373	6.0–8.0
	Coir	1.10–1.20	4–6	3–5	131–175	119–146	15–30
	Oil Palm	0.7–1.55	3–4	2–4	248	160–354	25.0

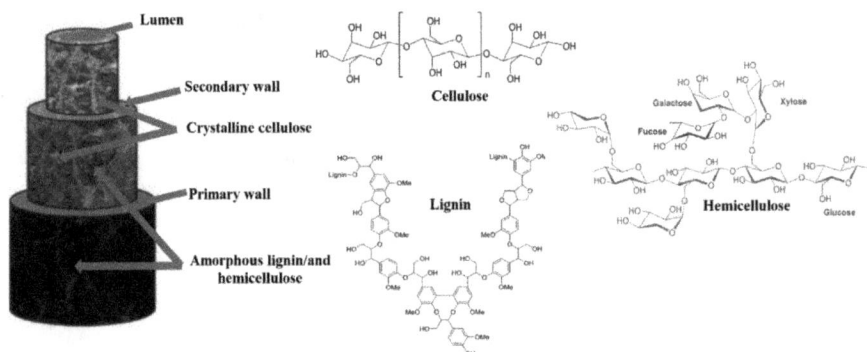

FIGURE 6.2 Schematic structure of the lignocellulosic natural fiber. (Reproduced from ref. [52].)

6.3 MAJOR COMPONENTS OF LIGNOCELLULOSIC FIBERS

The world's greatest source of renewable bioresources is lignocellulosic fibers. A schematic representation of the architecture of lignocellulosic fibers is shown in Figure 6.2. It is made up of a variety of chemical components, the composition of which varies according to the region where the crops are grown. It is made from several chemical constituents, which is also depending on the geographical area where the crops are sown [50]. Along with extracts like pectin, resins, and wax, the lignocellulosic components consist of 35–55 cellulose, 10–25 lignin, and 20–40 hemicellulose weight percentages, respectively [51].

Typically, fibers with higher cellulose content may have better mechanical and thermal qualities [50]. Lignin helps to improve the polymer matrix's thermal stability, reduce water absorption, and ensure proper cellulose dispersion in biopolymers, all of which contribute to the enhanced mechanical strength of composites [51]. Previous research demonstrates that the amount of cellulose, starch, and lignin in bio-composites affects their mechanical strength [53]. The flexural test also demonstrated that holo-cellulose modulus bio-composites of pecan nutshell fiber increased by 25% compared to pure biopolymer [53]. Another research covered the mechanical characteristics of cellulose-reinforced bioplastics made from maize starch with lignin acting as a plasticizer. It demonstrates that 1% cellulose and lignin will increase the maximum stress and elastic modulus by around 1400% and 250%, respectively, compared to the composite without cellulose and lignin [54]. High hemicellulose concentration, on the other hand, increases moisture absorption and accelerates fiber disintegration at low temperatures [50–52]. Lignin is different from cellulose and hemicellulose in that it has aromatic rings instead of lengthy molecular chains. It comprises a polyphenolic macromolecule made up of sinapyl alcohol, coniferyl alcohol, and three monomeric phenylpropane units. Hot water and low pH solutions do not cause lignin to dissolve (acid).

6.3.1 CELLULOSE

Cellulose is the major constituent of lignocellulosic fibers [50]. Long, unbranched chains comprised entirely of glucose and bound together by hydrogen bonds constitute cellulose. The cell shape of cellulose fibrils is a factor that affects the fiber's characteristics. The chemical formula for cellulose is $(C_6H_{10}O_5)_n$, whereby n represents the number of glucose groups. Cellulose has both crystalline and amorphous areas in its overall structure. Cellulose has 44.44% of carbon, 6.17% of hydrogen, and 49.39% of oxygen [55]. The number of chains of glucose molecules is between 4,000 and 8,000. At positions $C1$ and $C4$, cellulose's polymer chain was connected by glycosidic linkage. Three hydroxyl groups are present in each repeating unit. The hydroxyl group on cellulose gives it its hydrophilic property. It is worth highlighting that the physical properties, as well as the crystalline packing of cellulose, are significantly influenced by hydroxyl groups and the ability of hydrogen bond establishment. Numerous cellulose molecules make hydrogen bonds to create microfiber, which can then combine to form a fiber. More than 500,000 cellulose molecules are typically present in cellulose fiber. Hence, cellulose fiber has between 7,000 and 15,000 molecules of glucose per polymer [56]. The sturdy bond of hydrogen between the individual chains in cellulose microfibrils affects the overall mechanical properties of cellulose, particularly the tensile strength. Despite the fact that they might not be as strong as covalent bonds, 2.5 billion hydrogen bonds have a surprising amount of bonding energy.

6.3.2 HEMICELLULOSE

In contrast to cellulose, hemicelluloses are polysaccharides that include several sugar moieties. Glucose and monomers like hexoses (galactose, mannose), pentoses, and hexoses are among these sugars (xylose and arabinose). Hexoses and pentoses comprise the bulk of hemicellulose's structure, a branching carbohydrate with a low molecular weight [56]. Through hydrogen bonding, it is firmly attached to the cellulose fibrils. Both the hydroxyl and acetyl groups are present in its open structure and thus, making it hygroscopic and only partially soluble in water [57]. The natural fibers absorb much water due to this feature, making them weaker compared to cellulose. Hemicellulose also possesses an amorphous structure, and its breakdown temperature is also relatively low [58]. Hemicelluloses' interaction with cellulose and, in certain walls, lignin strengthens the cell wall, which is generally their most significant biological function. These characteristics are examined in connection to popular theories about the main wall.

6.3.3 LIGNIN

Lignin is a phenolic substance also known as the second essential source of organic material [59]. It is a complex copolymer of aliphatic and aromatic components that is amorphous in form, hydrophobic in nature, and very difficult. Hydroxyl, methoxyl, and carbonyl groups are in lignin [57,60]. Lignin is likely extremely unsaturated or aromatic due to its high carbon and low hydrogen contents. The presence of hydroxyls and several polar groups leads to the formation of strong intramolecular and intermolecular hydrogen bonds, causing lignin to be insoluble in solvents. Yet, lignin can dissolve in an alkaline solution due to the hydroxyl and carboxyl groups of phenols. Lignin is the cement that holds the fundamental fibers together with hemicellulose. Lignin covers the gap between hemicellulose and cellulose and gives cell walls their stiffness [61]. Lignin works as a water repellent, a defense against biological strikes, and a support and stiffener to the fibers because of its hydrophobic nature [62,63]. Additionally, lignin can potentially strengthen the interface between fibers and matrix by acting as a compatibilizer between hydrophilic fibers and hydrophobic polymers [64]. In a typical lignocellulosic, cellulose, hemicellulose, and lignin concentrations vary from 30% to 60%, 20% to 40%, and 15% to 25%, respectively. The mechanical characteristics of the fibers were greatly influenced by the lignocellulosic content, which also significantly impacted the mechanical and thermal performance of the polymer nanocomposites.

6.4 CURRENT PROSPECTS OF LIGNOCELLULOSIC FIBER-REINFORCED COMPOSITES

Lignocellulosic fiber-reinforced polymer composites are essential in creating environmentally responsible products. These fibers may be found in these plant parts, which are utilized to strengthen polymer composites. The current state of knowledge on lignocellulosic fiber-reinforced polymer composites is presented in this paper. Several researchers have investigated the manufacture and fabrication of concentration-ranging reinforced using lignocellulosic fiber-reinforced polymer composites and the mechanical, structural, and thermal characteristics of these composites [22,65–68]. For instance, Mittal et al. [69] discovered that the biodegradability of coir/epoxy composites is improved as modifications of lignocellulosic fibers which potentially can be applied to food packaging applications.

Generally, lignocellulosic fiber-reinforced polymer composites can be manufactured from thermoplastics and thermosets for various applications, such as agricultural and automotive. It can potentially lessen reliance on petroleum resources, which is linked to a slew of environmental issues. Natural fibers are renewable and biodegradable, which means they have a lower carbon footprint. The lignocellulosic fiber-based composites can be utilized in the construction and structural sectors, and some have been proven by Singh et al. [70] to have qualities equivalent to other mineral fillers. Other than that, industrial sectors such as construction, automotive, energy, and aerospace, are being pushed by society and authorities to produce more ecologically friendly goods to reduce the reliance on the usage of fossil fuels. In order to realize and emphasize the effort, the European Commission issued "European Guideline 2000/53/EG," aimed to increase automobile recycled content to 85% by weight in 2005. By the year 2015, the ratio had increased to 95% [71]. The regulation created a significant motivation for lignocellulosic fiber-based composites utilization.

In this situation, lignocellulosic fiber-based composites are a desirable option for businesses to address socioeconomic and environmental problems. Additionally, the use of lignocellulosic fiber-based composites would generate jobs in rural and underdeveloped areas, assisting in the achievement of the UN's sustainable development goals, which include reducing poverty level, fostering inclusive and sustainable industrialization, encouraging innovation, building sustainable cities and communities, and promoting responsible production and consumption [72,73]. As a result, natural fibers will be crucial to the socioeconomic growth of our civilization. The uses of natural fibers in many industries that are already in use, those that they may be employed in the future, and future prospects are shown and discussed in this part.

6.5 NANO-FILLERS IN LIGNOCELLULOSIC FIBER-REINFORCED POLYMER COMPOSITES

Nano-filler has emerged as promising candidates in lignocellulosic fiber-reinforced polymer composites because they provide specific properties to the composites besides improving interfacial adhesions of fiber/matrix. This kind of filler is considered material in solid form and exhibits various properties such as structures and compositions, which influence the final properties of the composites. Numerous shapes, including plates, laminas, and two-dimensional structures such as nanotubes and nanofibers. Other shapes such as iso-dimensional nanoparticles and are considered as three-dimensional shapes are possible for the nano-filler in terms of its physical structure. This nanomaterial can also be composed either from inorganic substances or organic., Nano-SiO_2 [23,24], MMT [74,75], nano-TiO_2 [76], and graphene [77,78] as well as CNT [24,79] are some of the different nano-fillers employed in lignocellulosic fiber-reinforced polymer composites. Other than these common traditional nano-fillers, the nano-fillers can also be extracted from biomass waste such as oil palm, known as oil palm nano-filler [80]. For a variety of cutting-edge applications, a tiny addition of these fillers can dramatically improve the mechanical, thermal, and water absorption qualities.

6.6 APPLICATIONS AND POTENTIAL USE OF NANO-FILLER/LIGNOCELLULOSIC FIBER-REINFORCED POLYMER NANOCOMPOSITES

Lignocellulosic fiber-based composites hybridized with nano-fillers have been the subject of substantial research to date and are becoming more well-liked as high-tech building and structural materials with excellent performance at an affordable price [33]. For example, Islam et al. [81] discovered that, when compared to wood/PP and coir/PP composites, hybrid composites (coir/wood/Polypropylene (PP)) had the highest tensile strength and modulus. The nano-filler enhanced the composites' mechanical characteristics by increasing the adhesion and interfacial contact of the fiber-polymer matrix. In addition, Majeed et al. [32] investigated the effect of the hybridization of nano-fillers such as montmorillonite (MMT) and rice husk ash (RHA) toward PP nanocomposites' mechanical and thermal properties. It was found that when the MMT and RHA were introduced into the PP matrix simultaneously, the tensile modulus of PP increased by 63%, and the flexural modulus increased by a staggering result of 92%. The research also found that the addition of RHA alone to the PP matrix caused a decrement in thermal stability. However, adding MMT to the PP/RHA composite improved its thermal stability. Based on the DSC experiment, adding RHA and MMT to the PP matrix enhanced the composite's crystallinity while constantly retaining the melting and crystallization temperatures. The introduction of hybrid nanotechnology triggers a new era in material science, assisting the creation and innovation of ultra-high-tech superior composites for future engineering applications. The material researcher intends to widen the utilization of MMT/lignocellulosic fiber nanocomposites in several fields, such as automotive, adsorbents, electronics, packaging, and outdoor.

6.6.1 PACKAGING APPLICATIONS

The rising need for environmentally friendly packaging materials, particularly for food packaging, reflects a change in lifestyle concerning the mobility and availability of goods from food companies. Therefore, the materials used for packaging must be trustworthy and safe enough to keep the food safe till it is consumed. It is projected that the use of nanotechnology in food packaging will increase dramatically due to the rising need for packaging that might lengthen the shelf life of goods brought on by globalization. During handling, shipment, and storage, packaging is intended to safeguard food products against physical damage such as crushing, abrasions, and shocks, chemical and ultraviolet radiations, as well as biological damage by microorganisms which lead to contamination [82]. Therefore, the optimal packaging material should possess outstanding mechanical properties, water resistance, thermal stability, and degradation resistance.

Lignocellulosic fiber and nano-filler are currently garnering attention as promising advanced composite materials due to their unique properties. High-performance natural fiber-based composites can be synthesized at low filler loadings, high aspect ratio, high surface area, high strength, high stiffness, rich intercalation chemistry, economic, and abundance in nature. Compared to nanocomposites with single fillers, hybridized nano-filler MMT with lignocellulosic fiber has synergistic effects that lead to superior mechanical performances, good water resistance, and decreased water absorption [83]. According to Huang et al. [84], the addition of the nanofillers improved the surface water contact angle of the nanocomposite, showing better hydrophobicity and making it a functional barrier material for packaging applications.

Separate investigations found that adding nano-filler to natural fiber hybrid composites improved application quality for existing packaging materials. Hemicellulose is one of the popular materials used as an environmentally friendly alternative to packaging materials. However, it has poor mechanical and thermal qualities. The hemicellulose-based film has been widely utilized to create polarising sheets and as materials for moisture and oxygen barriers [85]. Kassim et al. [86] reported on improving hemicellulose-based film-reinforced carboxymethyl cellulose (CMC) with nano-filler. The study discovered that the 60H-40CMC-MMT nanocomposite film had the best tensile strength and modulus values, which were 47.5 and 2.62 MPa, respectively. This can be attributed to the excellent compatibility between blended MMT/hemicellulose and CMC mixture ascribed to the electrostatic interaction and hydrogen bonding. With more hemicellulose loading, this mixture demonstrated the best results for good barrier characteristics. The thermal stability of the nanocomposite film is also improved by adding 2% of the nanofiller loading. The melting point of hemicellulose film, which was previously reported to be between 200°C and 250°C, increased to higher ranges between 250°C and 290°C with the addition of nano-filler.

6.6.2 AUTOMOTIVE APPLICATIONS

Manufacturing industries are transforming into more environmentallyfriendly and sustainable economic production as a result of considerable developments in science and technology. Researchers are concentrating on creating novel materials to aid society in this regard. With just a small percentage (5%) of nanoparticles, ligno-cellulosic fiber hybridization has the potential to outperform traditional composites in terms of properties like improved mechanical strength, higher elastic modulus, excellent thermal stability, and fire resistance [87]. Due to their simple production and general property enhancements, hybridized nanocomposites have created new opportunities in the automobile industry.

The main aspects affecting the adoption of clay polymer nanocomposites in the automotive industry are summarized by Galimberti et al. [88]. These factors include lighter vehicles, which result in lower fuel consumption and CO_2 emissions, enhanced safety, improved handling, and increased convenience. The polymer nanocomposites used in the car's construction can be given a nano-filler to achieve these goals. The nano-filler hybridized with polymer composites can be used in a variety of auto parts, including exhaust systems, suspension and braking systems, engine and powertrains, catalytic converters, body parts and frames, tyres, paints and coatings, lubrications, and electrical and electronic components. These materials provide better physical, chemical, and electrical properties than traditional materials because their nanoscale particles have these qualities. The aforementioned has a direct impact on the ability to produce vehicle parts in a lighter, safer, and more affordable way. More importantly, nanoparticles have the prospect of being applied in automotive applications for tribological, rheological, electrical, and optical functions. It results in improvements to the tyres, vision systems, surface coating, and powertrain and exhaust systems of the vehicle, which reduces the weight of the automobile, the amount of greenhouse gas it emits, and its overall carbon footprint. As indicated in Table 6.3, several thermoplastic or elastomeric polymers have been applied for hybridized nanocomposites, comprising around 80% clay particles

The automotive parts' flammability characteristics must be emphasized as another significant element. To make automobiles lighter and more fuel-efficient,

TABLE 6.3

Automotive Parts and Polymer Matrices for the Application of Clay Polymer Nanocomposites [88]

Automotive Parts	Polymer
Inner liner	Isoprene isobutylene copolymer, NBR
Timing belt/engine cover	Nylon-6
Internal compounds	NBR
Step-assist, doors, center bridge, sail panel, seat backs, and box-rail protector	PP, thermoplastic olefin
Thread	SBR, NBR, BR
Tire	SBR
Rear floor	Thermoset polymer matrix (with glass fiber)

SBR, styrene butadiene rubber; NBR, nitrile butadiene rubber; BR, butadiene rubber.

the majority of automotive parts have been replaced in contemporary times with plastic-based components. However, for some reason, it's possible that the severity of auto fires related to collisions and those that weren't. Despite the Federal Motor Vehicle Safety Standard and Flammability of Interior Materials standards that were introduced by the National Highway Traffic Safety Administration (NHTSA) in 1968 and 1972, respectively, fires have remained a serious risk. Electrical wiring, upholstery, and other flammable plastic components make up the majority of the items that ignite in car fires (by 47%), and their severity is increased by the presence of fuel (by 27% for ignitions). Plastic materials in motor vehicles have fire risks relating to the spreading speed of the smoke, flame, and combustion to the car's inner or passenger compartment, automobile industry has put forth massive efforts in finding ways to reduce the fire risk occurrence instead of depending on the rescue efforts when the fire ignites or started [89]. Bamboo/kenaf/epoxy nanocomposites were employed by Chee et al. [90] to strengthen several varieties of nano clay. The use of nano-filler to enhance the flammability qualities of lignocellulosic reinforced polymer composites is demonstrated by MMT and O-MMT. The UL94 horizontal burning test was passed by all hybrid nanocomposites in the study with an HB40 rating. Nonetheless, it was discovered that the limiting oxygen index (LOI) value of the nanocomposites increased from 20% to 28% with the addition of the nano-filler. In this instance, adding nanofiller to polymer composites might offer a high degree of grafting and leave behind a significant amount of active hydroxyl groups (CH-OH), which could covalently react with adscititious compatibilizer to encourage its dispersion and contact with polymer matrix. The resulting polymer nanocomposites demonstrated noticeably better flame retardancy and dramatically reduced smoke output.

6.6.3 OUTDOOR APPLICATIONS

Commonly, composite materials are created and manufactured for outdoor usage. Over the past few decades, various composite technology types have become available. However, not all are appropriate for incorporation or integration in building envelopes. In order to be suitable for outdoor applications, these materials must tolerate temperatures of at least 60°C [91]. Composite materials must adhere to strict guidelines such as good mechanical stability, great fire resistance, excellent acoustic properties, and good thermal insulation for construction or building materials implementation. Most polymers are photosensitive and exposed to weathering for almost their entire life cycle, making them vulnerable to deterioration in outdoor applications. Weathering aging such as photo-oxidation is one of the aging mechanisms that can significantly affect the lifespan of polymeric materials in elevated temperature settings, especially in outdoor applications [92]. Weathering aging happens due to chain crosslinking and chemical bond scission of the polymeric structure when exposed to the stated condition previously, which eventually leads to discoloration and further deterioration of mechanical properties when the polymer is exposed to UV sunlight, photos, and air [93]. All polymer materials ultimately suffer weathering aging, which first begins on the surface and depends on their chemical makeup, environmental conditions and variables, and of course, the time of exposure [94]. However, it must be emphasized and noted that even though the degradation can be postponed, it is inevitable that the process cannot be stopped completely. This exposure considerably decreases the overall thermomechanical characteristics of composite materials and jeopardizes their structural integrity or makes them catastrophically collapse [95].

According to Kord et al. [96], exposing weathered samples (virgin polymers) to an accelerated weathering environment for a more extended period increased the likelihood of color alteration, lightness, and rate of water absorption. Besides, hybrid nanocomposites only little changed regarding their ability to absorb water. Indirect weathering protection is provided by filling micro-voids and fiber lumens with nano-sized filler, which prevents water from further penetrating the composite and reduces oxidative activity and lightening [97]. They are therefore resistant to the sun, rain, and other weather elements. A similar discovery was made by Zahedi and his co-workers [98]. These materials can potentially be utilized for building enclosures such as walls, windows, roofs, and even shadings as the hybridization of the nano-filler with lignocellulosic fiber enhance the electricity chemical, heat, and flame resistance of the nanocomposite [99,100].

The overwhelming lignocellulosic fiber-based composites are applied in outdoor usages like building materials and construction items to meet fire safety standards like EN 13501-1-2009, which emphasize fire resistance qualities. It is essential to create and improve the materials at a laboratory scale in order to analyse the fire-retardant composite materials using a standard approach, such as TGA, Cone calorimetry based on ISO 5660-1 standard, LOI, burning test (UL 94 vertical or horizontal burning test) [101,102]. For outdoor applications, studies on composites as a replacement for more traditional materials, such as for components like roofing and door panels, have been widely publicized. Additionally, it is essential to ascertain the composite's thermal and flame resistance in the event of a fire accident. Rajini et al. [103] demonstrated how coconut sheath (CS) reinforced polyethelene (PE) composite has improved thermal and flammability qualities through a series of laboratory investigations. By using the char formation mechanism, it was found that CS/PE composites hybridized with 5 wt % MMT nanoclay significantly reduced their heat release and mass loss rate. The composites' flammability qualities were reduced due to the char development acting as a heat barrier. While treated, CS fiber also impacted the composites' thermal characteristics by lowering the temperature at which it degraded because it had more hydroxyl groups that attracted water. This research demonstrated using hybrid ligno-cellulosic/nano-filler polymer composites for outdoor applications.

6.7 CONCLUSIONS AND FUTURE OUTLOOK

This chapter delivers that the lignocellulosic fiber-reinforced polymer composites can be extended in their functionality by including the nano-fillers. The often applications of nanoparticles to modify the biocomposites especially in structural functions and usages are progressively growing. The nano-fillers occur in pure polymer and synthetic fiber-reinforced polymer composites and more toward biocomposites mainly made up of lignocellulosic fibers. Most of these kinds of materials have been employed to advance the potential of biocomposites toward better mechanical, thermal stability and water barrier properties. Practically, the addition of nanoparticles with well-dispersion can enhance its homogeneity, consequently, it provides better functionality and properties for the biocomposites toward automotive and outdoor products. Additionally, adding the nano-fillers in composites shows that it has low dielectric permittivity, low dielectric loss, and enhanced dielectric strength in comparison to the pure polymer. On top of that, the presence of nano-filler in lignocellulosic fiber/polymer composites could provide well in thermal stability and conductivity due to the compatibility of interfaces between nano-fillers and polymer matrix. In this point of view, the nano-filled biocomposites are suitable for packing applications, especially for food industries. In conclusion, the usage of nano-fillers in biocomposites could potentially advantage the structural industries, especially such as automotive and outdoor-based products such as electrical transmission towers. This is due to the nanofilled biocomposites can achieve great mechanical and thermal properties as well as promoted good electrical insulation properties in the biocomposites. In future studies, the authors would like to suggest on developing an optimum concentration or loading is critical to keep a balance of properties with economic consideration. This would open a gateway for future research of biomass-based materials to achieve the same quality as current synthetic materials.

ACKNOWLEDGEMENTS

The authors would like to express gratitude for the financial support received from the Universiti Teknologi Malaysia, the project "Characterizations of Hybrid Kenaf Fiber/Fiberglass Meshes Reinforced Thermoplastic ABS Composites for Future Use in Aircraft Radome Applications, grant number PY/2022/03758 — Q.J130000.3824.31J25".

REFERENCES

1. Asyraf, M. R. M., Ishak, M. R., Sapuan, S. M., & Yidris, N. (2021). Utilization of bracing arms as additional reinforcement in pultruded glass fiber-reinforced polymer composite cross-arms: Creep experimental and numerical analyses. *Polymers*, *13*(4), 620. https://doi.org/10.3390/polym13040620.

2. Ogin, S. L., Brøndsted, P., & Zangenberg, J. (2016). Composite materials: Constituents, architecture, and generic damage. In *Modeling damage, fatigue and failure of composite materials* (pp. 3–23). Woodhead Publishing, Sawston, Cambridge. https://doi.org/10.1016/B978-1-78242-286-0.00001-7.

3. Nurazzi, N. M., Asyraf, M. R. M., Khalina, A., Abdullah, N., Aisyah, H. A., Rafiqah, S. A., ... Sapuan, S. M. (2021). A review on natural fiber reinforced polymer composite for bullet proof and ballistic applications. *Polymers*, *13*(4), 646. https://doi.org/10.3390/polym13040646.

4. Asyraf, M. R. M., Ishak, M. R., Sapuan, S. M., & Yidris, N. (2021). Influence of additional bracing arms as reinforcement members in wooden timber cross-arms on their long-term creep responses and properties. *Applied Sciences*, *11*(5), 2061. https://doi.org/10.3390/app11052061.

5. Juneja, S., Chohan, J. S., Kumar, R., Sharma, S., Ilyas, R. A., Asyraf, M. R. M., & Razman, M. R. (2022). Impact of process variables of acetone vapor jet drilling on surface roughness and circularity of 3D-printed ABS Parts: Fabrication and studies on thermal, morphological, and chemical characterizations. *Polymers*, *14*(7), 1367. https://doi.org/10.3390/polym14071367.

6. Amir, A. L., Ishak, M. R., Yidris, N., Zuhri, M. Y. M., & Asyraf, M. R. M. (2021). Advances of composite cross arms with incorporation of material core structures: Manufacturability, recent progress and views. *Journal of Materials Research and Technology*, *13*, 1115–1131. https://doi.org/10.1016/j.jmrt.2021.05.040.

7. Asyraf, M. R. M., Ishak, M. R., Sapuan, S. M., Yidris, N., & Ilyas, R. A. (2020). Woods and composites cantilever beam: A comprehensive review of experimental and numerical creep methodologies. *Journal of Materials Research and Technology*, *9*(3), 6759–6776. https://doi.org/10.1016/j.jmrt.2020.01.013.

8. Asyraf, M. R. M., Ishak, M. R., Sapuan, S. M., & Yidris, N. (2021). Comparison of static and long-term creep behaviors between balau wood and glass fiber reinforced polymer composite for cross-arm application. *Fibers and Polymers*, *22*, 793–803. https://doi.org/10.1007/s12221-021-0512-1.

9. Ilyas, R. A., Sapuan, S. M., Harussani, M. M., Hakimi, M. Y. A. Y., Haziq, M. Z. M., Atikah, M. S. N., ... Asrofi, M. (2021). Polylactic acid (PLA) biocomposite: Processing, additive manufacturing and advanced applications. *Polymers*, *13*(8), 1326. https://doi.org/10.3390/polym13081326.

10. Asyraf, M. R. M., Syamsir, A., Zahari, N. M., Supian, A. B. M., Ishak, M. R., Sapuan, S. M., ... Rashid, M. Z. A. (2022). Product development of natural fibre-composites for various applications: Design for sustainability. *Polymers*, *14*(5), 920. https://doi.org/10.3390/polym14050920.

11. Ilyas, R. A., Sapuan, S. M., Asyraf, M. R. M., Dayana, D. A. Z. N., Amelia, J. J. N., Rani, M. S. A., ... Razman, M. R. (2021). Polymer composites filled with metal derivatives: A review of flame retardants. *Polymers*, *13*(11), 1701. https://doi.org/10.3390/polym13111701.

12. Chopra, L., Thakur, K. K., Chohan, J. S., Sharma, S., Ilyas, R. A., Asyraf, M. R. M., & Zakaria, S. Z. S. (2022). Comparative drug release investigations for diclofenac sodium drug (DS) by chitosan-based grafted and crosslinked copolymers. *Materials*, *15*(7), 2404. https://doi.org/10.3390/ma15072404.

13. Kyriazoglou, C., & Guild, F. J. (2005). Quantifying the effect of homogeneous and localized damage mechanisms on the damping properties of damaged GFRP and CFRP continuous and woven composite laminates-an FEA approach. *Composites Part A: Applied Science and Manufacturing*, *36*(3), 367–379. https://doi.org/10.1016/j.compositesa.2004.06.037.

14. Asyraf, M. R. M., Ishak, M. R., Norrrahim, M. N. F., Amir, A. L., Nurazzi, N. M., Ilyas, R. A., ... Razman, M. R. (2022). Potential of flax fiber reinforced biopolymer composites for cross-arm application in transmission tower: A review. *Fibers and Polymers*, *23*(4), 853–877. https://doi.org/10.1007/s12221-022-4383-x.

15. Asyraf, M. R. M., & Rafidah, M. (2022). Mechanical and thermal performance of sugar palm fibre thermoset polymer composites: A short review. *Journal of Fibers and Polymer Composites, 1*(1), 2.

16. Hamidon, M. H., Sultan, M. T., Ariffin, A. H., & Shah, A. U. (2019). Effects of fibre treatment on mechanical properties of kenaf fibre reinforced composites: A review. *Journal of Materials Research and Technology, 8*(3), 3327–3337. https://doi.org/10.1016/j.jmrt.2019.04.012.

17. Ali, S. S. S., Razman, M. R., Awang, A., Asyraf, M. R. M., Ishak, M. R., Ilyas, R. A., & Lawrence, R. J. (2021). Critical determinants of household electricity consumption in a rapidly growing city. *Sustainability, 13*(8), 4441. https://doi.org/10.3390/su1308444.

18. Ilyas, R. A., Sapuan, S. M., Atiqah, A., Ibrahim, R., Abral, H., Ishak, M. R., ... Ya, H. (2020). Sugar palm (*Arenga pinnata* [Wurmb.] Merr) starch films containing sugar palm nanofibrillated cellulose as reinforcement: Water barrier properties. *Polymer Composites, 41*(2), 459–467. https://doi.org/10.1002/pc.25379.

19. Ilyas, R. A., Sapuan, S. M., Atikah, M. S. N., Asyraf, M. R. M., Rafiqah, S. A., Aisyah, H. A., ... Norrrahim, M. N. F. (2021). Effect of hydrolysis time on the morphological, physical, chemical, and thermal behavior of sugar palm nanocrystalline cellulose (*Arenga pinnata* (Wurmb.) Merr). *Textile Research Journal, 91*(1–2), 152–167. https://doi.org/10.1177/0040517520932393.

20. Omran, A. A. B., Mohammed, A. A., Sapuan, S. M., Ilyas, R. A., Asyraf, M. R. M., Rahimian Koloor, S. S., & Petrů, M. (2021). Micro-and nanocellulose in polymer composite materials: A review. *Polymers, 13*(2), 231. https://doi.org/10.3390/polym13020231.

21. Alsubari, S., Zuhri, M. Y. M., Sapuan, S. M., Ishak, M. R., Ilyas, R. A., & Asyraf, M. R. M. (2021). Potential of natural fiber reinforced polymer composites in sandwich structures: A review on its mechanical properties. *Polymers, 13*(3), 423. https://doi.org/10.3390/polym13030423.

22. Bahrain, S. H. K., Masdek, N. R. N., Mahmud, J., Mohammed, M. N., Sapuan, S. M., Ilyas, R. A., ... Asyraf, M. R. M. (2022). Morphological, physical, and mechanical properties of sugar-palm (*Arenga pinnata* (Wurmb) merr.)-reinforced silicone rubber biocomposites. *Materials, 15*(12), 4062. https://doi.org/10.3390/ma15124062.

23. Bahrain, S. H. K., Rahim, N. N. C. A., Mahmud, J., Mohammed, M. N., Sapuan, S. M., Ilyas, R. A., ... Asyraf, M. R. M. (2022). Hyperelastic properties of bamboo cellulosic fibre-reinforced silicone rubber biocomposites via compression test. *International Journal of Molecular Sciences, 23*(11), 6338. https://doi.org/10.3390/ijms23116338.

24. Mohd Nurazzi, N., Asyraf, M. M., Khalina, A., Abdullah, N., Sabaruddin, F. A., Kamarudin, S. H., ... Sapuan, S. M. (2021). Fabrication, functionalization, and application of carbon nanotube-reinforced polymer composite: An overview. *Polymers, 13*(7), 1047. https://doi.org/10.3390/polym13071047.

25. Asyraf, M. R. M., Rafidah, M., Azrina, A., & Razman, M. R. (2021). Dynamic mechanical behaviour of kenaf cellulosic fibre biocomposites: A comprehensive review on chemical treatments. *Cellulose, 28*, 2675–2695. https://doi.org/10.1007/s10570-021-03710-3.

26. Amir, A. L., Ishak, M. R., Yidris, N., Zuhri, M. Y. M., & Asyraf, M. R. M. (2021). Potential of honeycomb-filled composite structure in composite cross-arm component: A review on recent progress and its mechanical properties. *Polymers, 13*(8), 1341. https://doi.org/10.3390/polym13081341.

27. Johari, A. N., Ishak, M. R., Leman, Z., Yusoff, M. Z. M., & Asyraf, M. R. M. (2020). Influence of CaCO3 in pultruded glass fiber/unsaturated polyester resin composite on flexural creep behavior using conventional and time-temperature superposition principle methods. *Polimery, 65*. https://doi.org/10.14314/polimery.2020.11.6.

28. Nurazzi, N. M., Harussani, M. M., Aisyah, H. A., Ilyas, R. A., Norrrahim, M. N. F., Khalina, A., & Abdullah, N. (2021). Treatments of natural fiber as reinforcement in polymer composites-a short review. *Functional Composites and Structures, 3*(2), 024002.

29. Zin, M. H., Abdan, K., Norizan, M. N., & Mazlan, N. (2018). The effects of alkali treatment on the mechanical and chemical properties of banana fibre and adhesion to epoxy resin. *Pertanika Journal of Science & Technology*, *26*(1).
30. Norizan, M. N., Abdan, K., Salit, M. S., & Mohamed, R. A. H. M. A. H. (2018). The effect of alkaline treatment on the mechanical properties of treated sugar palm yarn fibre reinforced unsaturated polyester composites reinforced with different fibre loadings of sugar palm fibre. *Sains Malaysiana*, *47*(4), 699–705. https://doi.org/10.17576/jsm-2018-4704-07.
31. Akil, H. M., Santulli, C., Sarasini, F., Tirillò, J., & Valente, T. (2014). Environmental effects on the mechanical behaviour of pultruded jute/glass fibre-reinforced polyester hybrid composites. *Composites Science and Technology*, *94*, 62–70. https://doi.org/10.1016/j.compscitech.2014.01.017.
32. Majeed, K., Ahmed, A., Abu Bakar, M. S., Indra Mahlia, T. M., Saba, N., Hassan, A., … Ali, Z. (2019). Mechanical and thermal properties of montmorillonite-reinforced polypropylene/rice husk hybrid nanocomposites. *Polymers*, *11*(10), 1557. https://doi.org/10.3390/polym11101557.
33. Ramesh, P., Prasad, B. D., & Narayana, K. L. (2020). Effect of fiber hybridization and montmorillonite clay on properties of treated kenaf/aloe vera fiber reinforced PLA hybrid nanobiocomposite. *Cellulose*, *27*(12), 6977–6993. https://doi.org/10.1007/s10570-020-03268-6.
34. Khan, A., Asiri, A. M., Jawaid, M., & Saba, N. (2020). Effect of cellulose nano fibers and nano clays on the mechanical, morphological, thermal and dynamic mechanical performance of kenaf/epoxy composites. *Carbohydrate Polymers*, *239*, 116248. https://doi.org/10.1016/j.carbpol.2020.116248.
35. Rozman, H. D., Musa, L., Azniwati, A. A., & Rozyanty, A. R. (2011). Tensile properties of kenaf/unsaturated polyester composites filled with a montmorillonite filler. *Journal of Applied Polymer Science*, *119*(5), 2549–2553. https://doi.org/10.1002/app.32096.
36. Asyraf, M. R. M., Ishak, M. R., Sapuan, S. M., Yidris, N., Ilyas, R. A., Rafidah, M., & Razman, M. R. (2020). Potential application of green composites for cross arm component in transmission tower: A brief review. *International Journal of Polymer Science*, *2020*, 1–15. https://doi.org/10.1155/2020/8878300.
37. Jia, Z. R., Gao, Z. G., Lan, D., Cheng, Y. H., Wu, G. L., & Wu, H. J. (2018). Effects of filler loading and surface modification on electrical and thermal properties of epoxy/montmorillonite composite. *Chinese Physics B*, *27*(11), 117806. https://doi.org/10.1088/1674-1056/27/11/117806.
38. Baihaqi, N. N., Khalina, A., Nurazzi, N. M., Aisyah, H. A., Sapuan, S. M., & Ilyas, R. A. (2021). Effect of fiber content and their hybridization on bending and torsional strength of hybrid epoxy composites reinforced with carbon and sugar palm fibers. *Polimery*, *66*(1), 36–43. https://doi.org/10.14314/polimery.2021.1.5.
39. Norrrahim, M. N. F., Nurazzi, N. M., Jenol, M. A., Farid, M. A. A., Janudin, N., Ujang, F. A., … Ilyas, R. A. (2021). Emerging development of nanocellulose as an antimicrobial material: An overview. *Materials Advances*, *2*(11), 3538–3551. https://doi.org/10.1039/D1MA00116G.
40. Sabaruddin, F. A., Paridah, M. T., Sapuan, S. M., Ilyas, R. A., Lee, S. H., Abdan, K., … & Abdul Khalil, H. P. S. (2020). The effects of unbleached and bleached nanocellulose on the thermal and flammability of polypropylene-reinforced kenaf core hybrid polymer bionanocomposites. *Polymers*, *13*(1), 116. https://doi.org/10.3390/polym13010116.
41. Norizan, N. M., Atiqah, A., Ansari, M. N. M., & Rahmah, M. (2020). Green materials in hybrid composites for automotive applications: Green materials. *Implementation and Evaluation of Green Materials in Technology Development: Emerging Research and Opportunities*, 56–76. https://doi.org/10.4018/978-1-7998-1374-3.ch003.

42. Zin, M. H., Abdan, K., Mazlan, N., Zainudin, E. S., Liew, K. E., & Norizan, M. N. (2019). Automated spray up process for pineapple leaf fibre hybrid biocomposites. *Composites Part B: Engineering*, *177*, 107306. https://doi.org/10.1016/j.compositesb.2019.107306.

43. Kumar, S., Manna, A., & Dang, R. (2022). A review on applications of natural Fiber-Reinforced composites (NFRCs). *Materials Today: Proceedings*, *50*, 1632–1636. https://doi.org/10.1016/j.matpr.2021.09.131.

44. Petroudy, S. D. (2017). Physical and mechanical properties of natural fibers. In *Advanced high strength natural fibre composites in construction* (pp. 59–83). Woodhead Publishing, Sawston, Cambridge. https://doi.org/10.1016/B978-0-08-100411-1.00003-0.

45. Neto, J., Queiroz, H., Aguiar, R., Lima, R., Cavalcanti, D., & Banea, M. D. (2022). A review of recent advances in hybrid natural fiber reinforced polymer composites. *Journal of Renewable Materials*, *10*(3), 561. https://doi.org/10.32604/jrm.2022.017434.

46. Subagyo, A., & Chafidz, A. (2018). Banana pseudo-stem fiber: Preparation, characteristics, and applications. *Banana Nutrition-Function and Processing Kinetics*, 1–19. https://doi.org/10.5772/intechopen.82204.

47. Feigel, B., Robles, H., Nelson, J. W., Whaley, J. M., & Bright, L. J. (2019). Assessment of mechanical property variation of as-processed bast fibers. *Sustainability*, *11*(9), 2655. https://doi.org/10.3390/su11092655.

48. Shah, D. U. (2013). Developing plant fibre composites for structural applications by optimising composite parameters: A critical review. *Journal of Materials Science*, *48*(18), 6083–6107. https://doi.org/10.1007/s10853-013-7458-7.

49. Zhong, Y., Kureemun, U., Tran, L. Q. N., & Lee, H. P. (2017). Natural plant fiber composites-constituent properties and challenges in numerical modeling and simulations. *International Journal of Applied Mechanics*, *9*(04), 1750045. https://doi.org/10.1142/S1758825117500454.

50. Komuraiah, A., Kumar, N. S., & Prasad, B. D. (2014). Chemical composition of natural fibers and its influence on their mechanical properties. *Mechanics of Composite Materials*, *50*, 359–376. https://doi.org/10.1007/s11029-014-9422-2.

51. Yang, J., Ching, Y. C., & Chuah, C. H. (2019). Applications of lignocellulosic fibers and lignin in bioplastics: A review. *Polymers*, *11*(5), 751. https://doi.org/10.3390/polym11050751.

52. Nurazzi, N. M., Khalina, A., Sapuan, S. M., Ilyas, R. A., Rafiqah, S. A., & Hanafee, Z. M. (2020). Thermal properties of treated sugar palm yarn/glass fiber reinforced unsaturated polyester hybrid composites. *Journal of Materials Research and Technology*, *9*(2), 1606–1618. https://doi.org/10.1016/j.jmrt.2019.11.086.

53. Agustin-Salazar, S., Cerruti, P., Medina-Juárez, L. Á., Scarinzi, G., Malinconico, M., Soto-Valdez, H., & Gamez-Meza, N. (2018). Lignin and holocellulose from pecan nutshell as reinforcing fillers in poly (lactic acid) biocomposites. *International Journal of Biological Macromolecules*, *115*, 727–736. https://doi.org/10.1016/j.ijbiomac.2018.04.120.

54. Miranda, C. S., Ferreira, M. S., Magalhães, M. T., Bispo, A. P. G., Oliveira, J. C., Silva, J. B., & José, N. M. (2015). Starch-based films plasticized with glycerol and lignin from piassava fiber reinforced with nanocrystals from eucalyptus. *Materials Today: Proceedings*, *2*(1), 134–140. https://doi.org/10.1016/j.matpr.2015.04.038.

55. Rahul, K. C. (2019). *Chemical analysis of spruce needles*. Centria University of Applied Sciences, Kokkola, Finland.

56. Brunner, G. (2014). Processing of biomass with hydrothermal and supercritical water. In *Supercritical fluid science and technology* (Vol. 5, pp. 395–509). Elsevier, Amsterdam. https://doi.org/10.1016/B978-0-444-59413-6.00008-X.

57. Mohanty, A. K., Misra, M. A., & Hinrichsen, G. I. (2000). Biofibres, biodegradable polymers and biocomposites: An overview. *Macromolecular Materials and Engineering*, *276*(1), 1–24. https://doi.org/10.1002/(SICI)1439-2054(20000301)276:1<1::AID-MAME1>3.0.CO;2-W.

58. Patel, J. P., & Parsania, P. H. (2018). Characterization, testing, and reinforcing materials of biodegradable composites. *Biodegradable and Biocompatible Polymer Composites*, 55–79.

59. Calvo-Flores, F. G., & Dobado, J. A. (2010). Lignin as renewable raw material. *ChemSusChem*, *3*(11), 1227–1235. https://doi.org/10.1002/cssc.201000157.

60. John, M. J., & Thomas, S. (2008). Biofibres and biocomposites. *Carbohydrate Polymers*, *71*(3), 343–364. https://doi.org/10.1016/j.carbpol.2007.05.040.

61. Liu, Q., Luo, L., & Zheng, L. (2018). Lignins: Biosynthesis and biological functions in plants. *International Journal of Molecular Sciences*, *19*(2), 335. https://doi.org/10.3390/ijms19020335.

62. Brebu, M., & Vasile, C. (2010). Thermal degradation of lignin-a review. *Cellulose Chemistry & Technology*, *44*(9), 353.

63. Ramamoorthy, S. K., Skrifvars, M., & Persson, A. (2015). A review of natural fibers used in biocomposites: Plant, animal and regenerated cellulose fibers. *Polymer Reviews*, *55*(1), 107–162. https://doi.org/10.1080/15583724.2014.971124.

64. Rozman, H. D., Tan, K. W., Kumar, R. N., Abubakar, A., Ishak, Z. M., & Ismail, H. (2000). The effect of lignin as a compatibilizer on the physical properties of coconut fiber-polypropylene composites. *European Polymer Journal*, *36*(7), 1483–1494. https://doi.org/10.1016/S0014-3057(99)00200-1.

65. Asyraf, M. R. M., Rafidah, M., Ishak, M. R., Sapuan, S. M., Yidris, N., Ilyas, R. A., & Razman, M. R. (2020). Integration of TRIZ, morphological chart and ANP method for development of FRP composite portable fire extinguisher. *Polymer Composites*, *41*(7), 2917–2932. https://doi.org/10.1002/pc.25587.

66. Asyraf, M. Z., Suriani, M. J., Ruzaidi, C. M., Khalina, A., Ilyas, R. A., Asyraf, M. R. M., … Mohamed, A. (2022). Development of natural fibre-reinforced polymer composites ballistic helmet using concurrent engineering approach: A brief review. *Sustainability*, *14*(12), 7092. https://doi.org/10.3390/su14127092.

67. Nurazzi, N. M., Asyraf, M. R. M., Fatimah Athiyah, S., Shazleen, S. S., Rafiqah, S. A., Harussani, M. M., … Khalina, A. (2021). A review on mechanical performance of hybrid natural fiber polymer composites for structural applications. *Polymers*, *13*(13), 2170. https://doi.org/10.3390/polym13132170.

68. Asyraf, M. R. M., Ishak, M. R., Syamsir, A., Nurazzi, N. M., Sabaruddin, F. A., Shazleen, S. S., … Razman, M. R. (2021). Mechanical properties of oil palm fibre-reinforced polymer composites: A review. *Journal of Materials Research and Technology*. https://doi.org/10.1016/j.jmrt.2021.12.122.

69. Mittal, M., & Chaudhary, R. (2019). Biodegradability and mechanical properties of pineapple leaf/coir fiber reinforced hybrid epoxy composites. *Materials Research Express*, *6*(4), 045301. https://doi.org/10.1088/2053-1591/aaf8d6.

70. Singh, S., Khairandish, M. I., Razahi, M. M., Kumar, R., Chohan, J. S., Tiwary, A., … Zakaria, S. Z. S. (2022). Preference index of sustainable natural fibers in stone matrix asphalt mixture using waste marble. *Materials*, *15*(8), 2729. https://doi.org/10.3390/ma15082729.

71. Chung, D.D.L. (2013). Introduction to carbon composites. *Properties and Characteristics of Polymer Composites*, 35, 227.

72. Asyraf, M. R. M., Syamsir, A., Supian, A. B. M., Usman, F., Ilyas, R. A., Nurazzi, N. M., … Rashid, M. Z. A. (2022). Sugar palm fibre-reinforced polymer composites: Influence of chemical treatments on its mechanical properties. *Materials*, *15*(11), 3852. https://doi.org/10.3390/ma15113852.

73. Asyraf, M. R. M., Rafidah, M., Ebadi, S., Azrina, A., & Razman, M. R. (2022). Mechanical properties of sugar palm lignocellulosic fibre reinforced polymer composites: A review. *Cellulose*, *29*(12), 6493–6516. https://doi.org/10.1007/s10570-022-04695-3.

74. Ashori, A., & Nourbakhsh, A. (2011). Herstellung und eigenschaften von polypropylen/holzmehl/nanoclay-verbundwerkstoffen. *European Journal of Wood and Wood Products*, *69*, 663–666. https://doi.org/10.1007/s00107-010-0488-9.

75. Alias, A. H., Norizan, M. N., Sabaruddin, F. A., Asyraf, M. R. M., Norrrahim, M. N. F., Ilyas, A. R., ... Khalina, A. (2021). Hybridization of MMT/lignocellulosic fiber reinforced polymer nanocomposites for structural applications: A review. *Coatings*, *11*(11), 1355. https://doi.org/10.3390/coatings11111355.

76. Vilakati, G. D., Mishra, A. K., Mishra, S. B., Mamba, B. B., & Thwala, J. M. (2010). Influence of TiO 2-modification on the mechanical and thermal properties of sugarcane bagasse-EVA composites. *Journal of Inorganic and Organometallic Polymers and Materials*, *20*, 802–808. https://doi.org/10.1007/s10904-010-9398-x.

77. Chaharmahali, M., Hamzeh, Y., Ebrahimi, G., Ashori, A., & Ghasemi, I. (2014). Effects of nano-graphene on the physico-mechanical properties of bagasse/polypropylene composites. *Polymer Bulletin*, *71*, 337–349. https://doi.org/10.1007/s00289-013-1064-3.

78. Sridharan, V., Raja, T., & Muthukrishnan, N. (2016). Study of the effect of matrix, fibre treatment and graphene on delamination by drilling jute/epoxy nanohybrid composite. *Arabian Journal for Science and Engineering*, *41*, 1883–1894. https://doi.org/10.1007/s13369-015-2005-2.

79. Nurazzi, N. M., Sabaruddin, F. A., Harussani, M. M., Kamarudin, S. H., Rayung, M., Asyraf, M. R. M., ... Khalina, A. (2021). Mechanical performance and applications of CNTs reinforced polymer composites-a review. *Nanomaterials*, *11*(9), 2186. https://doi.org/10.3390/nano11092186.

80. Saba, N., Paridah, M. T., Abdan, K., & Ibrahim, N. A. (2016). Effect of oil palm nano filler on mechanical and morphological properties of kenaf reinforced epoxy composites. *Construction and Building Materials*, *123*, 15–26. https://doi.org/10.1016/j.conbuildmat.2016.06.13.

81. Islam, M. S., Ahmad, M. B., Hasan, M., Aziz, S. A., Jawaid, M., Haafiz, M. M., & Zakaria, S. A. (2015). Natural fiber-reinforced hybrid polymer nanocomposites: Effect of fiber mixing and nanoclay on physical, mechanical, and biodegradable properties. *BioResources*, *10*(1), 1394–1407.

82. Qasim, U., Osman, A. I., Al-Muhtaseb, A. A. H., Farrell, C., Al-Abri, M., Ali, M., ... Rooney, D. W. (2021). Renewable cellulosic nanocomposites for food packaging to avoid fossil fuel plastic pollution: A review. *Environmental Chemistry Letters*, *19*, 613–641. https://doi.org/10.1007/s10311-020-01090-x.

83. Zakuwan, S. Z., & Ahmad, I. (2019). Effects of hybridized organically modified montmorillonite and cellulose nanocrystals on rheological properties and thermal stability of k-carrageenan bio-nanocomposite. *Nanomaterials*, *9*(11), 1547. https://doi.org/10.3390/nano9111547.

84. Huang, R., Zhang, X., Li, H., Zhou, D., & Wu, Q. (2020). Bio-composites consisting of cellulose nanofibers and Na+ montmorillonite clay: Morphology and performance property. *Polymers*, *12*(7), 1448. https://doi.org/10.3390/polym12071448.

85. Creighton, M. A., Zhu, W., van Krieken, F., Petteruti, R. A., Gao, H., & Hurt, R. H. (2016). Three-dimensional graphene-based microbarriers for controlling release and reactivity in colloidal liquid phases. *ACS Nano*, *10*(2), 2268–2276. https://doi.org/10.1021/acsnano.5b06963.

86. Haafiz, K. M., Taiwo, O. F., Razak, N., Rokiah, H., Hazwan, H. M., Rawi, N. F. M., & Khalil, H. A. (2019). Development of green MMT-modified hemicelluloses based nanocomposite film with enhanced functional and barrier properties. *BioResources*, *14*(4), 8029–8047. https://doi.org/10.15376/biores.14.4.8029-8047.

87. Kumar, S., Krishnan, S., & Samal, S. K. (2020). Recent developments of epoxy nanocomposites used for aerospace and automotive application. *Diverse Applications of Organic-Inorganic Nanocomposites: Emerging Research and Opportunities*, 162–190. https://doi.org/10.4018/978-1-7998-1530-3.ch007.

88. Galimberti, M., Cipolletti, V. R., & Coombs, M. (2013). Applications of clay-polymer nanocomposites. In *Developments in clay science* (Vol. *5*, pp. 539–586). Elsevier, Amsterdam. https://doi.org/10.1016/B978-0-08-098259-5.00020-2.

89. Lyon, R. E., & Walters, R. N. (2006). Flammability of automotive plastics. In *Society of automotive engineers (SAE) world congress*. SAE International. https://doi.org/10.4271/2006-01-1010.

90. Chee, S. S., Jawaid, M., Alothman, O. Y., & Yahaya, R. (2020). Thermo-oxidative stability and flammability properties of bamboo/kenaf/nanoclay/epoxy hybrid nanocomposites. *RSC Advances, 10*(37), 21686–21697. https://doi.org/10.1039/d0ra02126a.

91. Jiang, R., Michaels, H., Vlachopoulos, N., & Freitag, M. (2019). Beyond the limitations of dye-sensitized solar cells. In *Dye-sensitized solar cells* (pp. 285–323). Academic Press, Cambridge, Massachusetts. https://doi.org/10.1016/B978-0-12-814541-8.00008-2.

92. Beg, M. D. H., & Pickering, K. L. (2008). Accelerated weathering of unbleached and bleached Kraft wood fibre reinforced polypropylene composites. *Polymer Degradation and Stability, 93*(10), 1939–1946. https://doi.org/10.1016/j.polymdegradstab.2008.06.012.

93. Searle, N. D., McGreer, M., & Zielnik, A. (2002). Weathering of polymeric materials. *Encyclopedia of Polymer Science and Technology*. https://doi.org/10.1002/0471440264.pst401.pub2.

94. Brown, R. P., Kockott, D., Trubiroha, P., Ketola, W., & Shorthouse, J. (1995). *A review of accelerated durability tests*. National Physical Laboratory Teddington: Middlesex.

95. Raji, M., Zari, N., & Bouhfid, R. (2019). Durability of composite materials during hydrothermal and environmental aging. In Durability *and life prediction in biocomposites, fibre-reinforced composites and hybrid composites* (pp. 83–119). Woodhead Publishing, Sawston, Cambridge. https://doi.org/10.1016/B978-0-08-102290-0.00005-2.

96. Kord, B., Malekian, B., & Ayrilmis, N. (2017). Weathering performance of montmorillonite/wood flour-based polypropylene nanocomposites. *Mechanics of Composite Materials, 53*, 271–278. https://doi.org/10.1007/s11029-017-9660-1.

97. Eshraghi, A., Khademieslam, H., Ghasemi, I., & Talaiepoor, M. (2013). Effect of weathering on the properties of hybrid composite based on polyethylene, wood flour, and nanoclay. *BioResources, 8*(1), 201–210.

98. Zahedi, M., Pirayesh, H., Khanjanzadeh, H., & Tabar, M. M. (2013). Organo-modified montmorillonite reinforced walnut shell/polypropylene composites. *Materials & Design, 51*, 803–809. https://doi.org/10.1016/j.matdes.2013.05.007.

99. Zhang, T., & Yang, H. (2018). High efficiency plants and building integrated renewable energy systems: Building-integrated photovoltaics (BIPV). *Handbook of energy efficiency in buildings*. Elsevier: Amsterdam. https://doi.org/10.1016/B978-0-12-812817-6.00040-1.

100. Uddin, F. (2018). *Montmorillonite: An introduction to properties and utilization* (Vol. *817*). IntechOpen: London.

101. Nguyen, Q., Ngo, T., Mendis, P., & Tran, P. (2013). Composite materials for next generation building facade systems. *Civil Engineering and Architecture, 1*(3), 88–95. https://doi.org/10.13189/cea.2013.010305.

102. Nguyen, Q. T., Tran, P., Ngo, T. D., Tran, P. A., & Mendis, P. (2014). Experimental and computational investigations on fire resistance of GFRP composite for building façade. *Composites Part B: Engineering, 62*, 218–229. https://doi.org/10.1016/j.compositesb.2014.02.010

103. Rajini, N., Winowlin Jappes, J. T., Siva, I., Varada Rajulu, A., & Rajakarunakaran, S. (2017). Fire and thermal resistance properties of chemically treated ligno-cellulosic coconut fabric-reinforced polymer eco-nanocomposites. *Journal of Industrial Textiles, 47*(1), 104–124. https://doi.org/10.1177/1528083716637869.

7 Metal Nanofillers in Composite Structure

Soleha Mohamat Yusuff
Malaysian Nuclear Agency

Mohd Nor Faiz Norrrahim
Universiti Pertahanan Nasional Malaysia

Mohd Nurazzi Norizan
Universiti Sains Malaysia

7.1 INTRODUCTION

Polymers' properties are not particularly spectacular when compared to other materials, such as most metals. As a result, polymers are rarely utilized by themselves in the production of products and structures. However, by mixing a polymer (the matrix) and a reinforcing component (the filler), a composite is formed, and its properties can be upgraded (Delides, 2016). The properties include improved mechanical properties (Brostow et al., 2008) (such as strength, modulus, and dimensional stability), improved thermal stability, thermal conductivity (Vaggar et al., 2021), heat distortion temperature, improved chemical resistance, improved optical clarity, higher electrical conductivity, better surface appearance, reduced permeability to gases, water, and hydrocarbons, reduced thermal expansion coefficient, and reduced smoke emissions.

Minerals, diatomic earths, talc, chalk, silica and clay are examples of particle material that is added to polymers as fillers (Baird, 2003; Siraj et al., 2022). Fillers were initially used in composite manufacturing because they were cheap and employing them potentially bring down the price of polymers (Baird, 2003; Ilyas et al., 2021). They are also employed in matrix resins for processing improvements, density control, optical effects, thermal conductivity, thermal expansion control, electrical properties, magnetic properties, flame retardancy (Ilyas et al., 2021), and better mechanical properties like hardness and tear resistance (Park & Seo, 2011). In comparison to conventional fillers used in construction materials, a metal filler has a variety of benefits. Metal-based materials including metal oxide that have been used as metal fillers are zinc oxide (ZnO), cadmium oxide (CdO), nickel oxide (NiO) and titanium dioxide (TiO_2) (Munawar et al., 2020; Ullah et al., 2022). These metal fillers which had their own optical bandgap and exciton binding energy exhibit remarkable physical, optical, chemical, catalytic, magnetic, and electrical properties (Munawar et al., 2020; Tabib et al., 2017; Kumar et al., 2014; Paulose et al., 2017; Ullah et al.,

DOI: 10.1201/9781003400998-7

Particle size (µm)

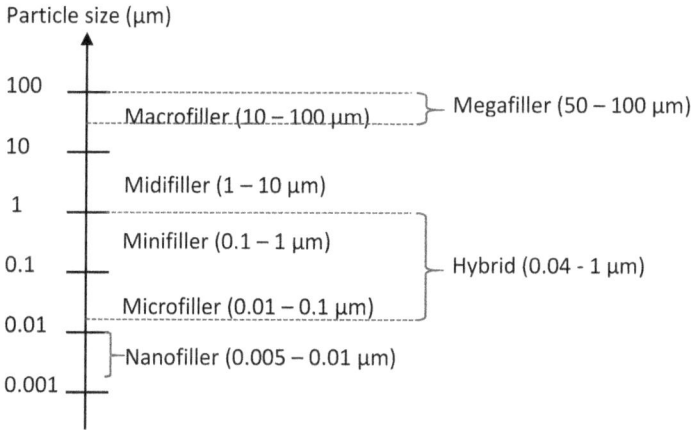

FIGURE 7.1 Composite filler ranges versus particle size. (Reproduced from ref Cangul & Adiguzel, 2017.)

2022; Tyagi et al., 2012). The benefits of metal fillers are related to a high level of plasticity, viscosity, and technical and casting qualities as well as high levels of strength parameters (Klyuchnikova & Lymar, 2005). The material performance is influenced by the type, shape, and the number of fillers, as well as the efficient coupling of fillers and matrix (Gajapriya et al., 2020). But, the size of the filler is the most important factor among filler characteristics (Delides, 2016). Figure 7.1 shows the scale meter of different particles.

In the past few decades, composites made with fillers of the micro- and nano-scale have become important materials because of their low weight, high specific stiffness and strength, high fatigue resistance, and excellent corrosion resistance (Fernandez et al., 2021; Basavaraj et al., 2013; Jacob et al., 2009; Palza et al., 2011). In contrast to nanofiller reinforcement, which offers more ductile fracture under loading, microscale filler-reinforced composites experience brittle fracture without having much energy absorption capacity (Fernandez et al., 2021). Ductile and brittle fractures are two different types of mechanical fractures in materials that result from the rupture of atomic or molecular bonds. The fracture is divided based on the size of the strain present at the time of fracturing (Wu, 2018). Crack development in materials that have minimal to no ductile deformation at the crack tip is known as brittle fracture. Contrarily, a ductile fracture involves a plastic deformation of the material at the crack tip. The preferred failure mode for materials that can tolerate damage is ductile fracture (Mouritz, 2012). Figure 7.2 shows the illustration of brittle and ductile fractures on the sample.

Filler with nano-scale referred to as nanomaterials. Four types (zero, one, two, and three dimensions) of nanomaterials or nano-objects exist (Joudeh & Linke, 2022). Zero-dimension nanometer-size objects such as quantum dots (Joudeh & Linke, 2022), one-dimension nanometer-size objects such as thin films, two-dimension nanometer-size objects such as nanowires, nanorods, and nanotubes, and three-dimension nanometer-size objects such as nanoparticles or nanoclusters (Domènech et al., 2012). Figure 7.3 depicts some illustrated nano-sized particles (Mukherji et al., 2018).

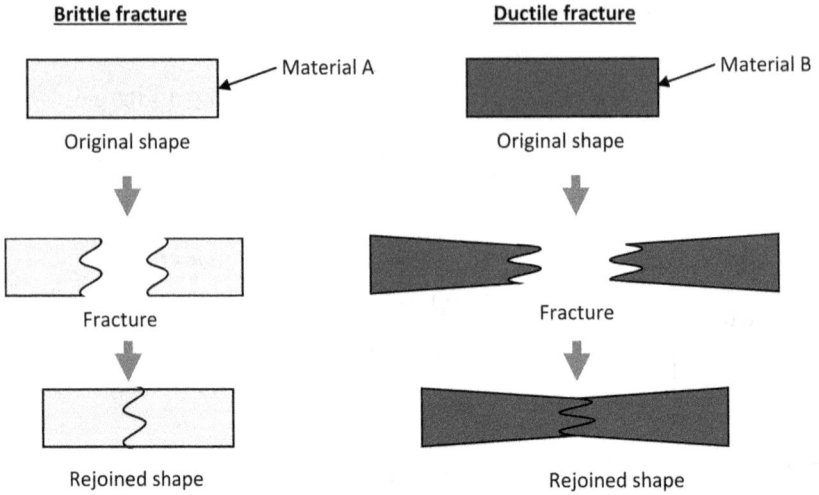

FIGURE 7.2 Illustration of brittle and ductile fracture of specimen.

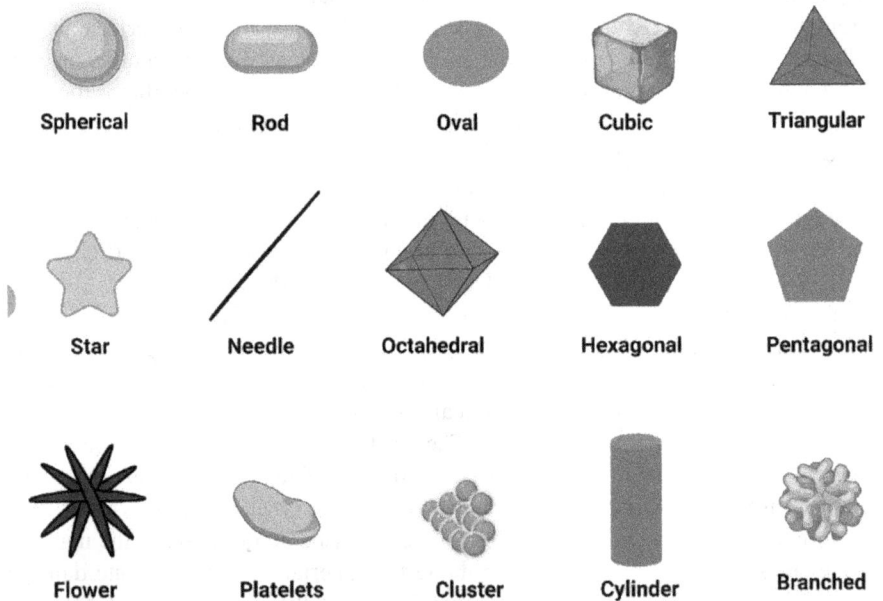

FIGURE 7.3 Various shapes of nano-scale particles. (Reproduced from ref Hamida et al., 2020.)

The nanomaterials include porous and polycrystalline materials with surface protrusions and nanometer-size metallic clusters (Domènech et al., 2012). Despite the unique electrical, optical, magnetic, and chemical properties of metal nanoparticles (Khan et al., 2019), this material has attracted scientists' and technologists' interest compared to other nano-sized materials (Domènech et al., 2012; Fritea et al., 2021).

Reduced in size from bulk to nanoscale, metal particles' behavior also undergoes notable behavioral changes, including surface-to-volume ratios, changes in the electronic structure of the particles, associations between the entities that make up the nanoparticles and their defects, as well as confinement and quantic effects (Domènech et al., 2012). In conclusion, different size of metal particles have their physiochemical characteristics, thus, a critical aspect of research. In this manner, metal nanoparticles were utilized to produce composite materials that performed better in various tests. Producing composites with metal nanofillers is simpler than manufacturing composites with metal nanoparticles. However, producing nano-size filler-reinforced composites is difficult because of their poor dispersion properties in matrices (Mouritz, 2012). The nanofiller existence and dispersion in the composite can be determined using spectroscopic techniques.

The objective of this chapter is to provide an overview of the characterization of metal nanofillers in composite structures using various spectroscopic techniques. Particular attention is on the distribution of metal nanoparticles inside the composite, morphology and other properties of the composite structure. In a large majority of cases, composite filled with metal nanoparticles shows better compatibility in terms of morphology size than pure composite. Moreover, the formation of a continuous structure is promoted by the presence of the metal nanofiller.

7.2 CHARACTERIZATION TECHNIQUES

The existence of metal nanofillers in the composite was detected and evaluated using a variety of techniques. There are X-ray diffraction (XRD), small angle X-ray absorption spectroscopy (XAS), small angle X-ray scattering (SAXS), X-ray photoelectron spectroscopy (XPS), Fourier Transform Infrared Spectroscopy (FTIR), nuclear magnetic resonance (NMR), Brunauer-Emmett-Teller surface area analysis (BET), thermogravimetric analysis (TGA), low-energy ion scattering (LEIS), UV-Vis, photoluminescence spectroscopy (PL), Dynamic Light Scattering (DLS), Nanoparticle Tracking Analysis (NTA), Inductively coupled plasma mass spectrometry (ICP-M), Secondary Ion Mass Spectrometer (SIMS), ToF-SIMS, matrix-assisted laser desorption/ionization (MALDI), Mössbauer, X-ray Magnetic Circular Dichroism (XMCD), Transmission electron microscopy (TEM), HRTEM, liquid TEM, cryo-TEM, electron diffraction, STEM, EELS, electron tomography, SEM-HRSEM, T-SEM-EDX, EBSD, AFM, Fluorescence techniques, UV-Vis spectroscopy, Solid-state NMR spectroscopy, WXRD, SAXS, infrared spectroscopy, Raman spectroscopy and many more (Mourdikoudis et al., 2018; Batool et al., 2022). Some of these techniques have been discussed on the basis of their accessibility, affordability, precision, non-destructiveness, ease of use, and affinity for particular compositions or materials (Mourdikoudis et al., 2018). These methods can be used separately or in combination to study a specific property, depending on the situation (Mourdikoudis et al., 2018). Furthermore, the development of nanotechnology had a significant impact on how analytical characterization techniques operated with samples of nanomaterials as opposed to samples of macroscopic materials (Mourdikoudis et al., 2018). This chapter covered the changes in morphology and surface texture by scanning electron microscopy (SEM), elemental content by Energy Dispersive X-ray spectroscopy (EDX), molecular structure by FTIR, thermal behavior by TGA, and crystallinity by XRD.

7.3 SURFACE MORPHOLOGY OF METAL NANOCOMPOSITES

Incorporating metal nanofillers gave significant change to the surface morphology of the composite compared to their matrix. The effect of the interaction between metal nanofillers and polymer matrix on the surface morphology of composites (Xu et al., 2022) can be identified using SEM at different magnifications (high and low respectively). This technique also could reveal the size distribution, elemental-chemical composition, dispersion of nanoparticles in the matrix and agglomeration state (Mourdikoudis et al., 2018). Under the microscope of SEM, the composite had crumpled microarchitectures and fishnet-like structures without the formation of any visible defects and bubbles (Mallya et al., 2022) (Figure 7.4). This surface is obviously different from the smooth surface with typical bubble-like structures and round

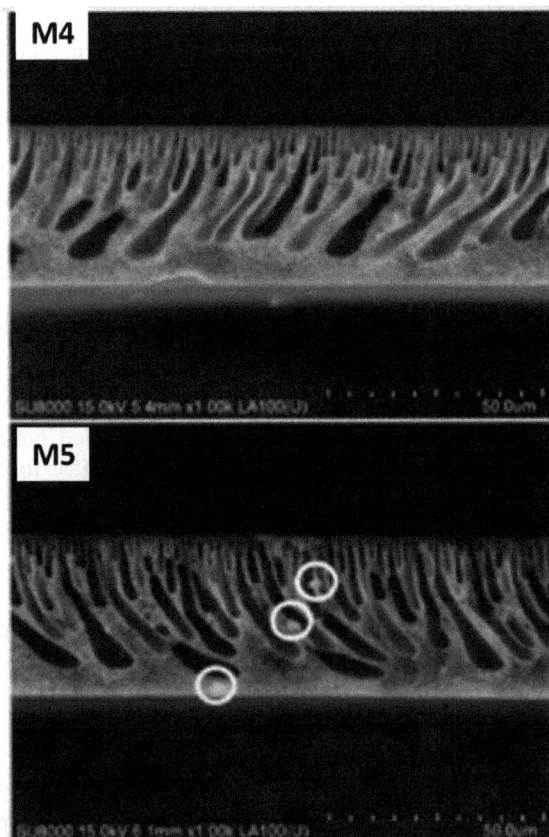

FIGURE 7.4 Scanning electron microscope of composite without (M4) and with metal nanofillers (M5). Metal nanoparticles were shown in a white circle. (Reproduced from ref Mallya et al., 2022.)

protrusions of the matrix (Mallya et al., 2022). Some distribution of metal nanofiller in the matrix also shows brighter, tiny spots corresponding to the presence of metal crystallites, compared to the homogeneous porous structure of the polymer matrix (Jamróz et al., 2020). The homogenous growth of metal nanofiller in the composite was also found from the formation of spherically shaped metal nanoparticles under the SEM technique (Sankar et al., 2020). SEM images of NiO-CdO-ZnO nanocomposite also have roughly spherical morphology (Munawar et al., 2020). From a cross-section view of the SEM image, some composites had porous finger-like structures with minimal macro voids (Mishra & Mukhopadhyay, 2021). SEM technique also captured the kidney stone monoclinic crystals of microwave-assisted synthesis of metal oxide CdO-CuO nanocomposite (Kannan et al., 2020).

The addition of metal nanofillers in the development of the composite caused the spherical crystal structure of the matrix to vanish; hence the smooth surface of the composite was obtained (Xu et al., 2022). Besides the morphology of the composite altered into stream-like patterns, a few white tiny ZnO nanoparticles may be seen distributed uniformly over the entire surface of the composite (Xu et al., 2022). Further addition of metal nanofillers caused the composite surface to become rough due to the creation of specific protrusions and the formation of agglomeration of metal nanofiller particles (Xu et al., 2022).

A composite prepared with 10 wt % silver nanoparticles (AgNPs) and polymer inclusion membranes (PIMs) have smooth, homogenous without obvious agglomeration of nanoparticles, obvious cracks and imperfections on the surface of the composite that implied that a uniform distribution of AgNPs on the composite (Maiphetlho et al., 2020). In addition, the results also confirm that the incorporation of AgNPs on PIM remains the composite stability (Maiphetlho et al., 2020). But, composite with higher than 10 wt % AgNPs and PIMs showed obvious cracks on the surface. These cracks became detrimental primarily to the composite stability and ion transport capacity performance (Maiphetlho et al., 2020). Increasing concentration of metal nanofiller also resulting nanofiller agglomeration in pore blockage (Mishra & Mukhopadhyay, 2021; van den Berg & Ulbricht, 2020) (Figure 7.5). The single, binary, and ternary nanocomposite also shows different surface characteristics. The surface roughness of nanocomposite increases as compositions of nanocomposites increase from single to binary to ternary (Domyati, 2022). The ternary nanocomposite had the lowest grains aggregation tendency compared to others (Domyati, 2022).

7.4 ELEMENTAL CONTENT IN METAL NANOCOMPOSITES

The existence and distribution of chemical components in material, including composite samples, have to be investigated using EDX since few cases exhibit no evident changes in the surface morphology of composite with a low and high concentration of metal nanofiller through SEM images (Liu et al., 2021).

The EDX approach (EDX elemental mapping and EDX analysis) will demonstrate the presence (Jeyabanu et al., 2019), pattern distribution (Mishra & Mukhopadhyay, 2021), and intensity of elements (Abu-Okail et al., 2021) of metal nanofillers on the surface of composite materials. Both EDX approaches also can

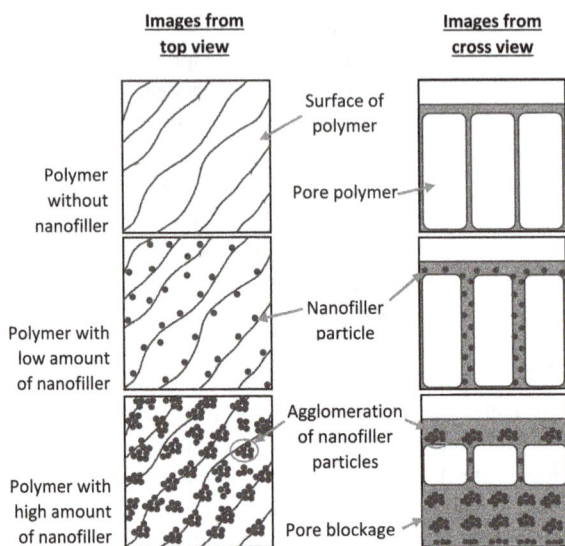

FIGURE 7.5   Schematic visualization of the nanofiller agglomeration in pore blockage.

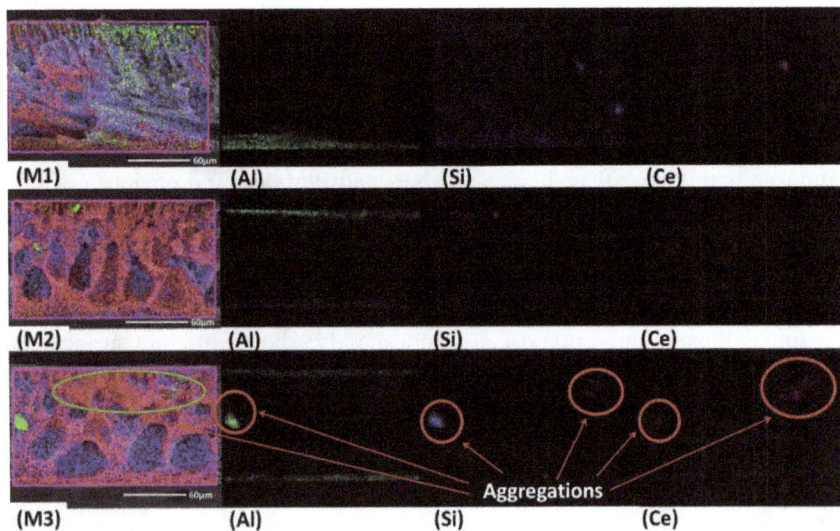

FIGURE 7.6   EDX mapping of CeO_2@HNTs/PVC membrane with samples labelled as M1, M2 and M3 show aggregations of Al, Si, and Ce element in sample M3. (Reproduced from ref Mishra & Mukhopadhyay, 2021.)

determine the atomic weight (%) of the constituents in the metal nanofiller composite (Mishra & Mukhopadhyay, 2021) (Figure 7.6).

As an illustration, the element mapping for Al and Ce in metal nanofillers was chosen, with the colour dots defining the corresponding elements in the 60 μm scan area (Mishra & Mukhopadhyay, 2021). The distribution of elements from the metal nanofiller was uniformly distributed based on EDX elemental mapping, indicating that the metal nanofiller is uniformly spread in the composite (Liu et al., 2021). The elements distribution in composite materials with a high metal nanofiller concentration is denser than that in composite materials with a low metal nanofiller concentration (Liu et al., 2021). Some works found few clusters and homogenous dispersion of low metal nanofiller content in the composites (Mishra & Mukhopadhyay, 2021). Increasing metal nanofiller content from low to optimum concentration gave increasing clusters in the matrix for composites (Mishra & Mukhopadhyay, 2021). Because further loading metal nanofillers exhibit blockage of pores and a reduction in pure water flux of composite, metal nanofiller agglomeration on the surface of the composite is obviously revealed using EDX mapping (Mishra & Mukhopadhyay, 2021). The percentage of elements displayed in EDX spectra of composite for example 0.44% of Al, 22.83% of O, and 76.73% of C from hybrid glass and carbon fibers composite samples confirm the successful loading of Al_2O_3 (metal nanofillers) and other materials in composite manufacturing (Abu-Okail et al., 2021). From the elemental data of EDX results of ZnS/ Al_2O_3/GO nanocomposite, 35.9 wt% of O and 28.93 wt% of Al originated from the Al_2O_3 nanomaterials (Domyati, 2022). High purity of ternary nanocomposite also identified based on the $K\alpha$ and $K\beta$ characteristic peaks of Al and O as well as Zn and S (Domyati, 2022). The chemical purity of the composite also confirms based on the occurrence of all elements in the nanocomposite without impurities peaks (Kannan et al., 2020).

7.5 FUNCTIONAL GROUPS OF METAL NANOCOMPOSITES

Changes in information in composite structure and matrix can be drawn from the materials' FTIR spectra (Kochkina & Butikova, 2019). The possible interactions between the matrix and other materials able to be interpreted (Jamróz et al., 2020) based on characteristic peaks of absorption bands appearing in the FTIR spectra (Jakubowska et al., 2023). Indirectly, it confirms the existence of metal nanofiller in the composite and the successful development of composite.

Every material had its characteristic peaks of FTIR spectra which indicate the chemical bonds available in that material (Xu et al., 2022). Therefore, comparing the FTIR spectra of pure metal nanofiller, pure matrix and composite are crucial and beneficial to identify the existence of metal nanofillers in the composite and possible interactions of metal nanofiller and other materials in composite structure (Xu et al., 2022). For example, the absorption bands located at 505 and 620 cm^{-1} of the CdO-CuO composite belongs to Cu-O and Cd-O stretching vibration, respectively (Kannan et al., 2020), which obviously confirm the existence of CuO and CdO in the composite. Other bands located at 1,113 and 1,416 cm^{-1} corresponded to carbonyl groups, 1,644 and 3,432 cm^{-1} corresponded to OH− group vibrations of water, and 860 and 779 cm^{-1} corresponded to CO_2 asymmetric stretching vibration happening in the CdO-CuO composite also revealed by FTIR spectra (Kannan et al., 2020). The nano copper-alumina (Cu-Al_2O_3) particles in the polypyrrole (PPy) chain of composite also confirms based on the existence of sharp and resolved absorption bands of nano copper-alumina (Cu-Al_2O_3) in FTIR spectrum (Sankar et al., 2020). The small alteration in the shape of the FTIR spectrum like the presence of a new band in composite which is attributed to the weak interaction of matrix and metal nanofillers is observed in nanoparticles (Bahrami et al., 2019; Jamróz et al., 2020). Furthermore, the appearance of new band and traces of the band (in the counterpart composites) in the composite were not observed in the matrix (Salari et al., 2018; Kanmani & Rhim, 2014; Jamróz et al., 2020). Besides, metal nanoparticles may immobilize in composite through physical retention (Menezes et al., 2019; Jamróz et al., 2020).

However, some works found the FTIR spectrum of the matrix and the metal nanofillers are nearly similar (Jeyabanu et al., 2019) which the characteristic peak of the matrix and the composite did not change in the location but gave different peak intensities (Xu et al., 2022). These nearly similar FTIR spectra prove that these materials had similar functional groups (Gaabour, 2020). Similar characteristic peaks were also observed from the FTIR spectrum of composite manufactured using low concentrations of metal nanofillers which is probably due to the overlapping bands (Kochkina & Butikova, 2019). The intensities of the bands grew as metal nanofiller concentration in the composite increased (Kochkina & Butikova, 2019). The introduction of metal nanofillers causes the other metal ion in the composite released and the decomposition of any molecule in the composite occur (Xu et al., 2022). As a result, the electrochemical performance of the composite improved (Xu et al., 2022). Further information about individual contributions can be measured using Fityk 0.9.8 software for example (Kochkina & Butikova, 2019). Moreover, a sufficient amount of metal nanofiller loading in the composite for the best complimentary also can be described from the minimal OH groups availability within the matrix network (Kochkina & Butikova, 2019).

FTIR spectra of composites with different concentration metal nanofillers may also show the shifting characteristics peaks due to the chemical interactions between metal nanofiller and the functional groups present in the matrix. For example, the strong electrostatic interaction of ZnO of metal nanofiller with the ether oxygen groups in the polymer matrix is displayed by the movement of the characteristic peak at $1,072\,cm^{-1}$ to left ($1,060\,cm^{-1}$) at the low contents of ZnO. But the electrostatic interaction decline at a high content of ZnO as shown by the movement of the characteristic peak at $1,072\,cm^{-1}$ to the right due to the uncontrol agglomeration of nanoparticles in composite structure (Xu et al., 2022). Addition of CuS nanoparticles in preparation of synthesized sodium alginate-assisted CuS composite also causes the shifted characteristic peaks at 1,412 and $1,020\,cm^{-1}$ which is due to the stabilizing CuS in the alginate groups (Jeyabanu et al., 2019).

7.6 CRYSTALLINITY OF METAL NANOCOMPOSITES

Crystallinity properties of composite with metal nanofillers might have different or similar to the matrix according to XRD pattern results. The crystal and amorphous structure of the composite were interpreted by the shape of the peak in XRD which a sharp and thin peak symbolized the crystal structure while a broad peak corresponded as an amorphous phase. The XRD patterns of metal nanofiller and composite greatly help in observing the effect of the existence and amount of metal nanofillers on the crystallinity properties of the composite.

Every metal nanofillers have its own crystallites which are identified from their XRD peaks. For example, the XRD pattern of ZnO shows hexagonal wurtzite crystallites (Sengwa & Dhatarwal, 2021) while TiO_2 shows a tetragonal anatase-rich crystal phase (Sengwa & Dhatarwal, 2021) (Figure 7.7). After the introduction of metal nanofiller in preparation for the composite, new sharp XRD peak positions signifies the metal nanofiller crystallite structure obviously detected in the XRD pattern of the composite (Sengwa & Dhatarwal, 2021) but the absence in the XRD pattern of the matrix (Xu et al., 2022). This confirms the crystallites of metal nanofillers maintained their original structures although the metal nanofillers were successfully dispersed in the matrix (Sengwa & Dhatarwal, 2021).

The amount of metal nanofillers in a composite structure influenced the peak intensity of crystal and amorphous properties of the metal nanofillers in the composite structure. Less peak intensity related to the low content of metal nanofillers

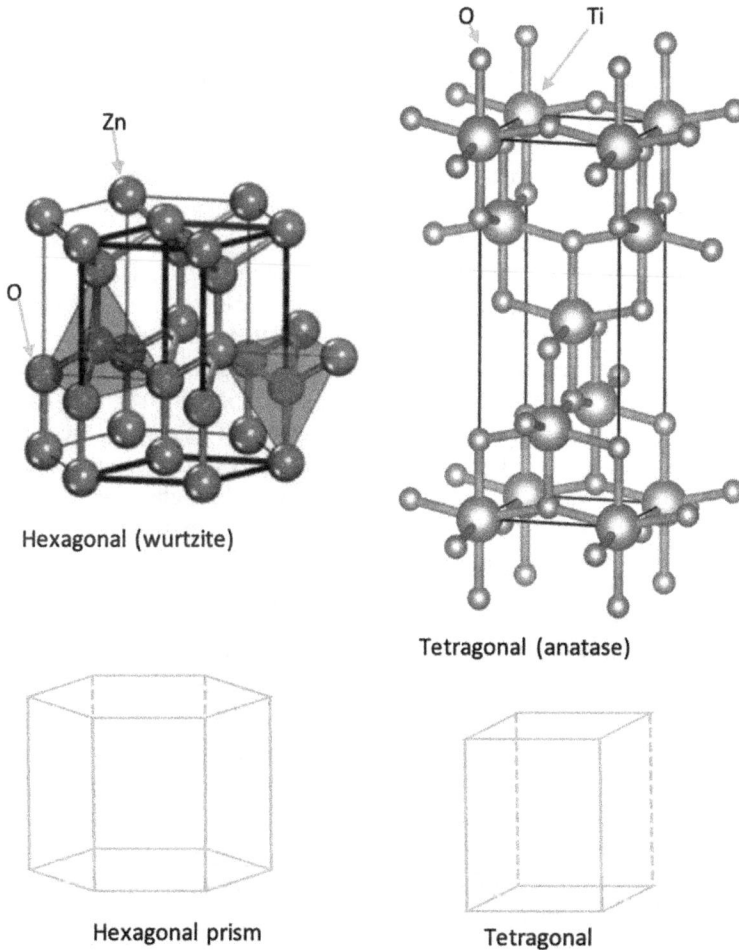

FIGURE 7.7 Hexagonal (wurtzite) and tetragonal (anatase) crystallite structures.

was found in XRD patterns of composite because the crystallinity value (Xu et al., 2022) and amorphous areas (Gaabour, 2020) of the polymer matrix are dominant. Composite loaded with very small amounts of metal nanofiller might also have similar XRD patterns to the matrix (Jamróz et al., 2020) which probably the crystalline structure of the polymer matrix did not affect by that amount of metal nanofiller (Kochkina & Butikova, 2019) or the crystalline structure of metal nanofillers in that composite still below the detection threshold in the diffraction method (Jamróz et al., 2020). No alteration and appearance of new diffraction peaks related to the structure of metal nanofillers in XRD patterns from both materials (composite with low metal nanofillers and matrix) are very beneficial information in indicating a very good distribution of metal nanofiller in the matrix (Jamróz et al., 2020). Nevertheless, this composite still has a good conductivity property compared to a composite without metal nanofiller content (Gaabour, 2020).

As the content of metal nanofiller increases, the crystallinity of metal nanofiller in the composite gradually increased (Xu et al., 2022) while the crystallinity and broadness of amorphous nature of the matrix decreased (Jeyabanu et al., 2019; Xu et al., 2022). As a result, semi-crystalline peaks of the matrix in XRD patterns belong to the composite became sharp (Gaabour, 2020). Reduction of semi-crystalline behavior may result from complex formation between polar groups in the polymer matrix with surface groups in the metal nanofiller (Gaabour, 2020). Increasing amounts of metal nanofillers offer more strong electrostatic interactions and Lewis's acid-base interaction between metal ions in metal nanofillers and the oxygen bonds in the polymer matrix (Xu et al., 2022). Electrostatic interactions depend on the number of metal ions. Because electrostatic interaction involves the opposite ions, increasing metal nanofillers offer a more positive charge of metal ions to promote electrostatic interactions. In addition, increasing amounts of metal nanofillers in composite inhibits polymer crystallization by increasing free volume and accelerating segment movement. The electrostatic interactions (Xu et al., 2022) capable of destroying the spherulites of the polymer matrix cause a reduction in crystallinity of the polymer matrix (Xu et al., 2022) while Lewis's acid–base interaction tuning the highest occupied molecular orbital (HOMO) thermodynamically which contributes to the improvement in the electrochemical window of polymer electrolytes and enhancing the anti-oxidative stability of polymer electrolytes. This clarifies that further metal nanofillers loading weakened the polymer matrix's crystalline zone ordering (Gaabour, 2020) and brought charge transfer complexes to evolve in composite, hence enhancing electrical conductivity property (Gaabour, 2020). The ion conductivity of the composite material also increases as metal nanofiller concentration increases (Xu et al., 2022). The peak intensity belongs to the crystal structure of the matrix may also disappear (Xu et al., 2022) when the amount of crystal structure of metal nanofillers is dominant in the composite. It can be concluded that the peak intensity of metal nanofiller in the composite has a linear relation to the concentration or amount of metal nanofiller which in principle confirms the successful composite formation (Sengwa & Dhatarwal, 2021). The evenly dispersed metal nanoparticles in the polymer matrix as shown by the amorphous nature of the polymer matrix confirmed the complexation between metal nanofiller and the matrix also takes place in the amorphous region (Jeyabanu et al., 2019).

However, an excess amount of metal nanofillers loaded in the matrix causes serious agglomeration of metal nanoparticles (Xu et al., 2022). The crystal structure belonging to matrix became dominant in the composite structure as confirmed by the reappearance of peaks related to crystallinity of the polymer matrix in the XRD pattern (Xu et al., 2022). The ion conductivity of the composite material loaded with excess metal nanofillers also reduced (Xu et al., 2022) drastically compared to the composite material loaded with the optimum amount of metal nanofillers. Therefore, the suitable addition of metal nanofillers is beneficial to obtain nanocomposite material with good crystallinity value (Xu et al., 2022).

The overlapping signals of peaks from matrix and metal nanofillers also occur in the XRD patterns of the composite. The difference in the crystallite sizes of the polymer matrix and simultaneous change in the crystalline structure is hard to quantify with any accuracy using XRD patterns (Kochkina & Butikova, 2019). Therefore, another method for example the DSC method can be applied to quantify the change in the crystalline structure with metal nanofillers based on the melting temperature data of bulk-like and filler-induced crystals (Kochkina & Butikova, 2019).

7.7 DECOMPOSITION OF METAL NANOCOMPOSITES

The decomposition patterns characteristics of samples obtained using TGA enable in the measurement of thermal stability and composition of the material. The incorporation of metal nanofiller in composite preparation gave significant changes in decomposition patterns and characteristics of composite materials. Although one main degradation stage is observed in the TGA curves, the addition of metal nanofillers affects the amount of thermal decomposition residues in the composite (Liu et al., 2021). The occurrence of metal nanofiller leads to attaching and retaining more char (Liu et al., 2021). As a result, the differences in residual masses of the composite are lower than those masses of the matrix (Liu et al., 2021). These also prove the potential of metal nanofillers in the composites to be acted as a mass transport barrier that retards the degradation of the product (Sankar et al., 2020).

In addition, the slight reduction in mass at temperatures above 400°C for the composite compared to the matrix was due to the concomitant loss of sub-functional surface groups of metal nanofillers (Liu et al., 2021; Han et al., 2016). For instance, the oxidation of metal nanofillers of composite in a CO_2 atmosphere (Liu et al., 2021; Feng et al., 2017; Sun et al., 2019) occurred at a temperature range between 150°C and 500°C.

7.8 VISIBLE COLOR OF METAL NANOCOMPOSITES

The presence of metal nanofillers in composite structures can be noticed from color changes between composite and matrix (Jamróz et al., 2020). Without metal nanofillers, the matrix is highly transparent (Figure 7.8). Introduction of metal nanofillers, giving opaque color to the composite as shown in Figure 9. For example, $Ti_3C_2T_x$/epoxy composites prepared by Liu et al. (2021) gave black and opaque color compared to the clear and transparent epoxy. Despite using various $Ti_3C_2T_x$ concentrations

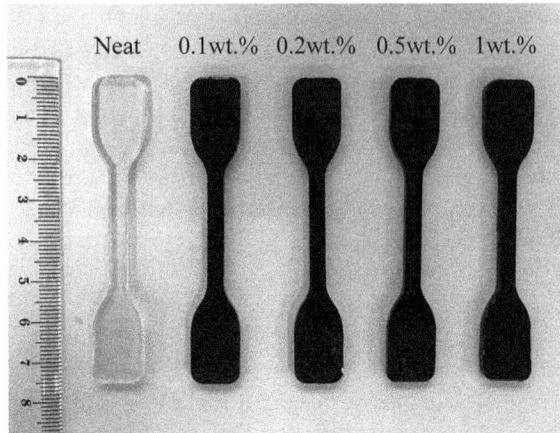

FIGURE 7.8 Photographs of matrix (neat) and $Ti_3C_2T_x$/epoxy composites at different $Ti_3C_2T_x$ concentration. (Reproduced from ref Liu et al., 2021.)

during preparation, the $Ti_3C_2T_x$/epoxy composites did not exhibit significant color variations (Liu et al., 2021). The uniform color indicates the well-dispersed loaded metal nanoparticles in the matrix (Sengwa & Dhatarwal, 2021). Moreover, the developed composite had a highly smooth surface compared to the matrix (Sengwa & Dhatarwal, 2021). A rising amount of metal nanofiller loading into the matrix reduces the relative transparency to visible light of the composite, thus limiting the optical uses of the composite (Sengwa & Dhatarwal, 2021).

7.9 CONCLUSION

Existence of metal nanofillers in composite structures able to be identified using various characterization tools including SEM, EDX, FTIR, and TGA. Similar to metal nanofillers themselves, their composites also have unique and variable structures. Although different methods were used to manufacture the composite using metal nanofillers, every metal nanofiller carries its own characteristic peaks of FTIR spectrum and XRD pattern. The existence and distribution of metal nanofillers were confirmed from SEM images, EDX analysis or EDX mapping and the visible color of the composites. These analytical instruments have been good enough before optimizing the performance of composite in their target applications. Therefore, suitable methods in metal nanofillers and composite developments should be vigorously performed for wide application of nanomaterials in every sector.

REFERENCES

Abu-Okail, M., Alsaleh, N. A., Farouk, W. M., Elsheikh, A., Abu-Oqail, A., Abdelraouf, Y. A., & Abdel Ghafaar, M. (2021). Effect of dispersion of alumina nanoparticles and graphene nanoplatelets on microstructural and mechanical characteristics of hybrid carbon/glass fibers reinforced polymer composite. *Journal of Materials Research and Technology, 14*, 2624–2637. https://doi.org/10.1016/j.jmrt.2021.07.158.

Bahrami, A., Rezaei Mokarram, R., Sowti Khiabani, M., Ghanbarzadeh, B., & Salehi, R. (2019). Physico-mechanical and antimicrobial properties of tragacanth/hydroxypropyl methylcellulose/beeswax edible films reinforced with silver nanoparticles. *International Journal of Biological Macromolecules*, *129*, 1103–1112. https://doi.org/10.1016/j.ijbiomac.2018.09.045.

Baird, D. G. (2003). Polymer processing. In R. A. Meyers, *Encyclopedia of Physical Science and Technology* (pp. 611–643). Academic Press. https://doi.org/10.1016/B0-12-227410-5/00593-7.

Basavaraj, E., Ramaraj, B., & Lee, J. H. (2013). Microstructure, thermal, physico-mechanical and tribological characteristics of molybdenum disulphide-filled polyamide 66/carbon black composites. *Polymer Engineering & Science*, *53*(8), 1676–1686. https://doi.org/10.1002/pen.23428.

Batool, M., Haider, M. N., & Javed, T. (2022). Applications of spectroscopic techniques for characterization of polymer nanocomposite: A review. *Journal of Inorganic and Organometallic Polymers and Materials*, *32*(12), 4478–4530. https://doi.org/10.1007/s10904-022-02461-3.

Brostow, W., Buchman, A., Buchman, E., & Olea-Mejia, O. (2008). Microhybrids of metal powder incorporated in polymeric matrices: Friction, mechanical behavior, and microstructure. *Polymer Engineering & Science*, *48*, 1977–1981. https://doi.org/10.1002/pen.21119.

Cangul, S., & Adiguzel, O. (2017). The latest developments related to composite resins. *International Dental Research*, *7*, 32–41. https://doi.org/10.5577/intdentres.2017.vol7.no2.3.

Delides, C. G. (2016). Everyday life applications of polymer nanocomposites. *e-RA*, *11*, 1–8.

Domènech, B., Bastos-Arrieta, J., Alonso, A., Macanás, J., Muñoz, M., & Muraviev, D. N. (2012). Chapter 3–Bifunctional polymer-metal nanocomposite ion exchange materials. In A. Kilislioglu, *Ion Exchange Technologies* (pp. 35–72). IntechOpen, Barcelona.

Domyati, D. (2022). Chemical and thermal study of metal chalcogenides (zinc sulfide), aluminum oxide, and graphene oxide based nanocomposites for wastewater treatment. *Ceramics International*, *49*(1), 1464–1472. https://doi.org/10.1016/j.ceramint.2022.09.190.

Feng, A., Yu, Y., Jiang, F., Wang, Y., Mi, L., Yu, Y., & Song, L. (2017). Fabrication and thermal stability of NH_4HF_2-etched Ti_3C_2 MXene. *Ceramics International*, *43*(8), 6322–6328. https://doi.org/10.1016/j.ceramint.2017.02.039.

Fernandez, M. J., Abirami, K., Swetha, S., Soundarya, S., Rasana, N., & Jayanarayanan, K. (2021). The effect of nano, micro and dual scale filler reinforcement on the morphology, mechanical and barrier properties of polypropylene composites. *Materials Today: Proceedings*, *46*(10), 5067–5071. https://doi.org/10.1016/j.matpr.2020.10.424.

Fritea, L., Banica, F., Costea, T. O., Moldovan, L., Dobjanschi, L., Muresan, M., & Cavalu, S. (2021). Metal nanoparticles and carbon-based nanomaterials for improved performances of electrochemical (bio)sensors with biomedical applications. *Materials (Basel)*, *14*(21), 6319.

Gaabour, L. H. (2020). Effect of selenium oxide nanofiller on the structural, thermal and dielectric properties of CMC/PVP nanocomposites. *Journal of Materials Research and Technology*, *9*(3), 4319–4325. https://doi.org/10.1016/j.jmrt.2020.02.057.

Gajapriya, M., Jayalakshmi, S., & Geetha, R. V. (2020). Fillers in composite resins-recent advances. *European Journal of Molecular & Clinical Medicine*, *7*(1), 971–977.

Hamida, R.S., Ali, M.A., Redhwan, A., & Bin-Meferij, M.M. (2020). Cyanobacteria-A promising platform in green nanotechnology: A Review on nanoparticles fabrication and their prospective applications. *International Journal of Nanomedicine*, *15*, 6033-6066. https://doi.org/10.2147/IJN.S256134.

Han, M., Yin, X., Wu, H., Hou, Z., Song, C., Li, X., Zhang, L. & Cheng, L. (2016). Ti_3C_2 MXenes with modified surface for high-performance electromagnetic absorption and shielding in the X-band. *ACS Applied Materials & Interfaces*, *8*(32), 21011–21019. https://doi.org/10.1021/acsami.6b06455.

Ilyas, R., Sapuan, S., Asyraf, M., Dayana, D., Amelia, J., Rani, M.S.A., Norrrahim, M.N.F., Nurazzi, N.M., Aisyah, H.A., Sharma, S. & Ishak, M.R. (2021). Polymer composites filled with metal derivatives: A review of flame retardants. *Polymers (Basel)*, *13*(11), 1–21. https://doi.org/10.3390/polym13111701.

Jacob, S., Suma, K. K., Mendaz, J. M., George, A., & George, K. E. (2009). Modification of polypropylene/glass fiber composites with nanosilica. *Macromolecular Symposia*, *277*(1), 138–143. https://doi.org/10.1002/masy.200950317.

Jakubowska, E., Gierszewska, M., Szydłowska-Czerniak, A., Nowaczyk, J., & Olewnik-Kruszkowska, E. (2023). Development and characterization of active packaging films based on chitosan, plasticizer, and quercetin for repassed oil storage. *Food Chemistry*, *399*, 1–14. https://doi.org/10.1016/j.foodchem.2022.133934.

Jamróz, E., Khachatryan, G., Kopel, P., Juszcak, L., Kawecka, A., Krzyściak, P., Kucharek, M., Bębenek, Z. & Zimowska, M. (2020). Furcellaran nanocomposite films: The effect of nanofillers on the structural, thermal, mechanical and antimicrobial properties of bio-polymer films. *Carbohydrate Polymers*, *240*(116244), 1–12. https://doi.org/10.1016/j.carbpol.2020.116244.

Jeyabanu, K., Sundaramahalingam, K., Devendran, P., Manikandan, A., & Nallamuthu, N. (2019). Effect of electrical conductivity studies for CuS nanofillers mixed magnesium ion based PVA-PVP blend polymer solid electrolyte. *Physica B: Condensed Matter*, *572*, 129–138. https://doi.org/10.1016/j.physb.2019.07.049.

Joudeh, N., & Linke, D. (2022). Nanoparticle classification, physicochemical properties, characterization, and applications: A comprehensive review for biologists. *Journal of Nanobiotechnology*, *20*(262), 1–29. https://doi.org/10.1186/s12951-022-01477-8.

Kanmani, P., & Rhim, J. W. (2014). Physical, mechanical and antimicrobial properties of gelatin based active nanocomposite films containing AgNPs and nanoclay. *Food Hydrocolloids*, *35*, 644–652. https://doi.org/10.1016/j.foodhyd.2013.08.011.

Kannan, K., Radhika, D., Nikolova, M. P., Andal, V., Sadasivuni, K. K., & Krishna, L. S. (2020). Facile microwave-assisted synthesis of metal oxide CdO-CuO nanocomposite: Photocatalytic and antimicrobial enhancing properties. *Optik*, *218*(165112), 1–7. https://doi.org/10.1016/j.ijleo.2020.165112.

Khan, I., Saeed, K., & Khan, I. (2019). Nanoparticles: Properties, applications and tox-icities. *Arabian Journal of Chemistry*, *12*, 908–931. https://doi.org/10.1016/j.arabjc.2017.05.011.

Klyuchnikova, N. V., & Lymar, E. A. (2005). The effect of metal Filler on Structure Formation of Composite Materials. *Glass and Ceramics*, *62*(9–10), 319–320. https://doi.org/10.1007/s10717-005-0103-4.

Kochkina, N. E., & Butikova, O. A. (2019). Effect of fibrous TiO_2 filler on the structural, mechanical, barrier and optical characteristics of biodegradable maize starch/PVA com-posite films. *International Journal of Biological Macromolecules*, *139*, 431–439. https://doi.org/10.1016/j.ijbiomac.2019.07.213.

Kumar, P. S., Selvakumar, M., Bhagabati, P., Bharathi, B., Karuthapandian, S., & Balakumar, S. (2014). CdO/ZnO nanohybrids: Facile synthesis and morphologically enhanced photocatalytic performance. *RCS Advances*, *4*, 32977–32986. https://doi.org/10.1039/C4RA02502D.

Liu, L., Ying, G., Wen, D., Zhang, K., Hu, C., Zheng, Y., Zhang, C., Wang, X. and Wang, C. (2021). Aqueous solution-processed MXene ($Ti_3C_2T_x$) for non-hydrophilic epoxy resin-based composites with enhanced mechanical and physical properties. *Materials & Design*, *197*, 109276. https://doi.org/10.1016/j.matdes.2020.109276.

Maiphetlho, K., Shumbula, N., Motsoane, N., Chimuka, L., & Richards, H. (2020). Evaluation of silver nanocomposite polymer inclusion membranes (PIMs) for trace metal transports: Selectivity and stability studies. *Journal of Water Process Engineering*, *37*(101527), 1–14. https://doi.org/10.1016/j.jwpe.2020.101527.

Mallya, D. S., Yang, G., Lei, W., Muthukumaran, S., & Baskaran, K. (2022). Functionalized MoS$_2$ nanosheets enabled nanofiltration membrane with enhanced permeance and fouling resistance. *Environmental Technology & Innovation*, 27(102719), 1–15. https://doi.org/10.1016/j.eti.2022.102719.

Menezes, M. d. L. L. R., Pires, N. d. R., da Cunha, P. L. R., de Freitas Rosa, M., de Souza, B. W. S., Feitosa, J. P. d. A., & Souza Filho, M. d. S. M. d. (2019). Effect of tannic acid as crosslinking agent on fish skin gelatin-silver nanocomposite film. *Food Packaging and Shelf Life*, 19, 7–15. https://doi.org/10.1016/j.fpsl.2018.11.005.

Mishra, G., & Mukhopadhyay, M. (2021). Well dispersed CeO$_2$@HNTs nanofiller in the poly (vinyl chloride) membrane matrix and enhanced its antifouling properties and rejection performance. *Journal of Environmental Chemical Engineering*, 9(1), 104734. https://doi.org/10.1016/j.jece.2020.104734.

Mourdikoudis, S., Pallares, R. M., & Thanh, N. T. (2018). Characterization techniques for nanoparticles: Comparison and complementarity upon studying nanoparticle properties. *Nanoscale*, 10, 12871–12934. https://doi.org/10.1039/C8NR02278J.

Mouritz, A. P. (2012). 18–Fracture processes of aerospace materials. In A. P. Mouritz, *Introduction to Aerospace Materials* (pp. 428–453). Woodhead Publishing. https://doi.org/10.1533/9780857095152.428.

Mukherji, S., Bharti, S., Shukla, G. M., & Mukherji, S. (2018). Synthesis and characterization of size- and shape-controlled silver nanoparticles. *Physical Sciences Reviews*, 4(1), 20170082. https://doi.org/10.1515/psr-2017-0082.

Munawar, T., Iqbal, F., Yasmeen, S., Mahmood, K., & Hussain, A. (2020). Multi metal oxide NiO-CdO-ZnO nanocomposite-synthesis, structural, optical, electrical properties and enhanced sunlight driven photocatalytic activity. *Ceramics International*, 46(2), 2421–2437. https://doi.org/10.1016/j.ceramint.2019.09.236.

Palza, H., Vergara, R., & Zapata, P. (2011). Composites of polypropylene melt blended with synthesized silica nanoparticles. *Composites Science and Technology*, 71(4), 535–540. https://doi.org/10.1016/j.compscitech.2011.01.002.

Park, S. J., & Seo, M. K. (2011). Chapter 6–Element and processing. In M. K. S. Soo-Jin Park, *Interface Science and Technology* (Vol. 18, pp. 431–499). Elsevier. https://doi.org/10.1016/B978-0-12-375049-5.00006-2.

Paulose, R., Mohan, R., & Parihar, V. (2017). Nano-structures & nano-objects nanostructured nickel oxide and its electrochemical behaviour – a brief review. *Nano-Structures & Nano-Objects*, 11, 102–111. https://doi.org/10.1016/j.nanoso.2017.07.003.

Salari, M., Sowti Khiabani, M., Rezaei Mokarram, R., Ghanbarzadeh, B., & Samadi Kafil, H. (2018). Development and evaluation of chitosan based active nanocomposite films containing bacterial cellulose nanocrystals and silver nanoparticles. *Food Hydrocolloids*, 84, 414–423. https://doi.org/10.1016/j.foodhyd.2018.05.037.

Sankar, S., Parvathi, K., & Ramesan, M. (2020). Structural characterization, electrical properties and gas sensing applications of polypyrrole/Cu-Al$_2$O$_3$ hybrid nanocomposites. *High Performance Polymers*, 32(6), 719–728. https://doi.org/10.1177/09540083198991.

Sengwa, R. J., & Dhatarwal, P. (2021). Polymer nanocomposites comprising PMMA matrix and ZnO, SnO$_2$, and TiO$_2$ nanofillers: A comparative study of structural, optical, and dielectric properties for multifunctional technological applications. *Optical Materials*, 113, 1–12. https://doi.org/10.1016/j.optmat.2021.110837.

Siraj, S., Al-Marzouqi, A., Iqbal, M., & Ahmed, W. (2022). Impact of micro silica filler particle size on mechanical properties of polymeric based composite material. *Polymers*, 14(4830), 1–18. https://doi.org/10.3390/polym14224830.

Sun, Y., Sun, Y., Meng, X., Gao, Y., Dall'Agnese, Y., Chen, G., Dall'Agnese, C. and Wang, X.-F. (2019). Eosin Y-sensitized partially oxidized Ti$_3$C$_2$ MXene for photocatalytic hydrogen evolution. *Catalysis Science & Technology*, 9, 310–319. https://doi.org/10.1039/C8CY02240B.

Tabib, A., Bouslama, W., Sieber, B., Addad, A., Elhouichet, H., Férid, M., & Boukherroub, R. (2017). Structural and optical properties of Na doped ZnO nanocrystals: Application to solar photocatalysis. *Applied Surface Science, 396*, 1528–1538. https://doi.org/10.1016/j. apsusc.2016.11.204.

Tyagi, M., Tomar, M., & Gupta, V. (2012). Influence of hole mobility on the response characteristics of p-type nickel oxide thin film based glucose biosensor. *Analytica Chimica Acta, 726*(13), 93–101. https://doi.org/10.1016/j.aca.2012.03.027.

Ullah, N., Shah, S. M., Erten-Ela, Ş., Ansir, R., Hussain, H., Qamar, S., & Usman, M. (2022). Dithizone, carminic acid and pyrocatechol violet dyes sensitized metal (Ho, Ba & Cd) doped TiO_2/CdS nanocomposite as a photoanode in hybrid heterojunction solar cell. *Ceramics International, 48*(21), 31478–31490. https://doi.org/10.1016/j. ceramint.2022.07.067.

Vaggar, G. B., Sirimani, V. B., & Sataraddi, D. P. (2021). Effect of filler materials on thermal properties of polymer composite materials: A review. *International Journal of Engineering Research & Technology, 10*(8), 1–5. https://doi.org/10.1007/s00289-022-04249-4.

van den Berg, T., & Ulbricht, M. (2020). Polymer nanocomposite ultrafiltration membranes: The influence of polymeric additive, dispersion quality and particle modification on the integration of zinc oxide nanoparticles into polyvinylidene difluoride membranes. *Membranes (Basel), 10*(197), 1–19. https://doi.org/10.3390/membranes10090197.

Wu, J. (2018). Chapter 5–Material interface of pantograph and contact line. In J. Wu, *Pantograph and Contact Line System* (pp. 165–191). Academic Press. https://doi. org/10.1016/B978-0-12-812886-2.00005-7.

Xu, Y., Li, J., & Li, W. (2022). Evolution in electrochemical performance of the solid blend polymer electrolyte (PEO/PVDF) with the content of ZnO nanofiller. *Colloids and Surfaces A: Physicochemical and Engineering Aspects, 632*(127773), 1–15. https://doi. org/10.1016/j.colsurfa.2021.127773.

8 Bio-oils as the Precursor for Carbon Nanostructure Formation

M. M. Harussani
Universiti Putra Malaysia
Tokyo Institute of Technology

S. M. Sapuan and Gwyth Muosso
Universiti Putra Malaysia

Shahriar Ahmad Fahim
Tokyo Institute of Technology

8.1 INTRODUCTION

Carbon nanostructures (CNSs) are the principal focus of discussions in the field of nanotechnology. Research has been conducted on different members of CNS from fullerenes to carbon nanotubes (CNTs) since their discovery, including carbon nanoscience – a discipline to understand their properties, structures, and interrelationships at a nanoscale level (Shenderova et al., 2002). A new carbon materials' era started with the discovery of Buckminster fullerenes, and Ijima's first observation of nanotubes in the 1980s (Iijima, 1980; Kroto et al., 1985). The next big step happened when Iijima (1991) introduced the world to a new type of molecular carbon, nanofibers, or CNTs of 4~30 nm diameter. This caught the attention of many researchers for its potential applications in various fields. Hence, as of now, the research on carbon nanomaterials has been progressing persistently (Goswami et al., 2021).

CNSs have been predicted to be useful across various fields like electronics, medical diagnostics, biosensors, energy storage, nanocomposites, and so on. However, their fruitful implementation is still somewhat challenged by their affordability in terms of costs. So, these nanomaterials are yet to be used to their full potential in the industries. Currently, most of the CNTs are produced from fossil fuel-based precursors, methane, benzene, xylene, acetylene, etc. (Azmina et al., 2012). Dependence on fossil fuels means increased raw materials costs due to their finite supply and limited availability geographically (Saputri et al., 2020) as well as leads to a non-environmental-friendly approach which leads to higher greenhouse gas (GHG) emission. Therefore, it is crucial to discover novel approaches to fabricating nanomaterials in an affordable and environmental-friendly way.

DOI: 10.1201/9781003400998-8

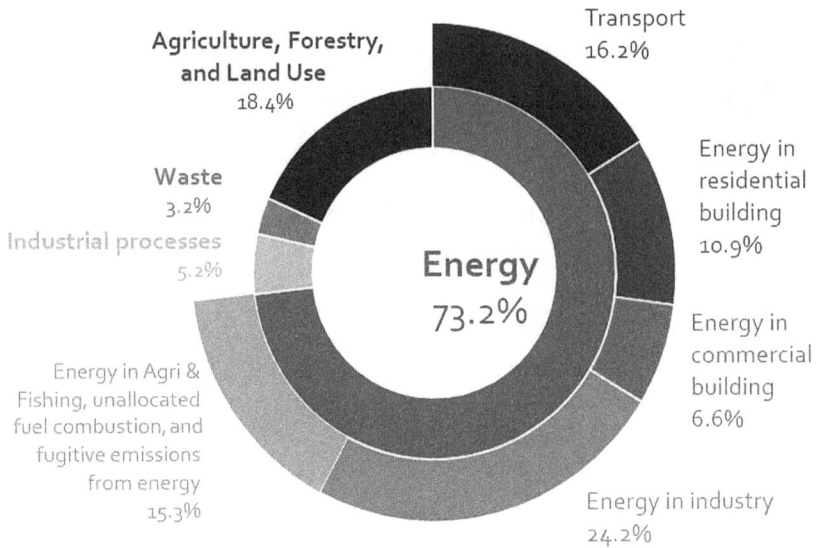

FIGURE 8.1 Carbon dioxide emissions according to the sectors globally. (Reproduced from ref Abdul Latif et al., 2021.)

According to Abdul Latif et al. (2021), out of 50 billion tons of GHG emissions globally, nearly three-fourths of them were coming from the energy sector, as illustrated in Figure 8.1. Ellabban et al. (2014) argued that the transition to renewable energy can help to reduce GHG emissions by reducing the effects of harsh weather and climate changes while also maintaining a consistent, affordable, and stable energy supply.

Hence, an alternative solution is to use renewable bioresources to derive CNSs because of their high availability, low cost, and sustainability. The usage of different bio-oils, such as camphor oil, castor oil, turpentine oil, olive oil, coconut oil, palm oil, etc., has been studied in recent years for synthesizing CNSs. As the precursor, bio-oils provide highly pure final products due to the lower thermal stability of bicyclic compounds present in them compared to their non-renewable counterparts. However, they might still have some limitations. For instance, Liu et al. (2011) found that catalytic decomposition of methane and benzene was possible at a lower temperature than camphor oil for producing nanotubes. Methane was found to be the best precursor to produce the best-quality single-walled CNT (SWCNT) in their research. Janas (2020) also argued that synthesized CNTs with bio-feedstocks (both solid and liquid) as precursors were of lower purity compared to the nanotubes produced with synthetic precursors.

Some recent studies related to the conversion of bio-oil, waste, and biomass into CNS with some of their applications had been studied. Kumar et al. (2013) synthesized CNTs and carbon–nitrogen nanotubes through spray pyrolysis technique from natural green precursor biodiesel derived from jatropha mixed with ferrocene and acetonitrile as catalysts in presence of argon (Ar) at a high temperature. To fabricate materials with desired electrical properties, the authors suggested that carbon–nitrogen nanotubes might be more useful. Tripathi et al. (2012) developed CNTs using natural non-edible castor oil. Three different catalysts, Ni, Zn, and Co, have been catalyzed, in a thermal catalytic vapor reactor to investigate the adsorption efficiency for removing arsenic (As) from water. Alancherry (2018) devised an environmental-friendly technique based on plasma-enhanced chemical vapor deposition (CVD) process to prepare CNS from bio-renewable precursors. *Citrus sinensis* essential oil from cold extraction of orange peels was used to prepare vertical graphene nanostructures followed by possible gas-sensing applications. In another study by Ghosh et al. (2007), *eucalyptus* oil was used to manufacture SWCNTs through a CVD method using Fe/Co zeolite catalyst. A possible route of using bio-oils to synthesize CNSs in the liquid phase was explored by Ballotin et al. (2019). In this study, H_2SO_4 was used to dehydrate the bio-oil complex to condense, aromatize, and carbonize for the formation of CNSs of different characteristics. Sankaranarayanan et al. (2019) presented a novel application of nanocarbon materials derived from a renewable precursor for the germination of seeds in different plants. Graphene oxide–carbon dot (GO-CD) composite was synthesized through hydrothermal treatment from *Macrocystis pyrifera*, brown macroalgal biomass. Spillage of waste cooking oil to water bodies is an inevitable phenomenon in daily life. Suriani et al. (2010) produced vertically aligned CNTs (VACNTs) from waste cooking palm oil for the first time that can be used for flat lamps and panel displays. They also concluded that similar waste cooking oils are feasible to be used as precursor for synthesis of CNT. Despite numerous studies already being done on the use of bio-oils as a precursor to make CNSs, Janas (2020) stated in his paper that a successful industrial implementation still requires further investigations.

In this chapter, state-of-the-art techniques for synthesizing CNS with bio-oils as the precursor are reviewed. Both solid and liquid precursors derived from biomass, waste, or bio-oils are evaluated with detailed fabrication methods to present a full understanding of the current development in this field. Next, promising applications of bio-oil-derived nanomaterials in electronics, sensing technologies, mechanical equipment, and batteries are discussed. Lastly, critical views on future steps and prospects to implement this idea into commercially feasible practice are provided in the conclusions, as recommendations for engineers and researchers.

8.2   CARBON NANOSTRUCTURES (CNSs)

The very basis of the structural formation of different CNSs is centered on different ways of bonding with other carbon atoms or other elements in sp, sp^2, and sp^3 hybridization states. In this section, different types of carbon, that is nanoallotropes, are discussed. Georgakilas et al. (2015) presented three different ways to classify CNSs. The first one is based on covalent bonds of the C atom, the second one is according to morphological characteristics, and the third one is in terms of dimensions.

Considering covalent bonds linking the C atoms, CNSs are classified into two types. The first type includes graphene, made of sp^2 hybridized carbon atoms closely packed together hexagonal crystal lattice. Examples of this type are graphene, nanosheets made of graphene, CNTs, carbon nano-onions, and carbon quantum dots. The second type of this classification includes CNSs with both sp^2 and sp^3 states, but in different proportions and mixtures of amorphous and crystalline graphitic regions. As of now, the only member of this type is the nanodiamond. The morphological classification is done on whether the carbon nanoallotrope has empty space internally or not. Fullerenes, CNTs, and nanohorns have empty internal spaces. On the other hand, graphene, nanodiamonds, carbon nano-onions, and carbon quantum dots have no internal spaces.

Lastly, the classification of CNSs in terms of dimension is defined as the number of dimensions that is outside the scope of the nanoscale range in the corresponding material. For example, the nanomaterial is 0D when none of the dimensions are outside the nanoscale, that is, all three dimensions can be measured at the nano-level. Such materials are fullerenes, carbon nano-onions, carbon quantum dots, and nanodiamonds. If one dimension of the nanomaterial cannot be measured in the nanoscale, it is a 1D nanomaterial, for instance, CNTs, carbon nano-fibers, and carbon nanohorns. These members are again classified into SWCNTs and multi-walled CNTs (MWCNTs). Finally, 2D nanostructures have two of three dimensions that are out of the scope of the nanoscale level, such as graphene, graphene nanoribbons, and few-layer graphene with their derivatives (graphene oxide (GO) and reduced GO) (Georgakilas et al., 2015; Nyholm & Espallargas, 2023). Figure 8.2 shows a diagram to overview the CNSs in terms of dimension, and Table 8.1 provides their basic structural characteristics. Figure 8.3 presents the structural diagrams of the carbon nanomaterials discussed in this section and their family structure.

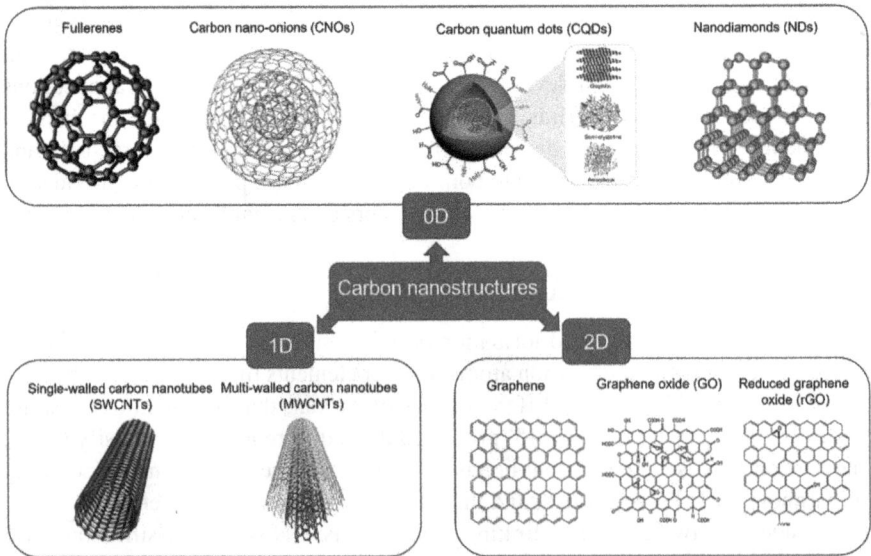

FIGURE 8.2 Classification of CNSs in terms of dimensions. (Reproduced from ref Nyholm & Espallargas, 2023.)

TABLE 8.1

Different Members of CNSs with Structural Characteristics

Dimension	CNS	Structural Characteristics
0D	Fullerenes	Hollow, close-caged, truncated icosahedral (sp^2 hybridized)
		Smallest possible: C$_{20}$
		Most abundant and smallest stable: C$_{60}$
		Outer diameter of 0.71 nm
	Carbon nano-onions	Made of concentric graphite shells
		Outer diameter: 10 nm–1 μm
		Inner diameter: 0.7–1 nm
	Carbon quantum dots	Mostly contain sp^3 hybridized and amorphous in nature carbon atoms while crystalline graphite structure results in strong photoluminescence
		Size: 2–10 nm
	Nanodiamonds	sp^3 hybridized carbon atoms in crystal with diamonded-like topology
		Size: 1–20 nm
1D	SWCNTs	Tubular nanostructured, single-layered graphene sheet
		Length: 50 nm–1 μm
		Diameter: Typical 0.4–2 nm
	MWCNTs	Tubular nanostructured, but consist of several layers arranged concentric cylinders of graphitic carbon
		Length: 10 nm–1 μm
		Outer diameter: 2.5–30 nm
2D	Graphene and their derivatives	Crystalline carbon film with the carbon atom of sp^2 hybridization arranged in a hexagonal honeycomb lattice
		Single graphene sheet is of 10–15 nm
3D	Graphite	Layer structure where carbon atoms are hexagonal in each layer, where each layer is a graphene layer

Source: Reproduced from refs Chung (2002), Georgakilas et al. (2015), Giubileo et al. (2018), Goswami et al. (2021), Norizan et al. (2020), Ren et al. (2023), and Shenderova et al. (2002).

8.3 FABRICATION OF CNS

8.3.1 CHEMICAL VAPOR DEPOSITION (CVD)

CVD, which may also be called catalytic growth, typically involves the decomposition of hydrocarbons (with the more commonly used ones including ethylene and acetylene) within a reactor at temperatures within the range of 550°C–750°C in the presence of a catalyst; this step is subsequently followed by a cooling process which fosters the formation of CNT structures (CNTs) as depicted in Figure 8.4. It is been found that Fe, Ni, and Co (all of which have been found to work just as well in other CNT synthesis methods) are the best catalysts for the facilitation of nanotubes growth with the presumption that with regards to the contact force between the catalyst particles and the substrate, nanotubes may emerge from the catalyst nanoparticles inserted in the pores as tip or base growth (Popov, 2004). The general principle

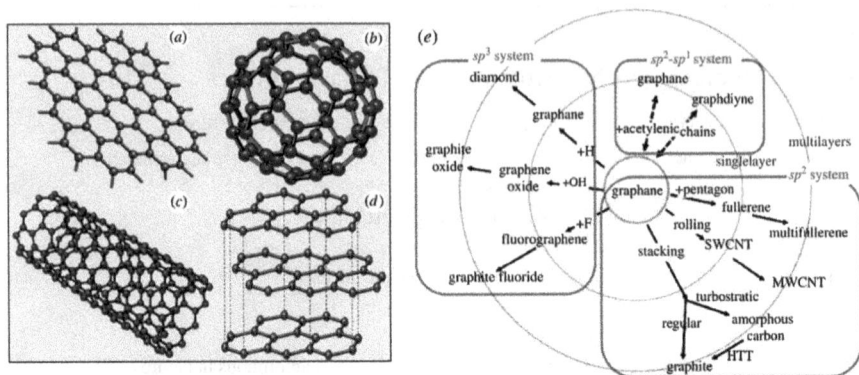

FIGURE 8.3   (a) 2D graphene layer, (b) 0D fullerene, (c) 1D CNT, (d) 3D graphite, (e) graphene and its derivatives. (Reproduced from ref Kumar et al., 2022.)

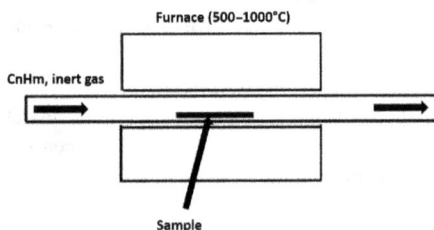

FIGURE 8.4   Simplified demonstration of the CVD process. (Reproduced from ref Popov, 2004.)

of this method begins with the introduction of hydrocarbon vapor via carrier gas (nitrogen and hydrogen) into the reaction chamber, where along with the metallic catalyst, the mixture is heated. The carbon is the separated from the hydrogen where the latter escapes while the former crystallizes in the form of a cylindrical framework. This reaction is sustained by the endothermic and exothermic nature of the hydrocarbon decomposition and the carbon crystallization accordingly. The formation of the nanotubes finally stops when the metal substrate is fully saturated with the crystallized carbon (Baker & Waite, 1975). Within this process, the production of SWCNTs and MWCNTs is typically decided by the particle size of the metal catalyst used in the process (Sinnott et al., 1999).

This methodology depends on several other parameters that also determine the properties of the product some of which include the processing time, pressure, gas flow rate, catalyst, and hydrocarbons used to initiate the process. For example, the structure of the resulting CNT is dependent on the structure of the precursor hydrocarbon as linear hydrocarbon typically produces straight and hollow CNT while cyclic hydrocarbons produced relatively curved CNTs (Nerushev et al., 2003; Pant et al., 2021); a notable trend within this process dictates that the density and the growth rate of the CNS are proportional to an increasing temperature, with this increase also resulting in a tendency for increased vertical alignment.

Furthermore, when it comes to the selection of the carrier gas, it is important to note that the individual components of the synthesis process must be in thermal equilibrium with the transport gas in order to prevent the formation of vapor condensation upon contact with any cool surface. Because of this, the surfaces and reactive chemical components of the system participating in the reaction are generally warmer than the supply of transport gas (Pant et al., 2021). Prior to the introduction of the decomposition reaction, the different phases that the precursor hydrocarbon might undertake undergo processing. Liquid hydrocarbons such as benzene and alcohol would first be heated with the vapor released being delivered by the carrier gas into the reaction chamber; on the other hand, solid precursors are placed directly into the reaction chamber where they undergo sublimation transition. Furthermore, the catalysts may also be used in different phases where they are typically fed via the floating catalyst method which involves conducting catalyst vapor pyrolysis at an appropriate temperature, where in-situ metal nanoparticles are released. In addition, substrates with catalyst coatings can be used in the reaction and put within the systemic reactor in order to start the CNT development.

This method has proven to be quite useful in the synthesis of CNTs as the advances made have resulted in the reduction of capital costs and resources used while producing high-quality CNTs in relatively large amounts (Dai, 2001). There are various studies that utilized CVD approach to synthesize CNTs, graphene, and other carbon-based nanomaterials (CNMs). A study by Kumar et al. (2013) was carried out using spray pyrolysis-assisted CVD method to grow CNTs using jatropha-derived biodiesel which was used as a carbon source, assisted with ferrocene ($Fe(C_5H_5)_2$) as a source of iron (Fe) which acts as catalyst. Using thermal CVD and with application of ferrocene catalysts, Suriani et al. (2013) had successfully synthesized waste chicken fat as the carbon source at a high temperature of 750°C (deposition temperature) for an hour processing time. However, CVD does not come without its detriments as the CNSs grown by this process are observed to have high defect densities; this was attributed to the relatively low growth temperature which prevents adequate annealing of the CNTs.

8.3.2   ARC DISCHARGE

Arc discharge (Figure 8.5) typically involves the use of graphite electrodes placed at a distance of 1mm in an inert atmosphere – which may be either an inert gas or a liquid environment – to produce CNTs. The anode is usually filled with a powdered carbon precursor along with a catalyst while the cathode is composed of pure graphite. The process uses a 20–25 V, 50–120 A current to generate high temperatures (around 5,000 K) which subsequently evaporates the precursor (via sublimation); the gaseous carbon proceeds to condense onto the cathode (due to the temperature difference) where along with the soot the sample is collected and purified to obtain the CNTs (Harris, 2004; Iijima, 1991).

While a number of experiments and observations have been made, the mechanism behind the growth of these CNTs is yet to be fully understood. A proposed outline of the synthesis suggests that, during the arc generation of the electrodes, the high temperatures generated turn the surrounding gas into plasma (within the area

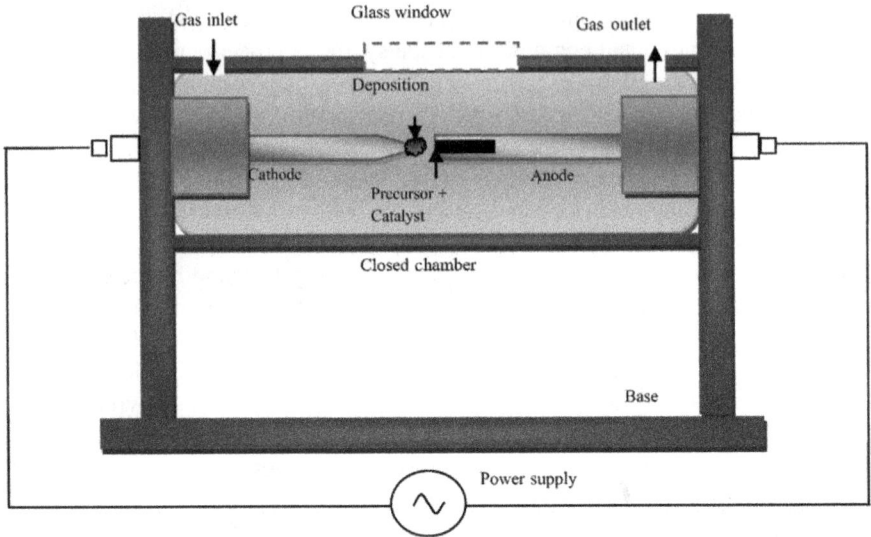

FIGURE 8.5   Setup in the arc discharge approach. (Reproduced from ref Arora & Sharma, 2014.)

between the electrodes). Due to the intense heat produced from the resistance of the electrodes, the carbon precursor undergoes sublimation and is then turned into carbon vapor which in turn decomposes into carbon ions. The carbon vapor then drifts toward the relatively cooler cathode where they undergo condensation into liquid carbon and due to the environmental conditions further crystallize to produce cylindrical structures on the cathodes surface (Arora & Sharma, 2014).

The CNTs produced were found to be in the form of tubes with diameters ranging from 4 to 30 nm (depending on the density of carbon vapor in the plasma) and 1 μm in length. Using transmission electron microscopy, MWCNTs were thus discovered and were described as needles comprised of coaxial tubes of graphitic sheets with layers ranging from 2 to roughly 50 in number. Moreover, it was reported that the carbon needles ended with a polygonal, cone-shaped, or curved caps formed due to the instability of the plasma, with the presumption that the tubes underwent spiral growth steps at these ends (Iijima, 1991). This process is governed by several parameters each playing a key role in the quantity and quality of the produced CNSs; these parameters include the operating current and arc properties, environmental conditions, electrode shape, and catalyst.

When an AC is used, the CNTs were observed to have been deposited onto the walls of the chamber as opposed to being on the cathode; this was attributed to the thermal effect created by this current. However, a method termed as pulsed arc method involves pulses (in the milliseconds) generated by an AC, striking the anode surface in order to assume the expected process of CNT collection at the cathode. On the other hand, DC current usually involves the continuous vaporization of the carbon precursor and along with the high temperature gradient (originating from the lower current density of the cathode due to its relatively smaller diameter) results in the aforementioned mechanism taking place. Between these operating currents, it is reported that the pulsed method is the most convenient as the DC method tends to impede the process due to the continuous flow of the ionized gas, while AC tends to have a reduced yield due to the CNS formation only occurring at the positive cycles.

Besides the operating currents flow, the strength also has its effect on the resulting CNTs as reported by multiple papers. There were also studies proposed different current densities for better CNT yield with Nishizaka et al. (2011) and Matsuura et al. (2009) reporting 250–270 and ~450 A/cm^2 accordingly. Moreover, it has also been reported that low currents typically result in lower yields (as per the correlation) due to an unstable arc with the current used having no effect whatsoever on the structure of the CNT (Matsuura et al., 2009; Scalese et al., 2010).

The catalyst used also plays a vital role in the resulting properties of the CNTs, where it has been suggested that the size and yield of the nanotubes is governed by the metal catalyst particle size and concentration, with Wang et al. (2004) reporting that the wt% of the catalyst (iron (Fe) in this context) should not be more that 10% as larger concentrations might hinder the flow of carbon vapors toward the cathode. Moreover, it was observed by Lin et al. (1994) that the catalysts tend to favor the production of SWCNTs rather than MWCNTs with the most common catalysts being Ni, Fe, and Co; however, other studies have disputed this with the formation of MWNTs in the presence of these catalysts (Jahanshahi et al., 2010; Kim & Kim, 2006; Song et al., 2007; Tsai et al., 2009).

Additionally, it has been shown that the creation of MWCNTs can be accomplished in an open vessel while liquid nitrogen is submerged beneath the surface (Ishigami et al., 2000) without the need to evacuate the reactor. The purpose of nitrogen gas is to completely purge all oxygen to allow for quick reactions. The ease of the experimental setup and the prospect of a straightforward approach to the reaction chamber during the operation made this method highly appealing. However, the excessive evaporation brought on by the arc discharge results in a relatively limited thermal exchange between the environment and the carbon material produced. As a result, the majority of the synthesized MWCNTs exhibit poor structural quality. Furthermore, it has been observed that MWCNTs and polyaromatic carbon shells can be created using arc discharge in deionized water (Lange et al., 2003; Sano et al., 2001; Zhu et al., 2002). In addition to offering a more favorable environment than liquid nitrogen, liquid water also offers the ideal temperature conditions needed to produce MWCNTs of high quality (Liu et al., 2014). This is due to the fact that no significant effects are seen when water and heated carbon react. To obtain highly pure MWCNTs, the generated amorphous carbon is likely simply eliminated by thermal oxidation.

8.3.3 Carbonization/Pyrolysis

8.3.3.1 Conventional Pyrolysis

This form of CNT synthesis takes the principles of pyrolysis to produce the nanostructures, where the temperature is maintained within the ranges of 300°C–600°C (and in some cases >600°C) to foster the thermal decomposition of coal (along with other carbon biomass sources) and the secondary cracking of products. A study by Das et al. (2016) highlighted the procedure where it is reported that a high-sulfur coal (collected from molten alkali leaching) was mixed with KOH and NaOH before undergoing catalytic pyrolysis at temperatures of 350°C. The solid products were then collected and mixed with HCL – at a ratio of 4:1 – before being heated for 15 min at 100°C, from which differing CNSs were obtained. Within this procedure, it was pointed out that the use of molten alkali leaching was hugely beneficial as the method allowed for the reduction of by-products and retardation of the coals binding structure (which ultimately aided in the formation of the nanomaterials).

Other instances report of synthesizing highly graphitic nanostructures – with coconut coir dust as the precursor via pyrolysis combined with hydrothermal carbonization pretreatment (Barin et al., 2014). Moreover, from the study carried out by Thompson et al. (2015), it was found that bamboo-like graphitic nanostructures could be produced from pyrolyzing softwood sawdust at 800°C with Fe as the catalyst, all of which could be carried out in a single step; the authors' research revealed the creation of catalytic Fe_3C particles, which was said to aid in the growth and formation of the graphitic structures. Similar methods were used by Sevilla et al. (2007) in order to create graphitic CNSs from pinewood sawdust. The biomass was then heated for 3 hours in an inert environment at a temperature between 900°C and 1,000°C after being infused with a nitrate solution in ethanol. The authors postulated two catalytic graphitization mechanisms: (1) dissolution into metal particles and subsequent graphitic carbon precipitation; and (2) creation of carbides and their thermal breakdown, which results in the formation of graphitic carbon. In a different study, in order to create GO-CD composites, Sankaranarayanan et al. (2019) used hydrothermal carbonization using deionized water and macroalgal bio-oil as the green carbon sources, whereby an autoclave reactor was used to heat the reaction mixture at 170°C for 4 hours.

8.3.3.2 Microwave Pyrolysis

Similar to the conventional pyrolysis process, microwave pyrolysis utilizes the heating of carbon precursors to produce CNS with it differing in the method of heating – which is instead carried by microwaves. Essentially, the precursor is uniformly heated within a microwave field where CNTs and CNFs are formed from the resultant plasma. This method has proven to be superior to the conventional method due to having better mass transfer of volatiles and a more uniform heating, all of which result in an increased CNT yield. Moreover, this allows for the procurement of CNTs from renewable biomass resources without the need for a catalyst or additional gases.

A study by Shi et al. (2014) reported the use of a mix of gumwood biomass and silicon carbide being used at 20:1 mixing ratio undergoing pyrolysis at temperatures of 500°C, where CNTs discovered to have formed on the biochar product. Another study reports using the same method (albeit at temperatures around 550°C) with palm kernel shells as the precursor to produce hollow carbon nanofibers with a crystalline structure.

8.3.3.3 Mechanical Exfoliation

Mechanical exfoliation is a relatively straightforward method whereby graphene is collected from the exfoliating of graphite. The graphite bulk is peeled layer by layer in which "mechanical energy" is utilized to break apart the Van der Wall forces between the adjacent layers. There are two ways that exfoliation may be carried out, these being via normal or shear force. The normal force route, as the name implies, uses a normal force that is otherwise dubbed as micromechanical cleavage which may simply be carried out by scotch tape; on the other hand, the shear force method, also being as intuitive as it sounds, essentially takes advantage of graphite self-lubricating ability to promote relative motion between the graphite layers (Novoselov et al., 2004). Moreover, despite the fact that a significant amount of GO monolayer could be created by chemically oxidizing graphite and then exfoliating the resultant material, structural defects on the GO were revealed by Raman spectra (Eda et al., 2008; Stankovich et al., 2007), with the high intensity of the D-band for reduced GO serving as a sign of the presence of a significant number of defects.

While these methods are relatively simple and inexpensive, it is also highly time-consuming and labor-intensive, thereby hindering its scalability potential; this drawback limits its utilization to laboratory research alone (Yi & Shen, 2015). Another hindrance of this method is that fragmentation may occur, as shown in Figure 8.6; this is when large graphite pieces break laterally which while making it easier to exfoliate (lower overall Van der Waal forces between the layers) also reduces the surface area of the layers which is undesirable (Figure 8.7).

Nonetheless, there have been other attempts at advancing the scalability of graphene exfoliation, as in one instance, Jayasena et al. (2014) created a lathe-like experimental apparatus to cleave highly orientated pyrolytic graphite (HOPG) samples for producing graphene flakes, where an extremely sharp crystal diamond wedge is held stationary while the work material (graphite embedded in epoxy) is fed onto the wedge resulting in cleaved graphite flakes – the working principle is similar to that of the industrial manufacturing method. Another method that enhances the scalability of this method is termed sonification, which involves dispersing graphite powder within an organic solvent and subsequent sonification and centrifugation; however, this method has been observed to have relatively more defects than the conventional method and may result in inferior graphene due to excessive cavitation (Hernandez et al., 2008; Polyakova et al., 2011).

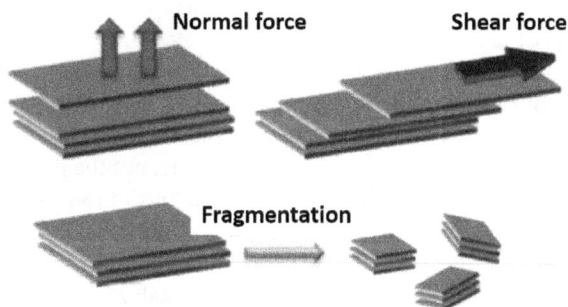

FIGURE 8.6 Depiction of the various manners in which the graphene may split apart (Yi & Shen, 2015).

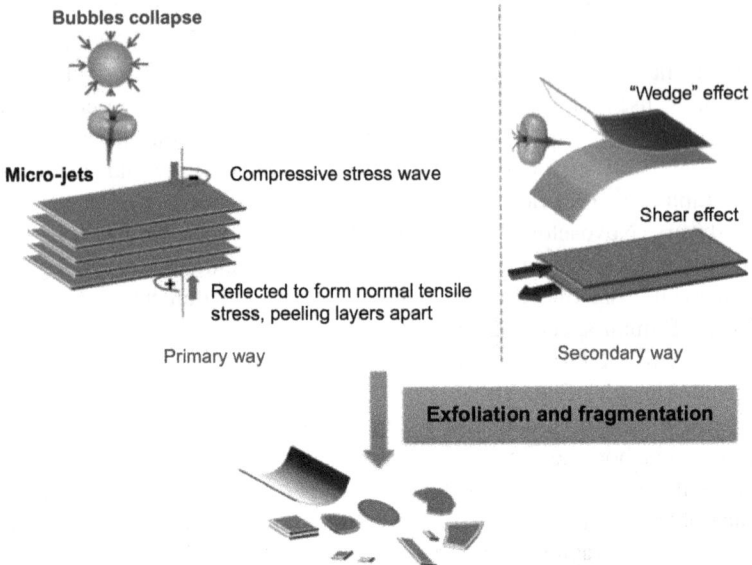

FIGURE 8.7   Mechanism behind the peeling via the sonification method. (Reproduced from ref Yi & Shen, 2015.)

8.3.4 LASER ABLATION

This process involves the vaporization of graphite target enclosed in a quartz glass tube to obtain CNTs. This method of CNT synthesis is similar to that of arc discharge in that they both involve the vaporization and condensation of a graphitic target – with the exception that laser ablation uses a laser beam to initiate its vaporization. This method also requires a controlled intensity of the laser beam, as below the required threshold, the beam simply heats the target, and above it then large chunks of the material are discharged as opposed to a small amount on the outer surface. The laser vaporizes a target that is enclosed within a quartz glass tube that is itself within a furnace with temperatures ranging from 800°C to 1,500°C (Figure 8.8). Moreover, an inert gas typically within the ranges of 500 Torr passes through the tube with the role of carrying the soot formed into a water-cooled Cu collector (Arepalli, 2004; Herrera-Ramirez et al., 2019; Popov, 2004).

As with most CNT synthesis methods, catalysts are used in the production of the nanostructures with Ni and Co being among the most commonly used. The use of catalysts is not necessarily a requirement as some studies report the production of CNTs from pure graphite; however, their presence dictates the production of SWNT and MWNT, as processes containing catalysts are known to produce only SWNTs (Thess et al., 1996). The function of catalysts has been surmised in the "scooter" mechanism, whereby a metal atom with a high electronegativity – which can retard the formation of fullerenes while being effective at catalyzing the nanotube growth – gets chemisorbed onto the open edge of a nanotube where it absorbs small carbon molecules and promotes the formation of a structure similar to that of graphitic

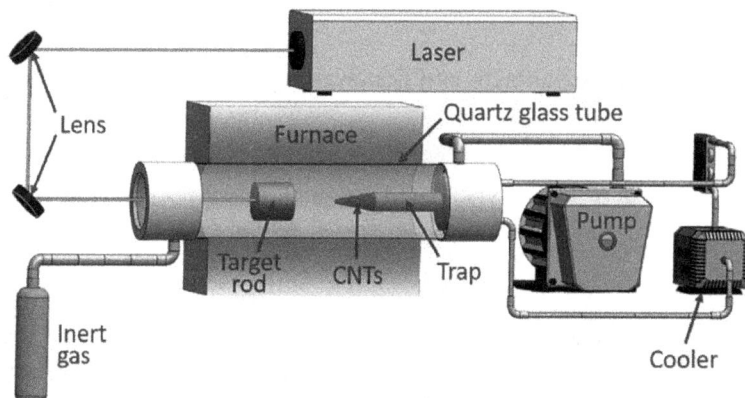

FIGURE 8.8 Laser ablation method where the laser beam is used to vaporize a graphitic target. (Reproduced from ref Herrera-Ramirez et al., 2019.)

sheets. This continues until the particle detaches or becomes over saturated, thus ending the catalysis and determining the tip structure of the nanotube (Popov, 2004).

While this process is known to have high conversion rates with yields in the realm of >70%, it also comes with the drawback of being very expensive to conduct. This is largely due to the amount of cost going into acquiring and powering the equipment with the laser taking up a large chunk of this cost. Another downside is that the CNTs are usually procured from the resultant soot where purification is required to separate the CNTs from the by-product, and this procedure more or less affects the quality of the final product. Table 8.2 shows the summary of comparison of the advantages and disadvantages of each fabrication approach for better understanding.

8.4 PLETHORA APPLICATION OF CNS

A wide range of prospects for the practical uses of CNMs generally dictate by its distinctive physical and chemical features, prompting a surge in their manufacture. Graphene and CNTs have been recorded to have the most extensive area of uses including in composites, coatings and textiles, energy storage as well as in sensors (De Volder et al., 2013). As of 2019, industrial manufacturing of CNTs and graphene has already surpassed several thousand tons where its production costs are significantly low (Taylor et al., 2021). Due to their mechanical properties with higher tensile strength, they are widely used as reinforcing agent in polymers and other materials in order to fabricate advanced composites with improved properties (Ahmad & Pan, 2015; Nurazzi et al., 2021).

From previous recently published review works, it can be summarized that due to that CNTs can be applied in producing strong and ultra-light materials which favors various sectors especially automotive sector (Peddini et al., 2014), aviation (Gohardani et al., 2014), wind (Loh & Ryu, 2014; Ma & Zhang, 2014) and marine turbines (Ng et al., 2013), sports equipment (De Volder et al., 2013; Tan & Zhang,

TABLE 8.2

Comparison of the Advantages and Disadvantages of Each CNS Synthesis Method

Method	Benefits	Drawback	References
CVD	• High yield of CNTs • Numerous sources of raw material • High purity of CNT • Relatively simple and flexible technology • Exceptional scalability potential • Relatively mild conditions required for CNT production	• High production cost, although can be considered relatively low cost in large-scale productions • High defect densities due to low temperature preventing adequate annealing	Pant et al. (2021), Pierson (1999), and Wu et al. (2022)
Arc discharge	• May be optimized for continuous production while providing high-quality MWNCT • High yield of CNTs may be easily produced • Selective (between SWNT and MWNT) by simply varying catalyst	• Numerous impurities within produce • Requirement of pure graphite target limits scalability • Energy extensive especially for large-scale production • Additional purification raises production costs	Mubarak et al. (2014)
Carbonization (pyrolysis)	• Wide variety and easily accessible raw material • Relatively simple process • High scalability potential	• High amount of impurities • Typically, a discontinuous process	Wu et al. (2022)
Exfoliation	• Conventional method is very inexpensive due to absence of machinery	• Low scalability potential • Probability of fragmentation (breaking apart into layers) remains an issue	Yi & Shen (2015)
Laser ablation	• Relatively ease of purification • May also be considered selective with varying catalyst use • Continuous and rapid production of CNTs is possible (positive scalability potential)	• Expensive equipment required for production • Energy extensive • Relatively low yield of CNTs • Graphitic target still limits scalability • Additional costs due to requirement of purification	Mubarak et al. (2014)

2012), and other structure materials. CNTs also can be employed in a variety of electrical applications (De Volder et al., 2013; Park et al., 2013) due to its better electrical conductivity, current carrying capacity, and thermal conductivity. Other than that, graphene and its derivatives offer a wide range of applications, including electronics, biochemical sensors, in solar cells as well as agriculture. Fullerene has numerous uses as well, as it was widely used in medicine (Shi et al., 2014; Tong et al., 2011), including gene and drug delivery (Uritu et al., 2015), as well as in cosmetics (Bergeson & Cole, 2012; Kato et al., 2010). Thus, in this section, various applications and possible uses of CNMs produced from bio-oil and other bio-sources will be summarized.

8.4.1 Electronic Devices

The significantly delocalized electronic structure of sp^2 hybridized CNMs implies that they are suitable for use as high-mobility electronic materials. The wide range of electronic properties of CNMs as a function of their chiral vector, along with their quasi-one-dimensional structure, opens up a number of intriguing possibilities for electronic applications.

8.4.1.1 Semiconductor

The growing demand of mobile electronics has encouraged researchers and industrial players to explore semiconductor materials that can be utilized in large-area, flexible macro electronics. Thus, according to Kumar et al. (2005), CNT-based semiconductors have demonstrated equivalent or greater field-effect mobility than the majority of commercialized organic and inorganic semiconductors in electronic application (Forrest, 2004). Furthermore, CNT thin films are chemically inert in the ambient environment and also have appealing mechanical and optical characteristics, making them ideal for flexible and transparent electronic devices (Rouhi et al., 2011). The incorporation of CNT thin films with flexible substrates, on the other hand, provides distinct production and processing obstacles that must be solved.

8.4.1.2 Lighting Application

Several researches have recently employed CNT emitters in lighting applications. Field emission flat-panel luminescent lamp incorporating CNT emitters might be viable option for this application (Bonard et al., 2001; Chen et al., 2003). Flat fluorescent light is commonly employed as a backlight in liquid crystal displays, and has the benefit of being mercury-free, controllable, able to necessitate a very significant brightness intensity of about 5,000 cd/m^2 with a long cycle life and good luminance uniformity. Bonard et al. (2001), for instance, effectively coated MWCNT on metallic wire and constructed a field emission luminescent tube with a brightness intensity of 10,000 cd/m^2. To provide greener solution, Suriani et al. (2010) employed leftover cooking palm oil as a raw material to fabricate VACNT as shown in Figure 8.9 which exhibited promising field electron emission potentials, which yield field enhancement, β, of 2,740, turn on electric field of 2.25 V/μm, threshold electric field of 3.00 V/μm, and maximum current densities of 6 mA/cm^2, as field emitter applications. There are also various recently published researches which had been done by other research groups (Liu et al., 2021; Park & Lim, 2022; Yoo & Park, 2022; Youh et al., 2020) to show the ability of CNTs in this field for further applications.

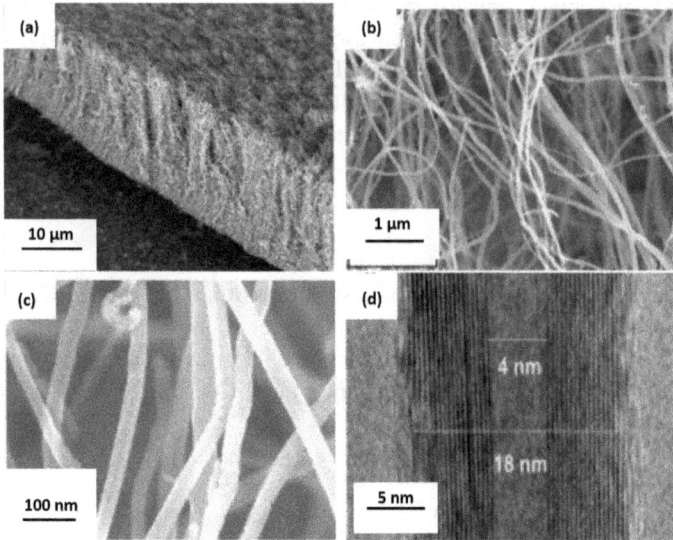

FIGURE 8.9 (a)–(c) FESEM images of VACNTs derived from waste chicken fat with different focus magnification, and (d) HRTEM image of MWCNT produced. Reproduced from ref Suriani et al., 2010.)

8.4.2 SENSOR APPLICATION

In order to facilitate production, CNT thin films and vertically grown CNT arrays have been studied for chemical-sensing applications. Recent studies employed as-grown heterogeneous CNT nanocomposites as chemiresistive sensors for detecting various volatile organic compounds (Kennedy et al., 2017; Sinha et al., 2021; Yoon et al., 2021), ammonia (NH_3) (Hamouma et al., 2019; Kim et al., 2022; Zhang et al., 2019), and dimethyl methyl phosphonate (Nurazzi et al., 2021; Tang et al., 2017). Selectivity difficulties were resolved to some extent by functionalizing these CNTs with both covalent (Bekyarova et al., 2004) and non-covalent (Novak et al., 2003) functionalization. Because each device efficiently averages the effects of the thin film's huge number of CNTs, device-to-device repeatability with great sensitivity may be accomplished over broad regions. Moreover, Alancherry (2018) studied the feasibility of CNS for gas-sensing applications by employing a plasma-assisted CVD process to manufacture CNSs from *Citrus sinensis* essential oil. *C. sinensis* essential oil, obtained by cold-extracting orange peels, is a rich source of naturally produced hydrocarbons, especially limonene. To examine the effectiveness of acetone sensing, *C. sinensis* oil was transformed into vertically oriented graphene nanostructures and combined into a sensor device.

Graphene has additionally been researched for employment in biosensors. Following in the footsteps of CNT biosensors, graphene biosensors have been employed for label-free protein detection (Ohno et al., 2009). While the earlier studies lacked sensing selectivity, recent investigations utilizing non-covalently functionalized graphene have demonstrated selective detection of biomolecules. A Pt nanoparticle-modified r-GO FET, for example, was utilized to detect DNA precisely (Huang et al., 2010), while r-GO functionalized with an Au nanoparticle–antibody conjugate was implemented to sense a protein exclusive to the antibody. Non-covalently functionalized CVD graphene may also detect glucose and glutamate in solution with good sensitivity and specificity (Huang et al., 2010).

8.4.3 Agriculture

CNMs have sparked considerable attention in agricultural applications. However, the current research shows that CNM exposure has a varied effect on plants, ranging from increased crop yield to acute cytotoxicity and genetic change. These apparent inconsistencies in study findings pose substantial barriers to the widespread application of CNMs in agriculture. Despite several challenges faced, the research and development regarding this field is still growing (Mukherjee et al., 2016; Zaytseva & Neumann, 2016) especially in several sectors–seed germination sector, target delivery of agrochemicals and fungicides.

8.4.3.1 Seed Germination

Seed germination is an intriguing field of study that plays an important function in plant physiology and facilitates quick agricultural plant production. Many nanomaterials, including carbon compounds, have been successfully used in the germination of seeds from diverse plant sources (Khodakovskaya et al., 2009). Carbon dots (CDs) have also been shown to be an active component in seed germination/plant growth experiments due to their better physiochemical capabilities (Chakravarty et al., 2015; Tripathi & Sarkar, 2015). Because of their potential toxicity, the use of non-renewable-source-derived CDs for seed germination may be restricted.

The use of CDs which is currently generated from non-renewable sources for seed germination has been hindered due to their possible toxicity. Thus, the generation of CDs from renewable resources for efficient seed germination is a greener method that can help with the maintenance of the ecosystem. Via hydrothermal treatment of bio-oil derived from brown macroalgal biomass (*Macrocystis pyrifera*) cultivated in a floating aquaculture system, a novel attempt was made to produce GO-CD composite. It was discovered that a low concentration of the composite enhanced plant growth as it was tested for mung bean seed germination. Sankaranarayanan et al. (2019) employed a hydrothermal technique to develop a GO-CD composite that emits green light that used brown macroalgal biomass as the carbon source. The manufacturing processes for the GO-CD are depicted in Figure 8.10, and the seed germination efficiency is then examined.

8.4.3.2 Agrochemical Delivery System

The advancement of smart agrochemical carrier system, a novel technology for agrochemical target delivery, has various significant benefits. Encapsulated agrochemicals display enhanced resilience and resistance to deterioration, with the goal of reducing the amount of agrochemicals used and increasing their performance (González-Melendi et al., 2008). According to Sarlak et al. (2014), fungicides encased in MWCNTs functionalized with citric acid were more harmful to the *Alternaria alternata* fungus than the bulk pesticide that was not even encapsulated.

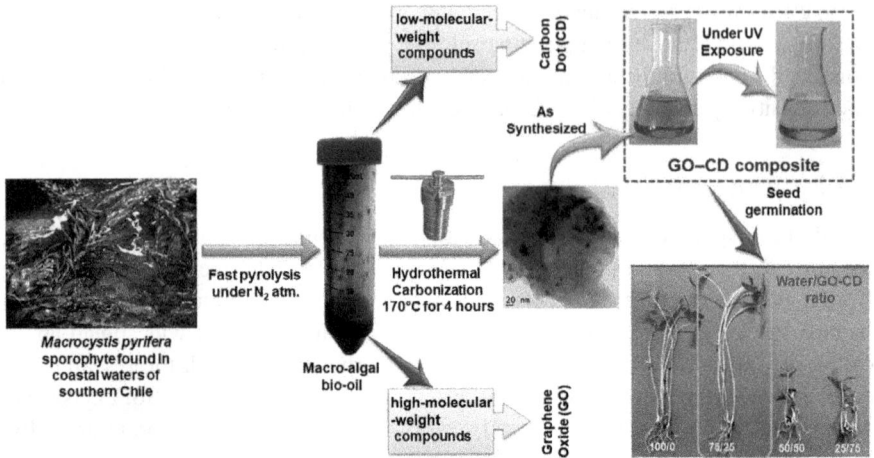

FIGURE 8.10 GO-CD production from *Macrocystis pyrifera*-derived bio-oil and its investigation for mung bean seed germination. (Reproduced from ref Sankaranarayanan et al., 2019.)

SWCNTs and ceria nanoparticles, according to a new study, may be transported into isolated chloroplasts. By adding electrons to the photosynthetic electron transport chain and gradually diffusing into the chloroplast membrane, these nanomaterials were able to alter photosynthetic activity (Giraldo et al., 2014). In addition to their use in agriculture, CNTs are being investigated for their potential as therapeutic molecular transporters in animal cells (Anzar et al., 2020; Das et al., 2013; Khan et al., 2013).

8.4.3.3   Fungicides and Herbicides

Carbon-based nanostructures are interesting candidates for the synthesis of new fungicides owing to their anti-fungal characteristics. From previous literature, it is recorded that SWCNTs outperformed other CNMs, studied against two plant pathogenic fungi, *Fusarium graminearum* and *Fusarium poae*. Fullerenes and AC, on the other hand, were generally ineffective for fungicide application. According to Wang et al. (2014), interaction between the nanoparticles and the fungal spores leads to plasmolysis, which is associated with decreased water content and growth inhibition, and appears to be essential for antifungal effectiveness. Other studies have connected the anti-microbial activity of GO to the creation of microbial membrane damage, disturbance of the membrane potential (Chen et al., 2014), electron transport (Liu et al., 2012), and oxidative stress via increased reactive oxygen species production (Hui et al., 2014). The above-discussed CNMs with antifungal and antimicrobial capabilities attracted a lot of attention since they might be used as cutting-edge fungicides and disinfectants that are suited for agricultural uses. It is challenging to foresee, however, how the mentioned in-vitro features would emerge when the CNMs or CNM-containing products are discharged into the environment due to the mostly unknown behavior of CNMs in complex environmental matrices.

8.4.4 Wastewater Treatment/Metal Adsorbent

Due to its vast surface area and capacity to adsorb a variety of organic and inorganic pollutants, activated carbon (AC) has been widely employed as a sorbent for traditional wastewater treatment. However, AC has a sluggish rate of adsorption, is a non-specific adsorbent, and has a poor level of efficiency against microbes. Pore obstruction in AC is frequently brought on by the presence of particles, oil, and grease in wastewater. Furthermore, AC must be changed on a regular basis since it is regularly removed along with the impurities it has adsorbed. With multiple instances in the literature, the use of CNMs in this context offers a great potential to enhance wastewater filtering systems (Das et al., 2014; Liu et al., 2013; Qu et al., 2013; Smith & Rodrigues, 2015).

It has been reported that the adsorption capacity of CNTs toward microcystins (cyanobacterial toxins) (Yan et al., 2006), lead (Li et al., 2002), and copper (III) (Dichiara et al., 2015) was even stronger than that of AC. Multi-walled nanotubes have been also used for sorption of antibiotics (Zhang et al., 2011), herbicides (Deng et al., 2012), or nitrogen and phosphorus in wastewater (Zheng et al., 2014). On the other hand, fullerenes as well as CNTs exhibit a mobilization potential for various organic pollutants, such as lindane (agricultural insecticide) (Srivastava et al., 2011) and persistent polychlorinated biphenyls (Wang et al., 2013). The primary advantages of CNMs are their enormous surface area, mechanical and thermal robustness, significant chemical affinity for aromatic compounds, and potential antibacterial properties (Yoo et al., 2011). Additionally, as pollutants may be desorbed from CNMs, filters made of CNMs are recyclable and greener (Choi et al., 2010; Wang et al., 2014). Tripathi et al. (2012) synthesized carbon nanospheres in a thermal catalytic vapor reactor using natural, non-edible castor oil as the carbon source and three distinct transition metals (Ni, Zn, and Co) as catalyst. Arsenic in water may be removed using carbon nanospheres, according to research. Li et al. (2004) studied the heavy metal adsorption by CNT. They found that CNTs are extremely effective in removing lead from water due to their exceptional potential for adsorption. Adsorption is greatly influenced by the pH of the solution and the state of the CNT's surface.

Commercial wastewater cleanup technology applications are still limited by issues with CNM's high production costs, the difficulty of finding CNTs with a uniform size and diameter distribution, uncertainty about the CNTs' potential for leaching, environmental safety issues, and issues to health.

8.4.5 Mechanical Applications

Due to the great variety of their morphologies, outstanding corrosion resistance, excellent mechanical behaviors, and high thermal conductivity, carbon materials have been the subject of extensive research in the fields of optics, electrochemistry, mechanics, and tribology for many years. CNTs and graphene are the allotropes of carbons with sp^2 hybridization that have drawn the most attention due to their exceptional charge carrier mobility (Avouris et al., 2007; Mittendorff et al., 2014; Norizan et al., 2020), electrochemical stability (Raccichini et al., 2015), mechanical stretchability/flexibility (Harussani et al., 2022; Lipomi et al., 2011; Park et al., 2013;

Rogers et al., 2010; Won et al., 2016), and so on. The widespread usage of CNTs and graphene in electronics, chemical separation, catalysis support, and super lubricity is a result of their exceptional features.

8.4.5.1 Composites in Automotive, Aerospace, and Marine Industries

According to a study by Lourie and Wagner (1998), one can see that the mechanical characteristics of CNTs rely on the sp^2 strength of the nanotubes' C–C bonds, making them suitable as reinforcement fibers for matrixes. It is interesting to note that these kinds of bonding are even more powerful than the sp^3 diamond bonds. Due to its electronegativity feature, the carbon atom in a nanotube creates a planar honeycomb lattice that is excellent at creating covalent connections with other elements. The assessment of how tightly an atom hangs onto the electrons in its orbit depends on this electronegativity feature.

The kind of bonding at the interface, the strength of the interface, and the mechanical load transmission from the surrounding matrix to the nanotubes all have a significant impact on the characteristics of CNT/polymer composites. The weak Van der Waals force between the polymer matrix and the CNT reinforcement is one category of the mechanism of interfacial load transfer from the matrix to nanotubes, according to the literature (Nurazzi et al., 2021). Better physical and mechanical characteristics of CNT composites need an efficient transmission of load stress from the matrix to the CNTs. The dispersion of CNTs in the polymer matrix by physical and chemical modifications has a major impact on the performance of CNT/polymer composites.

Because of their extreme lightweight and strong combination performance, CNTs have the potential to be used in highly structural applications for aeronautics, automotive parts, and marine ships. There are a lot of highly cited publications regarding this sectors which thoroughly studies the influence of CNT and its derivatives toward tensile, flexural, and impact strength of the reinforced polymer composites (Abazari et al., 2020; Harussani et al., 2022; Moghadam et al., 2015; Nguyen-Tran et al., 2018; Nurazzi et al., 2021).

8.4.5.2 Anti-Wear and Low-Friction Coatings

Tribological and mechanical investigations of graphene at the nanoscale reveal that friction rises as the number of atomic layers reduces (Balog et al., 2010; Wu et al., 2008) due to the strong electron–phonon interaction in single-layer epitaxial graphene (Kwon et al., 2012) as well as the puckering effect (Choi et al., 2011). The friction of graphene at the nanoscale may also be influenced by the sp3 functionalization of the surface. The potential for bigger band gaps and less Van der Waals interactions, which reduce adhesion forces and the number of free electrons, are the causes of this (Balog et al., 2010; Elias et al., 2009). It has been established that the outstanding monolayer-level protective stability and wear resistance of graphene as a covering originated there. The macroscopic frictional behavior and wear resistance of one-atom-thick graphene layer, which was created by chemically exfoliating HOPG, were initially described by Berman et al. (2014). Moreover, numerous studies have been conducted over the years to further enhance the anti-wear properties using various arrangements, compositions, functionalization, and modifications to the graphene structure (Gómez-Gómez et al., 2016; Kwon et al., 2012; Wang et al., 2015; Ye et al., 2016) which have demonstrated very promising wear and friction behavior.

A multistep self-assembly technique was created by Pu et al. (2014) to manufacture the hybrid coating with the fullerene C60 on silicon surfaces. On the surface of the amine-functionalized graphene, which was attached to the silicon surface via the nucleophilic addition procedure, the pure C60 was evenly chemisorbed (Figure 8.11). The fullerene C60's outer layer promoted microsphere contact during sliding, which would have reduced the contact area as a result of the cobblestone effect. In this research, fullerene C60 and amino-functionalized graphene were mixed to create a coating that included C60 and had benefits that went beyond their separate capabilities in terms of friction reduction and anti-wear improvement.

According to studies on the MWCNTs-reinforced coating, the introduction of MWCNTs may be a valuable strategy to increase the coating's durability and wear resistance due to their ability to inhibit fracture propagation (Li et al., 2007). The tribological behavior of an Al_2O_3 coating reinforced with MWCNTs that had been plasma-sprayed and had a thickness of around 400 m was examined by Keshri et al. (2010). This coating showed a falling trend as the testing temperature increased. The increase in wear resistance may be ascribed to the composite coating's continued high hardness, the protective layer's broad area of coverage over worn tracks at the elevated temperature, the bridging effect of MWCNTs between splats, and other factors. Umeda et al. (2015) synthesized MWCNT coatings with network topologies on a Ti substrate surface. By coating the substrate surface with MWCNT, the friction coefficient was lowered by about 80%. The MWCNTs' self-lubricating and nano-bearing characteristics allowed for the establishment of moderate sliding conditions due to the coatings on the surfaces during contact. Numerous additional works have also looked into effective lubrication coatings, including electroless MWCNTs/SiC coatings, in a manner similar to these researches on the CNT-containing composite

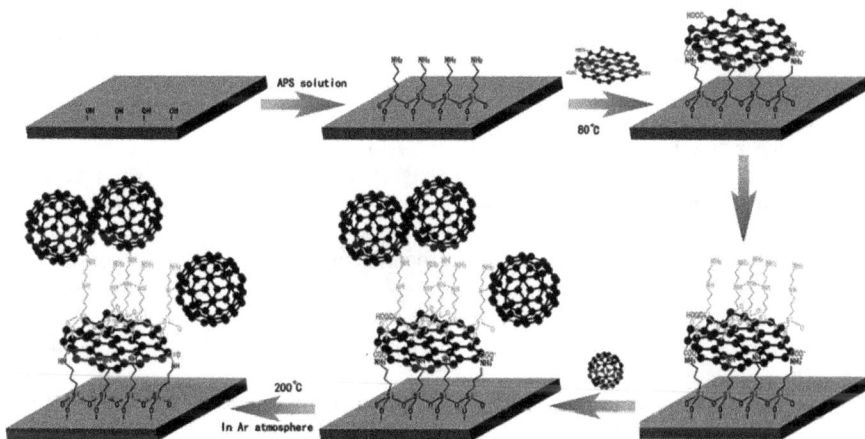

FIGURE 8.11 The schematic representation of the manufacturing methods for the hybrid coating that contains fullerene C60. (Reproduced from ref Pu et al., 2014.)

coatings (Li et al., 2007), MWCNTs/Cu composite coatings (Arai & Kanazawa, 2014), MWCNTs/epoxy coatings (Espósito et al., 2014), MWCNTs/polyurethane nanocomposite coatings (Song et al., 2011), dual-layer MWCNTs/Ag coatings (Lee et al., 2013), and MWCNTs/polyethylene oxide composite coatings (Ryu et al., 2014).

The nanodiamonds, unlike fullerene, CNTs, and graphene, include both sp^2 and sp^3 carbon atoms, with the rebuilding of sp^2 being crucial to the stability of the material (Mochalin et al., 2020). Currently, biocompatible and wear-resistant coatings using nanodiamonds have been investigated (Bao et al., 2012; Koizumi, 2011). They exhibit the desirable high hardness, surface areas, and customizable surface structures that allow for their special applications in cutting-edge fields including biomedicine and cosmetics. Similar to the CNTs, the nanodiamonds have the potential to be used on a genuinely large scale (Osswald et al., 2006).

8.5 CONCLUSIONS

It is carefully considered if renewable resources like plant shoots or essential oils may substitute petroleum-based feeds. Graphene, graphite-like structures, nanotubes, and closed shell nanoparticles are only a few examples of the CNSs that may be produced from bio-oil through reaction. The research demonstrates that because these resources are complicated, it is necessary to optimize the reaction conditions in order to produce products with the correct microstructure and chemical composition. A variety of proven high-performance applications for even high-purity CNS may be synthesized at minimal cost, nevertheless, with the right adjustment of the process parameters. The sheer volume of successful experiments that have been conducted on this front thus far and are detailed herein verifies that it is feasible to create strategies for the synthesis of such high-value products from common precursors.

REFERENCES

Abazari, S., Shamsipur, A., Bakhsheshi-Rad, H. R., Ismail, A. F., Sharif, S., Razzaghi, M., Ramakrishna, S., & Berto, F. (2020). Carbon nanotubes (CNTs)-reinforced magnesium-based matrix composites: A comprehensive review. *Materials*, *13*(19), 4421.

Abdul Latif, S. N., Chiong, M. S., Rajoo, S., Takada, A., Chun, Y.-Y., Tahara, K., & Ikegami, Y. (2021). The trend and status of energy resources and greenhouse gas emissions in the Malaysia power generation mix. *Energies*, *14*(8), 2200.

Ahmad, K., & Pan, W. (2015). Microstructure-toughening relation in alumina based multiwall carbon nanotube ceramic composites. *Journal of the European Ceramic Society*, *35*(2), 663–671.

Alancherry, S. (2018). Development of carbon nanostructures from non-conventional resources. (*Doctoral dissertation, James Cook University*).

Anzar, N., Hasan, R., Tyagi, M., Yadav, N., & Narang, J. (2020). Carbon nanotube-A review on synthesis, properties and plethora of applications in the field of biomedical science. *Sensors International*, *1*, 100003.

Arai, S., & Kanazawa, T. (2014). Electroless deposition and evaluation of Cu/multiwalled carbon nanotube composite films on acrylonitrile butadiene styrene resin. *Surface and Coatings Technology*, *254*, 224–229.

Arepalli, S. (2004). Laser ablation process for single-walled carbon nanotube production. *Journal of Nanoscience and Nanotechnology*, *4*(4), 317–325.

Arora, N., & Sharma, N. (2014). Arc discharge synthesis of carbon nanotubes: Comprehensive review. *Diamond and Related Materials*, *50*, 135–150.

Avouris, P., Chen, Z., & Perebeinos, V. (2007). Carbon-based electronics. *Nature Nanotechnology*, *2*(10), 605–615.

Azmina, M. S., Suriani, A. B., Salina, M., Azira, A. A., Dalila, A. R., Asli, N. A., Rosly, J., Nor, R. Md., & Rusop, M. (2012). Variety of bio-hydrocarbon precursors for the synthesis of carbon nanotubes. *Nano Hybrids*, *2*, 43–63. https://doi.org/10.4028/www.scientific.net/NH.2.43

Baker, R., & Waite, R. (1975). Formation of carbonaceous deposits from the platinum-iron catalyzed decomposition of acetylene. *Journal of Catalysis*, *37*(1), 101–105.

Ballotin, F. C., Perdigao, L. T., Rezende, M. V. B., Pandey, S. D., da Silva, M. J., Soares, R. R., Freitas, J. C., de Carvalho Teixeira, A. P., & Lago, R. M. (2019). Bio-oil: A versatile precursor to produce carbon nanostructures in liquid phase under mild conditions. *New Journal of Chemistry*, *43*(6), 2430–2433.

Balog, R., Jørgensen, B., Nilsson, L., Andersen, M., Rienks, E., Bianchi, M., Fanetti, M., Lægsgaard, E., Baraldi, A., & Lizzit, S. (2010). Bandgap opening in graphene induced by patterned hydrogen adsorption. *Nature Materials*, *9*(4), 315–319.

Bao, M., Zhang, C., Lahiri, D., & Agarwal, A. (2012). The tribological behavior of plasma-sprayed Al-Si composite coatings reinforced with nanodiamond. *Jom*, *64*(6), 702–708.

Barin, G. B., de Fátima Gimenez, I., da Costa, L. P., Souza Filho, A. G., & Barreto, L. S. (2014). Influence of hydrothermal carbonization on formation of curved graphite structures obtained from a lignocellulosic precursor. *Carbon*, *78*, 609–612.

Bekyarova, E., Davis, M., Burch, T., Itkis, M., Zhao, B., Sunshine, S., & Haddon, R. (2004). Chemically functionalized single-walled carbon nanotubes as ammonia sensors. *The Journal of Physical Chemistry B*, *108*(51), 19717–19720.

Bergeson, L. L., & Cole, M. F. (2012). Fullerenes used in skin creams. *Nanotechnology Law & Business*, *9*, 114.

Berman, D., Deshmukh, S. A., Sankaranarayanan, S. K., Erdemir, A., & Sumant, A. V. (2014). Extraordinary macroscale wear resistance of one atom thick graphene layer. *Advanced Functional Materials*, *24*(42), 6640–6646.

Bonard, J.-M., Stöckli, T., Noury, O., & Châtelain, A. (2001). Field emission from cylindrical carbon nanotube cathodes: Possibilities for luminescent tubes. *Applied Physics Letters*, *78*(18), 2775–2777.

Chakravarty, D., Erande, M. B., & Late, D. J. (2015). Graphene quantum dots as enhanced plant growth regulators: Effects on coriander and garlic plants. *Journal of the Science of Food and Agriculture*, *95*(13), 2772–2778.

Chen, J., Peng, H., Wang, X., Shao, F., Yuan, Z., & Han, H. (2014). Graphene oxide exhibits broad-spectrum antimicrobial activity against bacterial phytopathogens and fungal conidia by intertwining and membrane perturbation. *Nanoscale*, *6*(3), 1879–1889.

Chen, J., Zhou, X., Deng, S., & Xu, N. (2003). The application of carbon nanotubes in high-efficiency low power consumption field-emission luminescent tube. *Ultramicroscopy*, *95*, 153–156.

Choi, J. S., Kim, J.-S., Byun, I.-S., Lee, D. H., Lee, M. J., Park, B. H., Lee, C., Yoon, D., Cheong, H., & Lee, K. H. (2011). Friction anisotropy-driven domain imaging on exfoliated monolayer graphene. *Science*, *333*(6042), 607–610.

Choi, W. S., Yang, H. M., Koo, H. Y., Lee, H., Lee, Y. B., Bae, T. S., & Jeon, I. C. (2010). Smart microcapsules encapsulating reconfigurable carbon nanotube cores. *Advanced Functional Materials*, *20*(5), 820–825.

Chung, D. D. L. (2002). Review Graphite. *Journal of Materials Science*, *37*(8), 1475–1489. https://doi.org/10.1023/A:1014915307738.

Dai, H. (2001). Nanotube growth and characterization. In *Carbon Nanotubes* (pp. 29–53). Springer.

Das, M., Singh, R. P., Datir, S. R., & Jain, S. (2013). Intranuclear drug delivery and effective in vivo cancer therapy via Estradiol-PEG-appended multiwalled carbon nanotubes. *Molecular Pharmaceutics*, *10*(9), 3404–3416.

Das, R., Ali, M. E., Abd Hamid, S. B., Ramakrishna, S., & Chowdhury, Z. Z. (2014). Carbon nanotube membranes for water purification: A bright future in water desalination. *Desalination*, *336*, 97–109.

Das, T., Saikia, B. K., & Baruah, B. P. (2016). Formation of carbon nano-balls and carbon nano-tubes from northeast Indian Tertiary coal: Value added products from low grade coal. *Gondwana Research*, *31*, 295–304.

De Volder, M. F., Tawfick, S. H., Baughman, R. H., & Hart, A. J. (2013). Carbon nanotubes: Present and future commercial applications. *Science*, *339*(6119), 535–539.

Deng, J., Shao, Y., Gao, N., Deng, Y., Tan, C., Zhou, S., & Hu, X. (2012). Multiwalled carbon nanotubes as adsorbents for removal of herbicide diuron from aqueous solution. *Chemical Engineering Journal*, *193*, 339–347.

Dichiara, A. B., Webber, M. R., Gorman, W. R., & Rogers, R. E. (2015). Removal of copper ions from aqueous solutions via adsorption on carbon nanocomposites. *ACS Applied Materials & Interfaces*, *7*(28), 15674–15680.

Eda, G., Fanchini, G., & Chhowalla, M. (2008). Large-area ultrathin films of reduced graphene oxide as a transparent and flexible electronic material. *Nature Nanotechnology*, *3*(5), 270–274.

Elias, D. C., Nair, R. R., Mohiuddin, T., Morozov, S., Blake, P., Halsall, M., Ferrari, A. C., Boukhvalov, D., Katsnelson, M., & Geim, A. (2009). Control of graphene's properties by reversible hydrogenation: Evidence for graphane. *Science*, *323*(5914), 610–613.

Ellabban, O., Abu-Rub, H., & Blaabjerg, F. (2014). Renewable energy resources: Current status, future prospects and their enabling technology. *Renewable and Sustainable Energy Reviews*, *39*, 748–764. https://doi.org/10.1016/j.rser.2014.07.113.

Espósito, L. H., Ramos, J., & Kortaberria, G. (2014). Dispersion of carbon nanotubes in nanostructured epoxy systems for coating application. *Progress in Organic Coatings*, *77*(9), 1452–1458.

Forrest, S. R. (2004). The path to ubiquitous and low-cost organic electronic appliances on plastic. *Nature*, *428*(6986), 911–918.

Georgakilas, V., Perman, J. A., Tucek, J., & Zboril, R. (2015). Broad family of carbon nanoallotropes: Classification, chemistry, and applications of fullerenes, carbon dots, nanotubes, graphene, nanodiamonds, and combined superstructures. *Chemical Reviews*, *115*(11), 4744–4822. https://doi.org/10.1021/cr500304f.

Ghosh, P., Afre, R. A., Soga, T., & Jimbo, T. (2007). A simple method of producing single-walled carbon nanotubes from a natural precursor: Eucalyptus oil. *Materials Letters*, *61*(17), 3768–3770.

Giraldo, J. P., Landry, M. P., Faltermeier, S. M., McNicholas, T. P., Iverson, N. M., Boghossian, A. A., Reuel, N. F., Hilmer, A. J., Sen, F., & Brew, J. A. (2014). Plant nanobionics approach to augment photosynthesis and biochemical sensing. *Nature Materials*, *13*(4), 400–408.

Giubileo, F., Di Bartolomeo, A., Iemmo, L., Luongo, G., & Urban, F. (2018). Field Emission from Carbon Nanostructures. *Applied Sciences*, *8*(4), 526. https://doi.org/10.3390/app8040526.

Gohardani, O., Elola, M. C., & Elizetxea, C. (2014). Potential and prospective implementation of carbon nanotubes on next generation aircraft and space vehicles: A review of current and expected applications in aerospace sciences. *Progress in Aerospace Sciences*, *70*, 42–68.

Gómez-Gómez, A., Nistal, A., García, E., Osendi, M. I., Belmonte, M., & Miranzo, P. (2016). The decisive role played by graphene nanoplatelets on improving the tribological performance of Y2O3-Al2O3-SiO2 glass coatings. *Materials & Design*, *112*, 449–455.

González-Melendi, P., Fernández-Pacheco, R., Coronado, M. J., Corredor, E., Testillano, P., Risueño, M. C., Marquina, C., Ibarra, M. R., Rubiales, D., & Pérez-de-Luque, A. (2008). Nanoparticles as smart treatment-delivery systems in plants: Assessment of different techniques of microscopy for their visualization in plant tissues. *Annals of Botany, 101*(1), 187–195.

Goswami, A. D., Trivedi, D. H., Jadhav, N. L., & Pinjari, D. V. (2021). Sustainable and green synthesis of carbon nanomaterials: A review. *Journal of Environmental Chemical Engineering, 9*(5), 106118. https://doi.org/10.1016/j.jece.2021.106118.

Hamouma, O., Kaur, N., Oukil, D., Mahajan, A., & Chehimi, M. M. (2019). Paper strips coated with polypyrrole-wrapped carbon nanotube composites for chemi-resistive gas sensing. *Synthetic Metals, 258*, 116223.

Harris, P. J. (2004). Carbon nanotubes and related structures: New materials for the twenty-first century. *American Journal of Physics, 72*, 415. https://doi.org/10.1119/1.1645289.

Harussani, M., Sapuan, S., Nadeem, G., Rafin, T., & Kirubaanand, W. (2022). Recent applications of carbon-based composites in defence industry: A review. *Defence Technology, 18*(8), 1281–1230.

Hernandez, Y., Nicolosi, V., Lotya, M., Blighe, F. M., Sun, Z., De, S., McGovern, I. T., Holland, B., Byrne, M., & Gun'Ko, Y. K. (2008). High-yield production of graphene by liquid-phase exfoliation of graphite. *Nature Nanotechnology, 3*(9), 563–568.

Herrera-Ramirez, J. M., Perez-Bustamante, R., & Aguilar-Elguezabal, A. (2019). An overview of the synthesis, characterization, and applications of carbon nanotubes. *Carbon-Based Nanofillers and Their Rubber Nanocomposites*, 47–75.

Huang, Y., Dong, X., Shi, Y., Li, C. M., Li, L.-J., & Chen, P. (2010). Nanoelectronic biosensors based on CVD grown graphene. *Nanoscale, 2*(8), 1485–1488.

Hui, L., Piao, J.-G., Auletta, J., Hu, K., Zhu, Y., Meyer, T., Liu, H., & Yang, L. (2014). Availability of the basal planes of graphene oxide determines whether it is antibacterial. *ACS Applied Materials & Interfaces, 6*(15), 13183–13190.

Iijima, S. (1980). High resolution electron microscopy of some carbonaceous materials. *Journal of Microscopy, 119*(1), 99–111.

Iijima, S. (1991). Helical microtubules of graphitic carbon. *Nature, 354*(6348), 56–58.

Ishigami, M., Cumings, J., Zettl, A., & Chen, S. (2000). A simple method for the continuous production of carbon nanotubes. *Chemical Physics Letters, 319*(5–6), 457–459.

Jahanshahi, M., Shariaty-Niassar, M., Rostami, A., Molavi, H., & Toubi, F. (2010). Arc-discharge carbon nanotube fabrication in solution: Electrochemistry and voltammetric tests. *Australian Journal of Basic and Applied Sciences, 4*(12), 5915–5922.

Janas, D. (2020). From bio to nano: A review of sustainable methods of synthesis of carbon nanotubes. *Multidisciplinary Digital Publishing Institute, 12*, 4115.

Jayasena, B., Subbiah, S., & Reddy, C. (2014). Formation of carbon nanoscrolls during wedge-based mechanical exfoliation of HOPG. *Journal of Micro and Nano-Manufacturing, 2*(1), 011003.

Kato, S., Taira, H., Aoshima, H., Saitoh, Y., & Miwa, N. (2010). Clinical evaluation of fullerene-C60 dissolved in squalane for anti-wrinkle cosmetics. *Journal of Nanoscience and Nanotechnology, 10*(10), 6769–6774.

Kennedy, Z., Christ, J., Evans, K., Arey, B., Sweet, L., Warner, M., Erikson, R., & Barrett, C. (2017). 3D-printed poly (vinylidene fluoride)/carbon nanotube composites as a tunable, low-cost chemical vapour sensing platform. *Nanoscale, 9*(17), 5458–5466.

Keshri, A. K., Singh, V., Huang, J., Seal, S., Choi, W., & Agarwal, A. (2010). Intermediate temperature tribological behavior of carbon nanotube reinforced plasma sprayed aluminum oxide coating. *Surface and Coatings Technology, 204*(11), 1847–1855.

Khan, A. J., Khan, R. A., Singh, V. M., Newati, S. J., & Yusuf, M. (2013). In silico designed, self-assembled, functionalized single-walled carbon nanotubes and, deoxyribose nucleic acids (f-SW-CNT-DNA) bioconjugate as probable biomolecular transporters. *Journal of Bionanoscience, 7*(5), 530–550.

Khodakovskaya, M., Dervishi, E., Mahmood, M., Xu, Y., Li, Z., Watanabe, F., & Biris, A. S. (2009). Carbon nanotubes are able to penetrate plant seed coat and dramatically affect seed germination and plant growth. *ACS Nano*, *3*(10), 3221–3227.

Kim, H. H., & Kim, H. J. (2006). The preparation of carbon nanotubes by dc arc discharge using a carbon cathode coated with catalyst. *Materials Science and Engineering: B*, *130*(1–3), 73–80.

Kim, Y. G., Oh, B. M., Kim, H., Lee, E. H., Lee, D. H., Kim, J. H., & Koo, B. (2022). Trifluoromethyl ketone P3HT-CNT composites for chemiresistive amine sensors with improved sensitivity. *Sensors and Actuators B: Chemical*, 132076.

Koizumi, M. (2011). Nanodiamond composite plating layers and their tribological properties. *Journal of Japanese Society of Tribologists*, *56*(10), 621–626.

Kroto, H. W., Heath, J. R., O'Brien, S. C., Curl, R. F., & Smalley, R. E. (1985). C60: Buckminsterfullerene. *Nature*, *318*(6042), 162–163.

Kumar, N., Chamoli, P., Misra, M., Manoj, M. K., & Sharma, A. (2022). Advanced metal and carbon nanostructures for medical, drug delivery and bio-imaging applications. *Nanoscale*, *14*(11), 3987–4017. https://doi.org/10.1039/D1NR07643D

Kumar, R., Yadav, R., Awasthi, K., Shripathi, T., Sinha, A., Tiwari, R., & Srivastava, O. (2013). Synthesis of carbon and carbon-nitrogen nanotubes using green precursor: Jatropha-derived biodiesel. *Journal of Experimental Nanoscience*, *8*(4), 606–620.

Kumar, S., Murthy, J., & Alam, M. (2005). Percolating conduction in finite nanotube networks. *Physical Review Letters*, *95*(6), 066802.

Kwon, S., Ko, J.-H., Jeon, K.-J., Kim, Y.-H., & Park, J. Y. (2012). Enhanced nanoscale friction on fluorinated graphene. *Nano Letters*, *12*(12), 6043–6048.

Lange, H., Sioda, M., Huczko, A., Zhu, Y., Kroto, H., & Walton, D. (2003). Nanocarbon production by arc discharge in water. *Carbon*, *41*(8), 1617–1623.

Lee, H.-D., Penkov, O. V., & Kim, D.-E. (2013). Tribological behavior of dual-layer electroless-plated Ag-carbon nanotube coatings. *Thin Solid Films*, *534*, 410–416.

Li, Q.-L., Yuan, D.-X., & Lin, Q.-M. (2004). Evaluation of multi-walled carbon nanotubes as an adsorbent for trapping volatile organic compounds from environmental samples. *Journal of Chromatography A*, *1026*(1–2), 283–288.

Li, X., Zhou, Y., Sun, J., Zhang, H., & Wu, W. (2007). Tribological behavior of the electroless Ni-P-CNTs-SiC (nanometer) composite coating. *Xiyou Jinshu Cailiao Yu Gongcheng (Rare Metal Materials and Engineering)*, *36*, 712–714.

Li, Y.-H., Wang, S., Wei, J., Zhang, X., Xu, C., Luan, Z., Wu, D., & Wei, B. (2002). Lead adsorption on carbon nanotubes. *Chemical Physics Letters*, *357*(3–4), 263–266.

Lin, X., Wang, X., Dravid, V. P., Chang, R., & Ketterson, J. B. (1994). Large scale synthesis of single-shell carbon nanotubes. *Applied Physics Letters*, *64*(2), 181–183.

Lipomi, D. J., Vosgueritchian, M., Tee, B. C., Hellstrom, S. L., Lee, J. A., Fox, C. H., & Bao, Z. (2011). Skin-like pressure and strain sensors based on transparent elastic films of carbon nanotubes. *Nature Nanotechnology*, *6*(12), 788–792.

Liu, J., He, N., Li, X., Jiang, R., Yang, K., & Zeng, B. (2021). Alternated graphene/carbon nanotubes sandwich field emitter for making ultra-low-voltage field emission light directly driven by household electricity supply. *IEEE Electron Device Letters*, *42*(12), 1875–1877.

Liu, S., Hu, M., Zeng, T. H., Wu, R., Jiang, R., Wei, J., Wang, L., Kong, J., & Chen, Y. (2012). Lateral dimension-dependent antibacterial activity of graphene oxide sheets. *Langmuir*, *28*(33), 12364–12372.

Liu, W.-W., Aziz, A., Chai, S.-P., Mohamed, A. R., & Tye, C.-T. (2011). The effect of carbon precursors (methane, benzene and camphor) on the quality of carbon nanotubes synthesised by the chemical vapour decomposition. *Physica E: Low-Dimensional Systems and Nanostructures*, *43*(8), 1535–1542. https://doi.org/10.1016/j.physe.2011.05.012.

Liu, W.-W., Chai, S.-P., Mohamed, A. R., & Hashim, U. (2014). Synthesis and characterization of graphene and carbon nanotubes: A review on the past and recent developments. *Journal of Industrial and Engineering Chemistry*, *20*(4), 1171–1185.

Liu, X., Wang, M., Zhang, S., & Pan, B. (2013). Application potential of carbon nanotubes in water treatment: A review. *Journal of Environmental Sciences*, *25*(7), 1263–1280.

Loh, K., & Ryu, D. (2014). Multifunctional materials and nanotechnology for assessing and monitoring civil infrastructures. In *Sensor Technologies for Civil Infrastructures* (pp. 295–326). Elsevier.

Lourie, O., & Wagner, H. (1998). Evaluation of Young's modulus of carbon nanotubes by micro-Raman spectroscopy. *Journal of Materials Research*, *13*(9), 2418–2422.

Ma, P.-C., & Zhang, Y. (2014). Perspectives of carbon nanotubes/polymer nanocomposites for wind blade materials. *Renewable and Sustainable Energy Reviews*, *30*, 651–660.

Matsuura, T., Kondo, Y., & Maki, N. (2009). Selective mass production of carbon nanotubes by using multi-layered and multi-electrodes AC arc plasma reactor. *International Plasma Chemistry Society*.

Mittendorff, M., Winzer, T., Malic, E., Knorr, A., Berger, C., de Heer, W. A., Schneider, H., Helm, M., & Winnerl, S. (2014). Anisotropy of excitation and relaxation of photogenerated charge carriers in graphene. *Nano Letters*, *14*(3), 1504–1507.

Mochalin, V., Shenderova, O., Ho, D., & Gogotsi, Y. (2020). The properties and applications of nanodiamonds. *Nano-Enabled Medical Applications*, 313–350.

Moghadam, A. D., Omrani, E., Menezes, P. L., & Rohatgi, P. K. (2015). Mechanical and tribological properties of self-lubricating metal matrix nanocomposites reinforced by carbon nanotubes (CNTs) and graphene-a review. *Composites Part B: Engineering*, *77*, 402–420.

Mubarak, N., Sahu, J., Abdullah, E., Jayakumar, N., & Ganesan, P. (2014). Single stage production of carbon nanotubes using microwave technology. *Diamond and Related Materials*, *48*, 52–59.

Mukherjee, A., Majumdar, S., Servin, A. D., Pagano, L., Dhankher, O. P., & White, J. C. (2016). Carbon nanomaterials in agriculture: A critical review. *Frontiers in Plant Science*, *7*, 172.

Nerushev, O., Dittmar, S., Morjan, R.-E., Rohmund, F., & Campbell, E. (2003). Particle size dependence and model for iron-catalyzed growth of carbon nanotubes by thermal chemical vapor deposition. *Journal of Applied Physics*, *93*(7), 4185–4190.

Ng, K.-W., Lam, W.-H., & Pichiah, S. (2013). A review on potential applications of carbon nanotubes in marine current turbines. *Renewable and Sustainable Energy Reviews*, *28*, 331–339.

Nguyen-Tran, H.-D., Hoang, V.-T., Do, V.-T., Chun, D.-M., & Yum, Y.-J. (2018). Effect of multiwalled carbon nanotubes on the mechanical properties of carbon fiber-reinforced polyamide-6/polypropylene composites for lightweight automotive parts. *Materials*, *11*(3), 429.

Nishizaka, H., Namura, M., Motomiya, K., Ogawa, Y., Udagawa, Y., Tohji, K., & Sato, Y. (2011). Influence of carbon structure of the anode on the production of graphite in single-walled carbon nanotube soot synthesized by arc discharge using a Fe-Ni-S catalyst. *Carbon*, *49*(11), 3607–3614.

Norizan, M. N., Moklis, M. H., Demon, S. Z. N., Halim, N. A., Samsuri, A., Mohamad, I. S., Knight, V. F., & Abdullah, N. (2020). Carbon nanotubes: Functionalisation and their application in chemical sensors. *RSC Advances*, *10*(71), 43704–43732.

Novak, J., Snow, E., Houser, E., Park, D., Stepnowski, J., & McGill, R. (2003). Nerve agent detection using networks of single-walled carbon nanotubes. *Applied Physics Letters*, *83*(19), 4026–4028.

Novoselov, K. S., Geim, A. K., Morozov, S. V., Jiang, D., Zhang, Y., Dubonos, S. V., Grigorieva, I. V., & Firsov, A. A. (2004). Electric field effect in atomically thin carbon films. *Science*, *306*(5696), 666–669.

Nurazzi, N., Harussani, M., Zulaikha, N. S., Norhana, A., Syakir, M. I., & Norli, A. (2021). Composites based on conductive polymer with carbon nanotubes in DMMP gas sensors-an overview. *Polimery*, *66*(2), 85–97.

Nyholm, N., & Espallargas, N. (2023). Functionalized carbon nanostructures as lubricant additives - A review. *Carbon*, *201*, 1200–1228. https://doi.org/10.1016/j.carbon.2022.10.035.

Ohno, Y., Maehashi, K., Yamashiro, Y., & Matsumoto, K. (2009). Electrolyte-gated graphene field-effect transistors for detecting pH and protein adsorption. *Nano Letters*, *9*(9), 3318–3322.

Osswald, S., Yushin, G., Mochalin, V., Kucheyev, S. O., & Gogotsi, Y. (2006). Control of sp2/sp3 carbon ratio and surface chemistry of nanodiamond powders by selective oxidation in air. *Journal of the American Chemical Society*, *128*(35), 11635–11642.

Pant, M., Singh, R., Negi, P., Tiwari, K., & Singh, Y. (2021). A comprehensive review on carbon nano-tube synthesis using chemical vapor deposition. *Materials Today: Proceedings*, *46*, 11250–11253.

Park, K. C., & Lim, J. (2022). Direct grown vertically full aligned carbon nanotube electron emitters for X-ray and UV devices. In *Nanostructured Carbon Electron Emitters and Their Applications* (pp. 245–267). Jenny Stanford Publishing, Singapore.

Park, S., Vosguerichian, M., & Bao, Z. (2013). A review of fabrication and applications of carbon nanotube film-based flexible electronics. *Nanoscale*, *5*(5), 1727–1752.

Peddini, S., Bosnyak, C., Henderson, N., Ellison, C., & Paul, D. (2014). Nanocomposites from styrene-butadiene rubber (SBR) and multiwall carbon nanotubes (MWCNT) part 1: Morphology and rheology. *Polymer*, *55*(1), 258–270.

Pierson, H. O. (1999). *Handbook of Chemical Vapor Deposition: Principles, Technology and Applications*. Elsevier, Amsterdam.

Polyakova, E. Y., Rim, K. T., Eom, D., Douglass, K., Opila, R. L., Heinz, T. F., Teplyakov, A. V., & Flynn, G. W. (2011). Scanning tunneling microscopy and X-ray photoelectron spectroscopy studies of graphene films prepared by sonication-assisted dispersion. *ACS Nano*, *5*(8), 6102–6108.

Popov, V. N. (2004). Carbon nanotubes: Properties and application. *Materials Science and Engineering: R: Reports*, *43*(3), 61–102.

Pu, J., Mo, Y., Wan, S., & Wang, L. (2014). Fabrication of novel graphene-fullerene hybrid lubricating films based on self-assembly for MEMS applications. *Chemical Communications*, *50*(4), 469–471.

Qu, X., Alvarez, P. J., & Li, Q. (2013). Applications of nanotechnology in water and wastewater treatment. *Water Research*, *47*(12), 3931–3946.

Raccichini, R., Varzi, A., Passerini, S., & Scrosati, B. (2015). The role of graphene for electrochemical energy storage. *Nature Materials*, *14*(3), 271–279.

Ren, S., Cui, M., Liu, C., & Wang, L. (2023). A comprehensive review on ultrathin, multi-functionalized, and smart graphene and graphene-based composite protective coatings. *Corrosion Science*, *212*, 110939. https://doi.org/10.1016/j.corsci.2022.110939.

Rogers, J. A., Someya, T., & Huang, Y. (2010). Materials and mechanics for stretchable electronics. *Science*, *327*(5973), 1603–1607.

Rouhi, N., Jain, D., & Burke, P. J. (2011). High-performance semiconducting nanotube inks: Progress and prospects. *ACS Nano*, *5*(11), 8471–8487.

Ryu, B.-H., Barthel, A. J., Kim, H.-J., Lee, H.-D., Penkov, O. V., Kim, S. H., & Kim, D.-E. (2014). Tribological properties of carbon nanotube-polyethylene oxide composite coatings. *Composites Science and Technology*, *101*, 102–109.

Sankaranarayanan, S., Vishnukumar, P., Hariram, M., Vivekanandhan, S., Camus, C., Buschmann, A. H., & Navia, R. (2019). Hydrothermal synthesis, characterization and seed germination effects of green-emitting graphene oxide-carbon dot composite using brown macroalgal bio-oil as precursor. *Journal of Chemical Technology & Biotechnology*, *94*(10), 3269–3275.

Sano, N., Wang, H., Chhowalla, M., Alexandrou, I., & Amaratunga, G. A. (2001). Synthesis of carbon 'onions' in water. *Nature*, *414*(6863), 506–507.

Saputri, D. D., Jan'ah, A. M., & Saraswati, T. E. (2020). Synthesis of carbon nanotubes (CNT) by chemical vapor deposition (CVD) using a biogas-based carbon precursor: A review. *IOP Conference Series: Materials Science and Engineering*, *959*(1), 012019. https://doi.org/10.1088/1757-899X/959/1/012019.

Sarlak, N., Taherifar, A., & Salehi, F. (2014). Synthesis of nanopesticides by encapsulating pesticide nanoparticles using functionalized carbon nanotubes and application of new nanocomposite for plant disease treatment. *Journal of Agricultural and Food Chemistry*, *62*(21), 4833–4838.

Scalese, S., Scuderi, V., Bagiante, S., Gibilisco, S., Faraci, G., & Privitera, V. (2010). Order and disorder of carbon deposit produced by arc discharge in liquid nitrogen. *Journal of Applied Physics*, *108*(6), 064305.

Sevilla, M., Sanchís, C., Valdés-Solís, T., Morallón, E., & Fuertes, A. (2007). Synthesis of graphitic carbon nanostructures from sawdust and their application as electrocatalyst supports. *The Journal of Physical Chemistry C*, *111*(27), 9749–9756.

Shenderova, O. A., Zhirnov, V. V., & Brenner, D. W. (2002). Carbon Nanostructures. *Critical Reviews in Solid State and Materials Sciences*, *27*(3–4), 227–356. https://doi.org/10.1080/10408430208500497.

Shi, J., Wang, L., Gao, J., Liu, Y., Zhang, J., Ma, R., Liu, R., & Zhang, Z. (2014). A fullerene-based multi-functional nanoplatform for cancer theranostic applications. *Biomaterials*, *35*(22), 5771–5784.

Sinha, M., Neogi, S., Mahapatra, R., Krishnamurthy, S., & Ghosh, R. (2021). Material dependent and temperature driven adsorption switching (p-to n-type) using CNT/ZnO composite-based chemiresistive methanol gas sensor. *Sensors and Actuators B: Chemical*, *336*, 129729.

Sinnott, S., Andrews, R., Qian, D., Rao, A. M., Mao, Z., Dickey, E., & Derbyshire, F. (1999). Model of carbon nanotube growth through chemical vapor deposition. *Chemical Physics Letters*, *315*(1–2), 25–30.

Smith, S. C., & Rodrigues, D. F. (2015). Carbon-based nanomaterials for removal of chemical and biological contaminants from water: A review of mechanisms and applications. *Carbon*, *91*, 122–143.

Song, H., Qi, H., Li, N., & Zhang, X. (2011). Tribological behaviour of carbon nanotubes/polyurethane nanocomposite coatings. *Micro & Nano Letters*, *6*(1), 48–51.

Song, X., Liu, Y., & Zhu, J. (2007). Multi-walled carbon nanotubes produced by hydrogen DC arc discharge at elevated environment temperature. *Materials Letters*, *61*(2), 389–391.

Srivastava, M., Abhilash, P., & Singh, N. (2011). Remediation of lindane using engineered nanoparticles. *Journal of Biomedical Nanotechnology*, *7*(1), 172–174.

Stankovich, S., Dikin, D. A., Piner, R. D., Kohlhaas, K. A., Kleinhammes, A., Jia, Y., Wu, Y., Nguyen, S. T., & Ruoff, R. S. (2007). Synthesis of graphene-based nanosheets via chemical reduction of exfoliated graphite oxide. *Carbon*, *45*(7), 1558–1565.

Suriani, A., Dalila, A., Mohamed, A., Mamat, M., Salina, M., Rosmi, M., Rosly, J., Nor, R. M., & Rusop, M. (2013). Vertically aligned carbon nanotubes synthesized from waste chicken fat. *Materials Letters*, *101*, 61–64.

Suriani, A., Nor, R. M., & Rusop, M. (2010). Vertically aligned carbon nanotubes synthesized from waste cooking palm oil. *Journal of the Ceramic Society of Japan*, *118*(1382), 963–968.

Tan, D., & Zhang, Q. (2012). Research of carbon nanotubes/polymer composites for sports equipment. In *Future Computer, Communication, Control and Automation* (pp. 137–146). Springer, Berlin, Heidelberg.

Tang, R., Shi, Y., Hou, Z., & Wei, L. (2017). Carbon nanotube-based chemiresistive sensors. *Sensors*, *17*(4), 882.

Taylor, L. W., Dewey, O. S., Headrick, R. J., Komatsu, N., Peraca, N. M., Wehmeyer, G., Kono, J., & Pasquali, M. (2021). Improved properties, increased production, and the path to broad adoption of carbon nanotube fibers. *Carbon, 171*, 689–694.

Thess, A., Lee, R., Nikolaev, P., Dai, H., Petit, P., Robert, J., Xu, C., Lee, Y. H., Kim, S. G., & Rinzler, A. G. (1996). Crystalline ropes of metallic carbon nanotubes. *Science, 273*(5274), 483–487.

Thompson, E., Danks, A., Bourgeois, L., & Schnepp, Z. (2015). Iron-catalyzed graphitization of biomass. *Green Chemistry, 17*(1), 551–556.

Tong, J., Zimmerman, M. C., Li, S., Yi, X., Luxenhofer, R., Jordan, R., & Kabanov, A. V. (2011). Neuronal uptake and intracellular superoxide scavenging of a fullerene (C60)-poly (2-oxazoline) s nanoformulation. *Biomaterials, 32*(14), 3654–3665.

Tripathi, S., & Sarkar, S. (2015). Influence of water soluble carbon dots on the growth of wheat plant. *Applied Nanoscience, 5*(5), 609–616.

Tripathi, S., Sharon, M., Maldar, N., Shukla, J., & Sharon, M. (2012). Carbon Nano Spheres and nano tubes synthesized from Castor oil as precursor; for removal of Arsenic dissolved in water. *Archives of Applied Science Research, 4*(4), 1788–1795.

Tsai, Y., Su, J., Su, C., & He, W. (2009). Production of carbon nanotubes by single-pulse discharge in air. *Journal of Materials Processing Technology, 209*(9), 4413–4416.

Umeda, J., Fugetsu, B., Nishida, E., Miyaji, H., & Kondoh, K. (2015). Friction behavior of network-structured CNT coating on pure titanium plate. *Applied Surface Science, 357*, 721–727.

Uritu, C. M., Varganici, C. D., Ursu, L., Coroaba, A., Nicolescu, A., Dascalu, A. I., Peptanariu, D., Stan, D., Constantinescu, C. A., & Simion, V. (2015). Hybrid fullerene conjugates as vectors for DNA cell-delivery. *Journal of Materials Chemistry B, 3*(12), 2433–2446.

Wang, H., Ma, H., Zheng, W., An, D., & Na, C. (2014). Multifunctional and recollectable carbon nanotube ponytails for water purification. *ACS Applied Materials & Interfaces, 6*(12), 9426–9434.

Wang, L., Fortner, J. D., Hou, L., Zhang, C., Kan, A. T., Tomson, M. B., & Chen, W. (2013). Contaminant-mobilizing capability of fullerene nanoparticles (nC60): Effect of solvent-exchange process in nC60 formation. *Environmental Toxicology and Chemistry, 32*(2), 329–336.

Wang, Q., Bai, B., Li, Y., Jiang, Y., Ma, L., & Ren, N. (2015). Investigating the nano-tribological properties of chemical vapor deposition-grown single layer graphene on SiO 2 substrates annealed in ambient air. *RSC Advances, 5*(13), 10058–10064.

Wang, Y.-H., Chiu, S.-C., Lin, K.-M., & Li, Y.-Y. (2004). Formation of carbon nanotubes from polyvinyl alcohol using arc-discharge method. *Carbon, 42*(12–13), 2535–2541.

Won, S., Jang, J.-W., Choi, H.-J., Kim, C.-H., Lee, S. B., Hwangbo, Y., Kim, K.-S., Yoon, S.-G., Lee, H.-J., & Kim, J.-H. (2016). A graphene meta-interface for enhancing the stretchability of brittle oxide layers. *Nanoscale, 8*(9), 4961–4968.

Wu, L., Liu, J., Reddy, B. R., & Zhou, J. (2022). Preparation of coal-based carbon nanotubes using catalytical pyrolysis: A brief review. *Fuel Processing Technology, 229*, 107171.

Wu, X., Sprinkle, M., Li, X., Ming, F., Berger, C., & de Heer, W. A. (2008). Epitaxial-graphene/graphene-oxide junction: An essential step towards epitaxial graphene electronics. *Physical Review Letters, 101*(2), 026801.

Yan, H., Gong, A., He, H., Zhou, J., Wei, Y., & Lv, L. (2006). Adsorption of microcystins by carbon nanotubes. *Chemosphere, 62*(1), 142–148.

Ye, X., Liu, X., Yang, Z., Wang, Z., Wang, H., Wang, J., & Yang, S. (2016). Tribological properties of fluorinated graphene reinforced polyimide composite coatings under different lubricated conditions. *Composites Part A: Applied Science and Manufacturing, 81*, 282–288.

Yi, M., & Shen, Z. (2015). A review on mechanical exfoliation for the scalable production of graphene. *Journal of Materials Chemistry A, 3*(22), 11700–11715.

Yoo, J., Ozawa, H., Fujigaya, T., & Nakashima, N. (2011). Evaluation of affinity of molecules for carbon nanotubes. *Nanoscale*, *3*(6), 2517–2522.

Yoo, S. T., & Park, K. C. (2022). Extreme ultraviolet lighting using carbon nanotube-based cold cathode electron beam. *Nanomaterials*, *12*(23), 4134.

Yoon, B., Choi, S.-J., Swager, T. M., & Walsh, G. F. (2021). Flexible chemiresistive cyclohexanone sensors based on single-walled carbon nanotube-polymer composites. *ACS Sensors*, *6*(8), 3056–3062.

Youh, M.-J., Huang, C.-L., Wang, Y.-L., Chiang, L.-M., & Li, Y.-Y. (2020). Development of a high-brightness field-emission lighting device with ITO electrode. *Vacuum*, *181*, 109733.

Zaytseva, O., & Neumann, G. (2016). Carbon nanomaterials: Production, impact on plant development, agricultural and environmental applications. *Chemical and Biological Technologies in Agriculture*, *3*(1), 1–26.

Zhang, L., Song, X., Liu, X., Yang, L., Pan, F., & Lv, J. (2011). Studies on the removal of tetracycline by multi-walled carbon nanotubes. *Chemical Engineering Journal*, *178*, 26–33.

Zhang, W., Cao, S., Wu, Z., Zhang, M., Cao, Y., Guo, J., Zhong, F., Duan, H., & Jia, D. (2019). High-performance gas sensor of polyaniline/carbon nanotube composites promoted by interface engineering. *Sensors*, *20*(1), 149.

Zheng, X., Su, Y., Chen, Y., Wei, Y., Li, M., & Huang, H. (2014). The effects of carbon nanotubes on nitrogen and phosphorus removal from real wastewater in the activated sludge system. *RSC Advances*, *4*(86), 45953–45959.

Zhu, H., Li, X., Jiang, B., Xu, C., Zhu, Y., Wu, D., & Chen, X. (2002). Formation of carbon nanotubes in water by the electric-arc technique. *Chemical Physics Letters*, *366*(5–6), 664–669.

9 Nanoplastics in Environment
Environmental Risk, Occurrence, Characterization, and Identification

MohdSaiful Samsudin
Universiti Sains Malaysia

Azman Azid
Universiti Sultan Zainal Abidin

Nurul Latiffah Abd Rani
Universiti Malaysia Terengganu

9.1 INTRODUCTION

Given that plastic pollution is one of the most significant environmental issues and a developing environmental concern in recent decades, worries about it are growing on a global scale. The increase of plastic pollution globally serves as a warning that these pollutants may stay in the environment and have a negative impact on ecosystems and public health. The usage of plastics has significantly improved people's daily lives during the past 50 years. Plastics may be seen as a pillar material in a worldwide "throwaway culture" economy since they are frequently employed in basic everyday necessities (such as fresh food requirements, the transportation of shopping items, aseptic medical products, and meal preparation) (Oliveira & Almeida, 2019). The increased use of plastics has caused serious environmental degradation on a global scale, drawing growing attention from scientists. Large plastic waste is probably going to degrade into microplastics (less than 5 mm) and even nanoplastics (less than 1,000 nm) (Cai et al., 2021). The fact that nanoplastics, which are too tiny to be seen with the human eye, lack a defined size makes them comparable to microplastics. However, smaller plastic particles of less than 0.1 mm in size are more commonly used to describe nanoplastics. The breakdown of macro- and microplastics into the tiniest plastic particles, micro- and nanometer-sized particles, is the main cause of

DOI: 10.1201/9781003400998-9

nanoplastics. As a result, the physical characteristics of nanoplastics are extremely polydisperse and their composition is diverse (Gigault et al., 2018). Nanoplastics are challenging to define since they have not been sufficiently studied due to their small size. Classification of microplastics and nanoplastics in the environment is based on their morphological characteristics, such as size, shape, and color, as illustrated in Figure 9.1.

A daily discharge of microplastics and nanoplastics might contain 50,000–15,000,000 particles, according to estimates (Barcelo & Knepper, 2019). There are several primary sources of nanoplastics to take into account as follows:

i. manufactured polymer nanoparticles used in 3D printer ink, cosmetic items, and other applications.
ii. Plastic pieces breaking down due to physical stress, hydrolysis, UV light, or microbial action, commonly seen in marine environments.
iii. Wastewater treatment plant (WWTP) by-products such as biosolids and discharge water, found in sewage sludge.

Household and industrial waste, including personal care products, synthetic materials, and industry discharge all contribute to plastic pollution.

Besides, poor municipal solid waste segregation during disposal enhances soil contamination with plastic waste in landfills and open dump sites. Nanoplastics may accumulate up in soil and freshwater systems as a result of their disintegration. In comparison with primary sources, it is more difficult to detect, manage, and quantify nanoplastics in marine and terrestrial environments. Considerably, while plastic is an environmental hazard on its own, its characteristics can be even more damaging when mixed with additional substances. Other organic compounds are frequently added to plastic goods to improve or change them; in certain cases, they account for 50% of the final product such as bisphenol, phthalate acid esters, perfluoroalkyl substances, nonylphenol, and brominated flame retardants (BFRs) (Boyle & Örmeci, 2020). All of the BFRs are popular plastic additives and contaminants. The problem with plastic additives, also known as plasticides, is that they often have tiny

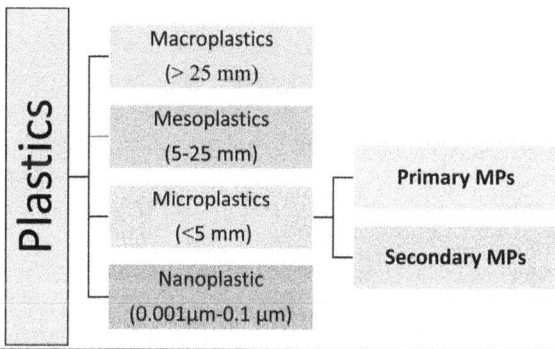

FIGURE 9.1 Classification of plastic in the marine environment. (Reproduced from ref Alprol et al. (2021)).

molecules and are not chemically attached to polymers, which means that under the right circumstances, they might readily separate from the polymer. For instance, substances (like phthalates) that aren't naturally lipophilic can rapidly dissociate when exposed to water. Once pollutants are floating freely, they can quickly bind to other polymeric materials and be transported over great distances.

9.2 OCCURRENCE

The presence of nanoplastic materials in various environmental matrixes was initially considered in the 1970s, but it is now recognized as an emerging environmental hazard. The recent finding of nanoplastics has prompted considerable debate about their prevalence and environmental fate (Gangadoo et al. 2020). The presence of nanoplastics in the environment is linked to human activity because they are a sort of plastic trash that is anthropogenic in origin. Higher population densities are typically thought to have more nanoplastics present (Zhang & Xu, 2022). Despite the numerous benefits that plastic provides across numerous industries, such as electronics, healthcare, packaging, construction, and automobiles, the improper disposal of plastic waste has become a pressing concern due to its detrimental effects on aquatic ecosystems and potential impact on human health over an extended period of time (Figure 9.2).

The main sources of primary nanoplastics in the environment include personal care items like cosmetics, sunscreens, and cleaning products, as well as electronics, paint, clothing washing, plastic tea bags, biomedical and drug products. Microbeads used in scrubs and shampoos contribute significantly to nanoplastics release upon fragmentation. Unfortunately, these consumer goods often end up in WWTPs after being rinsed off (Pohlmann et al., 2013; Hernandez et al., 2017).

The primary nanoplastics can transform into secondary nanoplastics through mechanical grinding, light degradation, and thermal oxidation (Zhang et al., 2020a). Additionally, the breakdown of microplastics can lead to the formation of nanoplastics, due to a combination of non-living and living processes (Koelmans et al., 2015). For instance, after undergoing full degradation into nanoplastics with an average size of 40 nm, the surface area of a conventional plastic bag ($0.2\,m^2$) would increase to $2,600\,m^2$ (Shen et al., 2019).

Primary nanoplastics that break down may result in decreased molecular weight and increased exposure to marine and land organisms. Secondary nanoplastics, produced by degradation or fragmentation of microplastics, can spread into the water through various means, like runoff and sewage discharge, posing potential harm. Smaller particles like nanoplastics are more mobile in water and can easily adhere to fish gills, putting fish and other aquatic life at risk. Additionally, nanoplastics can affect the growth and reproduction of certain aquatic plants. In conclusion, freshwater nanoplastics can have adverse effects on both fish and the environment.

Studies indicate that the actions of nanoplastics in the environment stem from features like clumping, redox reactions, compacting, and dissolving. These are due to the materials' physical and chemical attributes (Schirmer et al., 2013). Research discovered that nanoplastics maintain stability in deionized water and freshwater environments (Brandts et al., 2018). However, exposure to bacterial mediums, artificial

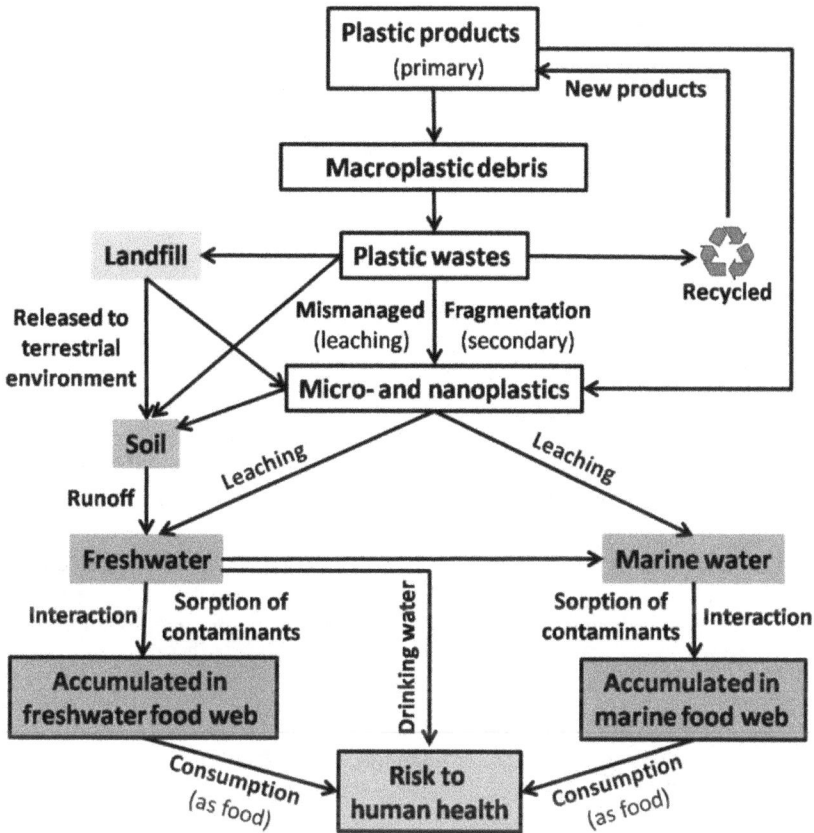

FIGURE 9.2 Plastic particle formation, dispersion into aquatic ecosystems, and interactions with other contaminants that pose risks to the environment and public health are all depicted schematically. (Reproduced from ref Cássio et al. (2022)).

and natural seawater, and biological fluids leads to an increase in particle size and distribution (Marques-Santos et al., 2018; Wegner et al., 2012).

Nanoplastics can spread through biodegradation or non-biodegradation processes in plastics. Since plastics are long-chain organic molecules, their capacity to interact with other pollutants and change in form, size, porosity, surface area, and crystallinity are what determine how quickly they degrade. Large bits of plastic are broken down into micro- and nanoplastics after being discarded in plastic trash, which then transforms chemically, biologically, and environmentally.

9.3   CHARACTERIZATION AND IDENTIFICATION

Further degradation of primary, secondary, and nanoplastics in particular would change their distinctive properties, such as size, shape, color, crystallinity, surface area, and densities. This would eventually have a negative effect on their physical and chemical activities in the ecosystem. Additives are added during plastic production to enhance its color, transparency, and handling qualities. These additions make the polymer more resistant to degradation caused by physical (e.g., temperature, ozone, light radiation), mechanical, electrical, and biological (e.g., fungi, bacteria) factors (Hale et al., 2020). The additives used may include colors, UV stabilizers, plasticizers, lubricants, flame retardants, as well as reinforcing fillers like wood, graphite, glass fibers, rock flour, kaolin, cotton flakes, jute, clay, linen, and cellulose pulp (Dobslaw et al., 2021). Though these additives may pose harm and contaminate the land, water, and air, they enhance the properties of plastic.

A comprehensive examination of the latest techniques for the preparation, recognition, and measurement of nanoplastics in the environment has been evaluated by Cai et al. (2021) as depicted in Figure 9.3. The research by Cai et al. (2021) analyzed 33 studies on the current state-of-the-art methodologies for the handling, isolation, identification, and quantification of nanoplastics. Despite many of the studies successfully discovering reference nanoplastics in controlled samples, they were unable to isolate and accurately quantify nanoplastics from actual field samples. When classifying microplastics, one of the key factors considered is its form. There is a wide variety of nanoplastics shapes available, including fragments, foam, paint, pellets, foil, spheres, fibers, films, lines, beads, flakes, sheets, granules, and nurdles (Campanale et al., 2020). The shape of nanoplastics is impacted by its original form, degradation, erosion processes, and exposure time in the environment (Murray & Örmeci, 2020). Wastewater and urban atmospheric deposits primarily consist of fibers and pieces, accounting for 52.7% (Annenkov et al., 2021). Microplastics fragments form from larger plastic products that undergo strain or UV light exposure, while fiber particles result from mechanical abrasion and chemical weathering (Zhou et al., 2020).

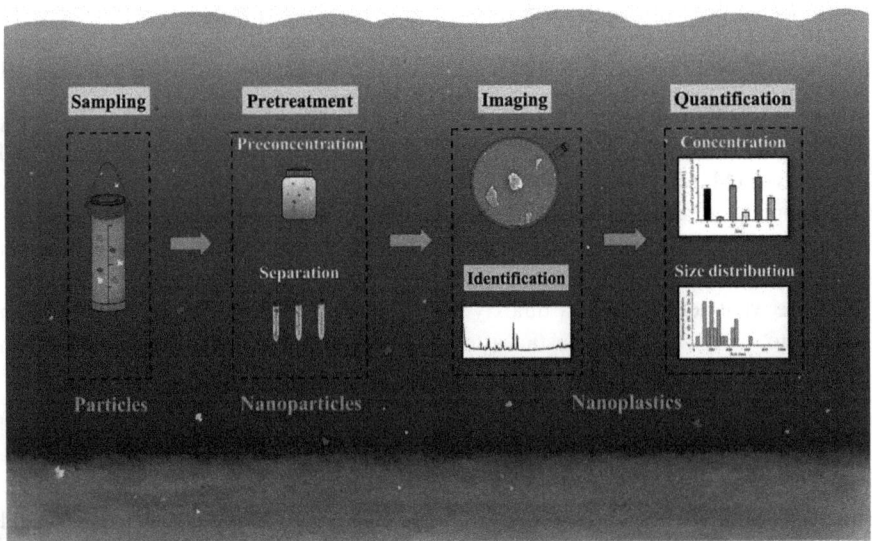

FIGURE 9.3 A generic technique of evaluating nanoplastics in environmental samples. (Reproduced from ref Cai et al. (2021)).

The size of nanoplastics is commonly determined through sampling and analytical techniques. Nanoplastics are defined as particles smaller than 100 nm, as determined by the National Oceanic and Atmospheric Administration. This definition was later adopted by the Scientific Aspects of Marine Environmental Protection (GESAMP, 2016) and the Joint Group of Experts (UNEP, 2014). Currently, the two most commonly used methods for nanoplastics classification are sieve retention and microscopic imaging. The most frequently found sizes in WWTP influent and effluent are 25, 100, and 500 nm (Mintenig et al., 2017).

Research by Murray and Örmeci (2020) and others has identified various nanoparticle colors, including red, orange, yellow, brown, tan, off-white, white, grey, blue, and green. There may be an undercount of particles that are dark, transparent, white, or translucent. Blue and red are the most frequently reported colors for fibers (Bergmann et al., 2019). Unlike polyethylene and low-density polyethylene, which were made opaque, polypropylene was produced with a clear and transparent appearance.

9.4 ENVIRONMENTAL RISK

Exposure to nanoplastics poses a major environmental threat and puts natural systems and environmental health at risk. Environmental risk assessment carefully evaluates the possibility that exposure to environmental stressors could harm ecosystems. The assessment of risk involves comparing exposure levels to ecological effects to determine the potential for negative impacts. The degree of exposure determines the impact of nanoplastics, and the risk assessment depends on how much exposure organisms receive in the environment. There has been a significant rise in scientific interest in the toxic effects of nanoplastics on living things in recent years, suggesting that these pollutants may be even more dangerous than microplastics (Masseroni et al., 2022).

The risk assessment process of nanoplastics is a four-step method, according to Koelmans et al. (2017), Besseling et al. (2019), and Masseroni et al. (2022):

    i. hazard identification
    ii. exposure assessment
    iii. effect assessment
    iv. risk characterization

Risk characterization is the final step, which involves determining if the predicted risk level is considered acceptable by combining data on exposure levels with the expected outcomes. The risk quotient (RQ) is calculated by dividing the substance's potential exposure by the amount where no negative effects are expected (RQ = exposure/toxicity) (Masseroni et al. 2022). If RQ < 1, the level of nanoplastics pollution is acceptable, while RQ > 1 signals a reason for concern.

The ecological risks of microplastics/nanoplastics in the environment will be evaluated using the RQ method, previously used to assess the risks of microplastics/nanoplastics in marine ecosystems (Zhang et al., 2020b).

The RQ was expressed as follows:

$$RQ = \frac{MEC}{PNEC}$$

where,

MEC: MP abundance at specific sampling points
PNEC: predicted no-effect concentration at which organisms were unlikely to
  experience any adverse effects

The calculated risk ratios were further classified into different risk levels: negligible risk (RQ < 1) and high risk (RQ > 1) (Zhang et al., 2020b).

The current environmental risk assessment utilizes a tiered and mostly deterministic methodology to efficiently identify low-concern compounds. The lowest tiers use highly conservative assumptions and high uncertainty, so the assessment factor (AF) is employed to factor in all potential risks. As more data become available, the AF decreases with each tier, reflecting reduced uncertainty. Higher tiers are more complex but also more ecologically realistic. A risk assessment of nanoplastics in aquatic environments has been conducted using available data:

$$RQ_{nps} = \frac{\text{Exposure}}{\text{Toxicity}} = \frac{1.92 - 2.82\,\mu g/L}{\left(5.4\,\mu g/L \Big/ 5\right)} = 1.7 - 2.6$$

The exposure data from Zhou et al. (2021) were the toxicity data, the exposure range is 1.92–2.82 µg/L, and the toxicity concentration is $\left(5.4\,\mu g/L/5\right)$. The present observed amounts of nanoplastics in aquatic environments provide an unacceptable risk for the wild populations, as shown by the RQ being > 1. The findings from this study indicating the potential harmful effects on aquatic life underline the importance of obtaining reliable information to conduct a thorough risk assessment of nanoplastics in the environment.

9.5 CONCLUSION

Nanoplastics have accumulated in the environment due to improper handling and usage of plastics, endangering the ecosystem and all living things. They are pervasive and are dispersed through rivers, lakes, stormwater runoff, sewage, sludge, and wastewater treatment facilities into many environmental compartments such as terrestrial, aquatic, and atmospheric depositions. The physicochemical characteristics of the plastics in soil and water have a significant impact on the fate and transport of nanoplastics. WWTPs are receiving particulate plastics from surface runoff, residential use, and industrial sources. The efficiency of the process is reduced as nanoplastics load increases in the sludge digestion system, and operational costs are also increased. To establish sustainable practices in plastic management, critical consideration of remediation technologies, prioritizing recycling, effective waste management, education and awareness, implementing circular models (such as reduce, reuse, recycle and recover), using bio-based materials (such as bioplastics), legislation, policy, and a clear roadmap are essential.

REFERENCES

Alprol, A. E., Gaballah, M. S., & Hassaan, M. A. (2021). Micro and Nanoplastics analysis: Focus on their classification, sources, and impacts in marine environment. Regional Studies in Marine Science, 42, 101625. https://doi.org/10.1016/j.rsma.2021.10162.

Annenkov, V. V., Danilovtseva, E. N., Zelinskiy, S. N., & Pal'shin, V. A. (2021). Submicro- and nanoplastics: How much can be expected in water bodies? *Environmental Pollution*, 278, 116910. https://doi.org/10.1016/j.envpol.2021.116910.

Barcelo, D., & Knepper, T. (2019). Analysis, fate and effects of microplastics in the environment: Barcelo, D., & Knepper, T. (2019). Analysis, fate and effects of microplastics in the environment: Preface to article collection. *TrAC Trends in Analytical Chemistry*, 121, 115671. https://doi.org/10.1016/j.trac.2019.115671

Bergmann, M., Mützel, S., Primpke, S., Tekman, M. B., Trachsel, J., & Gerdts, G. (2019). White and wonderful? Microplastics prevail in snow from the Alps to the Arctic. *Science Advances*, 5(8), eaax1157. https://doi.org/10.1126/sciadv.aax1157.

Besseling, E., Redondo-Hasselerharm, P., Foekema, E. M., & Koelmans, A. A. (2019). Quantifying ecological risks of aquatic micro-and nanoplastic. Critical Reviews in Environmental Science and Technology, 49(1), 32–80. https://doi.org/10.1080/106433 89.2018.1531688.

Boyle, K., & Örmeci, B. (2020). Microplastics and nanoplastics in the freshwater and terrestrial environment: A review. *Water*, 12(9), 2633. https://doi.org/10.3390/w12092633.

Brandts, I., Teles, M., Gonçalves, A. P., Barreto, A., Franco-Martinez, L., Tvarijonaviciute, A., Martins, M.A., Soares, A.M.V.M., Tort, L. & Oliveira, M. (2018). Effects of nanoplastics on Mytilus galloprovincialis after individual and combined exposure with carbamazepine. Science of the Total Environment, 643, 775–784. https://doi.org/10.1016/j. scitotenv.2018.06.257.

Cai, H., Xu, E. G., Du, F., Li, R., Liu, J., & Shi, H. (2021). Analysis of environmental nanoplastics: Progress and challenges. *Chemical Engineering Journal*, 410, 128208. https:// doi.org/10.1016/j.cej.2020.128208.

Campanale, C., Savino, I., Pojar, I., Massarelli, C., & Uricchio, V. F. (2020). A practical overview of methodologies for sampling and analysis of microplastics in riverine environments. *Sustainability*, 12(17), 6755. https://doi.org/10.3390/su12176755.

Cássio, F., Batista, D., & Pradhan, A. (2022). Plastic interactions with pollutants and consequences to aquatic ecosystems: What we know and what we do not know. *Biomolecules*, 12(6), 798. https://doi.org/10.3390/biom12060798.

Dobslaw, D., Woiski, C., Kiel, M., Kuch, B., & Breuer, J. (2021). Plant uptake, translocation and metabolism of PBDEs in plants of food and feed industry: A review. Reviews in Environmental Science and Bio/Technology, 20(1), 75–142. https://doi.org/10.1007/ s11157-020-09557-7.

Gangadoo, S., Owen, S., Rajapaksha, P., Plaisted, K., Cheeseman, S., Haddara, H., Truong, V.K., Ngo, S.T., Vu, V.V., Cozzolino, D. & Chapman, J. (2020). Nano-plastics and their analytical characterisation and fate in the marine environment: From source to sea. Science of the Total Environment, 732, 138792. https://doi.org/10.1016/j. scitotenv.2020.138792.

GESAMP (2016) Sources, fate and effects of microplastics in the marine environment (part 2). https://www.gesamp.org/publications/microplastics-in-the-marineenvironment-part-2.

Gigault, J., Ter Halle, A., Baudrimont, M., Pascal, P. Y., Gauffre, F., Phi, T. L., El Hadri, H., Grassl, B. & Reynaud, S. (2018). Current opinion: What is a nanoplastic?. *Environmental Pollution*, 235, 1030–1034. https://doi.org/10.1016/j.envpol.2018.01.024.

Hale, R. C., Seeley, M. E., La Guardia, M. J., Mai, L., & Zeng, E. Y. (2020). A global perspective on microplastics. Journal of Geophysical Research: Oceans, 125(1), e2018JC014719. https://doi.org/10.1029/2018JC014719.

Hernandez, L. M., Yousefi, N., & Tufenkji, N. (2017). Are there nanoplastics in your personal care products?. Environmental Science & Technology Letters, 4(7), 280–285. https://doi.org/10.1021/acs.estlett.7b00187.

Koelmans, A. A., Besseling, E., & Shim, W. J. (2015). Nanoplastics in the aquatic environment. Critical review (pp. 325–340). Marine Anthropogenic Litter. https://doi.org/10.1007/978-3-319-16510-3

Koelmans, A. A., Kooi, M., Law, K. L., & Van Sebille, E. (2017). All is not lost: Deriving a top-down mass budget of plastic at sea. Environmental Research Letters, 12(11), 114028. https://doi.org/10.1088/1748-9326/aa9500.

Marques-Santos, L. F., Grassi, G., Bergami, E., Faleri, C., Balbi, T., Salis, A., Damonte, G., Canesi, L. & Corsi, I. (2018). Cationic polystyrene nanoparticle and the sea urchin immune system: Biocorona formation, cell toxicity, and multixenobiotic resistance phenotype. Nanotoxicology, 12(8), 847–867. https://doi.org/10.1080/17435390.2018.1482378.

Masseroni, A., Rizzi, C., Urani, C., & Villa, S. (2022). Nanoplastics: Status and knowledge gaps in the finalization of environmental risk assessments. Toxics, 10(5), 270. https://doi.org/10.3390/toxics10050270.

Mintenig, S. M., Int-Veen, I., Löder, M. G., Primpke, S., & Gerdts, G. (2017). Identification of microplastic in effluents of waste water treatment plants using focal plane array- based micro-Fourier-transform infrared imaging. Water Research, 108, 365–372. https://doi.org/10.1016/j.watres.2016.11.015.

Murray, A., & Örmeci, B. (2020). Removal effectiveness of nanoplastics (< 400 nm) with separation processes used for water and wastewater treatment. Water, 12(3), 635. https://doi.org/10.3390/w12030635.

Oliveira, M., & Almeida, M. (2019). The why and how of micro (nano) plastic research. TrAC Trends in Analytical Chemistry, 114, 196–201. https://doi.org/10.1016/j.trac.2019.02.023.

Pohlmann, A. R., Fonseca, F. N., Paese, K., Detoni, C. B., Coradini, K., Beck, R. C., & Guterres, S. S. (2013). Poly (ε-caprolactone) microcapsules and nanocapsules in drug delivery. Expert Opinion on Drug Delivery, 10(5), 623–638. https://doi.org/10.1517/17425247.2013.769956.

Schirmer, K., Behra, R., & Sigg, L., (2013). Ecotoxicological aspects of nanomaterials in the aquatic environment, In Safety Aspects of Engineered Nanomaterials. Pan Stanford Publishing Pte. Ltd, Singapore.

Shen, M., Zhang, Y., Zhu, Y., Song, B., Zeng, G., Hu, D., Wen, X. & Ren, X. (2019). Recent advances in toxicological research of nanoplastics in the environment: A review. Environmental Pollution, 252, 511–521. https://doi.org/10.1016/j.envpol.2019.05.102

UNEP, U. (2014). Year Book 2014 emerging issues update. United Nations Environment Programme, Nairobi, Kenya.

Wegner, A., Besseling, E., Foekema, E. M., Kamermans, P., & Koelmans, A. A. (2012). Effects of nanopolystyrene on the feeding behavior of the blue mussel (Mytilus edulis L.). Environmental Toxicology and Chemistry, 31(11), 2490–2497. https://doi.org/10.1002/etc.1984.

Zhang, H., Liu, F. F., Wang, S. C., Huang, T. Y., Li, M. R., Zhu, Z. L., & Liu, G. Z. (2020a). Sorption of fluoroquinolones to nanoplastics as affected by surface functionalization and solution chemistry. Environmental Pollution, 262, 114347. https://doi.org/10.1016/j.envpol.2020.114347.

Zhang, M., & Xu, L. (2022). Transport of micro-and nanoplastics in the environment: Trojan-Horse effect for organic contaminants. Critical Reviews in Environmental Science and Technology, 52(5), 810–846. https://doi.org/10.1080/10643389.2020.1845531.

Zhang, X., Leng, Y., Liu, X., Huang, K., & Wang, J. (2020b). Microplastics' pollution and risk assessment in an urban river: A case study in the Yongjiang River, Nanning City, South China. *Exposure and Health*, 12(2), 141–151. https://doi.org/10.1007/s12403-018-00296-3.

Zhou, G., Wang, Q., Zhang, J., Li, Q., Wang, Y., Wang, M., & Huang, X. (2020). Distribution and characteristics of microplastics in urban waters of seven cities in the Tuojiang River Basin, China. Environ*mental* Res*earch*, 189, 109893. https://doi.org/10.1016/j.envres.2020.109893.

Zhou, X. X., He, S., Gao, Y., Li, Z. C., Chi, H. Y., Li, C. J., Wang, D. J. & Yan, B. (2021). Protein Corona-Mediated extraction for quantitative analysis of nanoplastics in environmental waters by pyrolysis gas chromatography/mass spectrometry. *Analytical Chemistry*, 93(17), 6698–6705. https://doi.org/10.1021/acs.analchem.1c00156.

10 Nanofillers in Pulp and Paper

Mohamad Nurul Azman Mohammad Taib
King Fahd University of Petroleum and Minerals

10.1 INTRODUCTION

The pulp and paper industries are among the important industries all over the world, and the demand for papers is now increasing by more than 423 million tons (1). Most consumers come from Asian countries, followed by the USA and European countries (2). Asia produces more than 40%, while an estimated 30% comes from Europe and 25% comes from North America (2). The primary stage in paper manufacturing involves several steps, starting with material preparation, pulp manufacturing, pulp bleaching, paper manufacturing, and fibers recycling (3–5). The use of nanofillers involves first using mineral additives to improve their intended properties for use in different paper applications (6). These mineral additives are also, at the same time, reducing the cost of paper manufacturing and making it much cheaper (1). In some cases, the use of functional nanofillers has been reported to increase opacity, brightness, water penetration control, wet and dry strength of paper, and some improvement on properties (7–9). However, this type of filler severely wears paper machine parts and printing cylinders. The abrasion that comes from this filler is influenced by particle hardness, finesses, and structures. Even at low content, it is very abrasive (8, 9). Therefore, the need to have nanofillers is important to avoid these bad effects. Nanofillers have become an important role in manufacturing the highest quality of printing paper and cost efficiency in papermaking (10). In the manufacturing process of papermaking, nanofillers are sometimes used together as a nano pigment for coating, nano polymer additives, nano sizing agents, nano retention systems, fiber nanocoating, and nano-based smart paper (11–14).

The area of nanotechnology has been explored for decades by researchers and scientists. It is becoming a trend to use nanotechnology in pulp and paper processing. The advantage that is provided by nanotechnology field is that nano size (at least one dimension of particles in the range of 1–100 nm) is used in many applications, especially in the pulp and paper industry (10, 15). Nanosize particles, including those from synthetic and natural resources such as clay or kaolin, cellulose, silica, calcium carbonate ($CaCO_3$), metal derivatives such as zinc oxide (ZnO), titanium oxide (TiO_2), and many more, are being studied by researchers to enhance performance as fillers, additives, reinforcement, and others (8, 14, 16, 17). Nanofillers have fine particles with tremendous properties such as has a high aspect ratio, good in mechanical strength, and easy modification (8). It can have new functionalities when added together in pulp and paper processing (14). The addition of nanofillers

DOI: 10.1201/9781003400998-10

is chosen by industries due to their enhancement of desired properties and application to new manufacturing and final products (8). These nanofillers can be added through various methods to pulps and papers. Some of the methods used are co-mixing with flocculation or fixation chemicals (this method is also considered as conventional loading method) or direct in-situ precipitation with various precursors (this method determines a high filler retention with more uniform fillers distribution within a sheet) (1, 15).

Even though the nanofillers are becoming a trend applying in pulps and papers, due to some limitations, they pose more challenges. The limitation is that some of the processing nanofillers involve some expensive processes to get nanosized materials for use in pulp and paper, and with this increase, the cost of manufacturing needs to be highly considered. Second, the surface area and aspect ratio are not uniformly distributed on the surface, making it hard to control the structure and performance. This happens especially when using natural nanofillers such as nanocellulose and nanoclay (15). Due to this, some modifications need to be made to provide some solution, such as the alteration of polar tails. Some modification occurs through the utilization of chemical treatments such as silylation, esterification, oxidation, and many more (18, 19). Some surfactant is also considered to be added with nanofillers to improve the properties (8, 9). Finally, the potential for detrimental interactions between the nanofiller-specific area and papermaking at the wet end of additives could affect the strength and sizing agents, and dyes also pose challenges (11, 15).

On top of that, this chapter will briefly discuss the uses of nanofillers and their application in the pulp and paper industries. Furthermore, the process of applying these nanofillers in the pulp and paper manufacturing process was also discussed to provide a clear understanding of the types of nanofillers and the parts to which they have been applied. The review for this chapter also elaborated on the application and function of nanofillers in the pulp and paper industries.

10.2 NANOFILLERS IN PULPS AND PAPERS

Nanofillers in pulps and papers can be added in a few ways and provide several functionalities depending on their applications and uses on the final end products. It can be added in bleaching as pigments to give brighter looks and as nanosizing agents to give some improvement in wear properties and adsorption characteristics. It can also be added for improving the strength and mechanical properties. Different nanofillers that are added will have different properties and functions. Usually, nanofillers are added as pigment agents. These pigment agents are one of the important components in the paper industry. It affected the final properties of the paper produced (15, 20). The paper pigments are used from inorganic compounds such as calcium carbonate and kaolin (8, 21). The use of nano fillers for pigments will help fill the voids and empty gap. It would also provide better barrier and retention properties (15). The other uses for nanofillers as for bleaching agents, retention agents, nano-minerals, nanosizing agents, improving the mechanical properties (tensile, tear, burst), superconductor agents (2, 22, 23) are discussed in detail in below sub-chapters.

10.2.1 Nano Calcium Silicate

Nano calcium silicate is the main compound of silica and calcium silicate (Ca_2O_4Si), also known as calcium orthosilicate. The addition of nano calcium silicate as nanofiller in pulp and paper results in reduction of paper quality and relative density, but at the same time increases in strength, bulk, smoothness, filler retention, opacity and brightness of the paper produced (24, 25). Using the conventional method, calcium silicate is prepared by a solid-state reaction of CaO or $CaCO_3$ and quartz (SiO_2) at higher temperature (1,150°C–1,200°C) for several hours (26, 27). Other methods are used such as chemical methods through combustion, sol-gel, co-precipitation followed by heat treatment at different temperatures used (28–30). The recent method for synthesizing nano calcium silicate is through a mechano-chemical route with the advantage of low cost, low processing energy (reaction carried out at room temperature), mass production with homogenous and uniform nano-sized component (31). It is also environmentally friendly and waste-free. In this method, it uses a high energy milling process that involves repeated mixing, deformation, commuting, welding, and re-welding of the reactant particles powder in a closed vial in ball milling (31).

10.2.2 Nano Calcium Carbonate ($CaCO_3$)

Nanofiller calcium carbonate ($CaCO_3$) is one of the common nanofillers used in paper production to give a brightness. It was reported that $CaCO_3$ provides higher brightness than normal clay and lower prices as compared with titanium oxide (8, 32). $CaCO_3$ is available as ground calcium carbonate and precipitated calcium carbonate. The ground calcium carbonate is produced from grinding limestone, marble, or chalk, whereas the precipitated calcium carbonate is produced by the carbonation of lime or with a special precipitation process (special calcium carbonate pigments are used) (33, 34). A wide variety of shapes and sizes is produced from precipitation process (34). The $CaCO_3$ nanofillers that are prepared by the carbonation method are applied for wet end papermaking applications (8, 15). The $CaCO_3$ nanofillers have a higher cost of production and poor retention properties. Usually, water-soluble additives are used to control the morphology, size, and surface properties. Additives such as chitosan and starch are used to improve the $CaCO_3$ nanofiller strength properties of papers (8, 35). Sometimes, surfactants (anionic and cationic surfactants) are added to the paper processing to improve the properties (8). A study by El-Sherbiny et al. (8) using $CaCO_3$ nanofillers in wet end papermaking showed higher brightness and opacity compared with commercial ones. It was also reported that there were decreases in mechanical strength such as burst index and tear strength, but for tensile strength, there was a slight increase from 1.52% to 6.53%, respectively. This is due to increase of filler retention and fiber-to-fiber bonds. Another report by Fortună et al. (1) showed that the addition of small amount of $CaCO_3$ nanofillers is not significant in reducing strength and, at the same time, gives higher opacity and retention. Another study was reported by Morsy et al. (12) on using hybrid $CaCO_3$ nanofillers mixed with nano silica via the sol-gel method in papermaking to improve the optical properties of paper, such as brightness, whiteness, and opacity. But the drawbacks were in the reduction of mechanical properties when compared with commercial $CaCO_3$ nanofillers.

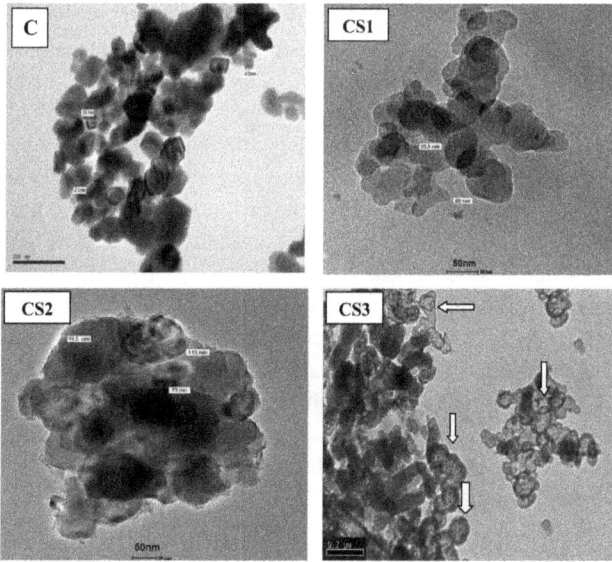

FIGURE 10.1   TEM images of CaCO$_3$ core particles (C) and SiO$_2$/CaCO$_3$ nanocomposites (CS1), (CS2) and (CS3). (Reproduced from ref (12).)

Figure 10.1 shows the TEM micrograph of silica and CaCO$_3$ nanofillers. The nano-silica has a particle size distribution range between 20 and 30 nm with a spherical shape, whereas the CaCO$_3$ nanofilers (C) and hybrid CaCO$_3$/SiO$_2$ (denoted by CS1, CS2, and C3) nanofillers have a sphere like with particle size ranging from 30 to 70 nm, 40 to 80 nm, respectively. The CaCO$_3$/SiO$_2$ in paper produced brighter paper as compared with CaCO$_3$ nanofillers only.

10.2.3   NANO CLAY

Nano clay can be added to pulp and paper manufacturing process to give some rigidity. The addition of this nanofiller could enhance the voids and gaps in pulp and paper while improving the mechanical strength. Furthermore, it can also enhance the gloss, opacity, and surface resistance of the coated paper (36, 37). The term clay refers to naturally occurring materials composed of fine-grained materials that have appropriate water contents and will harden when dried or fired (37). Clays and clay minerals have been widely utilized in many applications, such as agriculture, engineering, pulp and papers, and many more (37). The nanolcay from momontotrile (MMT) or bentonite is a very hydrophilic clay formed by layers of silica tetrahedral and alumina octahedral sheets in a ratio of 2:1 (38). Due to its abundant availablity and low cost, it was commonly studied for many applications, including in pulp and paper industries (39). It was reported that the addition of nanoclay, momontotrile (MMT), with an amount of 2% could improve the opacity, brightness, and tensile strength (40). This also indicated that the MMT would give positive results in paper processing. The nanoclay can also be used as a coating on paper to improve its mechanical and barrier properties (36, 39). The nano clay can be added in large amounts because of its abundantly available resources and ease of to obtaining the raw materials (38).

10.2.4 Nano Kaolin

Kaolin is one of most common types of fillers applied in many applications, including the pulp and paper industries (17, 34). It has a fine white color containing mineral kaolinite ($Al_2O_3 \cdot 2SiO_2 \cdot 2H_2O$) as main compound and ingredient (41). This kaolinite is a hydrous aluminum silicate with single silica tetra hedral layer that is linked through oxygen atoms to a single alumina octa hedral layer (41). Kaolin with a higher level of concentrations or loadings in papermaking affected the paper strength, stiffness, and bulking. On top of that, kaolin also has poor retention (34). Thus, making the fibers on paper is difficult to hydrophobize. There is a need for some modification in turn to improve the filler retention and negative impact of this filler (17). Filler modification with added bio-based modifier and poly-saccharide-based polymers is considered as a promising approach (42).

A study by Naijian et al. (17) on the application of three types of bio-based modified kaolin clay that are made from cationic starches, maize, and tapioca in papermaking as additives reported the improvement and well distributed of kaolin-starch fillers and bonded together to paper fibers. The kaolin–starch filler had a higher tendency to form cluster between 3.5 and 7.5 times larger compared with unmodified kaolin filler. Furthermore, the brightness and mechanical properties strength were better than unfilled papers as well paper filled with unmodified kaolin. In this research study, the stable hydrophobic properties were also observed for kaolin-starch filled paper.

Nano kaolin is modified using various methods to cover the disadvantages of kaolin such as from centrifuge or sedimentation process, chemical bleaching, acid activation, calcination or thermal treatment (17, 43–45). The surface modification is applied on kaolin using an intercalation process to modify the kaolin surface with the involvement and insertion of low molecular weight organic reagents that include guest molecules (dimethysulfoxide, formamide, potassium acetate, and urea) (46, 47). The insertion is applied on between layers consisting of two-dimensional arrangements of tetrahedral and octahedral sheets (46, 47). Bleaching using chemicals is a commonly used technique to enhance the brightness of nano kaolin for high-end paper products (17, 43). This process involves the discoloration of nano kaolin and the removal of iron particles in solution (43). Studies by Hassan et al. (48) as well as El-Gendy et al. (21) used different polymer materials together with nano kaolin after pulping. The results enhanced the mechanical properties and water absorption. Nano kaolin and nano calcium carbonate and are also used as coating in paper surface instead to brightness pigment on paper (2, 8, 17).

10.2.5 Nano Silica

Silica is one of the chemical compounds commonly used in paper coating, such as matte papers for in-jet printing (9). Silica can be synthesized using sol-gel method (49). The nano silica in papermaking improved whiteness, brightness, and opacity of

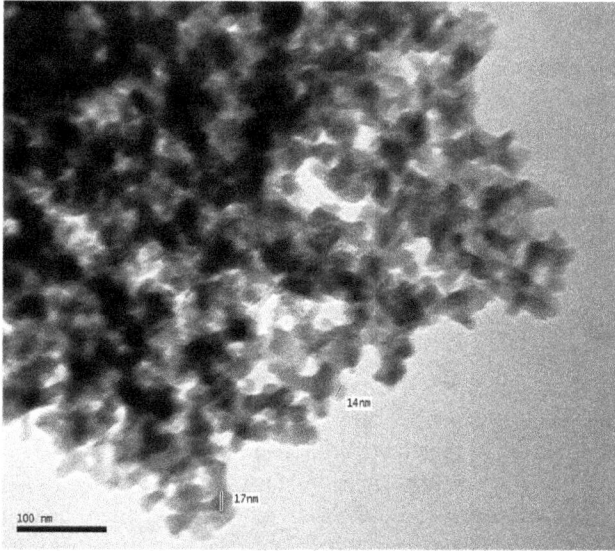

FIGURE 10.2 TEM micrograph of silica nano-particles. (Reproduced from ref. (12).)

the paper sheet as compared with unloaded or unmodified paper sheet (2, 9). This is due to the smaller particle size of nano silica, which provides a large surface area for higher light scattering on paper sheet as compared with unmodified or unloaded paper sheet (12). The introduction of nano silica resulted in the reduction of print through (2). And the nano silica also commonly provides paper with anti-bacterial properties that result in high degradation and high abatement (2). Figure 10.2 shows the TEM micrograph of nano silica with the nano silica has a narrow size distribution within a range of 20–30 nm. It also has a spherical shape and well dispersion. A study by Gamelas et al. (49) using silica in situ using sol-gel method improved the hydrogen-to-hydrogen bonding of cellulosic fibers of papers with silica. Another study by Lourenço et al. (9) using silica-modified ground calcium carbonate as nano-fillers, increased the tensile index from 16% to 20% and bulk from 7% to 13%. The enhanced fiber-to-filler bonding may be caused by the hydroxyl groups from silica coating and cellulosic fibers. The silica surface is mostly covered with chemical groups called as silanol groups (-Si-O) that have a very polar structure and are considered as chemically active (50).

The typical silica has semicrystalline phase. This was observed by XRD peaks in the study by Morsy et al. (12), with the existence of broad peak at $2\theta = 22.5°$ and also having higher retention properties. The study also concluded that the addition of nano silica in papermaking resulted in an improvement in burst and tear indexes due to fine particle size and higher retention. The small particles would cover a greater part of fiber surfaces and prevents inter-fiber contact over a larger fraction of surface area (33). But it was also reported that the tensile index was decreased with the addition of nano silica to 40.35 N m/g as compared with unmodified 72.74 N m/g (12).

10.2.6 Nano Titanium Dioxide

Nano titanum dioxide or nano TiO$_2$ is a fine white powder that is unreactive or chemically inert (20). Nano TiO$_2$ is classified based on crystalline arrangement. Both crystalline arrangements are known as anatase or rutile (20). The nano TiO$_2$ also known to have a significant number of hydroxyl groups on its surface. The low light absorption, high light reflectivity, and nano-size make this pigment ideal to obtain better opacity (13, 23). Thus, contribute to excellence performance of optical brightness and opacity of paper sheets. Nano TiO$_2$ is commonly used in pulp and paper industry to increase the brightness and opacity of paper (2, 13). Furthermore, it has antimicrobial properties that are non-toxic with potential bactericidal and fungicidal applications for packaging surfaces (51, 52).

Some studies also reported the use of nano TiO$_2$ for retention agent for supporting materials on paper sheet (53, 54). Nano TiO$_2$ is also used as a retention filter for end wet papermaking. It has excellent retention and filter effects. Nano TiO$_2$ suspension is possible to be coated on the paper sheet surface and wet end papermaking. Another possibility is to deposit nano TiO$_2$ onto individual fibers before the paper sheet formation. This process will have the bulking distribution of nano TiO$_2$ loading on the paper sheet. Furthermore, it effectively inhibits the pollution of white water by harmful substances (2). In a study by Huang et al. (7) that prepared a paper hand sheets with addition of modified nano TiO$_2$ to improve the opacity properties and hydrophobic properties of paper, the results had excellence hydrophobic properties with decreased tensile strength and increased opacity index. This is due to the uniform distribution obtained from modified nano TiO$_2$ with MPS ((3-trimethozysilyl) propyl methacrylate) coupling agent that were added before final paper sheet fabrication. In a different study by El-Sherbiny et al. (20) on nano TiO$_2$ as a special paper coating pigment showed that the addition of nano TiO$_2$ increased brightness and opacity of the coated paper. But at the same time, the paper roughness and air permeance decreased until they were stable at 50% level.

10.2.7 Nano Cellulose

Nanocellulose, also known as nano crystalline cellulose (NCC), nano fibrillated cellulose (NFC) or bacterial nanocellulose (BNC), is a potential material to be used in pulp and paper industries. Nanocellulose is made from wood and other bio-resources (kenaf, jute, pineapple leaves, and others) including bacteria that produce nanocellulose such *Gluconacetobater xylus* (55). Commonly, acid hydrolysis techniques are used for nano cellulose extraction (56). Figure 10.3 shows the hydrolysis of nanocellulose from cellulose using hydrochloric acid (HCl) acid. Due to its low cost, abundant availability, ease of surface modification, excellent mechanical properties, and biodegradability, it is one of the best nanofillers and additives to be added in pulp and papermaking through various methods such as in wet end papermaking, laminating, and coating (57). Application of nanocellulose in wet end application causes flocculation while at the same time strengthening the paper (57). The paper coating is done by covering the cellulosic fibers and filling the spaces between them with agents that are generally from binders, pigments, thickeners, dispersants, cross-linkers, lubricants,

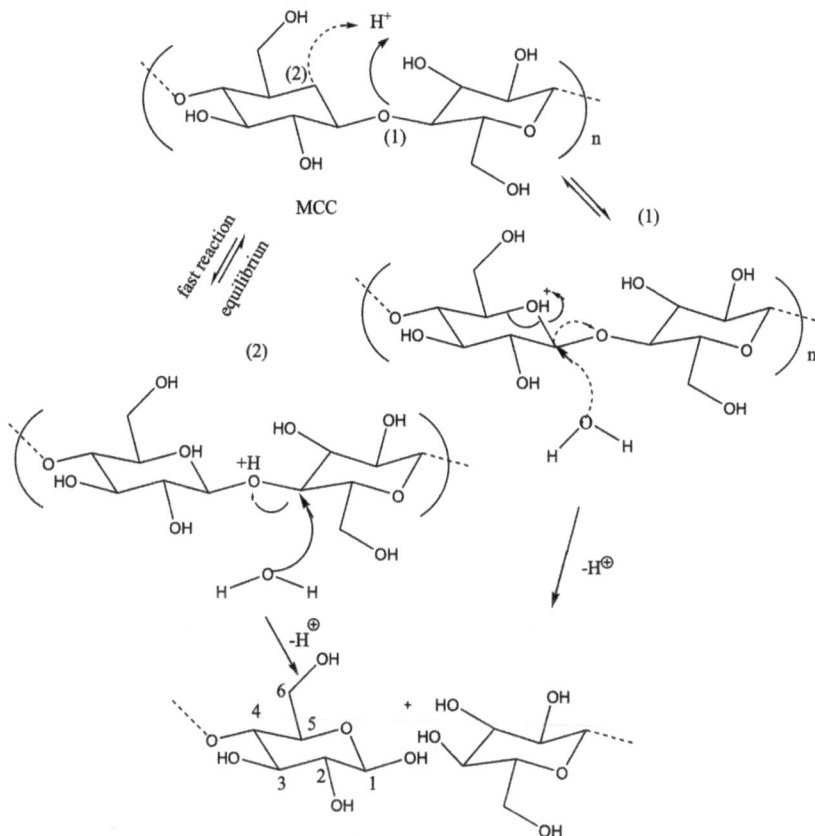

FIGURE 10.3 Acid hydrolysis of nanocellulose mechanism.

and optical brightening agents (11). Nanocellulose has excellent capacity to form an interconnected network due to OH bonds and plays a major role in influencing the coating performance with ease of surface modification or functionalization. With this modification, it can act as a gas barrier and improve the hydrophobicity, antibacterial, and UV protection behavior (57).

Nanocellulose is easily chemically functionalized with chemical treatments to change the polarity and improve the dispersion and bonding between fibers (58, 59). Chemical modifications or treatments are common ways to enhance compatibility and improve the dispersibility between polar and non-polar interactions. The OH groups are abundantly available in the nanocellulose structure, with three OH groups located at C2, C3, and primary OH group at C6, which is considered the most reactive and can be substituted with other functional groups such as acetyl, carbonyl, carboxyl, and others such as esterification and oxidation (Figure 10.4) (55, 58). The surface modification of nanocellulose can be done through TEMPO-mediated oxidation, acetylation/esterification, sulfonation, amination, carbamation, non-covalent crosslinking, grafting-onto and grafting from polymer backbone (55).

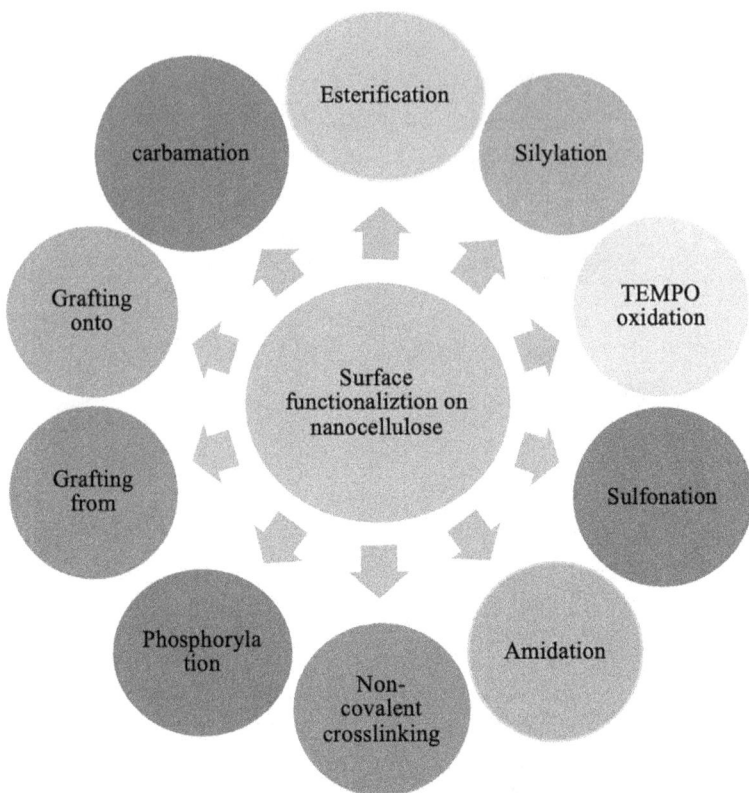

FIGURE 10.4   Surface functionalization methods on nanocellulose.

10.2.8   Nano Zinc Oxide (ZnO)

Nano zinc oxide or nano ZnO is a multifunctional mineral that is commonly used as a pigment and is suitable for paper coating (60). It exists as a mineral zincite and is non-toxic to human health (61). Nano ZnO has varied particle structures such as needles, rods, prims, sheets, triangles, squares, wires, ribbons, combs and tubes (61). It can also exist in two or three-dimensional structures. All these depend on methods of production of nanoscale ZnO such as vapour deposition micro-emulsion, sol-gel process, hydrothermal and precipitation (62–65). It is always a good choice for paper that needs greater printing properties and exposing brightness. On top of that, the nano ZnO also provides anti-fungal and UV protection due to its nano-size and zinc oxide properties (6, 61). Some studies used the nano ZnO/hybrid with starch composites as a stabilizing agent (6). The inorganic oxide such as nano ZnO has bigger advantages due to their robustness, high stability at elevated temperature, long shelf life, and ability to withstand harsh conditions (64, 65). Nano ZnO also shows microbial activity without photoactivation, as compared with TiO2, which needs photoactivation for antimicrobial activity to happen (66).

A study by Ma et al. (6) on nano ZnO mixed with starch for paper coating showed that the ZnO-starch nanocomposite and ZnO nanoparticles successfully blocked

UV radiation and had efficient antibacterial capability. Moreover, the coating on the paper surface could enhance the smoothness of the coated paper and the picking velocity as compared with the uncoated paper. The antibacterial properties were also investigated by another researcher Sobri et al. (61) who used nano ZnO on bamboo bleached pulp to produce antibacterial paper. The different percentages of zinc (15.08%–34.08%) and oxygen (17.45%–32.59%) were used. A higher percentage of precursors exhibited a more amorphous structure, and a measurement of more than 30% increment for the inhibition zone was reported for 10.00–25.00 mm against *S. aureus*, *S. choleraesuis*, and *E. coli*. Precursors more than 0.3M enhanced the growth of zinc oxide, providing better antibacterial properties.

10.3 CHALLENGES AND FUTURE RECOMMENDATIONS

Developing filler in nanosize for pulp and paper with intended desirable properties has become a major concern and challenge. The most challenging issue that needs to be reconsidered in the manufacturing process for industries is cost. The cost of pulp and paper manufacturing has always been an important issue in selecting the right nanofillers to be used during the manufacturing process. Table 10.1 shows the overall advantages and disadvantages of the nanofillers used in the pulp and paper process. The nanofillers that come from high-cost manufacturing process, mostly from inorganic sources such as nano titanium and zinc oxide, need to be used wisely, even though the improvement in properties is excellent compared with others. Moreover, the addition of inorganic nanofillers leads to non-environmentally friendly products. The used inorganic materials, such as nano calcium silicate, required high temperature and long processing time, resulting in only low yield of product. This also causes the cost of manufacturing paper and end products to be higher for consumers.

The need to look for more abundant natural resources, such as from organics and minerals, has become a priority for scientists and technologists. Commonly used natural resources such as nano calcium carbonate, nanoclay, and nano kaolin have good and comparable opacity, brightness, and mechanical properties when added in the pulp and paper process. The natural resources and organic nanofiller such as nanocellulose are becoming popular among researchers and scientists due to their high tensile, tear, and burst strength for papermaking; they are also environmentally friendly and not hazardous for human health. Whereas, the use of calcium silicate and calcium carbonate is both hazardous to health when inhaled by humans in a harsh environment for processing that involves making both nanofillers. Furthermore, the titanium dioxide when exposed to high concentrations, caused carcinogens. Organic nanofillers in pulp and paper will be in increasing demand for the next few decades to reduce the dependency on inorganic and mineral nanofillers. It's more environmentally friendly in terms of products and waste management. Nanofillers made from natural or bio-resources and organic materials have a broad future for potential use in pulp and papermaking. The nanofiller that has been added in papermaking was varied from 5% to 37% based on paper qualities and grades (67). Some of these nanofillers, even though they improve some properties such as brightness, have compromised the mechanical properties such as paper strength due to their ability to weaken the inter-fiber bonding (68). This is due to higher addition of fillers within

TABLE 10.1

The Advantages and Disadvantages of the Nanofillers Used in Pulp and Paper Process

Nanofiller	Advantages	Disadvantages
Nano calcium silicate	Improved for stiffness, whiteness and printing color density of paper.	Tightness, folding strength and tensile strength of paper decreases. Ash content increased with increases of amount of fillers. Required high temperature and long processing time, low yield and hazardous to human and environment also reduction in paper quality.
Nano calcium carbonate	Cost effective, abundantly available, give more brightness than clay,	Low in quality of paper produced
Nano clay	Enhancing the voids and gaps in pulp and paper and improving the mechanical strength. Improving gloss, opacity, brightness, tensile strength, barrier properties and surface resistance of the coated paper, abundantly available, low in processing cost.	Less brightness than calcium carbonate, high in hydrophilic properties resulting in agglomeration.
Nano kaolin	Used as paper coating and brightness pigment.	Poor retention, higher level loadings in papermaking affected the paper strength, stiffness and bulking.
Nano silica	Used as paper coating, provide whiteness, brightness and opacity of the paper sheet, higher light scattering, provide paper with anti-bacterial properties associated with high degradation and high abatement. Improve burst and tear indexes due to fine particle size and higher retention.	Higher in silanol and hydroxyl groups. Reduce the tensile strength of paper.
Nano titanium dioxide	Excellent brightness, good optical density, good light scattering, high refractive index, paper coating will have high strength, excellent retention and filter effects.	Expensive, exposure at high concentration can cause irritation to eyes, nose and throat, and carcinogen
Nano zinc oxide	Antioxidant, antimicrobial, safer than TiO_2, as pigment and coating paper, produce high quality paper and brightness, robustness, high stability at elevated temperature, long shelf life and withstand a harsh condition	Expensive process, inorganic fillers, not environmentally friendly.
Nano cellulose	Abundantly available resources, organic material, biodegradability, easy for surface modification, low in cost processing, high tensile, tear strengths and burst index.	Brittle effect due to high loading, poor dimensional stability and high in hydrophilic properties that lead to agglomerated form of nanocellulose

pulp and paper processes that led to poor filler interaction and loss of paper bulk (68). The addition of nano size fillers could improve the interaction and bonding within the pulp and paper processes. The tiny parts (pores) in pulp and paper would be fully covered, reducing the pulp retention process.

10.4 CONCLUSION

The nanofiller technology has been considered to be used in many applications, especially in the pulp and paper industries, but the need to look at and consider the cost and enhancement has become a concern for industrial main player as well as researchers and scientists. In addition, some nanofillers, such nano calcium silicate and nano CaCO3, improve brightness and opacity but at the same time reduce the mechanical properties. Other nanofillers, such as nanocellulose, increase the mechanical properties when incorporated into pulp and paper processing. Furthermore, the nano ZnO and nanoTiO$_2$ both have antimicrobial and antibacterial properties when added in pulp and paper processing. The addition of a small amount of nanofillers would have a positive effect on intended application uses. The future of nanofillers relies on the availability of the processing of the nanofillers to be added in pulp and paper processes. Nanofillers from natural resources have emerged, and the potential use of these nanofillers in the future is being given significant attention by researchers.

ACKNOWLEDGEMENTS

The author would like to thank you for the scientific and technical support provided by Universiti Sains Malaysia and M.N.A.M Taib profoundly acknowledges Universiti Sains Malaysia for the Postdoctoral Scheme.

REFERENCES

1. Fortună, M. E., Lobiuc, A., Cosovanu, L. M., & Harja, M. (2020). Effects of in-situ filler loading vs. conventional filler and the use of retention-related additives on properties of paper. *Materials, 13*(22), 5066. https://doi.org/10.3390/ma13225066.
2. Julkapli, N. M., & Bagheri, S. (2016). Developments in nano-additives for paper industry. *Journal of Wood Science, 62*, 117–130. https://doi.org/10.1007/s10086-015-1532-5.
3. Haile, A., Gelebo, G. G., Tesfaye, T., Mengie, W., Mebrate, M. A., Abuhay, A., & Limeneh, D. Y. (2021). Pulp and paper mill wastes: Utilizations and prospects for high value-added biomaterials. *Bioresources and Bioprocessing, 8*, 1–22. https://doi.org/10.1186/s40643-021-00385-3.
4. Eugenio, M. E., Ibarra, D., Martín-Sampedro, R., Espinosa, E., Bascón, I., & Rodríguez, A. (2019). Alternative raw materials for pulp and paper production in the concept of a lignocellulosic biorefinery. *Cellulose, 12*, 78.
5. Deshwal, G. K., Panjagari, N. R., & Alam, T. (2019). An overview of paper and paper based food packaging materials: Health safety and environmental concerns. *Journal of Food Science and Technology, 56*, 4391–4403. https://doi.org/10.1007/s13197-019-03950-z.

6. Ma, J., Zhu, W., Tian, Y., & Wang, Z. (2016). Preparation of zinc oxide-starch nanocomposite and its application on coating. *Nanoscale Research Letters*, *11*, 1–9. https://doi.org/10.1186/s11671-016-1404-y.

7. Huang, L., Chen, K., Lin, C., Yang, R., & Gerhardt, R. A. (2011). Fabrication and characterization of superhydrophobic high opacity paper with titanium dioxide nanoparticles. *Journal of Materials Science*, *46*, 2600–2605. https://doi.org/10.1007/s10853-010-5112-1.

8. El-Sherbiny, S., El-Sheikh, S. M., & Barhoum, A. (2015). Preparation and modification of nano calcium carbonate filler from waste marble dust and commercial limestone for papermaking wet end application. *Powder Technology*, *279*, 290–300. https://doi.org/10.1016/j.powtec.2015.04.006.

9. Lourenço, A. F., Gamelas, J. A., Sequeira, J., Ferreira, P. J., & Velho, J. L. (2015). Improving paper mechanical properties using silica-modified ground calcium carbonate as filler. *BioResources*, *10*(4), 8312–8324.

10. Mohieldin, S. D., Zainudin, E. S., Paridah, M. T., & Ainun, Z. M. (2011). Nanotechnology in pulp and paper industries: A review. *Key Engineering Materials*, *471*, 251–256. https://doi.org/10.4028/www.scientific.net/KEM.471-472.251.

11. Sangl, R., Auhorn, W., Kogler, W., & Tietz, M. (2013). Surface sizing and coating. In *Handbook of Paper and Board*. pp. 745–84.

12. Morsy, F. A., El-Sheikh, S. M., & Barhoum, A. (2019). Nano-silica and $SiO_2/CaCO_3$ nanocomposite prepared from semi-burned rice straw ash as modified papermaking fillers. *Arabian Journal of Chemistry*, *12*(7), 1186–1196. https://doi.org/10.1016/j.arabjc.2014.11.032.

13. Rawski, D. P. (2001). Pulp and paper: Nonfibrous components. *Encyclopedia of Materials: Science and Technology*, 7908–7910. https://doi.org/10.1016/B0-08-043152-6/01423-6.

14. Smook, G. A., & Kocurek, M. J. (1982). *Handbook for Pulp & Paper Technologists*. Canadian Pulp and Paper Association.

15. Shen, J., Song, Z., Qian, X., Yang, F., & Kong, F. (2010). Nanofillers for papermaking wet end applications. *BioResources*, *5*(3), 1328–1331.

16. Bajpai, P. (2018). *Biermann's Handbook of Pulp and Paper: Volume 1: Raw Material and Pulp Making*. Elsevier.

17. Naijian, F., Rudi, H., Resalati, H., & Torshizi, H. J. (2019). Application of bio-based modified kaolin clay engineered as papermaking additive for improving the properties of filled recycled papers. *Applied Clay Science*, *182*, 105258. https://doi.org/10.1016/j.clay.2019.105258.

18. Börjesson, M., & Westman, G. (2015). Crystalline nanocellulose-preparation, modification, and properties. *Cellulose-Fundamental Aspects and Current Trends*, *7*. https://dx.doi.org/10.57772/61899.

19. Chung, H., & Washburn, N. R. (2016). Extraction and types of lignin. *Lignin in Polymer Composites*, *10*, 13–25.

20. El-Sherbiny, S., Morsy, F., Samir, M., & Fouad, O. A. (2014). Synthesis, characterization and application of TiO 2 nanopowders as special paper coating pigment. *Applied Nanoscience*, *4*, 305–313. https://doi.org/10.1007/s13204-013-0196-y.

21. El Gendy, A., Khiari, R., Bettaieb, F., Marlin, N., & Dufresne, A. (2014). Preparation and application of chemically modified kaolin as fillers in Egyptian kraft bagasse pulp. *Applied Clay Science*, *101*, 626–631. https://doi.org/10.1016/j.clay.2014.09.032.

22. Lourenço, A. F., Gamelas, J. A., Sarmento, P., & Ferreira, P. J. (2019). Enzymatic nanocellulose in papermaking-The key role as filler flocculant and strengthening agent. *Carbohydrate Polymers*, *224*, 115200. https://doi.org/10.1016/j.carbpol.2019.115200.

23. Chauhan, V. S., & Chakrabarti, S. K. (2012). Use of nanotechnology for high performance cellulosic and papermaking products. *Cellulose Chemistry and Technology*, *46*(5), 389.

24. Liu, Q. X., Yin, Y. N., & Xu, W. C. (2013). Study on application of hydrated calcium silicate in paper from wheat straw pulp. In *Advanced Materials Research* (Vol. 774, pp. 1277–1280). Trans Tech Publications Ltd. https://doi.org/10.4028/www.scientific.net/AMR.774-776.1277.

25. Song, S., Zhang, M., He, Z., Li, J. Z., & Ni, Y. (2012). Investigation on a novel fly ash based calcium silicate filler: Effect of particle size on paper properties. *Industrial & Engineering Chemistry Research*, *51*(50), 16377–16384. https://doi.org/10.1021/ie3028813.

26. Yen, W. M., & Weber, M. J. (2004). *Inorganic Phosphors: Compositions, Preparation and Optical Properties*. CRC Press.

27. Shionoya, S., Yen, W. M., & Yamamoto, H. (Eds). (2018). *Phosphor Handbook*. CRC Press.

28. Khristov, T. I., Popovich, N. V., Galaktionov, S. S., & Soshchin, N. P. (1994). Calcium silicate phosphors obtained by the sol-gel method. *Glass and Ceramics*, *51*(9–10), 290–296.

29. Dhoble, S. J., Dhoble, N. S., & Pode, R. B. (2003). Preparation and characterization of Eu 3+ activated CaSiO3, (CaA) SiO 3 [A= Ba or Sr] phosphors. *Bulletin of Materials Science*, *26*, 377–382. https://doi.org/10.1007/BF02711179.

30. Nishisu, Y., Kobayashi, M., Ohya, H., & Akiya, T. (2006). Preparation of Eu-doped alkaline-earth silicate phosphor particles by using liquid-phase synthesis method. *Journal of Alloys and Compounds*, *408*, 898–902. https://doi.org/10.1016/j.jallcom.2005.01.101.

31. Singh, S. P., & Karmakar, B. (2011). Mechanochemical synthesis of nano calcium silicate particles at room temperature. *New Journal of Glass and Ceramics*, *1*(2), 49–52.

32. Shen, J., Song, Z., Qian, X., & Ni, Y. (2011). A review on use of fillers in cellulosic paper for functional applications. *Industrial & Engineering Chemistry Research*, *50*(2), 661–666. https://doi.org/10.1021/ie1021078.

33. Hubbe, M. A., & Gill, R. A. (2004). Filler particle shape vs. paper properties-A review. In *Proceedings of Spring Technology Conference*. TAPPI Press, Atlanta, GA.

34. Hubbe, M. A., & Gill, R. A. (2016). Fillers for papermaking: A review of their properties, usage practices, and their mechanistic role. *BioResources*, *11*(1).

35. Zhao, Y. (2005). Improvement of paper properties using starch-modified precipitated calcium carbonate filler. *TAPPI Journal*, *4*, 3–7.

36. Ghanbari, H., Kasmani, J. E., & Samariha, A. H. M. A. D. (2019). Improving printing and writing paper properties by coating with nanoclay montmorillonite (K10). *Cellulose Chemistry and Technology*, *53*, 395–403.

37. Gaikwad, K. K., & Ko, S. (2015). Overview on polymer-nano clay composite paper coating for packaging application. *Journal of Material Sciences and Engineering*, *4*(1), 151.

38. Floody, M. C., Theng, B. K. G., Reyes, P., & Mora, M. L. (2009). Natural nanoclays: Applications and future trends-a Chilean perspective. *Clay Minerals*, *44*(2), 161–176. https://doi.org/10.1180/claymin.2009.044.2.161.

39. de Oliveira, M. L. C., Mirmehdi, S., Scatolino, M. V., Júnior, M. G., Sanadi, A. R., Damasio, R. A. P., & Tonoli, G. H. D. (2021). Effect of overlapping eco-friendly cellulose nanofibrils and nanoclay layers on mechanical and barrier properties of spray-coated papers.

40. Kasmani, J. E., & Samariha, A. (2021). Effects of montmorillonite nanoclay on the properties of chemimechanical pulping paper. *Bioresources*, *16*(3), 6281–6291.

41. Olaremu, A. G. (2015). Physico-chemical characterization of Akoko mined kaolin clay. *Journal of Minerals and Materials Characterization and Engineering*, *3*(05), 353. https://doi.org/10.4236/jmmce.2015.35038.

42. Li, Q., Wang, S., Jin, X., Huang, C., & Xiang, Z. (2020). The application of polysaccharides and their derivatives in pigment, barrier, and functional paper coatings. *Polymers*, *12*(8), 1837. https://doi.org/10.3390/polym12081837.

43. González, J. A., & Ruiz, M. D. C. (2006). Bleaching of kaolins and clays by chlorination of iron and titanium. *Applied Clay Science*, *33*(3–4), 219–229. https://doi.org/10.1016/j.clay.2006.05.001.

44. Kassa, A. E., Shibeshi, N. T., & Tizazu, B. Z. (2022). Characterization and optimization of calcination process parameters for extraction of aluminum from Ethiopian kaolinite. *International Journal of Chemical Engineering*, *2022*. https://doi.org/10.1155/2022/5072635.

45. Tironi, A., Trezza, M. A., Irassar, E. F., & Scian, A. N. (2012). Thermal treatment of kaolin: Effect on the pozzolanic activity. *Procedia Materials Science*, *1*, 343–350. https://doi.org/10.1016/j.mspro.2012.06.046.

46. Cheng, H., Zhang, S., Liu, Q., Li, X., & Frost, R. L. (2015). The molecular structure of kaolinite-potassium acetate intercalation complexes: A combined experimental and molecular dynamic simulation study. *Applied Clay Science*, *116*, 273–280. https://doi.org/10.1016/j.clay.2015.04.008.

47. Martens, W. N., Frost, R. L., Kristof, J., & Horvath, E. (2002). Modification of kaolinite surfaces through intercalation with deuterated dimethylsulfoxide. *The Journal of Physical Chemistry B*, *106*(16), 4162–4171. https://doi.org/10.1021/jp0130113.

48. Hasan, A., & Fatehi, P. (2018). Stability of kaolin dispersion in the presence of lignin-acrylamide polymer. *Applied Clay Science*, *158*, 72–82. https://doi.org/10.1016/j.clay.2018.02.048.

49. Gamelas, J. A., Lourenco, A. F., & Ferreira, P. J. (2011). New modified filler obtained by silica formed by sol-gel method on calcium carbonate. *Journal of Sol-Gel Science and Technology*, *59*, 25–31. https://doi.org/10.1007/s10971-011-2456-1.

50. Sahakaro, K. (2017). Mechanism of reinforcement using nanofillers in rubber nanocomposites. In *Progress in Rubber Nanocomposites* (pp. 81–113). Woodhead Publishing. https://doi.org/10.1016/B978-0-08-100409-8.00003-6.

51. Rezić, I., Haramina, T., & Rezić, T. (2017). Metal nanoparticles and carbon nanotubes-Perfect antimicrobial nano-fillers in polymer-based food packaging materials. In *Food Packaging* (pp. 497–532). Academic Press. https://doi.org/10.1016/B978-0-12-804302-8.00015-7.

52. Corrales, M., Fernández, A., & Han, J. H. (2014). Antimicrobial packaging systems. In *Innovations in Food Packaging* (pp. 133–170). Academic Press. https://doi.org/10.1016/B978-0-12-394601-0.00007-2.

53. Adjimi, S., Roux, J. C., Sergent, N., Delpech, F., Thivel, P. X., & Pera-Titus, M. (2014). Photocatalytic oxidation of ethanol using paper-based nano-TiO2 immobilized on porous silica: A modelling study. *Chemical Engineering Journal*, *251*, 381–391. https://doi.org/10.1016/j.cej.2014.04.013.

54. Senadeera, G. K. R., Kitamura, T., Wada, Y., & Yanagida, S. (2006). Enhanced photoresponses of polypyrrole on surface modified TiO2 with self-assembled monolayers. *Journal of Photochemistry and Photobiology A: Chemistry*, *184*(1–2), 234–239. https://doi.org/10.1016/j.jphotochem.2006.04.033.

55. Mohamad Nurul Azman Mohammad, T., Sue Yee, T., & Hazwan Hussin, M. et al. Modification on nanocellulose extracted from kenaf (*Hibiscus cannabinus*) with 3-aminopropyltriethoxysilane for thermal stability in poly (vinyl alcohol) thin film composites, 03 November 2022, PREPRINT (Version 1) available at Research Square https://doi.org/10.21203/rs.3.rs-2143980/v1.

56. Taib, M. N. A. M., Yehye, W. A., & Julkapli, N. M. (2020). Synthesis and characterization of nanocrystalline cellulose as reinforcement in nitrile butadiene rubber composites. *Cellulose Chemistry and Technology*, *54*(1), 11–25.

57. Spagnuolo, L., D'Orsi, R., & Operamolla, A. (2022). Nanocellulose for paper and textile coating: The importance of surface chemistry. *ChemPlusChem*, *87*(8), e202200204. https://doi.org/10.1002/cplu.202200204.

58. Taib, M. N. A. M., Yehye, W. A., Julkapli, N. M., & Hamid, S. B. O. A. (2018). Influence of hydrophobicity of acetylated nanocellulose on the mechanical performance of nitrile butadiene rubber (NBR) composites. *Fibers and Polymers, 19*, 383–392. https://doi.org/10.1007/s12221-018-7591-z.

59. Habibi, Y. (2014). Key advances in the chemical modification of nanocelluloses. *Chemical Society Reviews, 43*(5), 1519–1542. https://doi.org/10.1039/C3CS60204D.

60. Sobri, Z., Ainun, Z. M. A., & Zainudin, E. S. (2018). Distribution of zinc oxide nanoparticles on unbleached and bleached bamboo paper via in-situ approaches. In *IOP Conference Series: Materials Science and Engineering* (Vol. 368, No. 1, p. 012046). IOP Publishing. https://doi.org/10.1088/1757-899X/368/1/012046.

61. Sobri, Z., Asa'ari, A. Z. M., Yacob, N., San H'ng, P., Abdullah, L. C., & Zainudin, E. S. (2021). In situ formation of zinc oxide on bamboo bleached pulp in preparation of antibacterial paper: Effect of precursors addition. *BioResources, 16*(3).

62. Cioffi, N., & Rai, M. (Eds.) (2012). *Nano-Antimicrobials: Progress and Prospects.*

63. Wang, Z. L. (2004). Zinc oxide nanostructures: Growth, properties and applications. *Journal of Physics: Condensed Matter, 16*(25), R829. https://doi.org/10.1088/0953-8984/16/25/R01.

64. Stoimenov, P. K., Klinger, R. L., Marchin, G. L., & Klabunde, K. J. (2002). Metal oxide nanoparticles as bactericidal agents. *Langmuir, 18*(17), 6679–6686. https://doi.org/10.1021/la0202374.

65. Kołodziejczak-Radzimska, A., & Jesionowski, T. (2014). Zinc oxide-from synthesis to application: A review. *Materials, 7*(4), 2833–2881. https://doi.org/10.3390/ma7042833.

66. Azizi-Lalabadi, M., Ehsani, A., Divband, B., & Alizadeh-Sani, M. (2019). Antimicrobial activity of Titanium dioxide and Zinc oxide nanoparticles supported in 4A zeolite and evaluation the morphological characteristic. *Scientific Reports, 9*(1), 17439. https://doi.org/10.1038/s41598-019-54025-0.

67. Song, S., Zhen, X., Zhang, M., Li, L., Yang, B., & Lu, P. (2018). Engineered porous calcium silicate as paper filler: effect of filler morphology on paper properties. *Nordic Pulp & Paper Research Journal, 33*(3), 534–541. https://doi.org/10.1515/npprj-2018-3045.

68. Lourenço, A. F., Gamelas, J. A., Zscherneck, C., & Ferreira, P. J. (2013). Evaluation of silica-coated PCC as new modified filler for papermaking. *Industrial & Engineering Chemistry Research, 52*(14), 5095–5099. https://doi.org/10.1021/ie3035477.

11 Design of Recycled Aluminium (AA7075)-Based Composites Reinforced with Nano Filler NiAl Intermetallic and Nano Niobium Powder Produced with Vacuum Arc Melting for Aeronautical Applications

Cagatay Kasar and Özgür Aslan
Atilim University

Fabio Gatamorta
University of Campinas

Ibrahim Miskioglu
Michigan Technological University

Emin Bayraktar
ISAE-Supmeca

11.1 INTRODUCTION

The development of NiAl intermetallic-reinforced composites is a useful solution for aeronautical and/or aerospace engineering due to their outstanding properties such as low density, high stiffness, high strength, high resistance to corrosion/oxidation, etc. In general, the application of TiAl, Ni-Al Nb$_2$Al and other intermetallics in the composites is structural materials that are being considered ideal new high-temperature

DOI: 10.1201/9781003400998-11

structural materials for civil and military applications [1,2]. For safety concerns in the aerospace area, the application of NiAl intermetallic requests a consistent manufacturing process such as diffusion bonding, with different materials to construct a new composite family [3–10].

Among them, nickel aluminide is generally used as a compound of NiAl regarding the compound of Ni_3Al due to its high corrosion resistance, low density and easy production [11]. NiAl shows good thermal conductivity, oxidation resistance and high melting temperature [12–17] which makes it very suitable for aeronautical applications. It means that it is ideal for high-temperature applications in gas turbines and jet engines.

In the frame of this work, NiAl intermetallic and niobium (Nb) were used in the recycled (fresh scrap) aluminium alloy, AA7075, for the high resistance composite production due to low manufacturing cost of certain parts of the turbo compressor. Due to the high and reliable mechanical properties of NiAl intermetallic-reinforced composites in the AA7075 alloy with a healthy and sound microstructure, a new design of these composites was developed in the frame of a joint research project with the French aeronautical society. A novel composite design was carried out with a special process in a vacuum arc melting oven (6,000°F). This process gives a strong chemical bonding diffusion compared to other processes such as sinter+ forging and/or 3D printing processes, etc. This process is cheaper and faster than the other processes if the final structure is compared. This process is followed by a second heat treatment to reduce the residual stresses and attain a relatively soft and ductile structure.

A perfect chemical diffusion bonding was carried out at the interface between matrix and reinforcements. By using this process in the frame of research collaboration with the French aeronautical society, detailed experimental tests were carried out to evaluate the static and cyclic properties of these composites. Microstructural analyses were carried out using scanning electron microscopy. A finite element method (FEM) was used based on the experimental results.

11.2   EXPERIMENTAL CONDITIONS

As a practical manufacturing process of these composites, two major reinforcements (received from VWR), fine nano NiAl (15 nm) and very fine Nb (<40 nm) powder, were added into the recycled fresh scrap AA7075 aluminium matrix. The recycled aluminium in the form of chips was supplied by the Brazilian aeronautical company. First, the recycled aluminium AA7075 chips were gas atomized and then mixed by high-energy milling in a planetary ball mill under an inert argon atmosphere to prevent oxidation of the powder (20/1 ball/powder ratio). Additionally, 3 wt% of zinc stearate was used as a lubricant during the preparation of the composite. After the milling operation, the thermal behaviour of the aluminium alloy (AA7075) powder was evaluated by differential scanning calorimetry–thermogravimetric analysis (DSC–TGA) and X-ray diffraction (XRD). Details of these experiments were given in former papers [1,3,4,6,7,9,11].

As minor reinforcements, molybdenum and copper (Mo 1 wt%, Cu 4 wt%, and GNPs 0, 15 wt%) were used. During the milling process, pure nano AA1050

(3–5 wt%) was added to homogenize the mixture of the recycled aluminium alloys. Biaxial compaction of the green compact specimens was done under 250 MPa. At the final stage, a novel composite design was carried out using a special process in a vacuum arc melting oven (6,000°F). The vacuum arc melting oven (6,000°F) was used for a short time (UNICAMP 2022), and the specimens were ready after cooling. This process is followed by second heat treatment to reduce the residual stresses and attain a relatively soft and ductile structure. For the static and cyclic, time-dependent properties of these composites, microstructural analyses were carried out using scanning electron microscopy. FEM was used based on the experimental results.

11.3 RESULTS AND DISCUSSION

Table 11.1 gives the composition of the four composites formulated for the innovative hybrid composite designed with two aluminium alloy AA7075 as a matrix. Besides the major reinforcements (Nb, NiAl), small amounts of Mo - Cu - GNPs increase the strength of the composites generating a strong cohesion of the reinforcements with the matrix mainly on the grain boundaries. For the fine distribution of the reinforcements and to obtain a fine grain size, nano Cu and nano Mo (<1 nm) and nano graphene platelets (GNPs, 500 m²/g with surface particle) were added to the matrix for each composite. Additionally, the presence of Mo and GNPs in the structure increases the mechanical resistance for toughening mainly due to a strong cohesion by chemical diffusion bonding at the interface between the matrix and reinforcements. We know that fine copper particles added to the composition even accelerated the chemical diffusion bonding in the matrix. The formation of the chemical diffusion bonding mechanism that will be presented in the next session is only indicative and should be improved with new measurements during the course of this research project.

Figure 11.1a shows an XRD diagram of NiAl intermetallic and Nb-reinforced AA7075-based composites indicating the phases, and additional information was given by "energy dispersive spectrometry (EDS)" chemical analysis for the composite used here. Figure 11.1b presents a mapping analysis of the microstructure for showing the distribution of the reinforced elements in the microstructure. Figure 11.1c gives a detailed analysis of the DSC diagram for AA7075 alloy and simulation of a fraction of solid depending on the temperature calculated with the software "Thermo-Calc" in the matrix to determine the critical transformation points during the heating and cooling stages.

TABLE 11.1(A)
Compositions of Four Composites (wt%)

Composite Name	AA7075	Nb	NiAl	GNP	Mo	Cu
E-I	B	25	35	0,15	1	1
E-II	B	35	25	0,15	1	1
E-III	B	20	25	0,15	1	1
E-IV	B	10	25	0,15	1	1

TABLE 11.1(B)
Chemical Composition of Scrap AA7075 Alloy (wt.%)

Element	Al	Cu	Fe	Mg	Mn	Si	Ni	Zn	Cr	Zr
wt.%	Balance	1.48	0.23	2.11	0.07	0.10	0.01	5.29	0.22	0.02

It seems that XRD patterns of the surface of the composite justify mainly an intermetallic NiAl phase which is supported by the EDS analyses of the composite structure. These analyses can justify the microstructure and mapping analyses showing the distribution of NiAl intermetallics and Nb particles. It is absolutely carried out with a strong chemical bonding diffusion between the matrix and reinforcements that give a high toughening mechanism.

Electrical conductivity levels were measured with an "Agilent 4338B Milli-ohm Meter". Three specimens were measured for each composite, and then, the mean values are given in Table 11.2. For the measurements, DC-regulated power supply voltage and current were set as 20V and 20A respectively. Data acquisition Card "NI9234" was connected in parallel with the output of the power to acquire the voltage data (voltage input accuracy was 24 bits). A high-precision multi-meter "Agilent U1253N" was connected in series to measure the current intensity (A). For the same specimens, thermal conductivity measurements carried out in our laboratory and also microhardness measurements taken from the samples produced under the same conditions for these composites are also presented in Table 11.2. All the data for the electrical and thermal measurements were revealed using the LabVIEW programme. These results obviously should be assumed as indicative data under the laboratory conditions. In Figure 11.2, the mechanical test device adapted on the Zwick test machine (ISAE-SUPMECA/Paris) for static compression test of the specimens and sub-size tensile test specimen was presented only as an example.

To understand the toughening mechanism, different types of comprehensive tests were conducted on the mechanical properties of the new hybrid aluminium-based composites. For the results of these tests, static and cyclic compression at a test speed of 1 mm/minute, time-dependent cyclic compression test with a test speed of 5 mm/second and also static tensile tests also were conducted, and all the results are presented in Figures 11.3a and b, respectively. All the static and cyclic compression tests were conducted using a pancake test specimen. The size of specimen E1 is width: 76,454 mm, depth: 102,108 and height: 205,232 mm and also the size of specimen E2 is width: 76,454 mm, depth: 102,108 and height: 205,232 mm.

The maximum stress generated in specimen E1 was about 1,200 MPa when the test was stopped due to the limit of the load cell, the Zwick test Machine. The other test results for specimens of E2, E3 and E4 were the same; we have stopped at a level of 1,000–1,200 MPa. Apparent Young's modulus estimated from these compression tests is variable between 90 and 96 GPa.

As for the comparison of the toughening mechanism of these composites, static tensile tests have been conducted according to DIN 50106 standards for only one composition containing 10 wt% Nb (E4) that has shown a typical stress-strain behaviour among other compositions. After that, one simulation has been carried out to

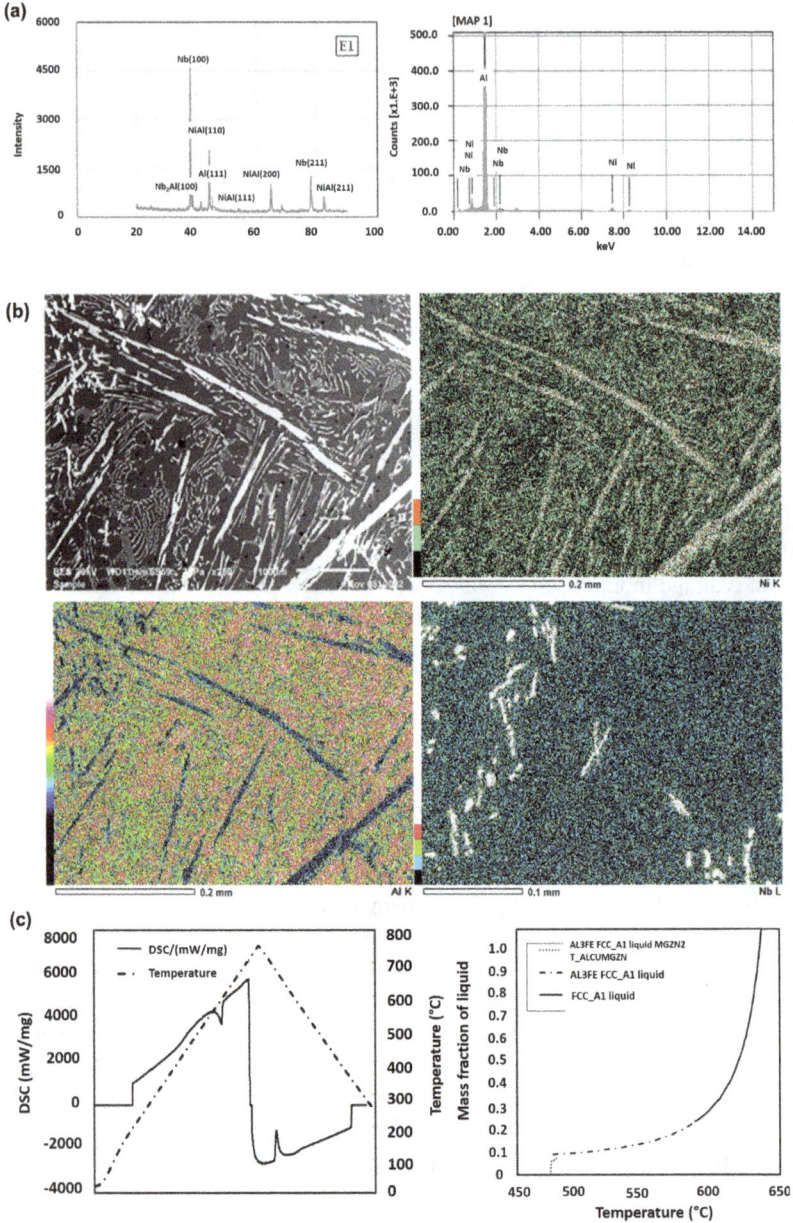

FIGURE 11.1 (a) XRD diagram of composite E1 indicating the phases and additional information was given by "EDS" chemical analysis for the composite used here, (b) Mapping analyses of the microstructure for showing the distribution of reinforced elements in the microstructure and (c) Differential scanning calorimetry (DSC) diagram measured for the AA7075 alloy with a heating rate of 5°C/minute and simulation of the fraction of solid depending on the temperature for AA7075.

TABLE 11.2

Electrical and Thermal Properties Measured for Four Composites with Micro Hardness Values

Composite Name	Electrical Conductivity at Ambient (S/m)	Thermal Conductivity (W/mK)	Microhardness $(HV_{0,1})$
E-I	5.35×10^9	4.330	585 ± 25
E-II	6.40×10^9	5.595	615 ± 30
E-III	7.20×10^9	9.115	315 ± 15
E-IV	9.10×10^9	8.755	285 ± 25

FIGURE 11.2 Vacuum arc melting oven (UNICAMP-CARAM's lab-Campinas/BR), static compression test on the test specimens (Michigan Tech-USA) and sub-size tensile test specimens. Mechanical test device adapted on the Zwick test machine (ISAE-SUPMECA/Paris).

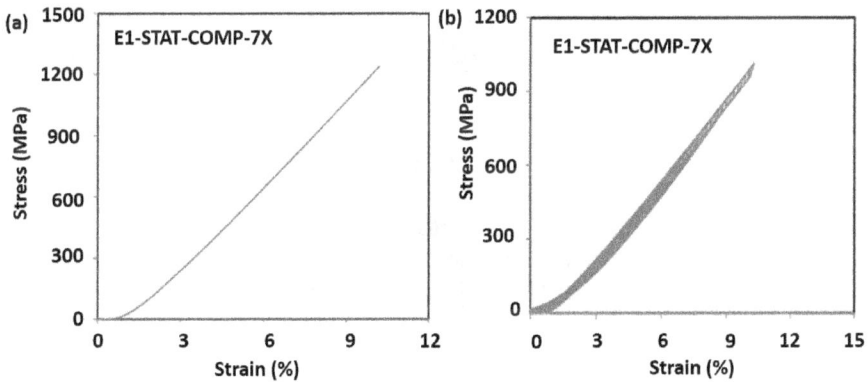

FIGURE 11.3 (a) Static compression test carried out on specimen E1 and (b) Cyclic compression test on specimen E1.

compare with the experimental result for this composition. For finite element analyses, a "large deformation theory" of rate-dependent plastic solids was used.

A multiplicative decomposition of the deformation gradient is used as a starting point:

$$F = F^e F^p$$

where F^e is the elastic distortion representing the local deformation of the material due to stretch and rotation and F^p is the plastic distortion representing the local deformation of the material due to the flow of defects (dislocations) through the microscopic structure. Hence, in a time integration process, F^e is directly calculated from:

$$F^e = FF^{p^{-1}}$$

where $\det(F^p) = 1$ and $\det(F^e) > 1$. After defining the elastic deformation gradient, constitutive equations for stress relations can be handled. First elastic rotation (R^e) and elastic stretch (U^e) tensors should be defined as:

$$F^e = R^e U^e$$

From the elastic stretch, Henky's logarithmic elastic strain is defined as:

$$E^e = \ln U^e$$

where Mandel's stress is defined as:

$$M^e = C : E^e$$

where C is the fourth-order isotropic elasticity tensor. Then, Cauchy stress is defined as:

$$T = J^{-1} F^e M^{e^T} F^{e-1}$$

where J^{-1} is the determinant of the deformation gradient. During a time integration process, the deformation gradient is known. Hence, only an evolution equation for the rate of plastic deformation F^p is required to calculate the stress and strain measures at the end of the step.

$$F^p = L^p F^p$$

where L^p is the plastic velocity gradient. Then, the stretching tensor is defined as:

$$D^p = \text{sym}\left(L^p\right) = \sqrt{\frac{3}{2}} \, \dot{j}^p N^p$$

where $\dot{\epsilon}^p$ is the equivalent plastic strain rate and N^p is the direction of flow which is defined as:

$$N^p = \sqrt{\frac{3}{2}}(T/\sigma)$$

where σ is equivalent stress which is defined as:

$$\sigma = \sqrt{\frac{3}{2}}|T|$$

During a time integration process, D^p can be directly calculated from the parameters given above with exponential mapping. During the analysis, rate dependency is expressed as power law quantity as follows:

$$Y(s) = Y^s \sqrt{\frac{3}{2}}\left(\frac{\dot{\epsilon}^p}{\epsilon_0}\right)^n$$

where ϵ_0 is the reference strain rate and Y^s is the yield resistance defined as:

$$Y^s = Y_0 + Q\left(1 - \exp\left(-b\int^p\right)\right)$$

where Q and b are isotropic hardening parameters and n is the rate sensitivity parameter.

A finite element analysis has been done using Abaqus/Standard. A special in-house user material code is developed for applying a large deformation rate-dependent plasticity theory. Results produced by this analysis are given below for the composition of E4 containing 10 wt% Nb in Figure 11.4.

As for time-dependent behaviour by means of nano-indentation, two basic tests have been carried out, wear and creep tests, for only two compositions, E1 and E2. For these tests, the creep compliance and the stress exponent were calculated by using data collection defined in Eq. (11.1) [13]:

$$\varepsilon(t) = \sigma_0 J(t) \tag{11.1}$$

where σ_0 the constant stress applied and $J(t)$ is calculated using Eq. (11.2)

$$J(t) = A(t) / (1 - v)P_0 \tan\theta \tag{11.2}$$

In Eq. (11.2), $A(t)$ is the contact area, P_0 constant applied load, θ is the effective cone angle which is $70.3°$ for a Berkovich indenter and the Poisson's ratio v is assumed to be 0.3.

This approach takes into account how the contact area under the Berkovich tip alters while displacing the surface changes.

The strain versus time behaviour during creep is characterized by a high strain rate in the primary stage of creep, and then in the secondary, steady-state stage of creep, the strain rate given in Eq. (11.3) can be written as:

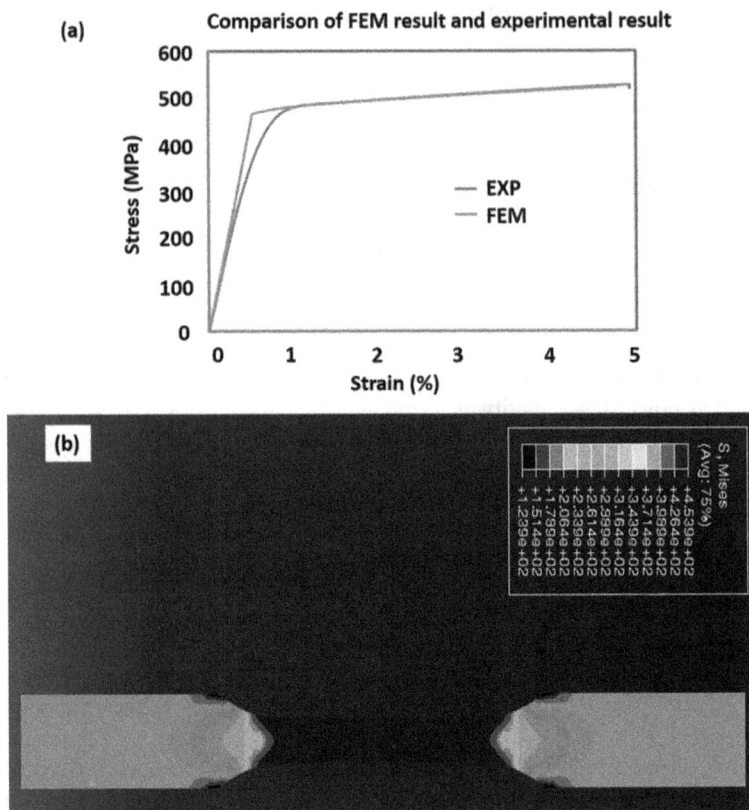

FIGURE 11.4 (a) FEM result in comparison with experimental static tensile test result for composition, E4–x10Nb. (b) Mises stress result of the FEM analysis of composition E4 containing 10 wt% Nb.

$$\dot{\varepsilon} = K\sigma^n \qquad (11.3)$$

where K is a constant and n is the stress exponent. The strain rate is calculated in the software and in turn n is obtained from the log-log plot of strain rate versus stress in the secondary stage of creep.

The materials under consideration are heterogeneous in nature, and it is the fact that the nano-indentation test is carried out over a small area/volume, a large scatter in the data is observed, and to overcome this, the sampling number was taken as large as possible.

Here, nano-indenter XP was used to perform the creep tests. Polished samples were loaded to a maximum load of 30 mN at a rate of 1 mN/s and the maximum load was applied for a duration of 500 seconds. A Berkovich tip was used for these tests. The stress exponent and the constant were calculated. The tests were done on

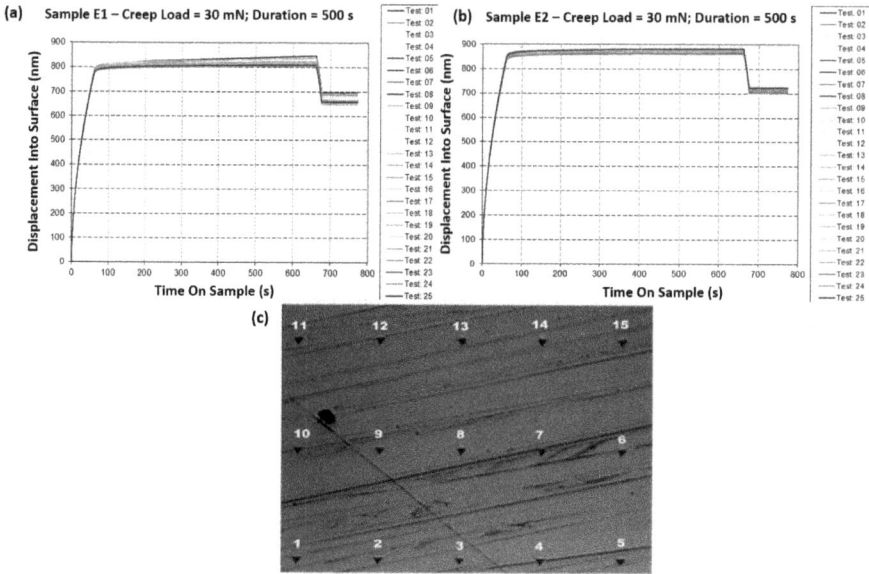

FIGURE 11.5 Mean creep test results carried out under the load of 30 mN during 500 seconds for the two test specimens (a) E1 and (b) E2. (c) the image of the first 15 indents that resulted from the creep tests with a horizontal spacing $= 50\,\mu m$ and vertical spacing $= 75\,\mu m$ (indents to 1,500 nm_500×).

TABLE 11.3

Summarize of Nano-Indentation Results for the Stress Exponent and Constant for Two Compositions, E1 and E2

Specimen	E1 Stress Exponent	E1 Constant	E2 Stress Exponent	E2 Constant
Mean value	0.0115	1.3628	0.0130	1.3617
Standard deviation	0.0007	0.0171	0.0012	0.0143
Max value	0.0135	1.3935	0.0159	1.3970
Minimum value	0.0104	1.3330	0.0115	1.3524

a 5X5 grid for a total of 25 tests. The results from samples E1 and E2 are shown in Figure 11.5a and b. The image of the first 15 indents that resulted from the creep tests is shown in Figure 5c. Stress exponent n and constant K results for each specimen and the averages of both for Sample E2 are given in Table 11.3.

Based on the average, the strain rate for samples E1 and E2 in the secondary steady-state stage creep is found as:

For the sample of the composition E1 $\dot{\varepsilon} = 1.3617\sigma^{0.0130}$, ($\sigma$ in MPa)

For the sample of the composition E1 $\dot{\varepsilon} = 1.3628\sigma^{0.0115}$, ($\sigma$ in MPa)

TABLE 11.4

The Results of Nano-Indentation Measurements of Two Compositions, E1 and E2

		Indentation Depth		Indentation Depth		Indentation Depth	
		250 nm		500 nm		1,500 nm	
		Modulus (GPa)	Hardness (GPa)	Modulus (GPa)	Hardness (GPa)	Modulus (GPa)	Hardness (GPa)
Sample E1	Average	91.9	4.99	93.7	5.33	92.4	4.92
	Standard Deviation	1.3	0.07	1.6	0.13	0.9	0.08
Sample E2	Average	93.0	4.92	92.0	4.80	91.6	5.01
	Standard Deviation	2.7	0.22	7.9	0.59	0.7	0.12

And also the mean values for the Young's modulus and hardness obtained by nano-indentation tests are for E1 89.4–91.9 GPa and 4.44–4.99 GPa and for E2 89.5–93.7 GPa and 4.52–4.80 GPa respectively.

The total mean values measured on at least three test specimens for each composition were summarized in Table 11.4. As indicated in the former section, nano-indentation tests were performed on a 5X5 grid at depths of 250, 500 and 1,500 nm, using the continuous stiffness measurement option of the nano-indenter. The modulus and hardness obtained are:

As for the wear test results, some details of the test results were given only for two compositions, E1 and E2. Wear tests were performed using the nano-scratch testing capability of the nano-indenter. Relatively fast wear tests can be performed to compare the wear behaviour of different samples. For the wear tests, a conical tip with a 90° cone angle was used. Wear tests were run under a normal load of 50 mN applied over a linear track of 500 mm for at least 50 cycles. Typical wear tracks for two compositions, E1 and E2, in Figure 11.6a and the corresponding initial and residual profiles of the track are given in Figure 11.6b. The wear is characterized as an area between the final and residual profiles and the results for the two samples were given in Table 11.5.

11.4 CONCLUSION

This manuscript gives partial results of our academic research collaboration project. We have developed an innovative microstructure by using high-level nano filler NiAl intermetallics and nano Nb powder as reinforcements in the recycled fresh scrap aluminium alloy AA7075. All the results and interpretations are based on the partial results of our research project. A strong and high toughness hybrid composite has been developed using vacuum arc melting sintering to carry out a strong chemical bonding diffusion between the matrix and reinforcement. This process is much more economical and very efficient for the hybrid composites for low-cost and efficient manufacturing regarding the 3D printing process.

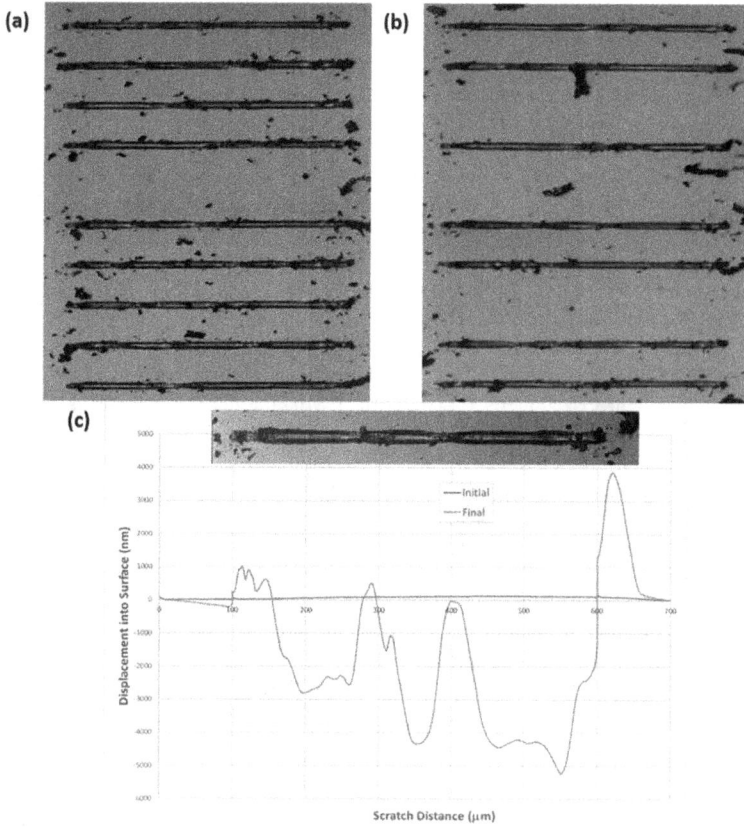

FIGURE 11.6   (a) Typical wear tracks for two compositions E1 and (b) E2 (c) the corresponding initial and residual profiles of the track.

TABLE 11.5
General Wear Results of Two
Compositions, E1 and E2

Values	Sample E1	Sample E2
Average (μm^2)	1029.578	1322.587
Standard deviation (μm^2)	437.1482	351.8577

ACKNOWLEDGEMENTS

This academic research has been carried out with the collaboration of ISAE-SUPMECA-Paris, UNICAMP-Campinas-BR, Atilim University-Ankara-TR and Michigan Tech Houghton, MI-USA. We acknowledge that Professor R. CARAM at UNICAMP, the Head of Physical Metallurgy Laboratory helped us to use the vacuum arc melting sintering device and discuss the results.

REFERENCES

[1] E. Bayraktar, F. Gatamorta, H. M. Enginsoy, J. E. Polis, I. Miskioglu, New Design of Composites from Fresh Scraps of Niobium for Tribological Applications, Mechanics of Composite, *Hybrid and Multifunctional Materials*, 6, 35–44 (2020).

[2] L.F.P. Ferreira, E. Bayraktar, M.H. Robert, I. Miskioglu, Chapter 17: Particles Reinforced Scrap Aluminium Based Composites by Combined Processing Sintering + Thixoforming, In *Mechanics of Composite and Multi-Functional Materials*, vol. 7, Springer, Berlin, pp. 145–152, 2016. 978-3-319-41766-0.

[3] F. Gatamorta, I. Miskioglu, D. Katundi, E. Bayraktar, Recycled Ti-Al-Cu Matrix Composites Reinforced with Silicon Whiskers and γ-Alumina (Al$_2$O$_3$) Fibres Through Sintering + Forging, In *SEM-Mechanics of Composite, Hybrid & Multi-Functional Materials*, vol. 5, Springer International Publishing, pp. 49–54, 2023, 10.1007/978-3-031-17445-2_6.

[4] U. Kaftancıoglu, G. Zambelis, F. Gatamorta, I. Miskioglu, Emin Bayraktar Development of Ni-Al/Nb2Al/ZrO2-Based Composites for Aircraft Engine Applications Produced by a Combined Method: Sintering + Forging, In *SEM-Mechanics of Composite, Hybrid & Multi-Functional Materials*, vol. 5, Springer International Publishing, pp. 55–60, 2023, 10.1007/978-3-031-17445-2_7.

[5] I. Miskioglu, G. Zambelis, F. Gatamorta, O. Aslan, E. Bayraktar, Toughening Mechanism of Silicon Whiskers and Alumina Fibers (γ-Al2O3) Reinforced Ni-Al-Cu Matrix Composites through "Sintering + Forging", *SEM-Mechanics of Composite, Hybrid & Multi-Functional Materials*, vol. 5, Springer International Publishing, pp. 29–35, 2022, 10.1007/978-3-031-17445-2_4.

[6] Y.X. Lu, C.H. Tao, D.Z. Yang. Improvement in Mechanical Properties of Nial Matrix Composites Fabricated by Reaction Compocasting. *Scripta Materialia*, 35, 10, 1243–1246 (1996). https://doi.org/10.1016/1359-6462(96)00276-x.

[7] D. Gu, Z. Wang, Y. Shen, Q. Li, Y. Li. In-Situ Tic Particle Reinforced Ti-Al Matrix Composites: Powder Preparation by Mechanical Alloying and Selective Laser Melting Behavior. *Applied Surface Science*, 255, 22, 9230–9240 (2009). https://doi.org/10.1016/j.apsusc.2009.07.008.

[8] V.A. Popov, E.V. Shelekhov, A.S. Prosviryakov, A.D. Kotov, M.G. Khomutov, Particulate Metal Matrix Composites Development on the Basisof In Situ Synthesis of TiC Reinforcing Nanoparticles During Mechanical Alloying. *Journal of Alloys and Compounds*, 707, 365–370 (2017).

[9] F. Gatamorta, I. Miskioglu, E. Bayraktar, M. L. Melo. Recycling of Aluminium-431 by High Energy Milling Reinforced with Tic-Mo-Cu for New Composites in Connection Applications. *Mechanics of Composite and Multi-Functional Materials*, 5, 41–46 (2019). https://doi.org/10.1007/978-3-030-30028-9_6.

[10] L.M.P. Ferreira, M.H. Robert, E. Bayraktar, D. Zaimova, New Design of Aluminium Based Composites through Combined Method of Powder Metallurgy and Thixoforming. *Advanced Materials Research*, 939, 1, 68–75 (2014).

[11] P.D. Srivyasa, M.S. Charoo, Role of Fabrication Route on the Mechanical and Tribological Behavior of Aluminium Metal Matrix Composites-A Review. *Materials Today: Proceedings*, 5, 20054–20069 (2018).

[12] L.-M.-P. Ferreira, E. Bayraktar, I. Miskioglu, M.-H. Robert. Influence of Nano Particulate and Fiber Reinforcements on the Wear Response of Multiferroic Composites Processed by Powder Metallurgy, *Advances in Materials and Processing Technologies*, 3, 1, 23 (2016).

[13] L.F.P. Ferreira, E. Bayraktar, M.H. Robert, I. Miskioglu, Chapter 17: Particles Reinforced Scrap Aluminium Based Composites By Combined Processing Sintering + Thixoforming, In *Mechanics of Composite and Multi-Functional Materials*, Springer, Berlin, 2016, pp. 145–152. 978-3-319-41766-0.

[14] L. Mihlyuzova, H.-M. Enginsoy, E. Bayraktar, S. Slavov, and D. Dontchev, Tailored Behaviour of Scrap Copper Matrix Composites Reinforced with "Zn-Ni-Al", Low Cost Shape Memory Structures, Mechanics of Composite, *Hybrid and Multifunctional Materials, Fracture, Fatigue, Failure and Damage Evolution*, 3, 49–54 (2021).

[15] L. Mihlyuzova, H. M. Enginsoy, D. Dontchev, and E. Bayraktar, Tailored Behaviour of Scrap Copper Matrix Composites Reinforced with Zinc and Aluminium: Low Cost Shape Memory Structures, Mechanics of Composite, *Hybrid and Multifunctional Materials*, 6, 27–34, (2020).

[16] F. Gatamorta, H. M. Enginsoy, E. Bayraktar, I. Miskioglu, and D. Katundi, Design of Recycled Alumix-123 Based Composites Reinforced with γ-Al2O3 through Combined Method; Sinter + Forging, Mechanics of Composite, *Hybrid and Multifunctional Materials*, 6, 9–18 (2020).

[17] D.K. Koli, G. Agnihotri, R. Purohit, Advanced Aluminium Matrix Composites: The Critical Need of Automotive and Aerospace Engineering Fields. *Materials Today Proceedings*, 2, 4–5, 3032–3041 (2015).

12 Performance Evaluation of Nanolignin in Polymer Composites

Aizat Ghani
Universiti Malaysia Sabah

Lee Seng Hua
Universiti Teknologi MARA (UiTM)

Syeed Saifulazry Osman Al-Edrus
Universiti Putra Malaysia

12.1 INTRODUCTION

The primary chemical components of raw plant fiber materials are cellulose, lignin, and hemicellulose [1]. Lignin is a biopolymer made from remnants of agro-industrial waste. The origin and pre-treatment method of lignin extraction affect its biological, morphological, and physicochemical properties [2]. Lignin works with the cell wall to alter its permeability and thermal stability; however, its main purpose is to act as a structural component that gives plant tissue more vigor and rigidity [3]. After cellulose, lignin is the second most prevalent natural polymer substance and has attracted a lot of interest recently [4].

Lignin is the most prevalent aromatic polymer on the earth and the second most prevalent organic polymer behind cellulose because it makes up 15%–40% of the dry weight of woody plants. According to annual biomass growth rates, the amount of lignin produced worldwide ranges from 5 to 36×108 tonnes [5]. Lignin has a structure of three-dimensional heterogeneous, randomly cross-linked structure that includes hydroxyl, carbonyl, and carboxylate groups on the surface in addition to a polyphenolic aromatic backbone [6].

The source from which lignin is extracted, the extraction techniques used, and any further treatments all significantly affect the mechanical and also physical characteristics of lignin [7]. Several pulping techniques can be used to extract lignin from several sources, including wood, cereal straws, pulp and paper, and sugarcane bagasse [3]. Intensive research is done to generate bio-based products from the deconstruction of lignin as well as numerous applications of lignin as an addition in composite materials [8]. The intricacy of the structures and the variety of lignin sources present both opportunities and obstacles for the study of lignin's possible applications [9].

DOI: 10.1201/9781003400998-12

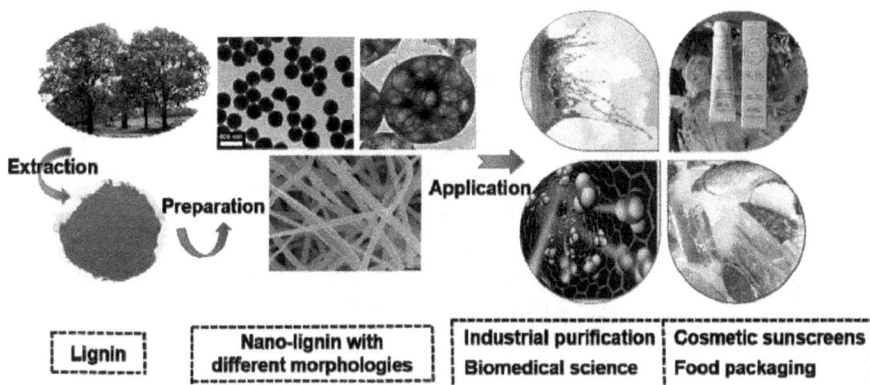

FIGURE 12.1 Nanolignins with various morphologies and their uses in various industries. Reproduced from ref. [4].

A lot of research is being done right now on using nanolignin (NL) for commercial purposes [10]. By manipulating the reaction mechanism of solvent/anti-solvent, lignin concentration, temperature and pH of the solution, and other factors, lignin nanoparticles (LNPs) with various morphologies as shown in Figure 12.1 (smooth colloidal, hollow, spherical, and quasi-spherical) have been successfully generated [11]. Due to their benefits of being nontoxic, resistant to the environment, having great thermal stability, and being biocompatible, LNPs have prospective applications as antioxidants, thermal/light stabilizers, reinforced materials, and nano-microcarriers [12].

12.2 PERFORMANCE EVALUATION OF NANOLIGNIN IN POLYMER COMPOSITES

One of the most extensive use of LNPs is the fabrication of bio-composites with better characteristics using micro- and nanoparticles of lignin [13]. NL is added to materials primarily as a reinforcing agent to improve the end product's mechanical qualities or as an additive to boost their antioxidant, antibacterial, and UV protection capabilities. A summary of some of the most recent and relevant studies on this topic will be given in the following discussion. The summary application research on NL is listed in Table 12.1.

Del Saz-Orozco [14] studied the influence of formulation variables on the mechanical properties and density of phenolic foams (PFs) and lignin nanoparticle-reinforced phenolic foams (LRPFs), on a stirring speed, blowing agent amount, and lignin nanoparticle weight fraction. LRPFs were obtained from calcium softwood lignosulfonates with an average diameter of 1.6 μm. The results showed that the density of LRPFs decreased as the weight percentage of LNPs and the amount of blowing agent increased. Lignosulfonates have surfactant characteristics [15], which reduce the surface tension of the formulation mixture, increase reagent compatibility, and promote bubble nucleation [16,17]. As a result, more bubbles form in the system,

TABLE 12.1

Summary of Application Research on Nanolignin

Sources of Nanolignin (Production Method)	Percent of Lignin Used (%)	Application of Nanolignin	Property Improvement and Advantages	References
Calcium softwood lignosulfates (Commercial lignotech ibéríca)	1.5–3.5	In resol resins for phenolic foam	• Modulus and strength improvement • Saving in blowing agent	[14]
Steam explosion lignin (hydrochloric acidolysis)	0.1–3	PLA biocomposites	• Bionanocomposites with extruded nanolignin had increased elongation at break	[20]
	1–3	Wheat gluten nanocomposite films	• Mechanical characteristics and thermal stability have been improved • Reduces water uptake of the bionanocomposites	[26]
	1–3	Binary and ternary PLA films with chitosan	• Lignin nanoparticles (LNP) have increased the PVA's Young's modulus and tensile strength	[30]
Kraft lignin (ultrasound treatment)	5–20	Waterborne PU-based (nano) composites	• Thermal stability has improved slightly • Mechanical characteristics have improved • The matrix contains a good distribution of nanoparticles	[32]
Alkali lignin (acidolysis)	Polyurethane	1–7	• Increased tensile strength and Young's modulus • Improved mechanical and thermal stability of PU • Improved hydrophobicity of nanocomposites	[33]
Steam explosion lignin (hydrochloric acidolysis)	Phenol–formaldehyde resol resin	5–10	• Favor the thermal cure process • Reduction of curing heat and curing temperature • The strength and shear strength of composites were improved, and their cross-link density increased with the addition of 5% nanolignin	[36]

Source: Adapted from ref. [13].

and the amount of gas in the foam increases, thus lowering its density. The compressive modulus and strength of LRPFs investigated ranged from 4.30–42.12 to 0.270–2.394 MPa, respectively. LRPFs have a higher compressive modulus than PFs at densities less than 155 kg/m³. With a density of 120 kg/m³, the compressive moduli of PF and LRPF, for example, were 14.70 and 18.86 MPa, respectively. The improvement in the mechanical characteristics obtained by using LNPs as a reinforcing agent in PFs is comparable to or better than that seen in other studies in which PFs were reinforced using non-biodegradable reinforcements such as aramid and glass fibers [18,19].

To create polylactic acid (PLA) bionanocomposites filled with 0, 1, and 3 wt% LNPs, Yang et al. [20] used two production processes, namely melt extrusion (E-PLA) and solvent casting (C-PLA). The materials were then tested for tensile strength. According to Gilca [21] and Frangville [22], when HCl acidolysis was used to create LNP suspension from lignin, it was discovered that the PLA film made using the solvent casting method (C-PLA) has a similar strength and elongation at break to the PLA film made using the melt extrusion method but a lower modulus due to the plasticizer effect left behind by leftover chloroform in the films (E-PLA). It was found that tensile strength and modulus (E) increased from 44.3 and 1955.8 MPa to 48.7 and 2153.2 MPa, respectively, with the addition of 1% LNPs (E-PLA/1LNP). Intriguingly, the elongation breaking point for E-PLA/1LNP was 26.7% as opposed to E-PLA (16.8%), which was different from several nanocomposites, including stiff nanofillers. This was mainly ascribed to the distinctive chemical structure of lignin, which contains a variety of functional groups (carboxyl, phenolic or aliphatic hydroxyls, and carbonyl), all of which greatly influenced the interactions (such as hydrogen bonding) between the PLA matrix and LNPs [23,24]. When LNP loading was raised to 3 wt% (E-PLA/3LNP), the elongation at the breaking point increased to 66.2%, with a decrease in tensile strength (41.0 MPa) and modulus (1,390.6 MPa) compared to E-PLA. This phenomenon was explained by the presence of lignin particles, which inhibited the development of a long-range continuous phase of PLA, resulting in a loss in mechanical characteristics [25]. Generally, this study found that the inclusion of LNPs had a favorable effect on mechanical properties, particularly elongation at breaks in extruded bionanocomposites, and this result was thought to be related to the chemical composition of lignin.

Yang et al. [26] studied the impact of incorporating LNPs at two different weight levels (1 and 3 wt%) on the performance of the bionanocomposite based on wheat gluten. Hydrochloric acidolysis was used to make LNP suspension from the lignin residue [21,22]. Then, the mechanical properties of the wheat gluten bionanocomposite were investigated through a tensile strength test. Initially, when LNPs are added to wheat gluten, it results in a significant increase in tensile strength and modulus, as well as a significant drop in the elongation breaking point. It was found that when the content of LNPs is increased from 0 to 3 wt%, the tensile strength and modulus improve from 5.5 to 13.3 MPa and 180.5 to 553.2 MPa, respectively while the elongation breaking point drops from about 300% to 27.5%. These findings are similar to those of Kunanopparat et al. [27] who found a substantial reinforcement (increase in tensile strength and modulus) in conjunction with gradual embrittlement (decrease in the elongation breaking point). According to Majer et al. [28] in comparison to a native polymer matrix, it is widely understood that adding hard fillers to a soft matrix increases the rigidity and embrittlement of the composite. It was found that LNPs may generate strong interactions with the wheat gluten matrix, such as hydrogen bonding and other physical forces; the results imply that adding LNPs can effectively improve the biocomposite's characteristics. In addition, when LNPs are added to wheat gluten, the moisture content and water uptake gradually decrease. The results show that, after soaking in distilled water for 10 minutes, the moisture content drops to 12.64% of its initial value for WG/0LNP and 10.14% for WG/3LNP, while the water uptake drops from 142.16% to 103.21%. Because of its aromatic structure, lignin is a hydrophobic molecule that naturally reduces the hydrophilicity of the wheat gluten bionanocomposite mixed with LNPs [29]. Overall, adding LNPs to wheat gluten-based bionanocomposites improves the mechanical characteristics (tensile strength and modulus), thermal stability, and bionanocomposite stability properties.

Yang et al. [30] in their study used solvent casting to create binary and ternary polymeric films containing chitosan (CH), LNPs, and polyvinyl alcohol (PVA), at two different concentrations (1 and 3 wt%), and PVA/CH/LNP nanocomposites, both binary and ternary, were investigated for their thermal and mechanical properties. The effects of LNP concentration and its interaction with PVA and CH on the elastic-plastic response of the produced nanocomposites were evaluated using tensile testing. When a binary PVA system with LNP is developed, it was discovered that the addition of LNP greatly improves the mechanical properties of the nanocomposites, as seen by increases in the tensile strength and Young's modulus, which go from 45.8 to 51.5 MPa and 1,100 to 2,100 MPa, respectively. Hence, 91% and 12.45% higher than pure PVA when 3 wt% LNP was incorporated in the PVA matrix. On the other hand, the elongation breaking point of PVA/1LNP and PVA/3LNP nanocomposites dropped from 163.4%–115.5% to 30.5%, respectively, while the PVA film displayed typical ductility. As opposed to PVA films, CH films were shown to be the least stretchy in binary CH-based systems. The addition of LNP did not affect the tensile strength, and the modulus dropped from 1,334 MPa (CH) to 995 MPa (CH/3LNP). However, with the increase in LNP loading, from 5.72% to 17.31%, the elongation at break increased gradually. The addition of CH to a ternary PVA/CH/LNP system resulted in a significant drop in tensile strength and Young's modulus. The findings showed that CH has no reinforcing impact on the PVA matrix, which was in line with prior research by Bonilla et al. [31]. Nanolignin (NL) particles were added to a waterborne thermoplastic polyurethane (PU) matrix at different concentrations to create bio-based nanocomposite materials [32]. LNPs were produced in this investigation using ultrasonography processing softwood kraft lignin to produce colloidally stable lignin–water dispersions. According to this study's findings, PU/NL nanocomposite materials gradually increase in elastic modulus as NL-6h loadings rise, PU/NL-5, PU/NL-10, and PU/NL-20 systems, respectively, resulting in E values of 45.4, 52.5, and 65.6 MPa when compared to the unfilled PU matrix. When compared to unfilled matrix materials, highly filled nanocomposites (20 wt% NL-6h) have a two-fold increase in elastic modulus (33.0 MPa). Recent studies on the strengthening effect of LNPs in polymer-based nanocomposites are consistent with this trend [20] and the polymer matrix may be responsible for the high surface-to-volume ratio of the nanoscale dimensions produced by the ultrasonication process.

Qi et al. [33] studied the impacts of LNPs, which operate as both eco-friendly bio-based polyols and cross-linkers, on the mechanical properties of PU nanocomposites. Furthermore, pre-polymerization of polyethylene glycol and diisocyanates in the presence of varying concentrations (1, 3, 5, and 7 wt%) of LNPs was used to make PU nanocomposites. Based on the outcomes and in comparison to the control PU (without the addition of LNP), the tensile strength of PU-1LNP increased from 5.9 to 12.4 MPa, Young's modulus increased from 105 to 128 MPa, and elongation at break increased from 24% to 1240%. From this study, it was found that when the amount of additional LNP into the PU nanocomposite increased to 7 wt%, the Young's modulus increased from 110 to 177 MPa and also the tensile strength rose from 6 to 17 Mpa from its original value. The increment in mechanical strength in PU composites mixed with LNP can be attributed to three different factors: a stiff benzene ring structure in LNP, which can be used as hard segments in the PU composites that result [34], the small LNP, which could operate as nanofillers, thus fills the space between molecular chains and enhancing the cross-linking density of PU composites. Moreover, the LNP itself contains hydroxyl groups that distribute well in the matrix and improve the interfacial bonding between nanoparticles and the PU composite matrix [35]. However, further increment of LNP amount will further reduce the value of break at the elongation point of PU composite. It was discovered that as the amount of lignin grew, the elongation at break was somewhat reduced. This phenomenon is possibly due to agglomeration occurring and lignin nanoparticle aggregation with the PU polymer, resulting in a reduction in the adhesion work of the composite.

Yang et al. [36] explored the effect of nanosized lignin (LNP) as an adhesive in bulk phenol–formaldehyde resol in contrast to microlignin (LMP) when incorporated at two distinct weight percentages (5 and 10 wt%), and then, investigations were conducted to determine how lignin-phenol-formaldehyde resol adhesive tensile shear strength was affected by micro- and nano-modification. According to this study, the addition of different sizes initially had a significant effect on the mechanical results of the interracial bonding between lignin and the phenolic resin. Field emission scanning electron microscopy was used to evaluate micrographs of the tested joints; the findings are shown in Figure 12.2 below.

The resol/10LNP surface, which has the weakest adhesion ability, showed the most detachment between the wood fiber and the nanomodified matrix. The shear strength of the composite was improved with 5 wt% micro- and nanolignin. Resol/5LMP and Resol/5LNP had shear strengths of 9.6 and 10.9 MPa, respectively, compared to 8.7 MPa for pure resol. This result suggests that nanosized lignin at lower weight contents could be a viable alternative to the phenolic resol adhesive [37]. This study's increased shear strength could be attributed to the fact that the copolymerization reaction between lignin and phenol-formaldehyde improved the cross-linking density and structural alignment of lignin and phenol by introducing a little amount of micro-/nanolignin [38]. This study also showed that the cross-link density of the glue will rise when lignin is added, increasing the adhesive's strength and shear strength.

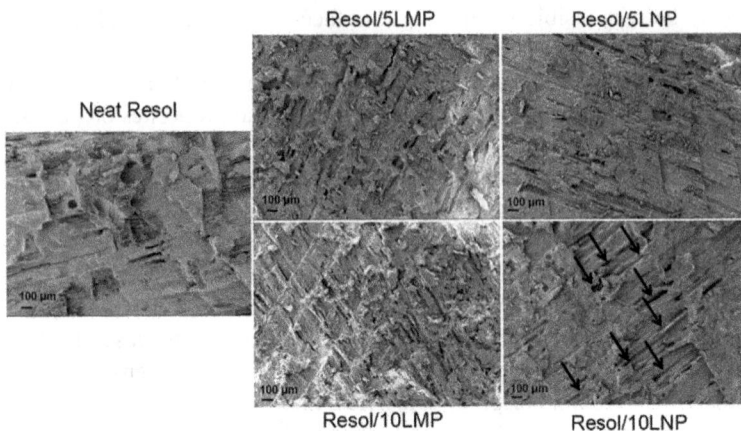

FIGURE 12.2   Fracture surface for lap joints with unmodified and lignin-modified adhesives, shown in FESEM micrograph. Reproduced from ref. [36].

12.3   POTENTIAL CHALLENGES AND FUTURE RECOMMENDATIONS OF LIGNIN IN BIOCOMPOSITES

Lignin micro- and nanoparticle production is a possible alternative to increase the range of use for this biopolymer. Recent interest in micro- and nanolignin has increased as a result of its superior qualities to the already available ordinary lignin [39]. An enormous rise in the interest in lignin micro- and nanoparticles for various applications can be attributed to the fact that they are an environmentally friendly material with properties such as antibacterial, antioxidant, ultraviolet absorption, biocompatibility, and biodegradability [4]. Other than that, lignin also exhibits a variety of special qualities, including UV absorption, high rigidity, resistance to oxidation and biological attack, and the capacity to delay and prevent oxidation reactions. Consequently, it can create high-value products from a vast amount of feedstock [40,41].

However, before lignin is widely used in biocomposites, several problems must be solved. The lack of uniformity in lignin production is one of the biggest problems. Depending on the source, extraction method, and processing conditions, the properties of lignin can vary significantly. For this reason, it is difficult to find uniform lignin of good quality for use in biocomposites. For example, different forms of biomass, such as wood, straw, or bagasse, can be used for lignin extraction, and each type of lignin has different qualities and chemical compositions. The properties of lignin can also be affected by the extraction process, for example, alkali extraction yields lignin with a high molecular weight, while acid extraction yields lignin with a low molecular weight [42,43].

Another challenge is the lack of compatibility between lignin and other bio-based polymers. Lignin's potential usage in biocomposites is constrained by its low compatibility with polymers like PLA [44]. Recent studies have nevertheless demonstrated that lignin may be made compatible with other bio-based polymers utilizing a variety of techniques, including chemical modification, mixing, or the addition of compatibilizers [45,46].

Moreover, the development of new lignin-based applications is hampered by the dearth of knowledge regarding the characteristics and behavior of lignin-based biocomposites. Because lignin is a complicated polymer with a heavily branched structure, it is challenging to anticipate its behavior and characteristics [6]. Therefore, additional studies are required to comprehend how lignin behaves in biocomposites, including its biodegradability, thermal stability, and mechanical characteristics.

In summary, lignin has enormous potential for its application in biocomposites as a bio-based material, but further studies are required to address the issues with its manufacture, compatibility, and characteristics. Future research should focus on standardizing lignin production, comprehending its characteristics and behavior in biocomposites, and creating techniques to make lignin compatible with another bio-based polymer.

12.4 CONCLUSION

The application of NL in the fabrication of polymer composites has been discussed in this chapter. The advantages of the application of NL in enhancing the properties of the polymeric composites were highlighted. NL is primarily served as a reinforcing agent to improve the end product's mechanical qualities or as an additive to boost the antioxidant, antibacterial, and UV protection capabilities of a polymer composite. NL has been reported to be added into a wide variety of polymers that proved the enormous potential of NL in the polymeric composite field. However, despite the advantages offered by NL, it has been identified to have low uniformity that has to be solved prior to expanding its utilization. In addition, the lack of compatibility between lignin and other bio-based polymers is also causing a negative effect on the properties of the polymer composites. Therefore, additional future studies are recommended to obtain a more comprehensive understanding of its performance.

REFERENCES

1. Ma, C., Kim, T. H., Liu, K., Ma, M. G., Choi, S. E., & Si, C. (2021). Multifunctional lignin-based composite materials for emerging applications. *Frontiers in Bioengineering and Biotechnology*, 9, 708976. https://doi.org/10.3389/fbioe.2021.708976.
2. Makri, S. P., Xanthopoulou, E., Klonos, P. A., Grigoropoulos, A., Kyritsis, A., Tsachouridis, K., Anastasiou, A., Deligkiozi, I., Nikolaidis, N. & Bikiaris, D. N. (2022). Effect of micro-and nano-lignin on the thermal, mechanical, and antioxidant properties of biobased PLA-lignin composite films. *Polymers*, *14*(23), 5274. https://doi.org/10.3390/polym14235274.
3. Haghdan, S., Renneckar, S. & Smith, G. D. (2016). *Sources of Lignin*, Elsevier Inc., ISBN 9780323355667.
4. Zhang, Z., Terrasson, V., & Guénin, E. (2021). Lignin nanoparticles and their nanocomposites. *Nanomaterials*, 11(5), 1336. https://doi.org/10.3390/nano11051336.
5. dos Santos, P. S., Erdocia, X., Gatto, D. A., & Labidi, J. (2014). Characterisation of Kraft lignin separated by gradient acid precipitation. *Industrial Crops and Products*, 55, 149–154. https://doi.org/10.1016/j.indcrop.2014.01.023.
6. Katahira, R., Elder, T. J., & Beckham, G. T. (2018). A brief introduction to lignin structure. *Lignin Valorization: Emerging Approaches*, Gregg T Beckham. https://doi.org/10.1039/9781788010351-00001.

7. Garcia, A., Amendola, D., Gonzalez, M., Spigno, G., & Labidi, J. (2011). Lignin as natural radical scavenger. Study of the antioxidant capacity of apple tree pruning lignin obtained by different methods. *Chemical Engineering Transactions*, *24*(925.10), 3303. https://doi.org/10.3303/CET1124155.

8. Liao, J. J., Abd Latif, N. H., Trache, D., Brosse, N., & Hussin, M. H. (2020). Current advancement on the isolation, characterization and application of lignin. *International Journal of Biological Macromolecules*, *162*, 985–1024. https://doi.org/10.1016/j.ijbiomac.2020.06.168.

9. Vishtal, A., & Kraslawski, A. (2011). Challenges in industrial applications of technical lignins. *BioResources*, *6*(3). https://doi.org/10.15376/biores.6.3.vishtal.

10. Richter, A. P., Brown, J. S., Bharti, B., Wang, A., Gangwal, S., Houck, K., … Velev, O. D. (2015). An environmentally benign antimicrobial nanoparticle based on a silver-infused lignin core. *Nature Nanotechnology*, *10*(9), 817–823. https://doi.org/10.1038/nnano.2015.141.

11. Mishra, P. K., & Ekielski, A. (2019). The self-assembly of lignin and its application in nanoparticle synthesis: A short review. *Nanomaterials*, *9*(2), 243. https://doi.org/10.3390/nano9020243.

12. Henn, A., & Mattinen, M. L. (2019). Chemo-enzymatically prepared lignin nanoparticles for value-added applications. *World Journal of Microbiology and Biotechnology*, *35*, 1–9. https://doi.org/10.1007/s11274-019-2697-7.

13. Ariyanta, H. A., Santoso, E. B., Suryanegara, L., Arung, E. T., Kusuma, I. W., Taib, M. N. A. M., Hussin, M.H., Yanuar, Y., Batubara, I. & Fatriasari, W. (2023). Recent Progress on the Development of lignin as future ingredient biobased cosmetics. *Sustainable Chemistry and Pharmacy*, *32*, 100966. https://doi.org/10.1016/j.scp.2022.100966.

14. Del Saz-Orozco, B., Oliet, M., Alonso, M. V., Rojo, E., & Rodríguez, F. (2012). Formulation optimization of unreinforced and lignin nanoparticle-reinforced phenolic foams using an analysis of variance approach. *Composites Science and Technology*, *72*(6), 667–674. https://doi.org/10.1016/j.compscitech.2012.01.013.

15. Ouyang, X., Qiu, X., & Chen, P. (2006). Physicochemical characterization of calcium lignosulfonate-A potentially useful water reducer. *Colloids and Surfaces A: Physicochemical and Engineering Aspects*, *282*, 489–497. https://doi.org/10.1016/j.colsurfa.2005.12.020.

16. Perkins, K. M., Gupta, C., Charleson, E. N., & Washburn, N. R. (2017). Surfactant properties of PEGylated lignins: Anomalous interfacial activities at low grafting density. *Colloids and Surfaces A: Physicochemical and Engineering Aspects*, *530*, 200–208. https://doi.org/10.1016/j.colsurfa.2017.07.061.

17. Hansen, R. (1993). *Handbook of Polymeric Foams and Foam Technology*, D. Klempner and K. C. Frisch, Hanser Publishers, Munich.

18. Shen, H., & Nutt, S. (2003). Mechanical characterization of short fiber reinforced phenolic foam. *Composites Part A: Applied Science and Manufacturing*, *34*(9), 899–906. https://doi.org/10.1016/S1359-835X(03)00136-2.

19. Desai, A., Auad, M. L., Shen, H., & Nutt, S. R. (2008). Mechanical behavior of hybrid composite phenolic foam. *Journal of Cellular Plastics*, *44*(1), 15–36. https://doi.org/10.1177/0021955X07078021.

20. Yang, W., Fortunati, E., Dominici, F., Kenny, J. M., & Puglia, D. (2015). Effect of processing conditions and lignin content on thermal, mechanical and degradative behavior of lignin nanoparticles/polylactic (acid) bionanocomposites prepared by melt extrusion and solvent casting. *European Polymer Journal*, *71*, 126–139. https://doi.org/10.1016/j.eurpolymj.2015.07.051.

21. Gilca, I. A., Ghitescu, R. E., Puitel, A. C., & Popa, V. I. (2014). Preparation of lignin nanoparticles by chemical modification. *Iranian Polymer Journal*, *23*, 355–363. https://doi.org/10.1007/s13726-014-0232-0.

22. Frangville, C., Rutkevičius, M., Richter, A. P., Velev, O. D., Stoyanov, S. D., & Paunov, V. N. (2012). Fabrication of environmentally biodegradable lignin nanoparticles. *ChemPhysChem*, *13*(18), 4235–4243. https://doi.org/10.1002/cphc.201200537.

23. Thunga, M., Chen, K., Grewell, D., & Kessler, M. R. (2014). Bio-renewable precursor fibers from lignin/polylactide blends for conversion to carbon fibers. *Carbon*, *68*, 159–166. https://doi.org/10.1016/j.carbon.2013.10.075.

24. Ferry, L., Dorez, G., Taguet, A., Otazaghine, B., & Lopez-Cuesta, J. M. (2015). Chemical modification of lignin by phosphorus molecules to improve the fire behavior of polybutylene succinate. *Polymer Degradation and Stability*, *113*, 135–143. https://doi.org/10.1016/j.polymdegradstab.2014.12.015.

25. Ouyang, W., Huang, Y., Luo, H., & Wang, D. (2012). Poly (lactic acid) blended with cellulolytic enzyme lignin: Mechanical and thermal properties and morphology evaluation. *Journal of Polymers and the Environment*, *20*, 1–9. https://doi.org/10.1007/s10924-011-0359-4.

26. Yang, W., Kenny, J. M., & Puglia, D. (2015). Structure and properties of biodegradable wheat gluten bionanocomposites containing lignin nanoparticles. *Industrial Crops and Products*, *74*, 348–356. https://doi.org/10.1016/j.indcrop.2015.05.032.

27. Kunanopparat, T., Menut, P., Morel, M. H., & Guilbert, S. (2012). Improving wheat gluten materials properties by Kraft lignin addition. *Journal of Applied Polymer Science*, *125*(2), 1391–1399. https://doi.org/10.1002/app.35345.

28. Majer, Z., Hutař, P., & Náhlík, L. (2013). Determination of the effect of interphase on the fracture toughness and stiffness of a particulate polymer composite. *Mechanics of Composite Materials*, *49*, 475–482. https://doi.org/10.1007/s11029-013-9364-0.

29. Duval, A., Molina-Boisseau, S., & Chirat, C. (2013). Comparison of Kraft lignin and lignosulfonates addition to wheat gluten-based materials: Mechanical and thermal properties. *Industrial Crops and Products*, *49*, 66–74. https://doi.org/10.1016/j.indcrop.2013.04.027.

30. Yang, W., Owczarek, J. S., Fortunati, E., Kozanecki, M., Mazzaglia, A., Balestra, G. M., Kenny, J.M., Torre, L. & Puglia, D. (2016). Antioxidant and antibacterial lignin nanoparticles in polyvinyl alcohol/chitosan films for active packaging. *Industrial Crops and Products*, *94*, 800–811. https://doi.org/10.1016/j.indcrop.2016.09.061.

31. Bonilla, J., Fortunati, E. L. E. N. A., Atarés, L., Chiralt, A., & Kenny, J. M. (2014). Physical, structural and antimicrobial properties of poly vinyl alcohol-chitosan biodegradable films. *Food Hydrocolloids*, *35*, 463–470. https://doi.org/10.1016/j.foodhyd.2013.07.002.

32. Garcia Gonzalez, M. N., Levi, M., Turri, S., & Griffini, G. (2017). Lignin nanoparticles by ultrasonication and their incorporation in waterborne polymer nanocomposites. *Journal of Applied Polymer Science*, *134*(38), 45318. https://doi.org/10.1002/app.45318.

33. Qi, G., Yang, W., Puglia, D., Wang, H., Xu, P., Dong, W., Zheng, T. & Ma, P. (2020). Hydrophobic, UV resistant and dielectric polyurethane-nanolignin composites with good reprocessability. *Materials & Design*, *196*, 109150. https://doi.org/10.1016/j.matdes.2020.109150.

34. Tavares, L. B., Boas, C. V., Schleder, G. R., Nacas, A. M., Rosa, D. S., & Santos, D. J. (2016). Bio-based polyurethane prepared from Kraft lignin and modified castor oil. *Express Polymer Letters*, *10*(11), 927. https://doi.org/10.3144/expresspolymlett.2016.86.

35. Hazarika, D., Gupta, K., Mandal, M., & Karak, N. (2018). High-performing biodegradable waterborne polyester/functionalized graphene oxide nanocomposites as an eco-friendly material. *ACS Omega*, *3*(2), 2292–2303. https://doi.org/10.1021/acsomega.7b01551.

36. Yang, W., Rallini, M., Natali, M., Kenny, J., Ma, P., Dong, W., Torre, L. & Puglia, D. (2019). Preparation and properties of adhesives based on phenolic resin containing lignin micro and nanoparticles: A comparative study. *Materials & Design*, *161*, 55–63. https://doi.org/10.1016/j.matdes.2018.11.032.

37. Turunen, M., Alvila, L., Pakkanen, T. T., & Rainio, J. (2003). Modification of phenol-formaldehyde resol resins by lignin, starch, and urea. *Journal of Applied Polymer Science*, *88*(2), 582–588. https://doi.org/10.1002/app.11776.

38. Khan, M. A., Ashraf, S. M., & Malhotra, V. P. (2004). Eucalyptus bark lignin substituted phenol formaldehyde adhesives: A study on optimization of reaction parameters and characterization. *Journal of Applied Polymer Science*, *92*(6), 3514–3523. https://doi.org/10.1002/app.20374.

39. Beisl, S., Friedl, A., & Miltner, A. (2017). Lignin from micro-to nanosize: applications. *International Journal of Molecular Sciences*, 18(11), 2367. https://doi.org/10.3390/ijms18112367.

40. Doherty, W. O., Mousavioun, P., & Fellows, C. M. (2011). Value-adding to cellulosic ethanol: Lignin polymers. *Industrial Crops and Products*, *33*(2), 259–276. https://doi.org/10.1016/j.indcrop.2010.10.022.

41. Laurichesse, S., & Avérous, L. (2014). Chemical modification of lignins: Towards biobased polymers. *Progress in Polymer Science*, *39*(7), 1266–1290. https://doi.org/10.1016/j.progpolymsci.2013.11.004.

42. Oriez, V., Peydecastaing, J., & Pontalier, P. Y. (2020). Lignocellulosic biomass mild alkaline fractionation and resulting extract purification processes: Conditions, yields, and purities. *Clean Technologies*, *2*(1), 91–115. https://doi.org/10.3390/cleantechnol2010007.

43. Zhang, Q., Li, H., Guo, Z., & Xu, F. (2020). High purity and low molecular weight lignin nano-particles extracted from acid-assisted MIBK pretreatment. *Polymers*, *12*(2), 378. https://doi.org/10.3390/polym12020378.

44. Esakkimuthu, E. S., DeVallance, D., Pylypchuk, I., Moreno, A., & Sipponen, M. H. (2022). Multifunctional lignin-poly (lactic acid) biocomposites for packaging applications. *Frontiers in Bioengineering and Biotechnology*, 10. https://doi.org/10.3389/fbioe.2022.1025076.

45. Muthuraj, R., Hajee, M., Horrocks, A. R., & Kandola, B. K. (2021). Effect of compatibilizers on lignin/bio-polyamide blend carbon precursor filament properties and their potential for thermostabilisation and carbonisation. *Polymer Testing*, 95, 107133. https://doi.org/10.1016/j.polymertesting.2021.107133.

46. Kun, D., & Pukánszky, B. (2017). Polymer/lignin blends: Interactions, properties, applications. *European Polymer Journal*, *93*, 618–641. https://doi.org/10.1016/j.eurpolymj.2017.04.035.

13 Natural Nanofillers in Biopolymer-Based Composites
A Review

Siti Hasnah Kamarudin
UiTM Shah Alam

Mohd Salahuddin Mohd Basri
Universiti Putra Malaysia

Syaiful Osman and So'bah Ahmad
UiTM Shah Alam

Intan Syafinaz Mohamed Amin Tawakkal
and Ummi Hani Abdullah
Universiti Putra Malaysia

13.1 INTRODUCTION

Nowadays, abundant wastes from nonbiodegradable types of plastics are rapidly becoming a serious issue. Hence, the production of materials with biodegradable properties is increasingly becoming the main study subject as suitable replacements for petroleum-based materials [1–3]. For the manufacture of such materials, biopolymers are regarded as superior raw materials for a wide variety of materials and product applications. A fascinating field of research has been focusing on biopolymer films reinforced with nanostructures derived from plants. The high demand for plastic and polymeric products, which continue to grow each year, has made them as an important sector for which sustainability is a key factor to be taken into account. Therefore, making it as a unique creature with safety properties, sustainability and eco-friendly business are necessary [4,5]. Properties such as renewability, biodegradability, and biocompatibility with low impact on the environment create a huge demand for biopolymer-based materials, especially for application such as packaging. Examples of biopolymers may include starch, pectin, chitosan, and gelatin [6].

Nanocomposite films are a class of materials composed primarily of bio-based natural polymers (e.g., chitosan, starch) and synthetic polymers (e.g., poly(lactic

DOI: 10.1201/9781003400998-13

acid)) and nanofillers (clay, organic, inorganic, or carbon nanostructures) with varying properties [7]. The interaction of environmental-friendly natural polymers and nanofillers improves the functionality of nanomaterials. The properties of nanocomposites can be improved depending on the properties of the nanofillers, including impermeability, enhanced mechanical strength, antimicrobial, antioxidant properties, or thermal stability [8].

Nanofillers are enabling ground-breaking advances in material science and polymer composites. In this process, a small quantity of substance is applied to several polymers and other materials, which can tremendously enhance the effectiveness and quality of materials, such as mechanical, thermal, flame retardancy, and many others [9,10]. Nanofillers are solid-form additives that can improve the properties of original materials without increasing their density. Natural fibers are frequently used as an alternative reinforcement of synthetic fibers in polymer composites owing to their economic feasibility, renewability, lighter weight, excellent mechanical properties, and sustainability. However, the compatibility issue between the hydrophobic polymer matrix and the hydrophilic fibers remains a concern, which leads inevitably to performance degradation.

There have been efforts made to enhance the compatibility of the polymer matrix and natural fibers. There is a huge opportunity for nanoparticles or nanofillers to be used as filler materials to enhance the properties of polymer composites. Researchers and scholars all over the world have used nanoclay, nanosilicon dioxide (SiO_2), carbon nanotubes, and a variety of other nanofillers to improve the properties of natural fiber-reinforced polymer composites [10]. Nanoclay has been shown to reduce sisal fiber-reinforced composites' water absorption, which is also very attractive. Carbon nanotubes have also been shown to improve natural fibers like kenaf, bamboo, sugar palm, and oil palm empty fruit bunch (OPEFB) fiber-reinforcement's mechanical and water absorption properties [11].

The performance of composites has been constantly being enhanced over the last few years. Due to their biocompatibility and biodegradability, biopolymers may be used to improve other biologically active molecules' capabilities in a product. Biopolymers made from a variety of natural resources have been regarded as desirable alternatives for uses in the biomedical engineering, food industry, manufacturing, and nonrenewable plastic packaging [7,12]. Carbohydrates' remarkable film-forming and mechanical properties have led to their use in the creation of polymer films. Other types of biopolymers, such as pullulan, proteins (gelatin), and poly(lactic) acid, lipids, poly(glycolic acid)/polyurethane (PGA/PU)-based, can also be utilized. Table 13.1 lists various types of biopolymer films along with their benefits and drawbacks.

Nanofillers come in a variety of forms and sizes, but their specific properties such as particle sizes are below 100 nm by nanomaterials' design. The mixture of materials focusing on biopolymers and nanofillers is currently gaining popularity. The size of nanofillers is very useful for nanocomposite materials because they have a large surface area, resulting in a significant interphase or barrier region in between matrix material and the nanofiller. The biopolymer matrix is modified because of this interaction, improving the bionanocomposite materials' physical, barrier, and thermal properties [7].

TABLE 13.1
Several Examples of Various Biopolymers Used as Films Together with Advantages and Limitations

Biopolymer-Based Films Types	Benefits	Limitations
Cellulose	Has no flavour or smell, is hydrophilic and high resistance to oil and fat, and has good thermal and chemical stability.	Hard to melt or disintegrate as a result of high crystallinity; nonantimicrobial activity
Chitin and chitosan	Antimicrobial activity with good carbon dioxide barrier properties	Lack of antioxidant and naturally infected; poor ability to block oxygen and water
Pectin	Super oxygen-resistance capacity	Strong absorption to water vapor; low mechanical performance
Starch	Tasteless, excellent oxygen, and carbon dioxide-barrier characteristics	Poor tensile and water vapor barrier
Gelatin	Excellent barriers and mechanical properties	Poor permeation of water vapor
Lipids	Great impediments to moisture migration	Affect the aesthetics and smooth finish of the food products with coatings
Bacterial cellulose	Superior mechanical properties and fluidity	Water-insoluble
PLA	Good biological compatibility, transparency, and also environmental-friendly	Poor strength and thermal stability, high hardness, and brittleness

This guide summarizes current developments and uses of natural nanofillers in biopolymer-based composites, as well as on the development of nanotechnology to produce novel, bio-based bionanocomposites with high efficacy. Importantly, the influence of enticing natural nanofiilers as an essential factor for end-user exposure is required before this implementation can be available for commercialization.

13.2 CLASSES OF NATURAL NANOFILLERS IN BIOPOLYMER-BASED COMPOSITES

Nanofiller has the potential in order to increase the tensile strength, thermal stability, and barrier characteristics of the materials used for packing [13]. Bionanocomposite could be formed by incorporating nanofiller with biopolymer.

13.2.1 TYPES OF NATURAL NANOFILLERS AND CLASSES OF BIONANOCOMPOSITES

Organic nanofillers include natural nanofillers such as cellulose, chitin, and chitosan. Nanocellulose is divided into three types based on size and shape: cellulose nanocrystals (CNC), cellulose nanofibers (CNF) or microfibrillated cellulose (MFC),

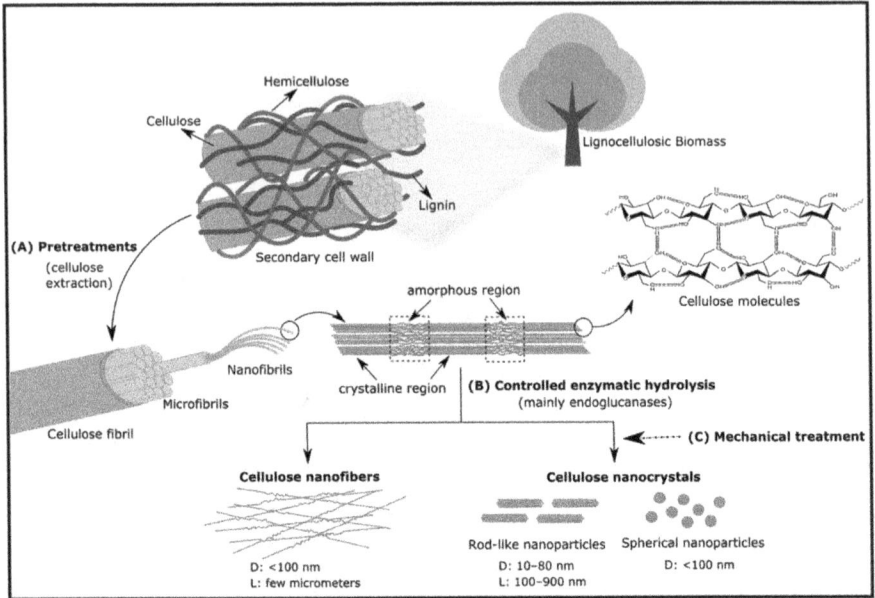

FIGURE 13.1 Enzymatic hydrolysis-based method for the synthesis of nanocellulose. Reproduced with permission from Ref. [4].

and bacterial nanocellulose (BNC) [14,15]. Nanocellulose is a natural nanomaterial derived from the cell walls of plants.

The enzymatic hydrolysis process that is involved in the synthesis of nanocellulose is shown in Figure 13.1. The procedure starts with lignocellulosic biomass pretreatments for cellulose extraction, then the process moves on to controlled enzymatic hydrolysis to produce cellulose nanofibers and cellulose nanocrystals (rod-like and spherical) and their relative sizes, as well as an indication of the feasibility of applying mechanical treatment after enzymatic hydrolysis, which is typically used to obtain more uniform particles [16].

Despite the fact that their chemical composition and molecular structures are same among all types (see Figure 13.2), they are distinct from one another in terms of their morphology, particle size, crystallinity, and other characteristics as a direct result of the various sources and extraction techniques [18,19]. Cellulose is rapidly crystallized and has a linear and repetitive structure with numerous hydroxyl groups. As a consequence, it is able to readily and swiftly combine with hydrogen bonds to generate structured crystalline structures that are kept together by hydroxyl groups. The hydroxyl groups of the cellulose polymer can produce hydrogen bondings between the different types of cellulose, which is known as an intermolecular hydrogen bond, and the polymer itself (intramolecular hydrogen bonds) [20].

Besides cereal straws [21] and banana plant [22], other crops and by-products have been studied as sources of nanocellulose in recent years including cotton linter [23] and olive tree pruning [24]. Other raw materials used to make cellulose include wood, herbaceous plants, grass, animal, algae, microbiological sources, as well as

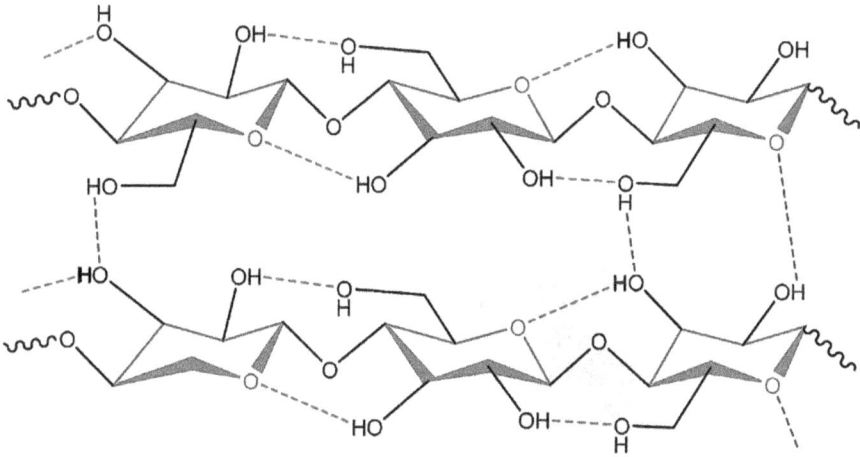

FIGURE 13.2   Molecular structure of cellulose. Reproduced with permission from Ref. [17].

TABLE 13.2
Various Sources for Nanocellulose Production

Source Group	Sources
Hardwood	Eucalyptus, Aspen, Balsa, Oak, Elm, Maple, Birch
Softwood	Pine, Juniper, Spruce, Hemlock, Larch, Cedar
Annual plants/ agricultural residues	Oil palm, Hemp, Jute, Agave, Sisal, Triticale straw, Soybean straw, Alfa, Kenaf, Coconut husk, Bagasse, Corn leaf, Sunflower, Bamboo, Canola, Wheat, Rice, Pineapple leaf and coir, Peanut shells, Potato peel, Tomato peel, Garlic straw residues, Mulberry fiber, Mengkuang leaves
Animal	Tunicates, *Chordata, Styela clava, Halocynthia roretzi*
Bacteria	*Gluconacetobacter, Salmonella, Acetobacter, Azotobacter, Agrobacterium, Rhizobium*
Algae	*Cladophora, Cystoseria myrica, Posidonia oceánica*

discarded paper [25], which is shown in Table 13.2. This promotes the circular economy by allowing for the integrated use of natural resources. Due to the nanoscale size of its cellulose fibers, nanocellulose possesses desirable qualities such as high strength, exceptional stiffness, and a large surface area [26,27].

An acid hydrolysis process is typically used to extract natural cellulose sources in order to produce cellulose nanocrystals (CNC) [28,29]. This process breaks down the hydrogen bonds in the fiber and cleaves the amorphous domains in order to produce well-defined crystalline rods from the natural cellulose. Common natural cellulose sources include wood pulp, cotton, algae, and bacteria. It possesses desirable characteristics such as high rigidity, large surface area, and superior mechanical qualities. It is crystalline in appearance and contains nanoparticles that are rod-like in form [30,31].

Due to the fact that they are biocompatible, biodegradable, durable, readily available, and possess exceptional mechanical characteristics, cellulose nanofibers (CNF),

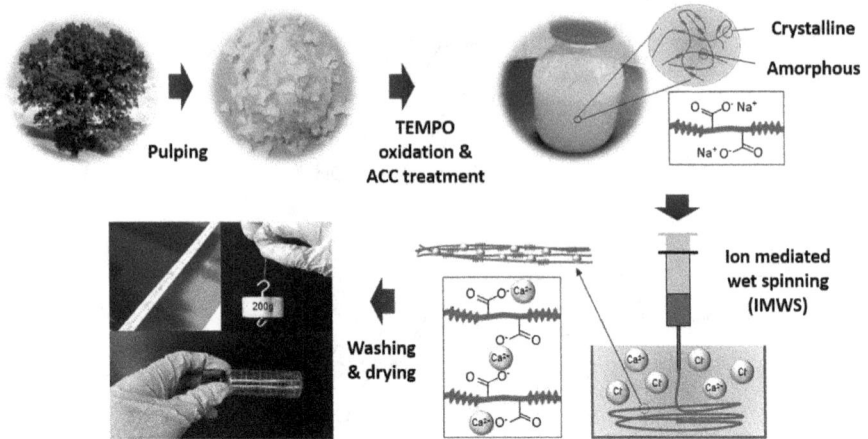

FIGURE 13.3 A schematic representation of the steps involved in the production of cellulose long fibers, including spinning, cross-linking, washing, drying, and stretching. Reproduced with permission from Ref. [35].

which are made up of various biopolymer molecules, are frequently used [32]. CNF generally ranges in width from 5 to 20 nm and ranges in length from a few micrometres to a few hundred micrometres. Extraction of CNF can be accomplished using a variety of processes, including homogenization, grinding, microfluidization, acid hydrolysis, and oxidation [33,34]. Kafy et al. [35] successfully fabricated cellulose long fiber from CNF via spinning, cross-linking, washing, drying, and stretching, as shown in Figure 13.3.

Bacterial cellulose, often known as BC, is a form of nanocellulose that has a variety of applications across a number of different sectors. However, because of the high manufacturing costs, its immense potential as a material with several uses has been severely restricted [36]. Many species of the genus Rhizobium, as well as Agrobacterium, Gluconacetobacter, and Sarcina, are responsible for the production of bacterial cellulose [37]. Bacterial cellulose is biodegradable and nontoxic, with a chemical structure similar to that of plant-based cellulose. The manufacturing and purification methods are simple, and the fact that bacterial cellulose can be shaped into a variety of forms both before and after the biosynthesis process makes it an intriguing biopolymer [38].

Chitin occurs naturally as a highly complex micro- and nanofibril structure that serves to maintain and preserve living systems, notably crustaceans, insects, and fungus, through the use of reinforcing and functional components [39–41]. Chitin is a component of a well-organized hierarchical structure that may be found in the exoskeletons of a wide variety of invertebrates, such as crabs, shrimps, lobsters, and krills. The dimensions of this structure range from nanometers to millimeters (Figure 13.4).

Chitin that has had some of its acetyl groups removed is the starting material to produce the linear polysaccharide known as chitosan. It is a copolymer of (1→4)-2-acetamido-2-deoxy-β-d-glucan (N-acetyl d-glucosamine) and (1→4)-2-amino-2-deoxy-β-d-glucan (d-glucosamine) units randomly or block distributed, based on the specific method of deacetylation, throughout the biopolymer

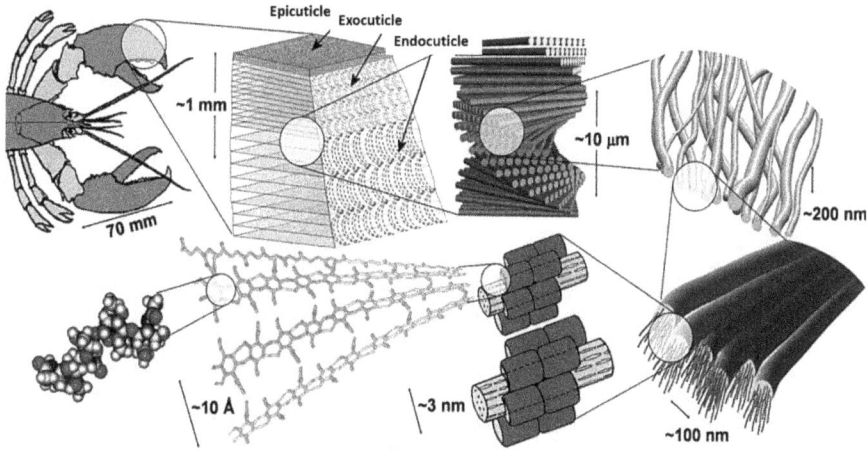

FIGURE 13.4 A hierarchical organization of the exoskeleton of an arthropod (H. americanus, American lobster) reveals different structural levels. Reproduced with permission from Ref. [42].

chain [43]. Chitosan dissolves in dilute organic acids such as acetic acid and formic acid. Nanoparticles of at least one dimension in the nanometer range (1–100 nm) are used to strengthen a biopolymer matrix in bionanocomposites, which results in greatly enhanced characteristics, thanks to the nanoparticles' high aspect ratio and surface area [44,45]. The most prevalent types of materials utilized as nanoparticles are layered clay minerals such as montmorillonite (MMT), hectorite, saponite, and laponite [46].

13.2.2 BIOPOLYMER USED AS THE MATRIX FOR THE FILMS WITH NATURAL NANOFILLERS

Growing interest is being shown in the use of nanocomposites to enhance the reinforcing and mechanical strength of polymeric composites. Nanocomposites use a matrix in which reinforcing components with diameters of less than 100 nm are dispersed [47]. Excellent interfacial contact and stress transfer between nanofillers and the polymer matrix are essential for the development of composites with excellent mechanical performance. A large number of studies on the biopolymer matrix combined with natural nanofillers in the production of film materials have been carried out. Table 13.3 shows some examples of natural nanofillers-reinforced biopolymer film matrices as well as the effects of the nanofillers.

13.3 ADVANTAGES AND DISADVANTAGES OF USING NATURAL NANOFILLERS IN BIOPOLYMER-BASED COMPOSITES

The physical, thermal, and mechanical characteristics of composites made from biopolymers have long been known to improve when nanofillers are used. Fauzi et al. [56] investigated the mechanical, physical, and chemical properties of sago

TABLE 13.3

The Effect of Nanofillers on Biopolymer Film Matrix

Nanofiller	Film Matrix	The Effect of the Nanofiller Reinforcement	References
Cellulose nanofibrils	Nanocellulose-tannin film	The air-barrier qualities of the cellulose films that included tannin were improved by a factor of six due to their high density and increased surface hydrophobicity.	[48]
Cellulose nanofibers and nanocrystals	Bionanocomposite films based on gelatin and nanocellulose (CNF and CNC)	Tensile strength is reduced when nanocellulose is added to a gelatin matrix. Nanocellulose helps to improve gelatin's excellent oxygen gas barrier properties.	[49]
Nanoclay and graphene oxide	Banana pseudostem nanocellulose films	Adding nanoclay and graphene oxide to plasticized films made of polyethylene glycol (PEG) increased their tensile strength and contact angle, respectively, compared to films made with only PEG.	[50]
Nanocellulose water hyacinth fiber	Yam Bean starch film	Using the highest proportion of nanocellulose, the highest values for tensile strength (5.8 MPa) and tensile modulus (403 MPa) were obtained (1 wt.%). The crystallinity index of bionanocomposite increased by more than 200% when nanocellulose was added at a concentration of less than 1 wt.%.	[51]
Cellulose nanocrystals	Cellulose nanocrystals/hydroxyapatite film	Hydrophilicity on the surface of the CNCs/HAP films was enhanced. Furthermore, the CNCs/HAP matrix outperformed the pure CNCs matrix in terms of its mechanical properties.	[52]
Crystalline nanocelullose (CNC)	Pectin-based bionanocomposite films reinforced with CNC	An increase in tensile strength of 84% was seen in a nanocomposite film containing 5% CNC, whereas a reduction in water vapor permeability of 40% was observed.	[53]
Cellulose nanocrystals	Thermoplastic starch films from tuber, cereal, and legume	Cellulose nanoparticles added to thermoplastic styrene (TPS) films helped in plasticization and enhanced stiffness, thermal stability, and moisture resistance.	[54]
Nanocellulose derived from microcrystalline cellulose	Polyvinyl alcohol/nanocellulose/silver nanocomposite films	The addition of 8 wt.% of NC to the PVA matrix resulted in an increase in the material's tensile strength. It was discovered that nanocomposite films exhibit significant antibacterial action against MRSA, which is a strain of *Staphylococcus aureus* and *Escherichia coli* (DH5-alpha).	[55]

starch-chitosan nanofillers film incorporated with chitosan nanofillers (CSN). Compared to sago starch film, which has a tensile strength of 46 MPa, sago starch/ chitosan nanofillers (SS/CSN) have a tensile strength of 88 MPa (SSF). The thermogravimetric measurement showed that the SS/CSN film was able to withstand temperatures up to 390°C with only a 60% loss in weight, whereas the SSF film lost 67% of its weight at 375°C.

Composite films made of nanocellulose/carboxymethyl cellulose (CMC) and nanochitosan/CMC were examined by Jannatyha et al. [57] in terms of their mechanical, barrier, and antibacterial properties. Casting methods were used to incorporate nanochitosan or nanocellulose in CMC film solutions at various concentrations (0.1%, 0.5%, and 1%). Both CMC/NCH and NCL were shown to have qualities that were less able to dissolve in water, contain more moisture, and absorb more moisture when the concentration was increased to 1 percent. The water vapor permeability of the polymer and nanofiller both reduced as the concentration of nanocomposite in the material was increased. In nanocomposite films, improvements in tensile strength and elongation at break were seen in conjunction with an increase in concentration.

Ren et al. [58] investigate the effects of various polyol plasticizers, such as glycerol, sorbitol, or a combination of the two, and clay content on potato starch/halloysite nanobiocomposites. Halloysite nanoclay was added to the nanobiocomposites, and regardless of the kind of plasticizer that was employed, this increased the nanobiocomposites' thermal stability and reduced their moisture absorption. Addition of halloysite to glycerol-plasticized starch increased its tensile strength by 47%, but doing the same for sorbitol-based and glycerol/sorbitol-based nanocomposites only increased it by 10.5% and 11%, respectively.

The effects of oil palm waste nanoparticles on the physical, mechanical, thermal, and morphological properties of hybrid plywood biocomposites are studied by Nuryawan et al. [59]. To use as adhesives, phenolic-formaldehyde (PFs) resins were synthesized and tested with varying amounts of oil palm ash (OPA) nanoparticles in layered fiber veneers and empty fruit bunch fiber mat. There were notable enhancements in water absorption and thickness swelling when the maximum loading of OPA nanoparticles in PF resin was employed. The flexural, shear, and impact properties of the biocomposites were improved. The panels of plywood were tested for their thermal stability, which demonstrated that the presence of PF-filled OPA nanoparticles contributed to the panels' thermal stability.

Ilyas et al. [60] systematically studied the water-barrier properties of sugar palm starch (SPS) bionanocomposites reinforced with sugar palm nanofibrillated cellulose (SPNFC). Solution casting methods are used to reinforce SPNFCs (0–1.0 wt.%) and SPS. The nanocomposites improved upon the control SPS film in terms of water resistance, moisture absorption, and light transmission. Addition of 1 wt% SPNFCs loading to the control SPS film increased water absorption by 24.13% and water solubility by 18.60% in the composite film.

Despite the benefits they provide in terms of physical, mechanical, and thermal properties, natural nanofillers have a few drawbacks that need to be taken into account. These include their susceptibility to moisture absorption, quality variations, low thermal stability, and incompatibility with hydrophobic polymer matrices [61,62].

It was also discovered that chitosan has poor mechanical properties, including a lack of elasticity, a high degree of fragility, low tensile strength, and high stiffness. Plasticizers, such as glycerol, sorbitol, or propylene glycol, have been commonly used to improve the properties [63,64]. Besides, nanofillers, such as nanoclays [65], carbon nanotubes [66], halloysite nanotubes [67], or inorganic nanoparticles [68,69], have also been used as additives in plasticized chitosan matrices.

13.4 MECHANISM INVOLVED OF USING NATURAL NANOFILLERS IN BIOPOLYMER-BASED COMPOSITES

Nanocellulose modification has the potential to enhance composite performance. For a deeper understanding of how chemical and surface modifications affect nanocellulose's reinforcing impact in polymer materials, more research into the material's reinforcing mechanism is required. However, no definitive conclusion has been reached regarding the best model for describing the reinforcement mechanisms of nanocomposites.

Huang et al. [70] incorporated unmodified nanofibrillated cellulose (UNFC), cationic nanofibrillated cellulose (CNFC), anionic nanofibrillated cellulose (ANFC), and cellulose nanocrystals (CNC) as the potential nanoadditives to make an insulating press paper. The effect of nanocellulose on the mechanical and electrical properties of prepared press paper was investigated. The tensile strength of composites containing 5% CNC-TEMPO was found to be greater than that of composites containing 5% CNC. They concluded that this was because the ionic bond between negatively charged CNC-TEMPO and positively charged chitosan was stronger. From the above discussion, it is evident that the size of the added nanocellulose and the electrical polarity of the nanocellulose have a significant impact on the mechanical properties of the nano-modified press paper.

Xiang et al. [71] studied the effects of bacterial cellulose's reinforcement mechanism on different types of paper sheets made from high-quality, medium-quality, and low-quality fibers such as softwood, hardwood, sugarcane bagasse, bamboo, wheat straw, and recycled fiber. This study shows that BC reinforcement effects were influenced by fiber properties such as fiber size and water retention value. Maintaining a low level of BC addition may be sufficient for effective reinforcement.

Toughening mechanisms, in addition to reinforcement mechanisms, have been extensively studied in epoxy and nanocomposite-related research. To improve the fracture toughness of epoxies or nanocomposite, different toughening mechanisms, such as soft particles and rigid fillers, can be mixed into epoxy resins at the same time to improve fracture toughness [72]. Lin et al. [73] investigated the toughening mechanism of polypropylene (PP) nanoparticles containing calcium carbonate ($CaCO_3$). The notched Izod impact strength of nanocomposites made of high-molecular-weight PP and 20 wt.% $CaCO_3$ nanoparticles with a monolayer coat of stearic acid was about 370 J/m, while that of unfilled PP was only 50 J/m. It has been found that the high-molecular-weight matrix's role is to produce high fracture stress, which stabilizes the plastic deformation prior to crack initiation.

Liang and Pearson [72] investigated the mechanism of toughening in hybrid epoxy-silica-rubber nanocomposites (HESRNs). A lightly cross-linked DGEBA/ piperidine epoxy system was blended with two different sizes of nanosilica (NS) particles, nominally 20 and 80 nm in diameter, as well as carboxyl terminated butadiene acrylonitrile (CTBN). The fracture toughness of CTBN-toughened epoxies was significantly increased by the addition of a minimal quantity of NS particles. This level of improvement was not possible by merely increasing the amount of CTBN content in the material. The effect of toughening is lessened when NS particles are grouped at high concentrations of CTBN.

13.5 SIGNIFICANT FIGURES FROM THE IMPROVEMENT INVOLVED

Nanofillers have been shown to improve the properties of biopolymer-based composites. The outcomes were determined by the technique, layering, direction, or treatment used. In order to synthesize highly ordered zein nanocomposites, Zhang and Wang [74] designed and studied a simple and low-cost approach. Making use of an external magnetic field, magnetic iron oxide (Fe_3O_4) nanofiller was synthesized, and a highly ordered structure was formed by reorienting the nanofiller particles in place. From the results, it showed that the tensile strength, elongation, and Young's modulus of Fe-Zein were all increased by 218%, 48%, and 264%, respectively, while the water vapor and oxygen permeability were reduced by 68% and 29%, respectively. More importantly, the mechanical and gas barrier properties of the highly ordered zein nanocomposites (Fe-Zein-Mag) were significantly improved.

Prashantha et al. [75] investigated the effect of nanofiller addition to both unmodified halloysites (HNT) and quaternary ammonium salt-treated halloysites (m-HNT). In order to create nanocomposites of poly(lactic acid) (PLA) and halloysite nanotubes (HNT), a melt extrusion technique with a masterbatch dilution procedure was used. It was found that halloysites are evenly distributed throughout the PLA matrix for both unmodified and modified nanotubes over the composition range studied. Halloysites have a greater nucleating impact on m-HNT, leading to a minor improvement in crystallinity. Halloysite addition to PLA also improves the material's stiffness, tensile strength, flexural strength, and impact resistance, with the latter four properties benefiting more from form-HNT than from HNT.

Cai et al. [76] investigated the effect of solvent treatment on morphology, crystallinity, and tensile properties of cellulose acetate nanofiber mats. After being immersed in a solvent, the fibers in a fibrous mat change both in diameter and orientation. Aligned nanofiber mats made from acetylated BC exhibited changes in fibrous shape and crystallinity after being soaked in an ethanol/acetone mixture.

Longer periods of submersion produced a denser, more compact fibrous structure with higher fiber diameters. Because of the solvent treatment, fibers get disoriented. Its tensile strength increased as a result of these changes.

13.6   CURRENT APPLICATIONS OF NATURAL NANOFILLERS IN BIOPOLYMER-BASED COMPOSITES

Appealing characteristics of natural nanofillers-based composites beyond synthetic fibers, as they are lighter, low-cost, nontoxic, and eco-friendly, make them predominant for modern applications, namely, automotive and other fields [77]. They have widespread utilizations in different fields due to their superlative properties in thermal, mechanical, and biodegradable. The applications of natural nanofillers in automotive, civil construction, sport and clothes, aerospace, biomedical, and pharmaceuticals are set out in Tables 13.4 and 13.5. From this data, it can be seen that plant fiber can be categorized into four, namely, leaf fibers, bast fibers, grass and reed fibers, and seed fibers. Each of these categories consists of its own type of fiber. This type of fiber has been detailed according to its use in modern applications as shown in Figure 13.5. From the chart, it has been shown that flax, hemp, and sisal were normally used in various applications related to automotive, civil constructions, sport and clothes, aerospace, biomedical, pharmaceutical, and others as detailed in Tables 13.4 and 13.5.

As can be seen from Table 13.4, the natural nanofillers-based composites grant lighter weight body parts, body panels, seat, carpets, and diverse interior part for cars, along with shelter against heat and any external crashes. These applications are also progressively decreasing the processing costs as well as designing an eco-friendly bio-based car concept as shown in Table 13.4. Data from this table also can break down into proportion of different categories of automotive compartments as shown in Figure 13.6. From the chart, it is apparent that natural nanofillers-based composites have greatest demand for door panel and seat parts, which leads to significantly eco-friendly bio-based car concept due to reducing total weight and promising fuel consumption savings. In addition, natural nanofillers-based composites are also extensively used to the same degree in civil constructions, sport and clothes, aerospace, biomedical, pharmaceutical, and others as shown in Table 13.5.

13.7   ADVANTAGES AND DISADVANTAGES OF USING NATURAL NANOFILLERS IN CURRENT FUNCTIONS

The characteristics of natural fibers are in accordance with their use. As shown in Table 13.6, each pure fiber has its own applications and has advantages and disadvantages based on its uses. In general, therefore, it seems that natural fibers have benefits in terms of lighter, less expensive, eco-friendly, high performance in firmness and rigidity, and recyclability. The major limitation of natural fiber, however, is water absorption. The evidence from these findings suggests that the importance of modern technology involved natural nanofillers in biopolymer-based composites to reveal its exceptional performance and competitive types of composites, as it claims supremacy over the mass of application from the largest field.

TABLE 13.4

Natural Fibers Utilizations in Automotive and Assembling Techniques

Origin Groups	Plant Fibers Category	Substances Used		Utilization	Assembling technique	References
		Fiber Reinforcement	Matrix/Binder Material			
Automotive						
Cellulose/ lignocellulose	Leaf fibers	Sisal	PP, PS, epoxy resin	Automobile body parts	Hand lay-up, compression moulding	[78]
	Bast fibers	Flax	PP, polyester, epoxy	Automotive industry		[79–82]
				Automotive	Resin transfer moulding, spray or hand lay-up, vacuum infusion	[83]
				Seatbacks, covers, rear parcel shelves, other interior trim, floor trays, pillar panels and central consoles, floor panels		[84]
		Hemp	PE, PP, PU	Automotive	Resin transfer moulding, compression moulding	[85]
		Ramie	cotton	Automotive		[86–88]
		Kenaf	PLA, PP, epoxy resin	Tooling, bearings, automotive parts	Compression moulding, pultrusion	[89]
				Automotive		[90–93]
				Door inner panel		[84]
	Grass and reed fibers	Rice Husk	PU, PE	Window/door frames, automotive structure.	Compression/injection moulding	[94]
	Seed fibers	Coir	PP, epoxy resin, PE	Building panels, roofing layers, lining boards, automotive structural components	Extrusion, injection moulding	[95]
				Car seat covers, mattresses, doormats, carpets		
			Glass fiber, epoxy	Wind turbine blades		[84]

TABLE 13.5

Natural Fibers with their Other Applications and Manufacturing Techniques

Origin Groups	Plant Fibers Category	Materials Used		Application	Manufacturing Technique	References
		Fiber Reinforcement	Matrix/Binder Material			
Civil Construction						
Cellulose/ lignocellulose fibers	Leaf fibers	Sisal	PP, PS, epoxy resin	Roofing sheet	Hand lay-up, compression moulding	[78]
			-	Civil constructions, used as fiber core of the steel wire cables of elevators	-	[79–82]
	Bast fibers	Flax	Polyester, epoxy, mortar	Masonry	-	[96]
			PP, polyester, epoxy	Structural	Resin transfer moulding, spray or hand lay-up, vacuum infusion	[83]
		Hemp	Polyester, epoxy, mortar	Masonry	-	[96]
			Polyethersulfone	Ignot bio- and polycal acoustic panel	-	[96]
			Polyester, epoxy, mortar	Masonry, Hemp chair furniture	-	[96]
			Polyethersulfone	Ignot bio- and polycal acoustic panel	-	[96]
		Jute	Polyester, PP	Roofing, covering of industrial buildings, setting of bungalows, prefabricated sheds, decorative materials	Hand lay-up, compression/ injection moulding	[97]
		Ramie	Polyester, epoxy, mortar	Masonry	-	[96]
			Glass fiber, vinyl ester	Deck panel	-	[96]
			PP, polyolefin, PLA	Civil	Extrusion with injection moulding	[98–99]

(Continued)

TABLE 13.5 (*Continued*)

Natural Fibers with their Other Applications and Manufacturing Techniques

Origin Groups	Plant Fibers Category	Fiber Reinforcement	Materials Used		Application	Manufacturing Technique	References
			Matrix/Binder Material				
			cotton		Constructions	-	[86–88]
			cement, metakaolin, fly ash		Cementitious materials	-	[96]
		Kenaf	PLA, PP, epoxy resin		Construction	Compression molding, pultrusion	[90–93]
			Plaster of Paris		Ceiling	-	[96]
	Seed Fibers	Coir	PP, epoxy resin, PE		Building boards, roofing sheets, insulation boards	Extrusion, injection moulding	[95]
			Polyester, epoxy, mortar		Masonry	-	[96]
Sport and clothes							
Cellulose/ lignocellulose	Leaf fibers	Sisal	Epoxy		Helmet shell	-	[96]
	Bast fibers	Flax	Epoxy/hemp		Racing bicycle	-	[96]
			Carbon fiber, epoxy		Bicycle frame		
		Jute	Polyester		Footwear, winter over coat		[96]
			Epoxy		Helmet shell		
		Hemp	Polyester		Footwear	-	[96]
		Kenaf	Kevlar, epoxy		Ballistic armor materials	-	[96]
			PLA		Mobile phone casing		[96]
			Epoxy		Recurve bow		[96]

(*Continued*)

TABLE 13.5 (Continued)

Natural Fibers with their Other Applications and Manufacturing Techniques

Origin Groups	Plant Fibers Category	Materials Used		Application	Manufacturing Technique	References
		Fiber Reinforcement	Matrix/Binder Material			
Aerospace						
Cellulose/ lignocellulose	Bast fibers	Hemp	Epoxy	Electronic shelfs for helicopter	-	[96]
		Ramie	Matrix	Aircraft wing boxes	-	[96]
		Kenaf	Glass fiber, matrix		-	[96]
Biomedical and pharmaceutical						
Cellulose/ lignocellulose	Leaf fibers	Sisal	Polycaprolactone	Orthoses materials	-	[96]
	Bast fibers	Flax	Poly(methyl methacrylate)	Drug delivery	-	[96]
			Epoxy	Bone grafting, orthopedic implants	-	[96]
		Jute	Plant extracts, polyester, PP, chitosan	Biomedical nanoparticles, antibiotics	-	[96]
			Plant extracts, polyester, PP, chitosan	Biomedical nanoparticles, antibiotics	-	[96]
		Hemp	Polycaprolactone	Orthoses materials	-	[96]
		Ramie	Epoxy	Bone grafting, orthopedic implants	-	[96]
		Kenaf	Plant extracts, polyester, PP, chitosan	Biomedical nanoparticles, antibiotics	-	[96]
	Seed Fibers	Coir	Polycaprolactone	Orthoses materials	-	[96]

(Continued)

TABLE 13.5 (*Continued*)
Natural Fibers with their Other Applications and Manufacturing Techniques

Origin Groups	Plant Fibers Category	Fiber Reinforcement	Matrix/Binder Material	Application	Manufacturing Technique	References
Others						
Cellulose/ lignocellulose	Leaf fibers	Sisal	PP, PS, epoxy resin	Roofing sheet	Hand lay-up, compression moulding	[78]
				Freighting industry (for mooring small craft and handling cargo), agricultural twine, or baler twine	–	[79–82]
			Polyester, PP, PE, PLA, epoxy, rubber, polyurethane, wheat gluten	Dielectric materials	–	[96]
			Glass fiber, epoxy	Wind turbine blades		
	Bast fibers	Flax	PP, polyester, epoxy	Textile	Resin transfer moulding, spray or hand lay-up, vacuum infusion	[83]
			Polyester, PP, PE, PLA, epoxy, rubber, polyurethane, wheat gluten	Dielectric materials	–	[96]
			Carbon nanotube	Electrodes	–	[96]
			Glass fiber, polyhydroxyalkanoates, epoxy	Marine materials	–	[96]
			Glass fiber, epoxy	Wind turbine blades	–	
			Aluminum, epoxy, polyester	Electromagnetic interference shields	–	

(*Continued*)

TABLE 13.5 (*Continued*)
Natural Fibers with their Other Applications and Manufacturing Techniques

Origin Groups	Plant Fibers Category	Materials Used		Application	Manufacturing Technique	References
		Fiber Reinforcement	**Matrix/Binder Material**			
Plant	Jute		Polyester, PP	Ropes, door panels and sanitary products (slab, ring, etc.), helmets, chest covers, leg covers, kitchen sinks, durable chairs, tables	Hand lay-up, compression/injection moulding	[97]
			Glass fiber, epoxy	Wind turbine blades	-	[78]
			Polyester, PP, PE, PLA, epoxy, rubber, polyurethane, wheat gluten	Dielectric materials	-	[96]
			Glass fiber, polyester	Solar parabolic trough collector	-	[96]
			Aluminum, epoxy, polyester	Electromagnetic interference shields	-	[84]
	Hemp		PE, PP, PU	Furniture	Resin transfer moulding, compression moulding	[96]
			Aluminum, epoxy, polyester	Electromagnetic interference shields		[98]
	Ramie		PP, polyolefin, PLA	Bulletproof vests, socket prosthesis, nanocomposites electrolytes	Extrusion with injection moulding	[86–88]
			cotton	Upholstery, gas mantle, fishing nets, and marine packings, furniture		[84]
				Sound-proofing, trunk panel, insulation		[90–93]
	Kenaf		PLA, PP, epoxy resin	Packaging, furniture, textiles, mats, paper pulp	Compression moulding, pultrusion	

(Continued)

TABLE 13.5 (*Continued*)
Natural Fibers with their Other Applications and Manufacturing Techniques

Origin Groups	Plant Fibers Category	Materials Used		Application	Manufacturing Technique	References
		Fiber Reinforcement	Matrix/Binder Material			
			Polyester, PP, PE, PLA, epoxy, rubber, polyurethane, wheat gluten	Dielectric materials	-	[96]
			Aluminum, epoxy, polyester	Electromagnetic interference shields	-	
	Grass and Reed Fibers	Rice husk	PU, PE	Window/door frames	Compression/injection moulding	[94]
	Seed fibers	Coir	Natural latex rubber	Containers, boxes, trays, packaging	-	[96]
			Glass fiber, epoxy	Wind turbine blades	-	

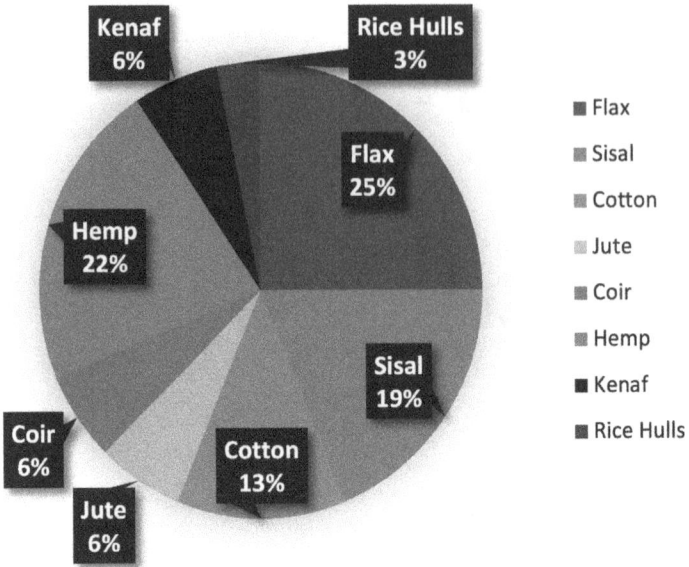

FIGURE 13.5   Natural fibers used in modern application.

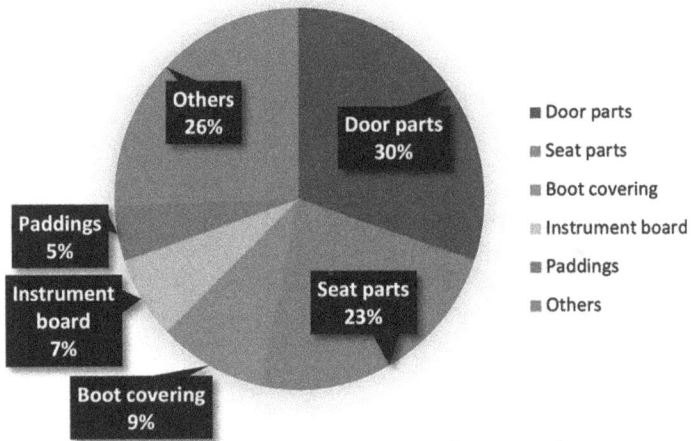

FIGURE 13.6   Natural nanofillers-based composites utilization in automotive compartments.

13.8   CURRENT MODERN TECHNOLOGY INVOLVING NATURAL NANOFILLERS IN BIOPOLYMER-BASED COMPOSITES

Engaging characteristics of natural nanofillers-based composites alike light weight, environmental friendliness, recyclability, and achievable mechanical performance attained from raw resources are due to the application of modern technology of natural nanofillers in biopolymer-based composites. The modern technology involved is

TABLE 13.6

Advantages and Disadvantages of Using Natural Nanofillers in Current Applications

Origin Groups	Plant Fibers Category	Materials Used		Advantages	Disadvantages	Application	References
		Fiber reinforcement	Matrix/binder material				
Cellulose/lignocellulose	Leaf fibers	Sisal	PP, PS, epoxy resin	- Product design flexibility, noise absorption, insulation; impact resistance in a crash followed by weight reduction contributes to saving fuel	- Water absorption	- AUTOMOBILE body parts, roofing sheet	[78]
	Bast fibers	Flax	PP, polyester, epoxy	- Good eco-friendly performance - High specific properties (strength and stiffness)	- Water absorption	- Structural, textile	[83]
		Jute	Polyester, PP	- Good physical setting and superlative work at lighter	/	- Ropes, roofing, door panels, and sanitary products (slab, ring, etc.), covering of industrial buildings, setting of bungalows, prefabricated sheds, decorative materials, helmets, chest guards, leg guards, kitchen sinks, durable chairs, tables.	[97]
		Hemp	PE, PP, PU	- Good specific firmness	- Poor adhesion to hydrophobic polymer matrix	- Furniture, automotive	[83]

(Continued)

TABLE 13.6 (Continued)
Advantages and Disadvantages of Using Natural Nanofillers in Current Applications

Origin Groups	Plant Fibers Category	Materials Used		Advantages	Disadvantages	Application	References
		Fiber reinforcement	Matrix/binder material				
				- Modulus	- An affinity for moisture absorption		
	Ramie	PP, polyolefin, PLA		- Low carbon footprint			[98]
				- Recyclability			
				- Lighter and less expensive than conventional bulletproof panels made from ceramic plate, Kevlar/aramid composites, or steel-based material	/	- Bullet-proof vests	
				- Ramie fiber was used to reduce the amount of Kevlar used and the production cost for economic armor production			
				- Improve safety and accessibility of prosthetic limb manufacture	/	- Socket prosthesis	
				- Lower density of the ramie/epoxy core	/	- civil	

(Continued)

TABLE 13.6 (*Continued*)
Advantages and Disadvantages of Using Natural Nanofillers in Current Applications

Origin Groups	Plant Fibers Category	Materials Used		Advantages	Disadvantages	Application	References
		Fiber reinforcement	Matrix/ binder material				
				- Cost of ramie fiber/aluminum laminate was only two-thirds of that of the aluminum	/	- Nanocomposite electrolyte	
				- Higher storage modulus at elevated temperatures than the unfilled PEO–LiTFSI sample			
				- Less decreases in ionic conduction which indicates very stable in mechanical properties at high temperature			
		Kenaf	PLA, PP, epoxy resin	- Motives of price, weight reduction - Low density and superior firmness properties	/	- Tooling, bearings, automotive parts	[84,89]
Grass and reed fibers		Rice husk	PU, PE	–	/	- Window/door frames, automotive structure.	[94]
Seed Fibers		Coir	PP, epoxy resin, PE	- Its tough protection to salt water and availability	/	- Boards, roofing sheets, insulation boards.	[84,95]

oriented toward improving the performance of natural nanofillers-based composites as well as an existing manufacturing process. Two methods are applied in improving the performance, namely, fiber treatment and modification as well as fiber hybridization, respectively [96]. Both techniques provide benefits in terms of improving the original structure of natural fiber as well as the overall quality, where through fiber treatment, it is able to boost the valid wetting and uniform dispersion to overcome incompatibility and inferior bonding of natural fiber in polymer and the latter, to adjust the deficiency of one specific fiber to achieve continuality, low-priced, and enhanced performance in terms of mechanical durability, chemical balance, nontoxicity, noncombustible, resistance to high temperature, and thermal or acoustic lining of natural fiber-reinforced composites [96].

Meanwhile, the modern technology involved relating to manufacturing technique can be divided into two leading focuses, namely, the advancement of conventional manufacturing process and automated manufacturing [100]. The methods involved in the advancement of conventional in manufacturing process are electrospinning [101,102], replacing the extrusion-injection in direct injection of molding process [96], applying the twin-screw processing [103], and adapting the additive manufacturing [103–105]. Next, robotic filament winding (RFW) technology has been introduced as one of the automated techniques in the manufacturing process [106,107]. The improvements that have been identified through this modern technology will be detailed in Table 13.7.

TABLE 13.7

Key Changes Related to Modern Technology Involving Natural Nanofillers in Biopolymer-Based Composites

Scope	Technique	Improvement	References
Advancement of conventional manufacturing processes	Electrospinning	- Enhance durability and resilience over process parameters, high surface area-to-volume ratio, and high porosity	[101,102]
	Extrusion-injection in molding process	- Boost strength achievement irrespective of the initial fiber length	[96]
	Twin-screw extrusion	- Decreased the amplification of shear stress and enhanced the distribute mixing greatly upgraded the overall fiber	[96]
	Additive manufacturing	- Rapid, simple manufacturing, improved performance of composites geometry, increased controllability of fiber volume and direction - Decrease thermal degradation/deconstruction of pure fiber, saving expensive moulds and tools - Capable to across design idea into the final product rapidly without the narrowing material and cycle time	[103–105]
Automated manufacturing	Robotic filament winding (RFW)	- Reproducing and decreased process hour by taking over a human operator	[106]

13.9   SIGNIFICANT FIGURES FROM THE IMPROVEMENT INVOLVED

The key changes related to modern technology involved for natural nanofillers in biopolymer-based composites are shown in Table 13.7. In general, it seems that there are two scopes involved in modern technology related to natural nanofillers, namely, advancement of conventional manufacturing processes and automated manufacturing, respectively. These data suggest that the involvement of modern technology of natural nanofillers in biopolymer-based composites significantly improves overall quality of natural nanofillers-based composites.

13.10   CHALLENGES AND CURRENT TRENDS OF NATURAL NANOFILLERS IN BIOPOLYMER-BASED COMPOSITES

The role and current challenges of natural nanofillers in biopolymer-based composites and their application trends are explained. Various types of natural nanofillers have been used in biopolymer-based composites to improve the performance and expand the applications of bionanocomposites in today's industry. Biopolymers have numerous and wide-ranging applications, especially in packaging and biomedical applications. Despite all the advantages of natural nanofillers in biopolymer-based composites, there are some limitations that need to be further investigated and overcome for a better future.

13.11   ISSUES AND CHALLENGES OF NATURAL NANOFILLERS IN BIOPOLYMER-BASED COMPOSITES

There are three main areas of deficiency with nanofillers and biodegradable polymers, namely, cost, processing, and safety, and health issues. In terms of cost, for example, the price of the cheapest biodegradable polymer types, poly(lactic acid), sold in the market is $1.9–$2.0/kg in 2019, compared to the widely used petroleum-based plastics polyethylene and polypropylene, which typically range from $1.2 to $1.3/kg [108]. In addition, other biodegradable polymer types such as PBS and PBAT cost about $3.5/kg in 2019 and PHAs cost $3.5/kg in 2018. In almost all studies, it was found that the slow growth in market demand and slow adoption of the materials by customers with annual sales of less than 500 tons for PHAs are the reasons for the high price of biodegradable polymers. In order to solve this problem and thus to boost the market demand, two crucial aspects that bring about important and significant changes in costs need to be considered, namely, exploring new strategies for high-purity and low-cost monomers from biomass. Therefore, as production capacity increases due to high demand in the market, the price of the product is expected to decrease. Over the next five years, the global bioplastics market is expected to grow at a compound annual growth rate of 21.7%, owing to consumer preference for eco-friendly plastic products and supportive government policies.

In terms of processing, the main obstacle for practical use is to obtain a super-high gas/moisture barrier that has good processability, balanced mechanical properties, and advanced composting that is either home-compostable, soil or marine biodegradable, and also industrially compostable, taking into account safety and health issues such as nanoparticle migration. If these bottleneck issues can be solved, nanocomposites with natural fillers can be industrially produced, becoming the next generation of environmental-friendly nanocomposites.

From a health and safety perspective, concerns have been raised about the health and environmental impacts of nanomaterials used in a variety of consumer and industrial applications. This is because when the nanoparticles are released, there is a risk of small particles becoming air-borne and being inhaled in the workplace, and the long-term health effects of exposure, particularly on the lungs, are largely unknown. The high aspect ratio of nanofillers can pose an environmental hazard due to the possibility of nanoparticles entering through skin nodules and being inadvertently ingested through the digestive tract. In addition, it is estimated that there are up to 10,000 reports of studies on toxicity. Han et al. [109] stated that some nanoparticles can cause intracellular damage, pulmonary inflammation, and vascular diseases.

13.12   CURRENT TRENDS OF NATURAL NANOFILLERS IN BIOPOLYMER-BASED COMPOSITES

Biodegradable composites enriched with nanofillers can solve the above problems and help to increase the efficiency of biopolymers. Biopolymers improved with nanofillers can benefit applications in biomedicine including such tissue engineering and drug delivery. The use of nanofillers can improve the properties of biopolymers and expand their applications and functions, leading to greater medical breakthroughs. As in future, biodegradable composites will be able to replace the majority of current materials, which is critical for our survival; consequently, this is an important challenge further to examine the significance of nanofillers in improving biodegradable composites. Biopolymers enriched with natural nanofillers have great potential for a sustainable future. To expand the applications and potential, researchers around the world continue to make progress in developing biopolymers with improved properties that meet industry and safety guidelines.

13.13   FUTURE RECOMMENDATIONS

Natural nanofiller-reinforced biodegradable composite materials are widely used, and the field has piqued the interest of numerous researchers. In the future, researchers will likely focus on natural nanofillers to improve composite processes. More research is needed on highly industrialized and cost-effective processes that are extremely difficult for this nanotechnology.

Furthermore, investigating new natural nanofillers with abundant availability can broaden sources and reduce material costs.

In practise, the development of natural nanofillers in biopolymer-based composites can provide economic advantages in rural regions, notably in much less developed regions in which these materials are abundant. A thorough life-cycle assessment (LCA) of biocomposites is required to maintain the potential significance in the lesson of emerging high-performance biocomposites. Despite the biocomposites' renewability and recyclability, researchers must contend with challenges related to standards and conformity assessment for these materials. Biocomposites designed for load-bearing applications must comply with regulations governing the disposal of massive amounts of waste. More research on product design and performance evaluation, as well as the effect of environmental ageing on the failure mechanics caused by the thermal properties processes of biocomposites, is required. Biocomposite materials must come to the realization with superior efficiency, reliability, consistency, and serviceability in order to expand their application.

In the future, bionanocomposite substances will be able to replace the majority of current materials, which is critical for our survival; therefore, it is an urgent task to investigate nanofillers to improve biodegradable composite materials.

13.14 CONCLUSION

Natural nanofillers used in biopolymer-based composites possess many advantages, including low density, biodegradability, excellent mechanical properties, being naturally renewable, minimal energy usage, and relatively high reactivity. As a result, they are the most versatile materials used in various aspects of our daily lives. Natural nanofillers can be fabricated by modern techniques including electrospinning, twin screw extruder, additive manufacturing, and can be served as an effective reinforcement in polymer matrices for various types of applications. This chapter further summarized the pros and cons of using natural nanofillers in biopolymer-based composites in terms of physical, thermal, and mechanical properties. In comparison with the virgin polymer, matrices alone, the results showed significant improvement in terms of the mechanical, barrier, and antimicrobial properties. However, despite their advantages, they also have some drawbacks, including moisture absorption, quality variations, poor thermal stability, and incompatibility with hydrophobic polymer matrices. For a clear understanding in particular reinforcing mechanism of nanofiller, it is necessary to further research on surface modification of nanocellulose in the polymer matrix. This chapter discussed on the potential applications of natural nanofillers in diverse industries such as automotive, civil construction, sport and clothes, aerospace, biomedical, and pharmaceuticals. Furthermore, the use of modern technology, notably in the production of natural nanofillers in biopolymer-based composites, might improve the material's overall quality. Although adding nanofillers to biopolymer-based composites could improve their properties and broaden their uses, safety and health concerns are among the primary challenges that require further research to determine whether they can be a viable alternative to conventional materials. To increase the compatibility of nanocellulose in the future, it will be important to discover suitable polymer matrices. It should be highlighted that the use of totally green nanocomposites in other applications, particularly the use of natural nanocellulose in biopolymer-based composites, should be extensively investigated. These studies also give an insight of the knowledge to advance product development, ensure compliance, and a successful market launch for the applications being developed on natural nanofillers from biodegradable-based polymer composites by 2025 for a promising future.

REFERENCES

[1] S. H. Kamarudin et al., "A Review on antimicrobial packaging from biodegradable," *Polymers*, vol. 14, no. 1, p. 174. doi: 10.3390/polym14010174

[2] N. M. Nurazzi et al., "A review on mechanical performance of hybrid natural fiber polymer composites for structural applications," *Polymers*, vol. 13, no. 13, pp. 1–47, 2021, doi: 10.3390/polym13132170.

[3] J. A. Sirviö, K. Hyypiö, S. Asaadi, K. Junka, and H. Liimatainen, "High-strength cellulose nanofibers produced: Via swelling pretreatment based on a choline chloride-imidazole deep eutectic solvent," *Green Chem.*, vol. 22, no. 5, pp. 1763–1775, 2020, doi: 10.1039/c9gc04119b.

[4] A. A. B. Omran et al., "Micro-and nanocellulose in polymer composite materials: A review," *Polymers*, vol. 13, no. 2. pp. 1–30, January 2021, doi: 10.3390/polym13020231.

[5] S. H. Kamarudin, L. C. Abdullah, M. M. Aung, and C. T. Ratnam, "Thermal and structural analysis of epoxidized jatropha oil and alkaline treated kenaf fiber reinforced poly(lactic acid) biocomposites," *Polymers*, vol. 12, no. 11, pp. 1–21, 2020, doi: 10.3390/polym12112604.

[6] D. S. Naidu and M. J. John, "Effect of clay nanofillers on the mechanical and water vapor permeability properties of xylan-alginate films," *Polymers*, vol. 12, no. 10, pp. 1–23, 2020, doi: 10.3390/polym12102279.

[7] E. Jamróz, P. Kulawik, and P. Kopel, "The effect of nanofillers on the functional properties of biopolymer-based films: A review," *Polymers*, vol. 11, no. 4, pp. 1–42, 2019, doi: 10.3390/polym11040675.

[8] E. Castro-Aguirre, R. Auras, S. Selke, M. Rubino, and T. Marsh, "Impact of nanoclays on the biodegradation of poly(lactic acid) nanocomposites," *Polymers*, vol. 10, no. 2, 2018, doi: 10.3390/polym10020202.

[9] N. M. Nurazzi et al., "Mechanical performance and applications of cnts reinforced polymer composites-a review," *Nanomaterials*, vol. 11, no. 9, pp. 1–25, 2021, doi: 10.3390/nano11092186.

[10] G. L. Devnani and S. Sinha, "Effect of nanofillers on the properties of natural fiber reinforced polymer composites," *Mater. Today Proc.*, vol. 18, pp. 647–654, 2019, doi: 10.1016/j.matpr.2019.06.460.

[11] N. M. Nurazzi et al., "A review on natural fiber reinforced polymer composite for bullet proof and ballistic applications," *Polymers*, vol. 13, no. 4, p. 646, 2021, doi: 10.3390/polym13040646.

[12] S. Mohan, O. S. Oluwafemi, N. Kalarikkal, S. Thomas, and S. P. Songca, "Biopolymers - application in nanoscience and nanotechnology," *Recent Adv. Biopolym.*, vol. 1, no. 1, pp. 47–66, 2016, doi: 10.5772/62225.

[13] D. A. Pereira de Abreu, P. Paseiro Losada, I. Angulo, and J. M. Cruz, "Development of new polyolefin films with nanoclays for application in food packaging," *Eur. Polym. J.*, vol. 43, no. 6, pp. 2229–2243, 2007, doi: 10.1016/j.eurpolymj.2007.01.021.

[14] A. Dufresne, *Nanocellulose: From nature to high performance tailored materials*, vol. 2, pp. 192–213, 2012, De Gruyter, Berlin. doi:10.1515/9783110254600.

[15] Dieter Klemm et al., "Nanocellulose: A new family of nature-based materials," *Angew. Chemie Int. Ed.*, vol. 50, pp. 5438–5466, 2011, doi: 10.1002/anie.201001273.

[16] M. Michelin, D. G. Gomes, and and J. A. Teixeira, "Nanocellulose production: Exploring the enzymatic route and residues of pulp and paper industry," *Molecules*, vol. 25, no. 3411, pp. 1–36, 2020, doi: 10.1002/anie.201001273.

[17] H. Shaghaleh, X. Xu, and S. Wang, "Current progress in production of biopolymeric materials based on cellulose, cellulose nanofibers, and cellulose derivatives," *RSC Adv.*, vol. 8, no. 2, pp. 825–842, 2018, doi: 10.1039/c7ra11157f.

[18] R. J. Moon, A. Martini, J. Nairn, J. Simonsen, and J. Youngblood, Cellulose nanomaterials review: Structure, properties and nanocomposites, vol. 40, no. 7. 2011.

[19] N. Lavoine, I. Desloges, A. Dufresne, and J. Bras, "Microfibrillated cellulose - Its barrier properties and applications in cellulosic materials: A review," *Carbohydr. Polym.*, vol. 90, no. 2, pp. 735–764, 2012, doi: 10.1016/j.carbpol.2012.05.026.

[20] G. Henriksson and H. Lennholm, "Cellulose and carbohydrate chemistry," In Monica Ek, Göran Gellerstedt, and Gunnar Henriksson (Eds.), *Wood Chemistry and Biotechnology*, vol. 1, 2009, pp. 71–100. De Gruyter, Berlin, New York.

[21] E. Espinosa, R. Sánchez, R. Otero, J. Domínguez-Robles, and A. Rodríguez, "A comparative study of the suitability of different cereal straws for lignocellulose nanofibers isolation," *Int. J. Biol. Macromol.*, vol. 103, pp. 990–999, 2017, doi: 10.1016/j.ijbiomac.2017.05.156.

[22] J. Velásquez-Cock et al., "Influence of the maturation time on the physico-chemical properties of nanocellulose and associated constituents isolated from pseudostems of banana plant c.v. Valery," *Ind. Crops Prod.*, vol. 83, pp. 551–560, 2016, doi: 10.1016/j.indcrop.2015.12.070.

[23] J. P. S. Morais, M. D. F. Rosa, M. D. S. M. De Souza Filho, L. D. Nascimento, D. M. Do Nascimento, and A. R. Cassales, "Extraction and characterization of nanocellulose structures from raw cotton linter," *Carbohydr. Polym.*, vol. 91, no. 1, pp. 229–235, 2013, doi: 10.1016/j.carbpol.2012.08.010.

[24] Ú. Fillat et al., "Assessing cellulose nanofiber production from olive tree pruning residue," *Carbohydr. Polym.*, vol. 179, pp. 252–261, 2018, doi: 10.1016/j.carbpol.2017.09.072.

[25] D. Trache et al., *Nanocellulose: From Fundamentals to Advanced Applications*, vol. 8, May 2020.

[26] A. Dufresne, "Nanocellulose: Potential Reinforcement in Composites," In *Natural Polymers, Volume 2: Nanocompites*, vol. 2, RSC Publishing, 2012, pp. 1–25.

[27] A. Dufresne, "Nanocellulose: A new ageless bionanomaterial," *Mater. Today*, vol. 16, no. 6, pp. 220–227, 2013, doi: 10.1016/j.mattod.2013.06.004.

[28] P. Criado, C. Fraschini, M. Jamshidian, S. Salmieri, A. Safrany, and M. Lacroix, "Gamma-irradiation of cellulose nanocrystals (CNCs): Investigation of physicochemical and antioxidant properties," *Cellulose*, vol. 24, no. 5, pp. 2111–2124, 2017, doi: 10.1007/s10570-017-1241-x.

[29] B. Dhuiège, G. Pecastaings, and G. Sèbe, "Sustainable approach for the direct functionalization of cellulose nanocrystals dispersed in water by transesterification of vinyl acetate," *ACS Sustain. Chem. Eng.*, vol. 7, no. 1, pp. 187–196, 2019, doi: 10.1021/acssuschemeng.8b02833.

[30] T. J. Bondancia, L. H. C. Mattoso, J. M. Marconcini, C. S. Farinas, R. X. V. De Novembro, and S. Carlos, "A new approach to obtain cellulose nanocrystals and ethanol from eucalyptus cellulose pulp via the biochemical pathway," *Am. Inst. Chem. Eng. Biotechnol. Prog.*, pp. 1–38, 2017, doi: 10.1002/btpr.2486.

[31] N. M. Moo-Tun, A. Valadez-Gonzalez, and J. A. Uribe-Calderon, "Thermo-oxidative aging of LDPE/stearoyl chloride-grafted cellulose nanocrystals blown films," *J. Polym. Environ.*, vol. 27, pp. 1226–1239, 2019, doi: 10.1007/s10924-019-01424-z.

[32] S. Ling et al., "Biopolymer nanofibrils: Structure, modeling, preparation, and applications," *Prog. Polym. Sci.*, vol. 85, pp. 1–56, 2018, doi: 10.1016/j.progpolymsci.2018.06.004.

[33] S. Iwamoto, K. Abe, and H. Yano, "The effect of hemicelluloses on wood pulp nanofibrillation and nanofiber network characteristics," *Biomacromolecules*, vol. 9, no. 3, pp. 1022–1026, 2008, doi: 10.1021/bm701157n.

[34] M. Pääkko et al., "Enzymatic hydrolysis combined with mechanical shearing and high-pressure homogenization for nanoscale cellulose fibrils and strong gels," *Biomacromolecules*, vol. 8, no. 6, pp. 1934–1941, 2007, doi: 10.1021/bm061215p.

[35] A. Kafy et al., "Cellulose long fibers fabricated from cellulose nanofibers and its strong and tough characteristics," *Sci. Rep.*, vol. 7, no. 1, pp. 1–8, 2017, doi: 10.1038/s41598-017-17713-3.

[36] M. Skočaj, "Bacterial nanocellulose in papermaking," *Cellulose*, vol. 26, no. 11, pp. 6477–6488, 2019, doi: 10.1007/s10570-019-02566-y.

[37] W. Hu, S. Chen, J. Yang, Z. Li, and H. Wang, "Functionalized bacterial cellulose derivatives and nanocomposites," *Carbohydr. Polym.*, vol. 101, no. 1, pp. 1043–1060, 2014, doi: 10.1016/j.carbpol.2013.09.102.

[38] O. Shezad, S. Khan, T. Khan, and J. K. Park, "Physicochemical and mechanical characterization of bacterial cellulose produced with an excellent productivity in static conditions using a simple fed-batch cultivation strategy," *Carbohydr. Polym.*, vol. 82, no. 1, pp. 173–180, 2010, doi: 10.1016/j.carbpol.2010.04.052.

[39] S. Ifuku and H. Saimoto, "Chitin nanofibers: Preparations, modifications, and applications," *Nanoscale*, vol. 4, no. 11, pp. 3308–3318, 2012, doi: 10.1039/c2nr30383c.

[40] S. Muthukrishnan, H. Merzendorfer, Y. Arakane, and K. J. Kramer, *Chitin Metabolism in Insects*, December 2012.

[41] S. Nikolov et al., "Robustness and optimal use of design principles of arthropod exoskeletons studied by ab initio-based multiscale simulations," *J. Mech. Behav. Biomed. Mater.*, vol. 4, no. 2, pp. 129–145, 2011, doi: 10.1016/j.jmbbm.2010.09.015.

[42] D. Raabe, C. Sachs, and P. Romano, "The crustacean exoskeleton as an example of a structurally and mechanically graded biological nanocomposite material," *Acta Mater.*, vol. 53, pp. 4281–4292, 2005.

[43] A. Di Martino, M. Sittinger, and M. V. Risbud, "Chitosan: A versatile biopolymer for orthopaedic tissue-engineering," *Biomaterials*, vol. 26, no. 30, pp. 5983–5990, 2005, doi: 10.1016/j.biomaterials.2005.03.016.

[44] J. W. Rhim and P. K. W. Ng, "Natural biopolymer-based nanocomposite films for packaging applications," *Crit. Rev. Food Sci. Nutr.*, vol. 47, no. 4, pp. 411–433, 2007, doi: 10.1080/10408390600846366.

[45] R. Zhao, P. Torley, and P. J. Halley, "Emerging biodegradable materials: Starch- and protein-based bio-nanocomposites," *J. Mater. Sci.*, vol. 43, no. 9, pp. 3058–3071, 2008, doi: 10.1007/s10853-007-2434-8.

[46] Q. H. Zeng, A. B. Yu, G. Q. Lu, and D. R. Paul, "Clay-based polymer nanocomposites: Research and commercial development," *J. Nanosci. Nanotechnol.*, vol. 5, no. 10, pp. 1574–1592, 2005, doi: 10.1166/jnn.2005.411.

[47] M. Murariu, A. L. Dechief, R. Ramy-Ratiarison, Y. Paint, J. M. Raquez, and P. Dubois, "Recent advances in production of poly(lactic acid) (PLA) nanocomposites: A versatile method to tune crystallization properties of PLA," *Nanocomposites*, vol. 1, no. 2, pp. 71–82, 2015, doi: 10.1179/2055033214Y.0000000008.

[48] A. L. Missio et al., "Nanocellulose-tannin films: From trees to sustainable active packaging," *J. Clean. Prod.*, vol. 184, pp. 143–151, 2018, doi: 10.1016/j.jclepro.2018.02.205.

[49] G. Mondragon, C. Peña-Rodriguez, A. González, A. Eceiza, and A. Arbelaiz, "Bionanocomposites based on gelatin matrix and nanocellulose," *Eur. Polym. J.*, vol. 62, pp. 1–9, 2015, doi: 10.1016/j.eurpolymj.2014.11.003.

[50] R. F. Faradilla, G. Lee, P. Sivakumar, M. Stenzel, and J. Arcot, "Effect of polyethylene glycol (PEG) molecular weight and nanofillers on the properties of banana pseudostem nanocellulose films," *Carbohydr. Polym.*, vol. 205, pp. 330–339, 2019, doi: 10.1016/j.carbpol.2018.10.049.

[51] M. Asrofi, H. Abral, A. Kasim, A. Pratoto, M. Mahardika, and F. Hafizulhaq, "Characterization of the sonicated yam bean starch bionanocomposites reinforced by nanocellulose water hyacinth fiber (WHF): The effect of various fiber loading," *J. Eng. Sci. Technol.*, vol. 13, no. 9, pp. 2700–2715, 2018.

[52] C. Huang et al., "Bio-inspired nanocomposite by layer-by-layer coating of chitosan/ hyaluronic acid multilayers on a hard nanocellulose-hydroxyapatite matrix," *Carbohydr. Polym.*, vol. 222, p. 115036, 2019, doi: 10.1016/j.carbpol.2019.115036.

[53] M. Chaichi, M. Hashemi, F. Badii, and A. Mohammadi, "Preparation and characterization of a novel bionanocomposite edible film based on pectin and crystalline nanocellulose," *Carbohydr. Polym.*, vol. 157, pp. 167–175, 2017, doi: 10.1016/j.carbpol.2016.09.062.

[54] B. Montero, M. Rico, S. Rodríguez-Llamazares, L. Barral, and R. Bouza, "Effect of nanocellulose as a filler on biodegradable thermoplastic starch films from tuber, cereal and legume," *Carbohydr. Polym.*, vol. 157, pp. 1094–1104, 2017, doi: 10.1016/j. carbpol.2016.10.073.

[55] M. S. Sarwar, M. B. K. Niazi, Z. Jahan, T. Ahmad, and A. Hussain, "Preparation and characterization of PVA/nanocellulose/Ag nanocomposite films for antimicrobial food packaging," *Carbohydr. Polym.*, vol. 184, pp. 453–464, 2018, doi: 10.1016/j. carbpol.2017.12.068.

[56] B. Fauzi, M. G. Mohd Nawawi, R. Fauzi, and S. N. L. Mamauod, "Physicochemical characteristics of sago starch- chitosan nanofillers film," *BioResources*, vol. 14, no. 4, pp. 8324–8330, 2019, doi: 10.15376/biores.14.4.8324-8330.

[57] N. Jannatyha, S. Shojaee-Aliabadi, M. Moslehishad, and E. Moradi, "Comparing mechanical, barrier and antimicrobial properties of nanocellulose/CMC and nanochitosan/CMC composite films," *Int. J. Biol. Macromol.*, vol. 164, pp. 2323–2328, 2020, doi: 10.1016/j.ijbiomac.2020.07.249.

[58] J. Ren, K. M. Dang, E. Pollet, and L. Avérous, "Preparation and characterization of thermoplastic potato starch/halloysite nano-biocomposites: Effect of plasticizer nature and nanoclay content," *Polymers*, vol. 10, no. 8, 2018, doi: 10.3390/polym10080808.

[59] A. Nuryawan et al., "Enhancement of oil palmwaste nanoparticles on the properties and characterization of hybrid plywood biocomposites," *Polymers*, vol. 12, no. 5, 2020, doi: 10.3390/POLYM12051007.

[60] R. A. Ilyas et al., "Sugar palm (*Arenga pinnata* [Wurmb.] Merr) starch films containing sugar palm nanofibrillated cellulose as reinforcement: Water barrier properties," *Polym. Compos.*, vol. 41, no. 2, 2020, doi: 10.1002/pc.25379.

[61] D. N. Saheb and J. P. Jog, "Natural fiber polymer composites: A review," *Adv. Polym. Technol.*, vol. 18, no. 4, pp. 351–363, 1999.

[62] S. T. Georgopoulos, P. A. Tarantili, E. Avgerinos, A. G. Andreopoulos, and E. G. Koukios, "Thermoplastic polymers reinforced with fibrous agricultural residues," *Polym. Degrad. Stab.*, vol. 90, no. 2 SPEC. ISS., pp. 303–312, 2005, doi: 10.1016/j. polymdegradstab.2005.02.020.

[63] M. Matet, M. C. Heuzey, E. Pollet, A. Ajji, and L. Avérous, "Innovative thermoplastic chitosan obtained by thermo-mechanical mixing with polyol plasticizers," *Carbohydr. Polym.*, vol. 95, no. 1, pp. 241–251, 2013, doi: 10.1016/j.carbpol.2013.02.052.

[64] N. E. Suyatma, L. Tighzert, A. Copinet, and V. Coma, "Effects of hydrophilic plasticizers on mechanical, thermal, and surface properties of chitosan films," *J. Agric. Food Chem.*, vol. 53, no. 10, pp. 3950–3957, 2005, doi: 10.1021/jf048790+.

[65] M. H. Lee, S. Y. Kim, and H. J. Park, "Effect of halloysite nanoclay on the physical, mechanical, and antioxidant properties of chitosan films incorporated with clove essential oil," *Food Hydrocoll.*, vol. 84, no. April, pp. 58–67, 2018, doi: 10.1016/j.foodhyd.2018.05.048.

[66] A. Aryaei, A. H. Jayatissa, and A. C. Jayasuriya, "Mechanical and biological properties of chitosan/carbon nanotube nanocomposite films," *J. Biomed. Mater. Res.*, vol. 102, no. 8, pp. 2704–2712, 2014, doi: 10.1002/jbm.a.34942.

[67] M. Liu, Y. Zhang, C. Wu, S. Xiong, and C. Zhou, "Chitosan/halloysite nanotubes bionanocomposites: Structure, mechanical properties and biocompatibility," *Int. J. Biol. Macromol.*, vol. 51, no. 4, pp. 566–575, 2012, doi: 10.1016/j.ijbiomac.2012.06.022.

[68] A. Shah, I. Hussain, and G. Murtaza, "Chemical synthesis and characterization of chitosan/silver nanocomposites films and their potential antibacterial activity," *Int. J. Biol. Macromol.*, vol. 116, pp. 520–529, 2018, doi: 10.1016/j.ijbiomac.2018.05.057.

[69] A. M. Youssef, H. Abou-Yousef, S. M. El-Sayed, and S. Kamel, "Mechanical and antibacterial properties of novel high performance chitosan/nanocomposite films," *Int. J. Biol. Macromol.*, vol. 76, pp. 25–32, 2015, doi: 10.1016/j.ijbiomac.2015.02.016.

[70] J. Huang, Y. Zhou, L. Dong, Z. Zhou, and R. Liu, "Enhancement of mechanical and electrical performances of insulating presspaper by introduction of nanocellulose," *Compos. Sci. Technol.*, vol. 138, pp. 40–48, 2017, doi: 10.1016/j.compscitech.2016.11.020.

[71] Z. Xiang, X. Jin, Q. Liu, Y. Chen, J. Li, and F. Lu, "The reinforcement mechanism of bacterial cellulose on paper made from woody and non-woody fiber sources," *Cellulose*, vol. 24, no. 11, pp. 5147–5156, 2017, doi: 10.1007/s10570-017-1468-6.

[72] Y. L. Liang and R. A. Pearson, "The toughening mechanism in hybrid epoxy-silica-rubber nanocomposites (HESRNs)," *Polymer*, vol. 51, no. 21, pp. 4880–4890, 2010, doi: 10.1016/j.polymer.2010.08.052.

[73] Y. Lin, H. Chen, C. M. Chan, and J. Wu, "The toughening mechanism of polypropylene/calcium carbonate nanocomposites," *Polymer*, vol. 51, no. 14, pp. 3277–3284, 2010, doi: 10.1016/j.polymer.2010.04.047.

[74] B. Zhang and Q. Wang, "Development of highly ordered nanofillers in zein nanocomposites for improved tensile and barrier properties," *J. Agric. Food Chem.*, vol. 60, no. 16, pp. 4162–4169, 2012, doi: 10.1021/jf3005417.

[75] K. Prashantha, B. Lecouvet, M. Sclavons, M. F. Lacrampe, and P. Krawczak, "Poly(lactic acid)/halloysite nanotubes nanocomposites: Structure, thermal, and mechanical properties as a function of halloysite treatment," *J. Appl. Polym. Sci.*, vol. 128, no. 3, pp. 1895–1903, 2013, doi: 10.1002/app.38358.

[76] J. Cai, H. Niu, Y. Yu, H. Xiong, and T. Lin, "Effect of solvent treatment on morphology, crystallinity and tensile properties of cellulose acetate nanofiber mats," *J. Text. Inst.*, vol. 108, no. 4, pp. 555–561, 2017, doi: 10.1080/00405000.2016.1174456.

[77] A. B. Nair and R. Joseph, Chapter 9 Eco-friendly bio-composites using natural rubber (NR) matrices and natural fiber reinforcements. In *Chemistry, manufacture and applications of natural rubber*, pp. 249–283, 2014, Woodhead Publishing. doi:10.1533/9780857096913.2.249

[78] M. Saxena, A. Pappu, R. Haque, and and A. Sharma "Chapter 22 Sisal fiber-based polymer composites and their applications," In Kalia, S., Kaith, B., and Kaur, I. (Eds.), *Cellulose Fibers: Bio- and Nano-Polymer Composites*, Springer, Berlin, no. November, 2011, pp. 1–22. doi: 10.1007/978-3-642-17370-7_22

[79] M. Mihai and M.-T. Ton-That, "Novel polylactic/triticale straw biocomposites: Processing, formulation, and properties," *Polym. Eng. Sci.*, pp. 446–458, 2014, doi: 10.1002/pen.

[80] M. Ramesh, K. Palanikumar, and K. H. Reddy, "Mechanical property evaluation of sisal-jute-glass fiber reinforced polyester composites," *Compos. Part B Eng.*, vol. 48, pp. 1–9, 2013, doi: 10.1016/j.compositesb.2012.12.004.

[81] U. Nirmal, J. Hashim, and M. M. H. Megat Ahmad, "A review on tribological performance of natural fibre polymeric composites," *Tribol. Int.*, vol. 83, pp. 77–104, 2015, doi: 10.1016/j.triboint.2014.11.003.

[82] M. Aslan, M. Tufan, and T. Küçükömeroğlu, "Tribological and mechanical performance of sisal-filled waste carbon and glass fibre hybrid composites," *Compos. Part B Eng.*, vol. 140, pp. 241–249, 2018, doi: 10.1016/j.compositesb.2017.12.039.

[83] S. Goutianos, T. Peijs, B. Nystrom, and M. Skrifvars, "Development of flax fibre based textile reinforcements for composite applications," *Appl. Compos. Mater.*, vol. 13, no. 4, pp. 199–215, 2006, doi: 10.1007/s10443-006-9010-2.

[84] P. Peças, H. Carvalho, H. Salman, and M. Leite, "Natural fibre composites and their applications: A review," *J. Compos. Sci.*, vol. 2, no. 4, p. 66, 2018, doi: 10.3390/jcs2040066.

[85] T. Sullins, S. Pillay, A. Komus, and H. Ning, "Hemp fiber reinforced polypropylene composites: The effects of material treatments," *Compos. Part B Eng.*, vol. 114, pp. 15–22, 2017, doi: 10.1016/j.compositesb.2017.02.001.

[86] T. G. Cengiz and F. C. Babalik, "The effects of ramie blended car seat covers on thermal comfort during road trials," *Int. J. Ind. Ergon.*, vol. 39, no. 2, pp. 287–294, 2009, doi: 10.1016/j.ergon.2008.12.002.

[87] E. Marsyahyo, Jamasri, H. S. B. Rochardjo, and Soekrisno, "Preliminary investigation on bulletproof panels made from ramie fiber reinforced composites for NIJ Level II, IIA, and IV," *J. Ind. Text.*, vol. 39, no. 1, pp. 13–26, 2009, doi: 10.1177/1528083708098913.

[88] T. Sen and H. N. J. Reddy, "Various industrial applications of hemp, kinaf, flax and ramie natural fibres," *Int. J. Innov. Manag. Technol.*, vol. 2, no. 3, pp. 192–198, 2011.

[89] B. Abdi, S. Azwan, M. R. Abdullah, and A. Ayob, "Flexural and tensile behaviour of kenaf fibre composite materials," *Mater. Res. Innov.*, vol. 18, pp. S6-184-S6-186, 2014, doi: 10.1179/1432891714Z.000000000954.

[90] T. Nishino, K. Hirao, M. Kotera, K. Nakamae, and H. Inagaki, "Kenaf reinforced biodegradable composite," *Compos. Sci. Technol.*, vol. 63, no. 9, pp. 1281–1286, 2003, doi: 10.1016/S0266-3538(03)00099-X.

[91] H. Anuar and A. Zuraida, "Improvement in mechanical properties of reinforced thermoplastic elastomer composite with kenaf bast fibre," *Compos. Part B Eng.*, vol. 42, no. 3, pp. 462–465, 2011, doi: 10.1016/j.compositesb.2010.12.013.

[92] A. Atiqah, M. A. Maleque, M. Jawaid, and M. Iqbal, "Development of kenaf-glass reinforced unsaturated polyester hybrid composite for structural applications," *Compos. Part B Eng.*, vol. 56, pp. 68–73, 2014, doi: 10.1016/j.compositesb.2013.08.019.

[93] E. Kipriotis, X. Heping, T. Vafeiadakis, M. Kiprioti, and E. Alexopoulou, "Ramie and kenaf as feed crops," *Ind. Crops Prod.*, vol. 68, pp. 126–130, 2015, doi: 10.1016/j.indcrop.2014.10.002.

[94] R. Arjmandi, A. Hassan, K. Majeed, and Z. Zakaria, "Rice husk filled polymer composites," *Int. J. Polym. Sci.*, vol. 2015, 2015, doi: 10.1155/2015/501471.

[95] D. Verma, P. C. Gope, A. Shandilya, A. Gupta, and M. K. Maheshwari, "Coir fibre reinforcement and application in polymer composites: A review," *J. Mater. Environ. Sci.*, vol. 4, no. 2, pp. 263–276, 2013.

[96] M. Li et al., "Recent advancements of plant-based natural fiber-reinforced composites and their applications," *Compos. Part B Eng.*, vol. 200, 2020, doi: 10.1016/j.compositesb.2020.108254.

[97] J. A. Khan and M. A. Khan, Chapter 1 The use of jute fibers as reinforcements in composites. In *Biofiber Reinforcements in Composite Materials* pp. 3–34, 2015, Woodhead Publishing. doi: 10.1533/9781782421276.1.3

[98] Y. Du, N. Yan, and M. T. Kortschot, The use of ramie fibers as reinforcements in composites. In *Biofiber Reinforcements in Composite Materials*, pp. 104–137, 2015, Woodhead Publishing. doi: 10.1533/9781782421276.1.104

[99] D. K. Rajak, D. D. Pagar, P. L. Menezes, and E. Linul, "Fiber-reinforced polymer composites: Manufacturing, properties, and applications," *Polymers*, vol. 11, no. 10, 2019, doi: 10.3390/polym11101667.

[100] N. Bhardwaj and S. C. Kundu, "Electrospinning: A fascinating fiber fabrication technique," *Biotechnol. Adv.*, vol. 28, no. 3, pp. 325–347, 2010, doi: 10.1016/j.biotechadv.2010.01.004.

[101] G. Wang, D. Yu, A. D. Kelkar, and L. Zhang, "Electrospun nanofiber: Emerging reinforcing filler in polymer matrix composite materials," *Prog. Polym. Sci.*, vol. 75, pp. 73–107, 2017, doi: 10.1016/j.progpolymsci.2017.08.002.

[102] G. D. Goh, Y. L. Yap, S. Agarwala, and W. Y. Yeong, "Recent Progress in Additive Manufacturing of Fiber Reinforced Polymer Composite," *Adv. Mater. Technol.*, vol. 4, no. 1, pp. 1–22, 2019, doi: 10.1002/admt.201800271.

[103] C. Hu, Z. Sun, Y. Xiao, and Q. Qin, "Recent patents in additive manufacturing of continuous fiber reinforced composites," *Recent Patents Mech. Eng.*, vol. 12, no. 1, pp. 25–36, 2019, doi: 10.2174/2212797612666190117131659.

[104] P. Parandoush, L. Tucker, C. Zhou, and D. Lin, "Laser assisted additive manufacturing of continuous fiber reinforced thermoplastic composites," *Mater. Des.*, vol. 131, pp. 186–195, 2017, doi: 10.1016/j.matdes.2017.06.013.

[105] L. Sorrentino et al., "Robotic filament winding: An innovative technology to manufacture complex shape structural parts," *Compos. Struct.*, vol. 220, pp. 699–707, 2019, doi: 10.1016/j.compstruct.2019.04.055.

[106] J. Frketic, T. Dickens, and S. Ramakrishnan, "Automated manufacturing and processing of fiber-reinforced polymer (FRP) composites: An additive review of contemporary and modern techniques for advanced materials manufacturing," *Addit. Manuf.*, vol. 14, pp. 69–86, 2017, doi: 10.1016/j.addma.2017.01.003.

[107] F. Wu, M. Misra, and A. K. Mohanty, "Challenges and new opportunities on barrier performance of biodegradable polymers for sustainable packaging," *Prog. Polym. Sci.*, vol. 117, p. 101395, 2021, doi: 10.1016/j.progpolymsci.2021.101395.

[108] S. O. Han, M. Karevan, M. A. Bhuiyan, J. H. Park, and K. Kalaitzidou, "Effect of exfoliated graphite nanoplatelets on the mechanical and viscoelastic properties of poly(lactic acid) biocomposites reinforced with kenaf fibers," *J. Mater. Sci.*, vol. 47, no. 8, pp. 3535–3543, 2012, doi: 10.1007/s10853-011-6199-8.

14 Effect of Dispersion and Interfacial Functionalization of Multiwalled Carbon Nanotubes in Epoxy Composites
Structural and Thermogravimetric Analysis Characteristics

H. M. Mohammed, N. M. Nurazzi, N. F. M. Rawi, M. H. M. Kassim, and K. M. Salleh
Universiti Sains Malaysia

N. Abdullah and M. N. F. Norrrahim
Universiti Pertahanan Nasional Malaysia (UPNM)

14.1 INTRODUCTION

Carbon nanotubes (CNTs) are among the most promising in delivering excellent multifunctional characteristics, mechanical properties, and appreciated nanomaterials in the current and previous decades [1–3]. For instance, adding CNTs as a structural component to a polymer matrix can significantly enhance the mechanical properties with a relatively low filler content of less than 5 wt.%. CNTs as a conducting filler can modify the transportation properties (thermal conductivities and electrical) of polymer-based composites. However, appropriate dispersion and strong interfacial interaction between the carbon nanotubes and polymer matrix must be achieved for these materials to be used as efficient reinforcements in polymer composites [4,5]. Furthermore, CNTs are typically inert, easily entangled, and aggregate due to their size and high aspect ratio. Therefore, the major challenge in polymer composites

DOI: 10.1201/9781003400998-14

fabrications is to modify the CNTs chemicals structure and hence to improve their interfacial adhesion with polymer matrix [6]. Types of CNTs provide excellent thermal conductivity, pore size, dimensions, and mechanical, electrical, and magnetic characteristics, making them one of the most used materials in engineering technologies [7]. Since the study discovered fullerene-related carbon nanotubes in 1991 by the Japanese scientist "Sumio Iijima," CNTs have increased global interest because of their attractive and practical mechanical, electrical, thermal, and chemical characteristics and outstanding stability under typical environmental conditions [8].

Presently, two primary varieties of carbon nanotubes can be found: (1) single-walled carbon nanotubes (SWCNTs), with a diameter of 1–10 nm and a length of several micrometres, are constructed from one layer of graphite sheet and (2) multi-walled carbon nanotubes (MWCNTs) comprised many layers of the graphene sheet and 3–30 nm about the outer diameter range [9]. MWCNTs are mainly created on an industrial scale using chemical vapor deposition. These MWCNTs are highly entangled and agglomerated in their original conditions. These agglomerates are challenging to disperse completely through melt mixing, and the dispersion state is sensitive to the chosen processing parameters [10,11]. Poor dispersion made it more challenging to use CNTs as a proper reinforcement [12–14]. Weak interfacial bonding among the epoxy matrix and CNTs results in inefficient stress transfer [15]. Furthermore, it is noticed that there is a 10%–20% decrease in both the elastic modulus and tensile strength of the composite due to their poor interaction among the epoxy matrix and pristine MWCNTs [14].

Amino-functionalized multiwalled carbon nanotubes (MWCNT-NH$_2$) and carboxyl-functionalized carbon nanotubes (CNT-COOH) have been prepared to reinforce epoxy resin and carbon fiber/epoxy composites due to the enhanced dispersion and interfacial adhesion with an epoxy matrix compared to the pristine CNTs. The comparison between CNT-COOH and untreated carbon nanotubes to reinforce diglycidyl ether bisphenol A-epoxy resin was made by Montazeri et al., and they found that the CNT-COOH had higher Young's modulus values than untreated CNTs as a result [16]. MWCNT-NH$_2$ was investigated by Srikanth et al. [17] as a reinforcement material for carbon fiber/epoxy composites. The results revealed that mechanical parameters, such as flexural strength and interlaminar shear strength (ILSS), were increased by adding MWCNT-NH$_2$ at a loading of 0.5 wt.%.

According to Guadagno et al. [18], the existence of MWCNT that has been MWCNT-COOH did not significantly enhance the storage modulus of epoxy nanocomposites. Nevertheless, when the amount of MWCNT-COOH increased, the value remained almost constant or declined. As compared to MWCNT-NH$_2$, Cui et al. [19] discovered that the tensile strength of MWCNT-COOH/epoxy composites tended to decline more quickly as the MWCNT-COOH content rose. Compared to pristine MWCNT/epoxy composites, incorporated as unaltered MWCNT/epoxy composites, the addition of CTBN-grafted MWCNTs to the epoxy matrix significantly improved thermal stability and toughness. Furthermore, the functionalized MWCNTs increased interfacial bonding and homogenized the MWCNT dispersion in the matrix, improving the thermal stability properties of the composites. The fabricated MWCNTs epoxy related to the chemical bonding between the nanotube and epoxy matrix increases the interfacial strength.

14.2 DISPERSION OF MWCNTs WITH EPOXY MATRIX

The MWCNTs have been dispersed and functionalized with epoxy matrix through mechanical stirring [20], ultrasonication [20,21], calendaring (three-roll mill) methods [21], and ball milling methods [20,21].

14.2.1 MECHANICAL STIRRING

A Heidolph RZR-2102 stirrer was utilized for the mechanical stirring. The impact of the flow produced by the impeller on particle dispersion was assessed using three distinct mixing elements: a helix blade, a viscojet with three cones, and a viscojet with two cones. The primary distinction between a viscojet and a helix blade stir bar is the flow produced into the matrix which affects how heat is dissipated is presented in Figure 14.1 [20]. A Hielscher-type UP200S (24 kHz, 200 W) ultrasonic machine was used for ultrasonication. A 3 mm tip sonotrode and a 22 mm tip sonotrode were utilized for quantities ranging from 5 to 200 mL and 100 to 1,000 mL of the dispersion [20].

14.2.2 ULTRASONICATION

Ultrasonication is a significant method of MWCNTs dispersing and deagglomerating where high-intensity ultrasound waves cause cavitation in liquids, even though it is often very difficult to achieve homogenous MWCNTs dispersion in the polymeric and epoxy matrix. The ultrasonic bath and the ultrasonic horn are the two main processing methods to introduce ultrasonic energy into liquids. Lu et al. [22] stated that solids are predominantly dispersed by ultrasonication through the nucleation and bursting of microbubbles. Compared to cell dismembrator horns (25 kHz), the ultrasonication bath has a greater frequency (40–50 kHz). Three physical processes are brought on by ultrasonication of fluids: cavitation of the fluid, localized heating, and production of free radicals. The dispersion may result from cavitation and bubbles' production and implosion [22,23].

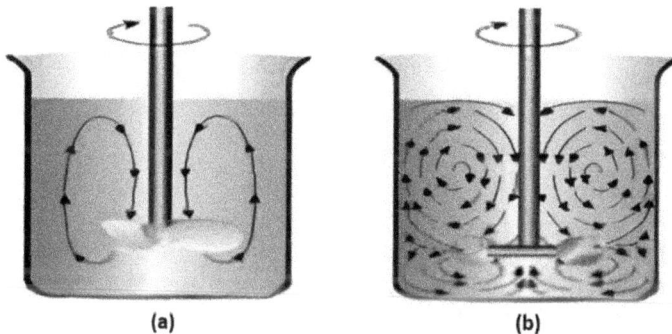

(a)   (b)

FIGURE 14.1 Flow produced from viscojet and a helix blade stir bar: (a) helix blade stir bar and (b) viscojet stir bar. (Reproduced from ref. [20].)

14.2.3 CALENDARING (THREE-ROLL MILL) METHODS

According to Hwang et al. [21], the calendaring method is generally called three-roll milling. This dispersion process is used to mix the nanotubes into epoxy or polymer and distribute MWCNTs. It uses both shear and extensional flow produced by rotating rolls at varying speeds. The center roller rotates opposite to the first and third rollers, typically called the feed and apron rolls. High shear rates are produced when the center roll's rotational velocity exceeds the feed roll's ($\omega2 > \omega1$). The liquid mixture flows down (basically coating) the surrounding rolls through its surface tension. It is subjected to strong shear stresses as the resin suspension is fed into the small space (δ) between the feed and centre rolls. The processed resin suspension is collected by placing a scraper blade in contact with the apron roll at each desired dwell period's conclusion. To enhance dispersion, this milling operation might be repeated numerous times. Viswanathan et al. [24] explained that one of the main remarkable benefits of this method is the ability to mechanically or hydraulically alter and maintain the gap width between the rollers, making it simple to get a regulated and narrow size distribution of particles in viscous fluids. In some procedures, the appropriate amount of particle dispersion can be achieved by gradually reducing the width of gaps. According to a recent research, using a calendar to disperse CNTs in a polymer matrix has emerged as a promising method to obtain a comparatively excellent CNTs dispersion [25].

14.2.4 BALL-MILLING METHODS

Ball milling is a technique that is frequently used to grind large materials to a fine powder. Due to the impact of the rigid balls in a sealed container during milling, high pressure is created locally. The ball's cascading impact changes the material into a fine powder. Typically, by using ceramic, flint pebbles and stainless steel manufacture the balls. According to Awasthi et al. [21], CNTs dispersion into polymer or epoxy matrices has been effectively accomplished using ball milling. Ball milling is a beneficial technique for obtaining CNTs with narrow length and diameter distributions and opening the nanotubes to boost their capacity for gas sorption. Jia et al. [26] has also noted that a significant quantity of amorphous carbon is produced, proving that the tubes are seriously damaged in many ways and that ball milling is a harmful process.

14.3 PREPARATION OF MWCNTs/EPOXY COMPOSITES

According to Cui et al. [19] findings on grafting carboxyl and amino groups onto the surfaces of MWCNTs, COOH-MWCNTs, respectively, and amino-MWCNTs initially, for 10 hours at 40°C, 4 g of pristine MWCNTs were treated by combining concentrated sulfuric and nitric acid in a volume ratio of 3:1 and were sonicated. Then the mixture was filtered, and the remaining material was thoroughly washed with deionized water several times. Finally, the sample was dried for 24 hours in a vacuum oven at 80°C to produce COOH-MWCNTs. Second, 2 g of COOH-MWCNTs were dispersed in the mixture of 20 mL of dimethyl formamide and 200 mL of thionyl chloride ($SOCl_2$), followed by 24 hours of 70°C magnetic stirring during reflux. After the reaction, the vacuum distillation eliminated the unreacted $SOCl_2$. The MWCNTs functionalized by acyl chloride were then added to TETA and refluxed with magnetic stirring at 120°C for 96 hours. Finally, the mixture was filtered, and the leftover material was repeatedly cleaned with ethanol until it was clear and amino-MWCNTs were produced [19].

Finally, the three distinct types of MWCNTs (pristine, COOH–, and amino-MWCNTs) were dispersed with epoxy using a three-roller machine with water-bath heating. Thereafter, TETA was progressively included in the mixture to minimize the formation of bubbles. The mixture was then poured into a mould and baked at 40°C for 0.5 hours, 80°C for 0.5 hours, and 110°C for 1 hour to cure it. Throughout the curing phase, the heating rate was 2°C/minute.

Amidation process involves two reaction steps by converting the commercial MWCNT-COOH into $MWCNT-NH_2$. First, 300 mL $SOCl_2$ and 5 mL DMF were mixed with 15 g MWCNT-COOH to perform the acylation process, which was then heated to 70°C for 24 hours while being mechanically stirred. The mixture was then cooled, and the supernatant liquid was washed with anhydrous THF under N_2 until clear. For the later application, the surface-acylated MWCNT's dark powder was dried at room temperature. Next, a flask containing 10 g of surface-acylated MWCNT and 300 mL of EDA was heated at reflux for 72 hours at 110°C. Finally, the remaining residues of adsorbed EDA on the surface of $MWCNT-NH_2$ will be eliminated, and the excess EDA was dried at 70°C under vacuum after cooling at room temperature [27].

Konnola et al. [28] prepared MWCNT-OH and MWCNT-g-CTBN, MWCNT-OH by adding nitrene formed by thermolysis of 2-(2-azidoethoxy) ethanol to strained double bonds of MWCNT. First, 100 mg of MWCNTs were dispersed in 15 mL of NMP in an R.B. flask using ultrasonication for 2 hours to prepare a homogenous solution. After the mixture was at ambient temperature, 5 g of 2-(2-azidoethoxy) ethanol was added to conduct the functionalization. The mixture was then heated at 160°C for 18 hours in a nitrogen environment. To thoroughly eliminate the unre-acted 2-(2-azidoethoxy) ethanol, the liquid was cooled to room temperature, filtered using a 0.22-m PTFE membrane, and then repeatedly washed with acetone (100 mL) and distilled water (100 mL). After being vacuum-dried at 60°C overnight at room temperature, the functionalized MWNTs with hydroxyl groups were obtained in quantities of 108 mg. Then, Konnola et al. [28] manufactured MWCNT-g-CTBN by sonicating the 500 mg of MWCNT-OH and dispersed in 125 mL anhydrous dimethyl formamide (DMF) for 30 minutes. A two-necked flask, which had the 5-g solution of CTBN in 125 mL of DMF, was then filled with the MWNT-OH solution. After that, under nitrogen, a mixture of DCC (20.6 g, 100 mmol) and DMAP (1.2 g, 10 mmol) in DMF (250 mL) was added. The reaction was then allowed to continue at 40°C for 68 hours. To get MWCNT-g-CTBN, the finished product was filtered, washed with methanol, and dried under a vacuum for 24 hours at 60°C.

Che et al. [29] investigated dispersion techniques that affected the performance of MWCNTs/epoxy nanocomposites in the X-band microwave range. The disper-sion method by ball milling was used to create composites with different carbon nanotube-loading contents [19]. The MWCNT/epoxy nanocomposites were pro-duced with a thickness of 3 mm and various CNT compositions (0.25, 0.5, 0.75, 1, and 1.25 wt.%). First, MWCNTs and epoxy resin were added. The combination was then put through a ball-milling process utilizing a porcelain vertical-type ball mill jar with a 1 L capacity, one pivot, and 0.5 kg of porcelain balls ranging from 10 to 20 mm. Each batch weighed 300 g and was milled at a speed of 300 rpm for an optimum of 60 minutes. After ball milling, the hardener triethylenetetramine (TETA) was added, and the matrix was then cured for 24 hours in ambient conditions before being characterized [29].

14.4 FUNCTIONALIZATION OF CNTs

The functionalization of CNTs is to increase the interfacial interaction between CNTs with polymers, improve CNTs dispersion, and for the purpose of desired applications [30]. Covalent and noncovalent functionalization are two methods used to functionalize CNTs [31–34].

14.4.1 COVALENT FUNCTIONALIZATION

Norizan et al. [35] stated that covalent functionalization of CNTs can be accomplished by suitable surface-bound functional group modification on the nanotube end either by adding functional groups or other "active agents" directly to the sidewalls of nanotubes. Covalent functionalization of CNTs is most frequently started by the oxidation process, which causes the production of carboxyl groups (–COOH) on the surface of nanotubes. Since it adds carboxyl groups via oxidation, oxidation has become essential in the functionalization process. Molecular covalent coupling is made possible by carboxyl groups forming by amide and ester bonds. Covalent functionalization is frequently accomplished using two acid treatments: the first involves nitric acid solution (HNO_3) and the second involves sulfuric acid (H_2SO_4) and a nanotube sample (1:3 by volume) for up to 6 hours with high-intensity sonication [36]. Figure 14.2 depicts the covalent functionalization of MWCNT using the H_2SO_4/HNO_3 oxidation method in order to form MWCNT-COOH. The functionalization subsequently produces stable CNTs dispersions in various polar solvents, including

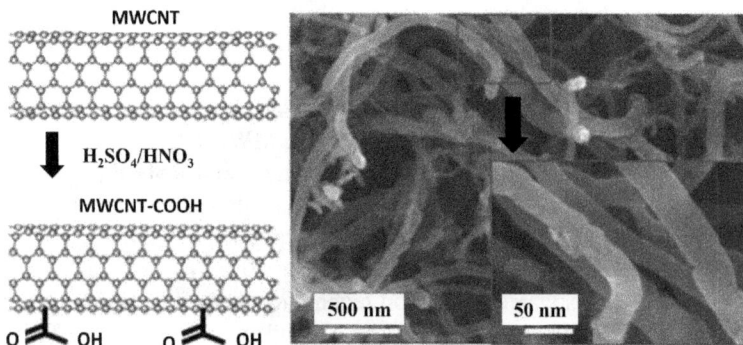

FIGURE 14.2 Covalent functionalization of nanotubes with the H_2SO_4/HNO_3 oxidation method. (Reproduced from ref. [38].)

with water. Additionally, the reduction of van der Waals interactions between the CNTs due to the covalent attachment of functional groups changes the hydrogen bonding, which in turn affects the stacking and layering characteristics of CNTs. This facilitates the phase separation of nanotube bundles into individual tubes [37].

14.4.2 NONCOVALENT FUNCTIONALIZATION

CNTs and the desired conductive polymer interact via a process known as noncovalent functionalization. According to Bose et al. [39], noncovalent functionalization is a useful technique for modifying the CNTs and polymer interface while keeping the tubes' dependability. This approach is particularly appealing since it allows for the adsorption of different groups of ordered structures on the CNT surface without interfering with the nanotubes' extended p-conjugation. The deagglomeration of CNTs in the presence of a modifier has been addressed extensively in recent years, and several different techniques have been put out [40].

Norizan et al. [40] stated that the idea of noncovalent functionalization has been well-known and frequently used due to the flexible design of the polymers and the enhancement of composites' natural features. However, due to the lack of covalent interaction between the CNTs and the polymer, this modification results in less stable functionalization and poor polymer reinforcement than covalent functionalization. On the other hand, noncovalent modification does not damage the conjugated system of the CNTs' sidewalls and end cap. Hence, it does not affect the material's fundamental structural characteristics. As a result of π-π stacking and van der Waals interactions between the polymer chains made up of aromatic rings and the exteriors of CNTs, conjugated polymers wrapped successfully using noncovalent functionalization of CNTs are then produced [40]. Saeb et al. [41] demonstrated a rise in thermal effects while curing epoxy nanocomposites, including MWCNTs with amino functionality. Epoxy composites with MWCNTs that have been amino-functionalized had calculated curing heat values that were greater than systems with unmodified MWCNTs and even higher than epoxy composites without MWCNTs.

14.5 STRUCTURAL CHARACTERISTICS OF MWCNT/EPOXY COMPOSITES

To confirm any significant structural modifications in the epoxy resin brought on by the change with CNTs, an infrared analysis is performed on the composites that have been manufactured. According to Konnola et al. [28], a distinctive C-Cl absorption peak may be seen in 2-(2-chloroethoxy)ethanol at about 745 cm^{-1}. In addition, peaks at 2,870 and 2,930 cm^{-1}, attributed to the C–H stretching vibrations, as well as a broad band with a center at 3,400 cm^{-1}, referred to as the stretching of OH groups were also seen. The C–O stretching is what causes chloroethoxyethanol to peak at 1,121 cm^{-1}. The FTIR spectrum of 2-(2-azidoethoxy)ethanol shows that most chloride groups have been replaced with azide groups due to the absence of the C–Cl stretch following the reaction with sodium azide and the development of the typical azido group absorption at around 2,094 cm^{-1}. Furthermore, the FTIR spectrum of the MWCNT-OH shows the characteristic absorption bands due to stretching vibration of the C–O at 1,075 cm^{-1}. In addition, the MWCNT-OH shows absorption peaks between 2,920 and 2,860 cm^{-1}, which is related to the C–H stretching absorption band. On the other hand, MMWCNT-g-CTBN exhibits three characteristic peaks related to vibrations of the CTBN: the peak at 963 cm^{-1} corresponds to =C–H out-of-plane bending vibration of 1,4 trans olefin in CTBN, peak at 1638 cm^{-1} is due to C=C stretching, and peak at 1,715 cm^{-1} corresponds to C=O stretching of carbonyl group in CTBN. The presence of small stretching band at 1,740 cm^{-1} clearly reveals the ester bands that came from the CTBN grafted to the MWCNT.

Qingjie et al. [42], in their experimental study, stated that the MWCNT surface was grafted with EDA. Therefore, the amino groups' covalent attachment to the MWCNT surface, as opposed to their simple adsorption, had to be verified. The surface functionalities were identified by FTIR spectroscopy as shown in Figure 14.3, to separate the adsorbed and chemically bonded EDA with a prominent peak of epoxy casting at 1,570 cm^{-1}. The sample of MWCNT-NH$_2$ experienced a heating rate of 5°C/minute from room temperature to observe, but MWCNT-COOH experienced no heating, which was consistent with both MWCNT stretch and NH bend vibrations. Indicating the existence of nonamide carbonyl species in MWCNT-NH$_2$, both samples peaked at 1,718 cm^{-1}, which can be attributed to the CO stretching vibration of carboxylic groups [42].

Peaks in the XPS spectrum indicate the proportion of electrons with a specific binding energy. Shorter peaks indicate fewer electrons. It can be understood that half as many electrons were detected with the binding energy at peak A as at peak B if, for instance, peak A is half as tall as peak B. As a result, the peak intensities reveal a material's percentage makeup. As can be observed in the initial image, the O1s have the highest peak, indicating that oxygen has the highest atomic composition. According to the researchers [43,44], the stronger attraction of the electron to the nucleus, the stronger the binding energy. For example, electrons in the first state will have peaks with higher energy than electrons in the second state. The energy of electrons in 2s will be higher than that of 2p. While some instruments offer peak identification characteristics, it is still possible to identify peaks and lines on spectra by comparing them to standards made of various materials. Moulder provides examples of these standards to assist in interpreting the spectra [44].

Results obtained by Che et al. [28] show the strength of the chemical bonds generated on the nanotube's surface before and after its functionalization with CTBN was

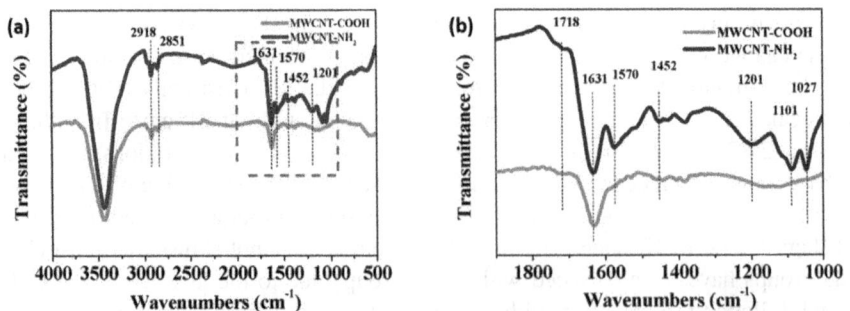

FIGURE 14.3 FTIR spectra: (a) MWCNT-COOH and MWCNT-NH$_2$ and (b) enlargement square area in (a). (Reproduced from ref. [42].)

tested using XPS. Pristine MWCNTs show a solid peak at 285 eV owing to C 1s and a weak peak at 532 eV due to O 1s from the nanotube defects. Upon functionalization, the O1s peak of MWCNT-OH and MWCNT-g-CTBN significantly increases, and a new peak at 400 eV is seen (N1s). The presence of organic moieties, which supported the accomplishment of the alteration, is what causes the elements' enhanced intensity [28].

Che et al.'s [28] experimental results shown in Figure 14.4a–c display the high-resolution data of the CNTs' C1s area. These CNTs are made up of MWCNT-OH and MWCNTg-CTBN. The primary peak at 284.5 eV for pristine MWCNT is ascribed to the sp^2 hybridized carbon atoms (C=C), and sp^3 hybridized carbon atoms produced by flaws in the nanotube structure give out a peak at 285.1 eV. (C–C), a peak at 286.2 eV associated with alcohols, phenols, and ethers' carbon-oxygen single bonds (C–O), a peak at 288.9 eV attributable to carboxylic acids' carbon-oxygen double bonds, esters (O–C=O) and carboxylic anhydrides, and ultimately, a peak at 291.6 eV, the typical position of the sp^2-hybridized carbon atoms 13, 34, and 35 satellite peak caused by the π–π* shake-up.

As per the Che et al. [28] study in the C1s spectra of the MWCNT-OH, compared to pristine MWCNT, the peak intensity of hydroxyl groups is much greater. Proving that the organic moiety and the nanotube have reacted, the elimination of the π–π* shake-up transition can be attributed to the electron system's more significant disruption, demonstrating a substantial alteration to the CNT sidewalls' electronic structure. Regarding the MWCNT-g-CTBN high-resolution spectrum on C1, the O–C=O groups in the band at 288.9 eV, which correlate to a considerable intensity increase, point to the covalent grafting of MWCNT and CTBN. Figure 14.4d deconvolution

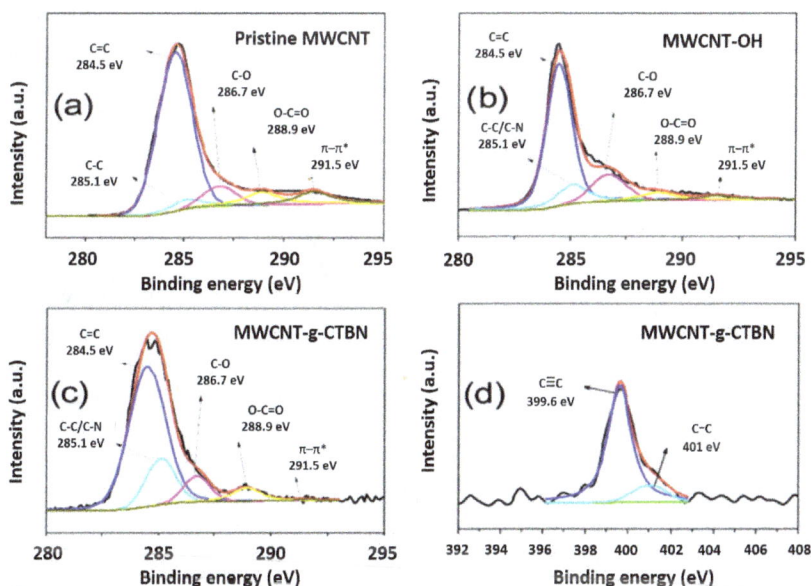

FIGURE 14.4 High-resolution C1s spectra of (a) pristine MWCNT, (b) MWCNT-OH, (c) and (d) MWCNT-g-CTBN in addition to high-resolution N1s spectra of MWCNT-g-CTBN. (Reproduced from ref. [28].)

of the N 1s spectrum reveals two contributions with corresponding binding energies of 399.6 eV and 401 eV. The polymer's CN is blamed for the former. C-N causes the former, indicating that CTBN36 successfully covalently functionalized MWCNT.

According to Figure 14.4, general XPS spectra of COOH-MWCNTs, pristine MWCNTs, and amino-MWCNTs have been presented to determine the chemical composition of the MWCNT surfaces. In every spectrum, there are peaks of C 1s and O 1s. Yet, the strength of the O 1s peak of COOH– MWCNTs significantly rises compared to the pristine MWCNTs. It is believed to be caused by the H_2SO_4/HNO_3 treatment, which grafted many carboxyl groups (–COOH) onto the surface of the pristine MWCNTs. Differences in the atomic configurations of the three samples are compatible with the findings above. Furthermore, spectra of amino-MWCNTs also showed an N 1s peak and a reduction in the strength of the O 1s peak. The fact that TETA was effectively grafted onto the surface of amino-MWCNTs indicates that TETA has partially replaced some of the hydroxide groups (–OH).

The distinctive absorption bands caused by the C-stretching O's vibration at 1,075 cm^{-1} are visible in the MWCNT-FTIR OH's spectra. The MWCNT-OH also exhibits absorption peaks in the range of 2,920–2,860 cm^{-1}, which are connected to the C–H stretching absorption band. The peak at 963 cm^{-1}, however, is due to the =CH out of plane bending vibration of the 1,4 trans olefin in CTBN, peak at 1,638 cm^{-1} is brought on by C=C stretching, and the C=O stretching of the carbonyl group in CTBN brings on the rise at 1,715 cm^{-1}. The ester bands from the CTBN grafted to the MWCNT are visible because of a tiny stretching band at 1,740 cm^{-1}. Consequently, compared to pristine MWCNT/epoxy composites, the DH of functionalized MWCNT/epoxy is substantially lower. Furthermore, adding MWCNTs to these epoxy matrices causes Tp to drop slightly, transferring it to a lower temperature. This suggests that adding MWCNTs lowers the reaction's activation energy and accelerates the curing process because they have large p-bonds on their surfaces.

XPS analysis by Qingjie et al. [42] was used to characterize the functional groups of the MWCNTs based on the chemical shift observations [45]. The XPS spectrum corresponds to C1s, N1s, and O1s on the surface of MWCNT-COOH and MWCNT-NH$_2$. MWCNT-NH$_2$ showed the N1s peak at 400.2 eV after amination, which confirmed the successful amino-group functionalization on the surface of MWCNTs. Carbon, oxygen, and nitrogen atomic concentrations for MWCNT-COOH and MWCNT-NH$_2$ were 94.1%, 5.9%, and 0%, and 87.8%, 7.4%, and 4.5%, respectively. The N1s peak of MWCNT-NH$_2$ may also be fitted to three component peaks at 399.9 eV (CNC), 401.0 eV (NCH$_2$), and 402.3 eV (CNOH) [46]. The results verified the FT-IR analysis's findings and further demonstrated that acylamide groups had been inserted into the surface of MWCNTs.

Two key features often distinguish the MWCNTs Raman spectra, which is tangential or G-band is the name given to the band at a wavelength of 1,572 cm^{-1}, resulted from the in-plane vibrations of sp^2-hybridized graphitic carbon. The so-called disorder (D) band, which corresponds to sp^3-carbon and is centered approximately 1,340 cm^{-1}, is caused by flaws and functional groups on the walls or ends of CNTs or amorphous carbon. The D' band, which at higher frequencies is a weak shoulder of the G-band, is likewise a double resonance characteristic brought on by disorder and faults. A benchmark for determining the flaws in CNTs is the intensity area ratio of the D to G bands or I_D/I_G. The higher the defect, which denotes the presence of more groups on the surface of MWCNTs, the larger the ratio. The I_D/I_G ratios for MWNT-OH (1.87) and MWCNT-g-CTBN (1.89), which are higher than those of pristine MWNTs (1.60), show that chemical functionalization has increased the number of defects in MWCNTs. The lack of a clear rise in D-band intensity following the successive functionalization of MWCNT-g-CTBN is probably due to the polymer's indirect attachment to the CNTs' carbon atoms. Similar events have been described in the past.

14.6 TGA CHARACTERISTICS OF MWCNTs/EPOXY COMPOSITES

TGA is an analytical method to examine the thermal stability of the material and the fraction of volatile components by observing the weight change that occurs due to heating at a constant rate. The measurement carried out by TGA is divided into three types: (1) dynamic TGA, (2) static or isothermal TGA, and (3) TGA curve [47,48].

14.6.1 DYNAMIC TGA

With a particular heating rate, the sample's temperature will steadily rise in this type. As a result, the sample's temperature changes linearly over time.

14.6.2 STATIC OR ISOTHERMAL TGA

The sample is maintained at a constant temperature during a predetermined time. The sample's mass change will be monitored throughout this time.

14.6.3 STATIC TGA

In this type, the sample is held at constant heat with increasing temperature. This is also known as isothermal TGA.

14.6.4 TGA CURVE

This represents single-stage decompositions. TGA curve commonly shows the plot of change in weight concerning temperature or time. For example, consider a TGA curve, "T_i" shows the initial temperature, and "T_f" represents the max temperature on completion [47,48]. According to Bom et al. [49], the weight loss vs temperature plot TGA acquires is called a "thermogram." A typical thermogram can be used to determine the following:

1. Thermal stability: The capacity of a material to preserve its properties as nearly unchanged by exposure to heat is used by TGA to assess a substance's thermal stability. Engineers must thoroughly understand thermal stability to forecast the temperature range of polymers, alloys, and nanomaterials.
2. Sample composition: It is a given that a sample will lose weight when the temperature is raised. For chemists to establish a sample composition and comprehend the reactions that occur throughout the decomposition process, it is important to know how much weight is lost. The weight loss profile may also be used to determine the presence of an unidentified drug in the sample

or to calculate the quantity or percentage of a certain component among various other compounds.

3. Procedural decomposition temperature: A thermogram may use to determine the procedural decomposition temperature, which displays a substance's evaporation or breakdown processes.

14.6.5 DEGRADATION

According to Acvi et al. [50], thermal degradation is a term used to describe the physical deactivation of materials by sintering, chemical reactions, and evaporation. It has a positive correlation with temperature. High temperature in the presence or absence of an appropriate chemical environment, and the change of high surface area catalysts which are not thermodynamically favoured into smaller surface area agglomerates, is the sintering process. Sintering might result in a decrease in the active phase's or the support's accessible regions.

Maron et al. [51] determine the decomposition behavior of the pristine MWCNTs, functionalized MWCNTs coated with zinc sulphide particles (MWCNTf-ZnS), and functionalized MWCNTs (MWCNTf). Up to 510°C, the pristine MWCNT displayed thermal resilience before entirely degrading at 580°C. When functional groups attached to the carbonic structure were present, the acid-treated MWCNTf and MWCNTf-ZnS began to lose weight at lower temperatures. The MWCNTf had reduced stability, mainly as a result of structural flaws during the functionalization process [52]. The MWCNT-ZnS performed the best out of the three evaluated samples where it is thermally stable up to 580°C, with full carbonic structural degradation occurring at 620°C. Only the sulfide particles have not degraded at higher temperatures while being exposed to heat. This may be explained by the fact that the CNTs are coated with more thermally stable particles, as can be observed by transmission electron microscopy.

According to Amit et al. [53], the TGA is used to analyze the heat degradation behavior of several epoxy adhesive formulations. The MWCNT/3% epoxy adhesive composite demonstrated lowest heat stability than the MWCNT/3% epoxy adhesive composite. The uniform dispersion and development of a better interfacial adhesion between the MWCNTs nanoparticles and the epoxy are the primarily factors in increasing the thermal stability. In addition, high interfacial adhesion reduced the mobility of the polymer molecular chains, increasing significantly the MWCNT/epoxy/vapor grown carbon fibers (VGCF) adhesive system's heat stability [54].

All three formulations of epoxy adhesive were found to degrade in two steps. With epoxy, the beginning decomposition is between 313°C and 380°C, while the complete decomposition is between 450°C and 610°C. Also, a 3 wt.% epoxy addition to MWCNT reveals a considerable improvement in thermal stability, with the first signs of disintegration appearing between 310°C and 450°C. This behaviour suggests that the epoxy resin network has been pyrolyzed, and the hydroxyl groups have been degraded [55]. The complete decomposition of the unreacted long-chain epoxy resin particles was clearly seen during the final decomposition of the MWCNT/3% epoxy adhesive, which was detected at 480°C–700°C.

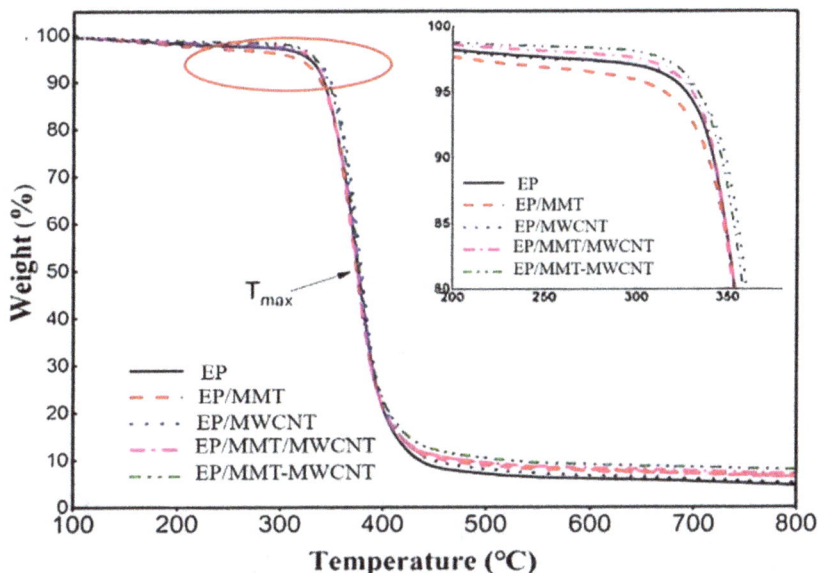

FIGURE 14.5 TGA thermogram of pure epoxy and epoxy composites in the atmosphere. (Reproduced from ref. [56].)

The first-stage decomposition of MWCNT containing 3 wt.% epoxy and VGCF begins between 298°C and 470°C, which is ascribed to the evaporation and decay of unreacted minor species, such as carbonyl, methoxy, and ester. The deterioration of the three-dimensional network structure of epoxy with a greater molecular weight created after curing may cause the second-stage disintegration at 440°C. The small degradation with increased char residue is connected to the second-stage decomposition over 500°C. In contrast to the MWCNT/epoxy/VGCF adhesive composite, the first severe deterioration of the MWCNT integrated with 3 wt.% of epoxy starts at a higher temperature and even the degradation occurs at a higher temperature [53].

The TGA characteristics of the as-prepared epoxy resin/montmorillonite (EP/MMT), epoxy resin/multiwalled carbon nanotubes (EP/MWCNT), epoxy resin/MMT/MWCNT (EP/MMT/MWCNT), and epoxy resin/montmorillonite-multiwalled carbon nanotubes (EP/MMT-MWCNTs) with neat epoxy acting as the control sample have been studied (Figure 14.5 and Table 14.1) [56]. The results found that the T5% and T50% of all but EP/MMT were lower than that of neat epoxy which is because of the presence of overlapped layered MMT decreases the cross-linking density of epoxy resin and thus reduces the thermal stability at the beginning. As the temperature increases, the negative effect of MMT on the thermal stability of EP alleviates, as indicated by the narrowing gap of T50% between EP and EP/MMT. The increase of T5% and T50% for EP/MWCNT exposes that tangled MWCNT can play a positive role in retarding thermal degradation of epoxy resin. Synergistic effect of MMT and MWCNT offsets the negative effect of MMT, leading to increased T5% and T50% of EP/MMT/MWCNT and EP/MMT-MWCNT. The char yields of all the nanocomposites at 800°C were higher

TABLE 14.1
TGA Data of Pure Epoxy and Epoxy MWCNT

Sample	Residue at 800°C (wt.%)	T(5%) (°C)	T(50%) (°C)
EP	4.6	326.8	375.7
EP/MMT	6.4	315.2	373.8
EP/MWCNT	5.1	333.0	378.8
EP/MMT/MWCNT	7.0	330.2	374.5
EP/MMT-MWCNT	7.9	333.5	377.2

than those of neat epoxy and follow an order of EP < EP/MWCNT < EP/MMT < EP/MMT/MWCNT < EP/MMT-MWCNT. This means MMT has better blocking effect at higher temperature, and MWCNT will act as charring agent by retarding thermal degradation of epoxy resin. The synergistic effect of self-assembled MMT-MWCNT could even make EP form more residues. The results concluded that the thermal resistance of EP/MMT/MWCNT is better than EP/MMT-MWCNT due to better distribution of MMT/MWCNT in epoxy resin.

With nonisothermal heating, the weight losses of MWCNT may be seen, indicating the existence of functional groups covalently linked. Around 150°C is when the chemical groups began to gradually lose heat, and also weight losses 3.8 and 15.3 wt.%, respectively, with graft ratio 1.1%–3.6% for the functionalized MWCNT between 150°C and 800°C. As a result of the higher temperature, the carboxylic groups on the MWCNT-COOH had a little weight loss of 3.8 wt.% in this temperature range. According to Zhang et al. [42], due to the breakdown of the amino groups, MWCNT-NH$_2$/epoxy composites displayed a quick weight loss of 10.2% wt.% from 150°C to 500°C, which was followed by a slower weight loss rate of 15.3% as the temperature rose to 800°C. Hence, it can be proven that EDA molecules have been chemically bound to the surface of MWCNT-NH$_2$/epoxy composites by integrating the results of FTIR, XPS, TGA, and particle size tests.

14.7 CONCLUSION

Based on the analysis conducted, it can be concluded that the dispersion and interfacial functionalization of multiwalled carbon nanotubes in epoxy composites have a significant impact on the structural and thermogravimetric analysis characteristics of the resulting materials. The dispersion of carbon nanotubes in the epoxy matrix is crucial for improving the mechanical properties of the composite, while the interfacial functionalization can enhance the interfacial bonding between the nanotubes and the matrix, leading to improved thermal stability. Overall, the findings suggest that the dispersion and interfacial functionalization of carbon nanotubes can be optimized to achieve the desired properties in epoxy-based composites for various applications.

ACKNOWLEDGEMENTS

This work was supported by a Universiti Sains Malaysia, Short-Term Grant with Project No: 304/PTEKIND/6315733.

REFERENCES

1. Han, Z.J., et al., Biological application of carbon nanotubes and graphene, in K. Tanaka and S. Iijima (Eds.), Carbon Nanotubes and Graphene. 2014. Elsevier. pp. 279–312. https://doi.org/10.1016/B978-0-08-098232-8.00012-7
2. Bokobza, L.J.P., Multiwall carbon nanotube elastomeric composites: A review. *Polymer.* 2007. 48(17): 4907–4920.
3. Hu, Y., et al., Carbon nanostructures for advanced composites. *Reports on Progress in Physics.* 2006. 69(6): 1847.
4. Fiedler, B., et al., Fundamental aspects of nano-reinforced composites. *Composites Science and Technology.* 2006. 66(16): 3115–3125.
5. Schadler, L.S., et al., Designed interfaces in polymer nanocomposites: A fundamental viewpoint. *MRS Bulletin.* 2007. 32(4): 335–340.
6. Kim, J., et al., Fabrication and mechanical properties of carbon fiber/epoxy nanocomposites containing high loadings of noncovalently functionalized graphene nanoplatelets. *Composites Science and Technology.* 2020. 192: 108101.
7. Aqel, A., et al., Carbon nanotubes, science and technology part (I) structure, synthesis and characterisation. *Arabian Journal of Chemistry.* 2012. 5(1): 1–23.
8. Iijima, S., Synthesis of carbon nanotubes. *Nature,* 1991. 354(6348): 56–58.
9. Gao, J., et al., Large-scale fabrication of aligned single-walled carbon nanotube array and hierarchical single-walled carbon nanotube assembly. *Journal of the American Chemical Society.* 2004. 126(51): 16698–16699.
10. Krause, B., Pötschke, P., and Häußler, L., Influence of small scale melt mixing conditions on electrical resistivity of carbon nanotube-polyamide composites. *Composites Science and Technology. Composites Science and Technology,* 2009. 69(10): 1505–1515.
11. Chung, C., et al., Biomedical applications of graphene and graphene oxide. *Accounts of Chemical Research.* 2013. 46(10): 2211–2224.
12. Roy, S., et al., Improved polymer encapsulation on multiwalled carbon nanotubes by selective plasma induced controlled polymer grafting. *ACS Applied Materials & Interfaces.* 2014. 6(1): 664–670.
13. Yourdkhani, M. and P.J.C. Hubert, A systematic study on dispersion stability of carbon nanotube-modified epoxy resins. *Carbon.* 2015. 81: 251–259.
14. Zhou, Y., et al., Experimental study on the thermal and mechanical properties of multi-walled carbon nanotube-reinforced epoxy. *Materials Science and Engineering: A.* 2007. 452: 657–664.
15. Lau, K.-T. and D.J.C. Hui, Effectiveness of using carbon nanotubes as nano-reinforcements for advanced composite structures. *Carbon.* 2002. 40(9): 1605.
16. Montazeri, A., et al., Mechanical properties of multi-walled carbon nanotube/epoxy composites. *Materials & Design.* 2010. 31(9): 4202–4208.
17. Srikanth, I., et al., Effect of amino functionalized MWCNT on the crosslink density, fracture toughness of epoxy and mechanical properties of carbon-epoxy composites. *Composites Part A: Applied Science and Manufacturing.* 2012. 43(11): 2083–2086.
18. Guadagno, L., et al., Effect of functionalization on the thermo-mechanical and electrical behavior of multi-wall carbon nanotube/epoxy composites. *Carbon.* 2011. 49(6): 1919–1930.
19. Cui, L.-J., et al., Effect of functionalization of multi-walled carbon nanotube on the curing behavior and mechanical property of multi-walled carbon nanotube/epoxy composites. *Materials & Design.* 2013. 49: 279–284.
20. Barra, G., et al., Different methods of dispersing carbon nanotubes in epoxy resin and initial evaluation of the obtained nanocomposite as a matrix of carbon fiber reinforced laminate in terms of vibroacoustic performance and flammability. *Materials.* 2019. 12(18): 2998.
21. Hwang, S.-H., et al., Smart materials and structures based on carbon nanotube composites. *Carbon Nanotubes-Synthesis, Characterization, Applications.* 2011. 514: 371–396.

22. Lu, K., et al., Mechanical damage of carbon nanotubes by ultrasound. *Carbon*, 1996. 34(6): 814–816.

23. Warner, J.H., et al., *Graphene: Fundamentals and Emergent Applications*. 2012. Newnes.

24. Viswanathan, V., et al., Challenges and advances in nanocomposite processing techniques. *Materials Science and Engineering: R: Reports*. 2006. 54(5–6): 121–285.

25. Thostenson, E.T. and T.-W.J.C. Chou, Processing-structure-multi-functional property relationship in carbon nanotube/epoxy composites. *Carbon*. 2006. 44(14): 3022–3029.

26. Jia, Z., et al., Production of short multi-walled carbon nanotubes. *Carbon*, 1999. 37(6): 903–906.

27. Abdellaoui, H., et al., *Mechanical Behavior of Carbon/Natural Fiber-Based Hybrid Composites, in Mechanical and Physical Testing of Biocomposites, Fibre-Reinforced Composites and Hybrid Composites*. 2019. Elsevier. pp. 103–122.

28. Konnola, R. and K.J.R.a. Joseph, Effect of side-wall functionalisation of multi-walled carbon nanotubes on the thermo-mechanical properties of epoxy composites. *RSC Advances*. 2016. 6(28): 23887–23899.

29. Che, B.D., et al., Effects of carbon nanotube dispersion methods on the radar absorbing properties of MWCNT/epoxy nanocomposites. *Macromolecular Research*. 2014. 22: 1221–1228.

30. Luo, Y., et al., Effect of amino-functionalization on the interfacial adhesion of multi-walled carbon nanotubes/epoxy nanocomposites. *Materials & Design*. 2012. 33: 405–412.

31. Blondeau, P., et al., Covalent functionalization of single-walled carbon nanotubes with adenosine monophosphate: Towards the synthesis of SWCNT-Aptamer hybrids. *Materials Science and Engineering: C*. 2011. 31(7): 1363–1368.

32. Xu, G., et al., Covalent functionalization of multi-walled carbon nanotube surfaces by conjugated polyfluorenes. *Polymer*. 2007. 48(26): 7510–7515.

33. Zhang, A., et al., Effect of percolation on the electrical conductivity of amino molecules non-covalently coated multi-walled carbon nanotubes/epoxy composites. *Applied Surface Science*. 2012. 258(22): 8492–8497.

34. Li, X., et al., Non-covalent functionalization of multi walled carbon nanotubes and their application for conductive composites. *Carbon*. 2008. 5(46): 829–831.

35. Norizan, M.N., et al., Carbon nanotubes: Functionalisation and their application in chemical sensors. *RSC Advances*, 2020. 10(71): 43704–43732.

36. Reddy, K.R., et al., A new one-step synthesis method for coating multi-walled carbon nanotubes with cuprous oxide nanoparticles. *Scripta Materialia*, 2008. 58(11): 1010–1013.

37. Mohd Saidi, N., et al., Characterizations of MWCNTs nanofluids on the effect of surface oxidative treatments. *Nanomaterials*, 2022. 12(7): 1071.

38. Hoa, L.T.M., Characterization of multi-walled carbon nanotubes functionalized by a mixture of HNO3/H2SO4. *Diamond and Related Materials*, 2018. 89: 43–51.

39. Bose, S., R.A. Khare, and P.J.P. Moldenaers, Assessing the strengths and weaknesses of various types of pre-treatments of carbon nanotubes on the properties of polymer/carbon nanotubes composites: A critical review. *Polymer*. 2010. 51(5): 975–993.

40. Norizan, M.N., et al., Carbon nanotubes: Functionalisation and their application in chemical sensors. *RSC Advances*. 2020. 10(71): 43704–43732.

41. Saeb, M.R., et al., Highly curable epoxy/MWCNTs nanocomposites: An effective approach to functionalization of carbon nanotubes. *Chemical Engineering Journal*. 2015. 259: 117–125.

42. Zhang, Q., et al., Dispersion stability of functionalized MWCNT in the epoxy-amine system and its effects on mechanical and interfacial properties of carbon fiber composites. *Materials & Design*. 2016. 94: 392–402.

43. Hanawa, T. and M.J.A.S.S. Ota, Characterization of surface film formed on titanium in electrolyte using XPS. *Applied Surface Science*. 1992. 55(4): 269–276.

44. Moulder, J.F., et al., *Handbook of X-Ray Photoelectron Spectroscopy: A Reference Book of Standard Spectra for Identification and Interpretation of XPS Data*. Perkin-Elmer Corporation (pp. 51–52). 1992.

45. Yang, K., et al., Effects of carbon nanotube functionalization on the mechanical and thermal properties of epoxy composites. *Carbon*. 2009. 47(7): 1723–1737.

46. Ma, P.-C., et al., Dispersion, interfacial interaction and re-agglomeration of functionalized carbon nanotubes in epoxy composites. *Carbon*. 2010. 48(6): 1824–1834.

47. Rajisha, K. R., B. Deepa, L. A. Pothan, and S. Thomas. Thermomechanical and spectroscopic characterization of natural fibre composites. In E. Zafeiropoulos (Ed.), *Interface engineering of natural fibre composites for maximum performance* (pp. 241–274). 2011. Woodhead Publishing Limited, Cambridge. https://doi.org/10.1533/9780857092281. frontmatter.

48. Thomas, S., et al., *Thermal and Rheological Measurement Techniques for Nanomaterials Characterization*. Vol. 3. 2017. Elsevier, Netherlands.

49. Bom, D., et al., Thermogravimetric analysis of the oxidation of multiwalled carbon nanotubes: Evidence for the role of defect sites in carbon nanotube chemistry. *Nano Letters*, 2002. 2(6): 615–619.

50. Avci, A.K. and Z.I. Önsan, 2.16 Catalysts. *Comprehensive Energy Systems* (pp. 475–523). 2018. Elsevier.

51. Maron, G., et al., Carbon fiber/epoxy composites: Effect of zinc sulphide coated carbon nanotube on thermal and mechanical properties. *Polymer Bulletin*. 2018. 75: 1619–1633.

52. Yoonessi, M., et al., Carbon nanotube epoxy nanocomposites: The effects of interfacial modifications on the dynamic mechanical properties of the nanocomposites. *ACS Applied Materials & Interfaces*. 2014. 6(19): 16621–16630.

53. Singh, A.K., et al., Aligned multi-walled carbon nanotubes (MWCNT) and vapor grown carbon fibers (VGCF) reinforced epoxy adhesive for thermal conductivity applications. *Journal of Materials Science: Materials in Electronics*. 2017. 28: 17655–17674.

54. Jagtap, S. and D.J.E.P.L. Ratna, Preparation and characterization of rubbery epoxy/multiwall carbon nanotubes composites using amino acid salt assisted dispersion technique. *Express Polymer Letters*. 2013. 7(4): 329–339.

55. Sahoo, S., S. Mohanty, and S. Nayak, Toughened bio-based epoxy blend network modified with transesterified epoxidized soybean oil: Synthesis and characterization, *RSC Advances*, 2015. 5: 13674–13691.

56. Xue, Y., et al., A novel strategy for enhancing the flame resistance, dynamic mechanical and the thermal degradation properties of epoxy nanocomposites. *Materials Research Express*. 2019. 6(12): 125003.

15 Natural Nanofillers in Polyolefins-Based Composites A Review

H. A. Aisyah
Universiti Putra Malaysia

E. S. Zainudin and F. A. B. Balkis
Universiti Putra Malaysia

S. H. Lee
Universiti Putra Malaysia
Universiti Teknologi MARA Pahang Branch Campus Jengka

15.1 INTRODUCTION

Excellent properties and rising popularity mean that composites are finding increasing use in a diverse set of contexts such as transportation, construction, automotive part, sport as well as medicine. This opens up a wealth of opportunities. Because of its properties including high stiffness, strength, lightweight, dynamic properties, as well as chemical and corrosion resistance, composites are being used in a variety of industries. Generally, composite is made up from synthetic or natural polymer, and it was identified that synthetic polymers are widely employed in daily usage such as in packaging materials and food containers [1]. The growth in synthetic polymer manufacturing and usage created environmental difficulties because of non-biodegradable polymers that emit toxic vapors and hazardous substances, as well as accumulation disposal of wastes on land. Due to increased ecological issues and environmental concerns grow, the formation of biodegradable polymers in composite industry that are biocompatible and environmentally friendly is employed. Incorporating biodegradable nanostructure materials such as natural nanofiller in composite production offers degradation properties and has a good impact on fiber–matrix interaction [2,3]. The addition of various types of nanofiller did not only improve the physico-mechanical properties of the composite, it did enhance its thermal properties as well.

DOI: 10.1201/9781003400998-15

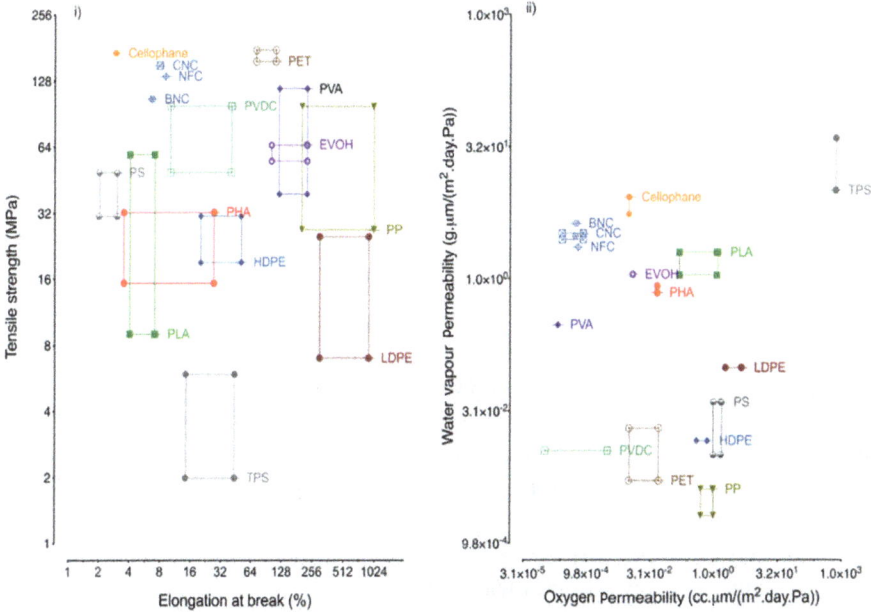

FIGURE 15.1 Mechanical and barrier properties of conventional plastics and polymers [4]. (i) Plot of tensile strength (in MPa) against elongation at break (in %) and (ii) Graph illustrating water vapor permeability compared to oxygen permeability.

15.2　NANOFILLER

Nanofillers have recently become popular for use in polymer composites due to the vast improvements in properties and biodegradability they may bring to the materials. The strength of polymer and biopolymer is comparable, but the elasticity is lower and the vapor permeability is higher for biopolymer. Figure 15.1 shows the mechanical and barrier properties of conventional plastics and biopolymers. Hence, for biopolymers to be applied and substitute conventional plastics, reinforcement fillers must be added into it. Ultra-low volume is the only need give significant changes to better composite's properties. Nanofillers are added to composites to boost their mechanical, thermal, optical, and electrical performance. In addition, the aim for using nanofillers is that they enhance the number of contacts between matrix and reinforcement material by increasing the surface-to-volume ratio, which raises the total material potentiality. Despite nanofiller reinforcement to improve mechanical properties, nanoclay is often applied to enhance thermal stability and barrier properties

Nanofillers are classified based on their dimensions, namely one dimensional (1D), two dimensional (2D), and three dimensional (3D) [5]. 1D nanofillers have dimensions of less than 100nm and take the form of thin sheets with thicknesses ranging from a few nanometers to hundreds of thousands of nanometers in length. Montmorillonite clay, nanoplatelets, nanosheets, and carbon nanowalls are examples of 1D nanofillers [6]. This type of nanofillers is usually used in applications such as medical, electronic, and sensors, because of their dimensionality, besides excellent

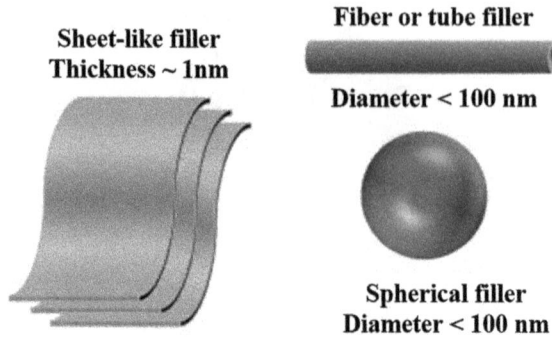

FIGURE 15.2 Nanofiller in the form of sheet, tube, and spherical [9].

electrical, magnetic, and optic properties. 2D nanofillers are nanofillers having two dimensions smaller than 100 nm and in form of tubes, threads, and filaments. Common examples of 2D nanofillers are carbon nanotubes, cellulose whiskers, 2D graphene, clay nanotubes, carbon fiber, cellulose fiber, and graphene [7]. On the other hand, 3D nanofillers are particles with all three dimensions in the nanometer scale and often in the form of spherical or cubical. Carbon black, nanosilica, nanoalumina, silicon carbide, and silica are all examples of 3D nanofillers [8]. Figure 15.2 shows nanofiller in different forms, and Figure 15.3 shows example of nanofillers.

Nanofillers also can be classified into organic and inorganic nanofillers as shown in Figure 15.4. The examples of organic nanofillers are natural biopolymers, while inorganic nanofillers are metal and clays. The natural nanofillers are made up of biopolymer molecules, likes cellulose, chitin, and starch. The primary properties of natural nanofillers that make it unique are their compatibility, degradability, and abundantly available. There are three main types of nanofillers from cellulose, and they can be differentiated by their morphology, crystallinity, and particle size. Also, their original sources and methods of extraction play an important role. One of the major types is nanocrystalline cellulose (NCC), the second one is nanofibrillated cellulose, and the third one is bacterial nanocellulose (NC) [10]. A derivative from chitin, called chitosan, is typically extracted from the skeleton of shellfish. Chitosan is known for a non-toxic substance, and it also has appealing surface area, biocompatibility as well as the ability to form film. The most common application of chitosan nanofiller is in biomedical applications, film, packaging, and coating due to its excellent barrier and antibacterial properties, along with prolonging perishable food's shelf life [11].

15.3 POLYOLEFINS

Polyolefins or polyalkenes are the largest class of thermoplastics and formed by the polymerization of olefinic monomers or alkenes monomer units. The number of olefinic monomers determined the polyolefins category into homopolymers, copolymers, or terpolymers [13]. Polyolefins are classified as ethylene-based, propylene-based, higher polyolefins, and polyolefin elastomers depending on their monomeric unit and chain structure [14]. They also can be categorized as crystalline or amorphous based on chain conformation, configuration, and processing circumstances. Examples of

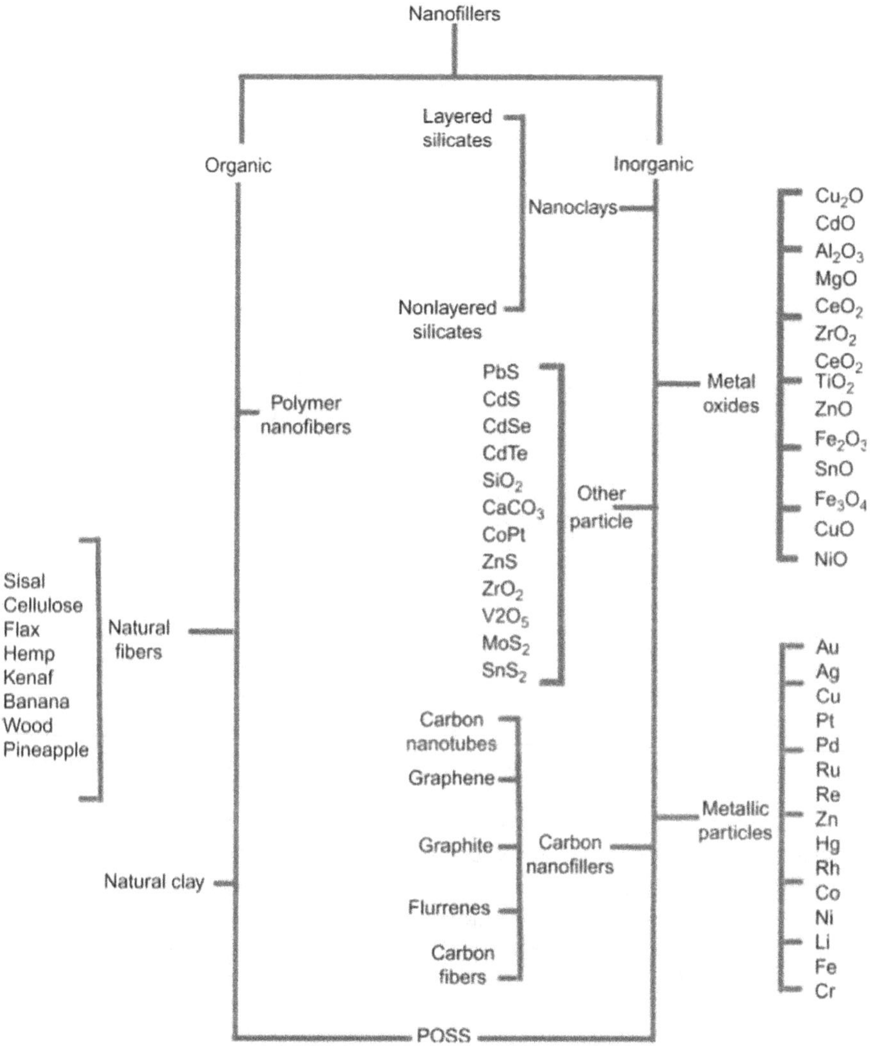

Nanofillers

Organic | Inorganic

Layered silicates

Nanoclays

Nonlayered silicates

Polymer nanofibers

Metal oxides: Cu_2O, CdO, Al_2O_3, MgO, CeO_2, ZrO_2, CeO_2, TiO_2, ZnO, Fe_2O_3, SnO, Fe_3O_4, CuO, NiO

Other particle: PbS, CdS, CdSe, CdTe, SiO_2, $CaCO_3$, CoPt, ZnS, ZrO_2, $V2O_5$, MoS_2, SnS_2

Natural fibers: Sisal, Cellulose, Flax, Hemp, Kenaf, Banana, Wood, Pineapple

Carbon nanofillers: Carbon nanotubes, Graphene, Graphite, Flurrenes, Carbon fibers

Metallic particles: Au, Ag, Cu, Pt, Pd, Ru, Re, Zn, Hg, Rh, Co, Ni, Li, Fe, Cr

Natural clay

POSS

FIGURE 15.3 An example of a various types of nanofillers [5].

polyolefins polymers are polyethylene (PE), polypropylene (PP), poly(styrene) (PS), polybutene-1, polyisobutylene (PIB), and ethylene propylene rubber, with PE and PP have been the most frequently used polyolefins in a wide range of applications such as plastic bags, wraps, carpets, food packaging, and plastic food containers. The advantages and disadvantages of some selected polyolefins are shown in Table 15.1. PEs are further classified as density, very-low-density PE (VLDPE), low-density PE (LDPE), medium-density PE (MDPE), and high-density PE (HDPE). The densities of VLDPE, LDPE, MDPE, and HDPE are 0.880–0.915 g/cm³, 0.910–0.925 g/cm³, 0.926–0.940 g/cm³, and 0.941–0.965 g/cm³, respectively [15]. Among these types of PE, HDPE has higher degree of crystallinity and chain packing. Meanwhile, LDPE has higher permeability compared to HDPE due to its high degree of chain branching

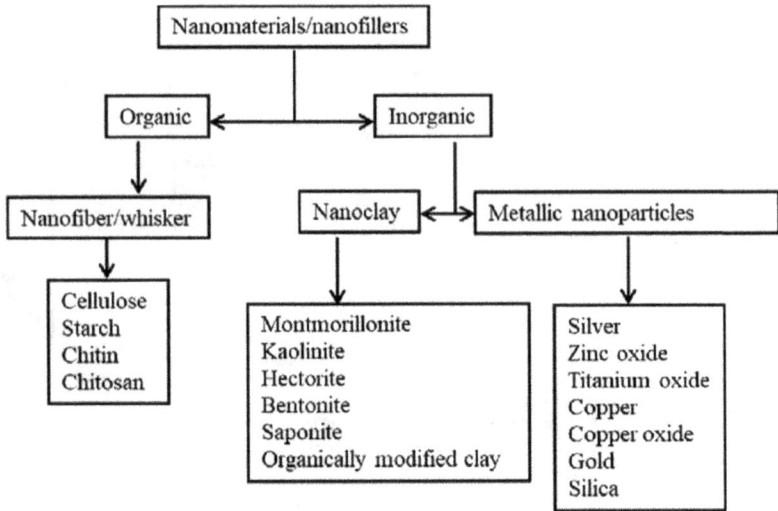

FIGURE 15.4   Organic and inorganic classification of nanofillers [12].

and lower crystallinity [16]. The extensive use of polyolefin polymer was because of their low cost and favorable properties, that is, low cost, recyclability, variety of design, and ease of manufacture.

15.4 NANOFILLER-REINFORCED POLYOLEFIN COMPOSITES

There are a few limitations of polyolefin polymer-based composite, including lower stiffness, and the performance decreased at elevated temperatures [22]. In food packaging industry, the main disadvantage is the inherent permeability to gases and other small molecules. In the composite industry, polyolefin is frequently added with nanofillers to provide a suitable stiffness and impact resistance, as well as improved dimensional stability, water resistance, and increased thermal stability. Figure 15.5 illustrates the concept of nanofillers in the polymer matrix production.

15.4.1 POLYETHYLENE

PE is a light, versatile synthetic resin derived from ethylene polymerization. PE is an important member of the polyolefin resin family. It is the most widely used plastic in the world, with applications ranging from food packaging to automobiles. Its monomer, ethylene (C_2H_4), is a gaseous hydrocarbon that is commonly produced by ethane cracking. Its double bond will be broken under the influence of polymerization catalysts, and the excess single bond will be linked with another ethylene molecule. The properties of PE plastic are determined by this simple structure, which is repeated thousands of times in a single molecule. Long, chain-like arrangement with linear or branching hydrogen atoms is connected to a carbon backbone. LDPE, HDPE, and ultra-high-molecular-weight PE (UHMWPE) are examples of branched PE. This section focuses solely on nanofillers reinforced in LDPE and HDPE polymers.

TABLE 15.1

Advantages and Disadvantages of Some Selected Polyolefins

Polyolefin Type	Advantages	Disadvantages	References
PE	• Highly ductile and satisfactory impact strength • Water resistant and durable • Superior electric insulation properties • Suitable for packaging due to its transparency and opacity • Recyclability • Good heat-resistant properties	• Hard to biologically degrade • Non-renewable • High energy consumption for production • Many types of PE and make the sorting for recycling difficult	[17]
PP	• Inert to acids and ideal containers to hold acidic liquids • Highly resistant to corrosion and chemical leaking, ideal for piping systems • Resists to freezing • High moldability • Good electrical insulator • Highly impermeable by water (waterproof) • Malleable • Low density but high tensile strength	• Prone to UV degradation • Limited use in high temperatures • Poor bonding properties • Extremely flammable	[18]
PIB	• Satisfactory ozone resistance • High weathering resistance • Extremely low gas permeability • Heat resistance • Flexible under low temperature • Compoundable with superior tensile strength	• Moderate abrasion and compression set • Inferior tensile strength • Lack of resiliency • Easy flammable • Inefficient curing rate	[19]
Polybutadiene	• Cheaper than isoprene, better • Flexible under low temperature • High compatibility with other polymers • Can be adhered finely with metals	• Inferior tensile strength • Low tear resistance • Low tack	[20]
PS	• Cost-effective • Strong and rigid • High transparency • Able to be molded in various forms • Dimensionally stable • Excellent electrical properties	• Brittle • Inferior resistance toward chemicals • Highly flammable • Poorly withstand UV degradation	[21]

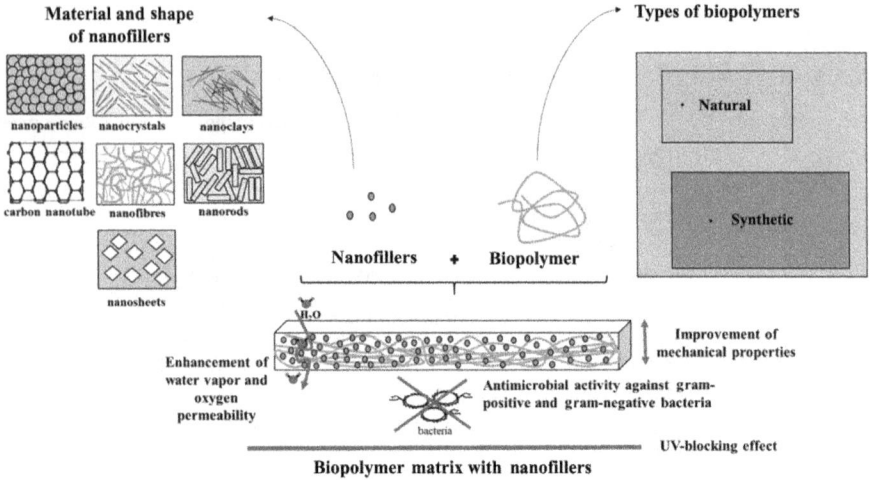

FIGURE 15.5 Illustration of nanofiller in the nanocomposite films and their functional properties [23].

15.4.2 NANOFILLERS IN HIGH-DENSITY POLYETHYLENE

HDPE polymer has superior properties in stiffness, electrical properties, and strength, especially high-insulation dielectric strength, making it highly popular in cable insulator materials. However, poor flame retardancy and high flammability are drawbacks that need to be considered. Although halogenated flame-retardant additives such as aluminum hydroxide ($Al(OH)_3$) and magnesium hydroxide ($Mg(OH)_2$) were good enough to improve fire resistivity with low cost, a very high loading (recommends in 50–70 wt%) of these additives is always needed to perform and consequently scarify product's strength. At this moment, a later generation of flame-retardant material, layered double hydroxide (LDH), reacts similar to the $Al(OH)_3$ and $Mg(OH)_2$ additives, but in nanoscales. The LDH/HDPE nanocomposite found significant improvement in thermal stability and delayed flammability [24].

Other than enhancing flame-retardancy properties of HDPE composites, nanofillers shall act as reinforcements in the HDPE polymer, to further strengthen its strength performances. In this moment, green nanofillers such as NCs are often applied. Two weight percent of organo-flouromica nanofillers in HDPE composite possess the satisfactory tensile strength values close to 24 MPa. Good interfacial conditions are the main factor of good strength [25]. Additionally, nanometric dispersion of silicate layers and enhancement in nanofiller–matrix interactions can improve HDPE tearing resistance.

However, deterioration of strength may be occurred for nanofiller insertion, due to poor interfacial conditions between nanofiller and HDPE, because of its hydrophobic and weakly polar nature, whereas NCs are hydrophilic in nature and polar, making them highly incompatible. Nanofillers aggregation is a frequent phenomenon found on the composite as brown spots. To workout this issue, many efforts have been done, such as copolymers, treatments, type, and/or nanofiller processing [26–28].

Partial stable suspension of NC in organic solvent also concluded to solve the potential aggregation [29]. While TEMPO-NC fillers reinforced HDPE had resulted with increased yield stress and modulus of the composite [30]. Besides, PE-grafted maleic anhydride coupling agent is also a frequently used agent in HDPE composites, to improve reinforcing effect of nanofillers. Contrarily, reduced impact strength by strong bonding restricted polymer chain's movements [31].

15.4.3 Nanofillers in Low-Density Polyethylene

LDPE has a lower "density" than HDPE by the meaning of having more branches of chains. The LDPE polymer is often applied in similar applications as HDPE polymer. Yet, LDPE has higher impact resistance, moisture, and chemicals barrier. It is found that the nanoclay insertion observed is very effective in enhanced morphology. The oxygen permeability of LDPE decreases in the presence of nanoparticles. Furthermore, the results show that specimens containing nanofiller have greater thermal stability than pure LDPE [32].

Furthermore, a number of surface modification techniques have been developed to reduce nanofiller agglomeration. Another emerging method is plasma treatment, which has the potential to improve nanofiller compatibility with polymers by modifying the surface of the nanofillers. It showed that 30-minute plasma-treated nanofillers had the greatest enhancement in composite analysis results [33]. Components' surface energy and thermodynamic parameters at the interface were treated as critical factors for nanofiller as a compatibilizer at the interface. Besides, nanofiller's specific surface area, particle geometry, chemical modifications, and potential for ordering into 3D networks shall also influence the reinforcement degree [34]. The orientation of nanofillers shall give different dispersion and reinforcement effects. It has been found that nanofillers create percolating network within the polymer and provide strong enhancing effects, at low fiber contents. Oil palm mesocarp fiber NC-reinforced LDPE composites give significant increment of mechanical properties, at just 3 wt% of nanofillers [35]. However, percolating network may hardly be formed when there is fiber agglomerated and consequently reduced mechanical properties.

As the circular economy of plastic composites gains traction, one of the major and difficult tasks is composite recycling. The majority of post-consumer plastic waste recycling occurs through mechanical recycling or energy recovery. LDPE polymers are commonly suitable for mechanical recycling. The purpose of recycling is to reuse and recreate. The recovered materials can be reused to recreate something new and useful. By doing so, the plastic waste issues can be partially solved and mitigated. A concept has been proposed to recycle two-layer packaging films with cellulosic barrier coatings into valuable new products. The cellulosic coatings were completely blended as microscale agglomerates in the LDPE matrix during the mechanical recycling process [36]. Barrier and heat seal ability properties were not affected by these agglomerates, and its mechanical properties were merely changed. However, from the point of economical view, optimization is utmost important for mechanical recycling of multilayer packaging films containing NC to be employed cost-effectively for commercialization. However, replacing unrecyclable barrier films with recyclable materials will eventually result in a more circular packaging economy in which even multilayer films are designed to be recycled and thus prevent pollution from entering the environment.

15.4.4 NANOFILLERS IN POLYPROPYLENE

PP has low density but excellent stiffness and tensile strength. In addition to that, PP is also inactive toward acids, alkalis, and many other solvents [37]. These attributes made PP an important polyolefin that has been used industrially. The utilization of PP brings beneficial effects to many industries including packaging and automotive components. In textiles industry, PP has been used to make ropes, thermal under-wear, and carpets. Nevertheless, physicochemical properties of PP are insufficient for advanced uses. Therefore, enhancement is required. One of the fast and most direct ways is by the addition of nanoscale fillers to the PP matrix. By doing so, the perfor-mance of the polymer is often improved instantly by enhancing the composition and properties of nanoscale fillers.

PP, which is used in food packaging, automobiles, and other industrial sectors, is non-toxic, is chemically resistant, and is low cost, has lately gained attention. However, the weak mechanical qualities and ageing resistance of PP prevented its use in many applications. Mechanical characteristics and ageing resistance of PP must be increased for advanced applications [38]. Natural reinforcement fibers have been widely used to improve the mechanical and thermal characteristics of polymer matrices, resulting in materials with superior mechanical and thermal capabilities.

Owing to its unique nature including renewability, cost effective, easily acces-sible, NCC has sparked a lot of interest, not to mention its superior strength, low density, and high surface reactivity. A study showed that a proportion of 3 wt% NCC/attapulgite (AT) was able to increase the tensile strength of the PP by almost 15% com-pared to that of pure, unmodified PP [39]. Moreover, the modified PP retained almost 99% of its tensile strength even after a 15-day thermal ageing period. Interestingly, the retention rate was 52% higher than pure PP. Meanwhile, modified PP retained 95% of its initial impact strength. The retention rate is twofold higher than that of pure PP. Another study found that incorporating NCC into LDPE nanocomposites increased mechanical parameters such as strength, modulus, and hardness by 6.5, 19, and 150%, respectively. The FTIR data revealed evidence of bonding between NCC and PP, indicating that the mechanical properties had improved. The differential scanning calorimetry and thermogravimetric analysis results demonstrated that the PP-NCC nanocomposite exhibited stable thermal behavior. The DSC Tm was within the same range for all samples.

15.5 APPLICATIONS

A wide range of interest in NC as a filler material for polyolefin and other polymers is due to its advantages as compared to the other fillers. Some of the well-established advantages of NC fillers are low density, high purity, high aspect ratio, high sur-face area, high tensile strength and modulus, low toxicity, etc. These characteris-tics are suitable to be integrated into polyolefin composite products as it improves the composite properties in general. In addition, NC filler is a greener option for the environment as the green global market is emerging over the years. With some modifications, NC filler will be more efficiently used in various applications such as packaging, automotive components, biomedicals, electronics, aerospace, and other industries. This section will focus on the two most applications of NC/PO, which are packaging and automotive panels.

15.5.1 Packaging

Food packaging has gone material revolution through progress in human civilization, from using pumpkins, shells, and leaves for food storage to ceramic vessels in the Mediterranean region in 1,500 BC–500 AD, then to glass-based packing in 1,100 BC. In 200 BC–220 AD, China has started to use paper packaging, and fast forward, to the twentieth century when people started to use plastic for packaging in the 1950s [34]. Plastic became an important element in packaging starting in the 1970s, and it has replaced most paper-based packaging products since then. The revenue generation of plastic packaging in 2020 was predicted to reach a value of $998 billion [40].

Most of the plastic-based used in the packaging come from petroleum-based known as polyolefin that include PP, LDPE, LLDPE, and HDPE. The usage of these plastics in packaging is due to their large availability, relatively low cost, good mechanical properties, and thermal stability. However, they inherited the main disadvantages which are non-renewable resource, non-degradable, and increase the accumulation of environmental pollution [41–44]. Increasing awareness and demand toward environmentally friendly products has led to the development of bio-based or greener packaging materials. One of the ways to promote this motivation is by incorporating natural resources such as cellulose-based in the polyolefin plastic packaging. Evaluation of biodegradability by soil burial test is one of the methods to study the degradation process of a material. Ahmadi et al. [41] explored the effect of integrating NC fiber into LDPE/thermoplastic starch (TPS) on their degradation properties. The rate of weight change percentage after 30 days of exposure to soil showed that composites with NC fibers were faster than the control samples. Rapid weight loss represents a faster degradation process that has occurred. This may be due to the hydrophilic nature of NC fibers that leads to a higher chance of microorganism attack, thus faster the degradation process [45,46].

A similar material combination of LDPE/TPS composites was performed by Gray et al. [44] to investigate the effect of adding 1% and 2% of NC crystals on the composites. Several tests for packaging criteria were conducted, namely water vapor permeability coefficient, water vapor transmission, and moisture absorption test. The results showed that this composite had improved with the addition of 1% NC crystals. Additionally, the composite mechanical properties such as tensile and hardness had shown positive increments. From this study, the results have shown that introducing NC crystal in the combination of LDPE/TPS/NC composites can compensate for weaknesses in the former formulation and can be considered as a sustainable alternative to replace LDPE packaging material.

In general, adding NC as fillers in polymer packaging improves oxygen transmission rate, water vapor permeability, oil, grease, and liquid barrier [39,42,47–50].

These advantages not only value-added the packaging properties when compared to the original formulation but also can be associated with a high level of optical transparency with a suitable polymer. Improved barrier properties of NC/polyolefin products are due to the high crystallinity of NC [47]. Increasing crystal region in NC/polyolefin composites resulted in impervious surface to liquid and gas transmission. This property is contradicted to the amorphous regions that allow water and gas molecules to diffuse easily. For instance, slow penetration of oxygen molecules into NC film was observed due to high fibril entanglements within the film and hence a lower permeability [51]. In general, the improvement of barrier properties by NC-based composites can be attributed to the dense network formed by these nanofibrils by having smaller and more uniform dimensions [50,51]. This characteristic made NC a promising material in packaging including food, drug delivery membranes/filters, and wound dressing.

15.5.2 AUTOMOTIVE INDUSTRY

The need for lightweight material to replace steel in automotive industries has started over 20 years ago [52]. This is because weight plays a major role in the energy consumption of a vehicle, and it was approximated that a 7% of fuel economy increases with a 10% weight reduction in automotive [53]. The use of polyolefin, which is a lightweight material in automotive industries, has been dramatically increased due to these drivers, reducing fuel consumption and low emissions for manufacturing and transportation sectors. The major advantages of polyolefin matrices that benefited this industry are they can be easily shaped, have a smooth surface, low cost, and lighter than metals and glass [53–55]. Parts that have been replaced by polyolefin material are bumpers, wheel housings, containers, door trim, windshield fluid containers, and many more. As mentioned before, there is a need to incorporate bio-based materials to promote greener and environmentally friendly products.

Composite made of a mixture of NC and polyolefin material for automotive parts is expected to improve manufacturing speed, promote recycling, reduce weight, and enhance environmental and thermal stability [43,56]. Some of the additional automotive components based on these materials are hood, trunk lid, roof side rail, seatback, and door trim of a vehicle. This nanocomposite offers an average of 25% weight reduction if compared to pure plastic, and this reduction provides a potential of substantial energy savings for the users [43]. For instance, the density of panels, made from 1.5% of NC in the PP matrix, resulted in 20 kg/m³ weight reduction from its average original weight of 625 kg/m³ when used 100% PP [57]. Moreover, adding NC improves the transparency and mechanical properties of polyolefin while maintaining the original qualities of matrices [48,55–59]. For instance, a study examined the effect of various percentage concentrations of NC crystals to fabricate NC/PP composites and tested on their mechanical and thermal properties. The results showed that adding 1% of NC crystals had improved the overall properties of the composite in terms of bending, hardness, and thermal properties. This improvement is due to the dense composite structure resulted from the effect of adding NC crystals that may lead to better bonding interaction between NCs in the matrix that was shown from FTIR spectra [60].

Aside from the common composite panels that use melting compounding, solvent casting, and in-situ polymerization methods, a pressure-quenching physical foaming method of carbon dioxide can be performed to fabricate ultra-high-strength and lightweight materials for automotive industries [61–64]. Ito et al. [65] had conducted a study to investigate the mechanical properties of the foam and solid NC/PP composites. On average, the NC/PP foams resulted had approximately 68 MPa which was slightly lower than the NC/PP solid panel that had nearly 90 MPa in flexural stress which is corresponding to their average relative densities by having mean values of 0.9 and 1.05, respectively. Although foam composite properties were slightly lower than the solid panel, the foam composite still had better flexural modulus and stress when compared to pure PP. In general, NC-based polyolefin has shown a positive effect on several properties of the composites including better flexibility, toughness, and thermal at a lower density which will benefit the automotive industries. This improvement is owing to their high-surface-to-volume ratio that can interact more strongly with the matrix, meaning better adhesion when compared to conventional microreinforcements or pure matrix [60,65,66].

15.5.3 OTHERS

The stringent requirement is obligatory for material specifications in the medical field especially when it is used inside the body like bone and joint replacement, cartilage replacement, sutures, cardiovascular applications, as well as disposal items such as syringes and tubes. The most important requirement is the biocompatibility of the polymers that is they are compatible and resistant to tissue, cells, enzymes, and biological fluids based on the applications. Furthermore, they should not degrade or aggravate thrombosis, breakdown of tissues, or harmful, immune, toxicological, or allergenic effects [56,67]. Wang et al. [68] investigated the effect of adding NC crystals into UHMWPE for artificial joint applications. Several tests were conducted to study the tribological property and biocompatibility of the composites. The results showed that the coefficient of friction and the number of wear debris of the composites decreased when compared to the pure plastic. Having smaller size debris (less than 200 nm) managed to reduce severe inflammatory responses and lower cell death when adding NC crystals in the matrix [69]. In contrast, larger size debris (100 nm–3 μm) was observed when used pure plastic. They concluded that there is a bright opportunity in using NC crystals with UHMWPE for wear resistant as well as it functions as good biocompatible material for lubricant.

Polymeric foams are getting distinct interest from a variety of fields such as tissue engineering, thermal and sound insulation, construction, and transportation. Another foam processing method is foam injection molding (FIM). FIM is a method to prepare the polymeric foam with a feasible to fabricate a complex 3D geometry. Modified NC fibers were used with PP to fabricate PP/CNF foams, and this product has potential in construction and transportation [70]. In their study, modified and 0.4° of substitution of NC exhibited dramatic improvement when compared to the foams that employed unmodified NC and pure plastic. The study concluded that the combination of PP/NC foams presented notable improvement in mechanical properties due to crystallization promotion effect that produced smaller cell sizes and higher cell densities [62,71,72]. Aside from these examples, many other applications have been studied and can be further explored to fully optimize the benefits from NC as well as polyolefin materials.

15.6   CHALLENGES AND FUTURE OUTLOOK

The main objective of filling polyolefin with NC material is to enhance their eco-friendly aspects to meet growing demand toward environmental and greener approaches. At the same time, many properties inherited from NC, such as low density, renewable, excellent mechanical properties, and electrical insulation properties to name a few, have benefited the end products [56,57,73]. Subsequently, the resulted NC/polyolefin composites had better or even overpass conventional composite properties such as had greater performance in mechanical and thermal, improves liquid and gas barriers, and biodegradable. These improvements can benefit several industries including the packaging in food, biomedical, automotive, etc. However, NC also inherited some disadvantages; thus, several challenges need to be considered and focused on. The two most challenging aspects of using NC into polyolefin are their compatibility due to different polarities and the production of the NC material itself.

Considering different characteristics of NC and polyolefin, it is important to have good compatibility and uniform dispersion of NC materials in the composite. This is due to that these intrinsic aspects will determine the overall composite properties [74–76]. There are many studies have been conducted to improve the interfacial bonding between hydrophilic NC and hydrophobic matrix. Having good interfacial bonding indicates that these two components have good compatibility. Not only do they need to be compatible with each other, but NC also needs to be uniformly dispersed in the matrix to promote a good bonding and better load distribution that may also lead to better composite performances. For instance, some of the modifications were conducted via chemical modifications either on NC or on polyolefin, using compatibilizers, or modifying the fabrication processes [41,77–80]. Although applying this modification can improve some of the composite properties, it is still necessary to further explore the best way to have good interfacial bonding that can reduce the processing time and cost in fabricating the composites. For instance, aside from the chemical modification efforts, Sapkota et al. [80] have implemented an organic-solvent-free, two-step process to produce NC crystal/PE composites as an alternative approach without any treatment on the NC. The composite produced resulted in a significant improvement of storage modulus by a factor of 2.5 with the addition of 15% of NC when compared to direct melt-mixing by extrusion method. This two-step process has shown a good dispersion of NC in the PE matrix, and this method is claimed to be readily applicable and scalable to industrial production scale.

The benefits of using NC have been well recognized by many studies. However, the more environmentally friendly and low energy production of NC is needed to keep up with the growing demands [43,81]. This production cost may come from the preparation of the NC itself or the processing part of the composites. For instance, high expenses are spent on the pretreatment techniques of the production of NC material such as enzymatic treatment [82]. On average, it was reported that pretreatment processing consumes more than 40% of the total processing costs [83]. Additionally, there is also a cost for by-products management, for instance, on a large scale for NC crystals that use the sulfuric acid hydrolysis technique for the isolation process. It was reported that using sulfuric acid is one of the cheapest and recyclable methods to produce NC crystals [84]. However, further processing to convert the by-products degraded sugar, for biofuel may require additional cost and energy. Moreover, transportation of low solid content of wet NC fiber can contribute to the major issue in the commercial advancement of NC. This is due to that NC crystal needs to be used immediately or else it needs to be preserved [85]. This may require an additional step to redisperse the material and may raise the end-use price of NC products. Therefore, it was suggested that we need to exploit the application of NC-based products by designing attractive composites to optimize the functionality and performances [86].

15.7 CONCLUSIONS

Because of its excellent mechanical properties, good processing properties, and superior electrical insulation process, conventional polyolefin is in high demand. However, adding nanofillers is one of the most common ways to improve their properties. Existing inorganic fillers such as mineral powder, carbon black, and silicon dioxide have a strong potential, but their density is much higher than the polyolefin itself. The problem of density causes a rise in transportation cost and labor intensity during the preparation and use of the composites. NC has a promising influence as a filler in the polyolefin composites. It has given an enormous opportunity in using the polyolefins in a more environmentally friendly approach without compromising the original qualities, and even, has improved most of the composite qualities. However, given the inherent properties of NC, research and innovation on the modifications of the material, or the composite processing, need to be further investigated, and the properties of the composite need to be evaluated. This modification is required to enhance the dispersion and interfacial adhesion between NC and polyolefin as it appears to determine the overall nanocomposite properties. By improving these characteristics, many more conventional polyolefins products can be integrated with NC material that can promote wider application use in sustainability, environmentally friendly, and greener approaches.

REFERENCES

1. Zhong, Y., Godwin, P., Jin, Y., & Xiao, H. (2020). Biodegradable polymers and green-based antimicrobial packaging materials: A mini-review. *Advanced Industrial and Engineering Polymer Research*, *3*(1), 27–35.
2. Jagadeesh, P., Puttegowda, M., Mavinkere Rangappa, S., & Siengchin, S. (2021). Influence of nanofillers on biodegradable composites: A comprehensive review. *Polymer Composites*, *42*(11), 5691–5711.
3. de Carvalho, A. P. A., & Junior, C. A. C. (2020). Green strategies for active food packaging: A systematic review on active properties of graphene-based nanomaterials and biodegradable polymers. *Trends in Food Science & Technology*, *103*, 130–143.
4. Silva, F. A., Dourado, F., Gama, M., & Poças, F. (2020). Nanocellulose bio-based composites for food packaging. *Nanomaterials*, *10*(10), 2041.
5. Akpan, E. I., Shen, X., Wetzel, B., & Friedrich, K. (2019). Design and synthesis of polymer nanocomposites. In Krzysztof Pielichowski and Tomasz M. Majka (Eds.), *Polymer Composites with Functionalized Nanoparticles* (pp. 47–83). Elsevier.

6. Devnani, G. L., & Sinha, S. (2019). Effect of nanofillers on the properties of natural fiber reinforced polymer composites. *Materials Today: Proceedings*, *18*, 647–654.

7. Kisiel, M., & Mossety-Leszczak, B. (2020). Development in liquid crystalline epoxy resins and composites-A review. *European Polymer Journal*, *124*, 109507.

8. Kumar, A. P., Depan, D., Tomer, N. S., & Singh, R. P. (2009). Nanoscale particles for polymer degradation and stabilization-trends and future perspectives. *Progress in Polymer Science*, *34*(6), 479–515.

9. Fu, S., Sun, Z., Huang, P., Li, Y., & Hu, N. (2019). Some basic aspects of polymer nano-composites: A critical review. *Nano Materials Science*, *1*(1), 2–30.

10. Ilyas, R. A., Sapuan, S. M., Atikah, M. S. N., Ibrahim, R., Hazrol, M. D., Sherwani, S. F. K., ... & Syafiq, R. (2020). Natural fibre: A promising source for the production of nanocellulose. In Sapuan S.M., Ilyas R.A. (Eds.), *Proceedings of the 7th Postgraduate Seminar on Natural Fibre Reinforced Polymer Composites* (pp. 14–21)). Institute of Tropical Forestry and Forest Products (INTROP).

11. Ilyas, R. A., Aisyah, H. A., Nordin, A. H., Ngadi, N., Zuhri, M. Y. M., Asyraf, M. R. M., ... & Ibrahim, R. (2022). Natural-fiber-reinforced chitosan, chitosan blends and their nanocomposites for various advanced applications. *Polymers*, *14*(5), 874.

12. Shankar, S., & Rhim, J. W. (2018). Bionanocomposite films for food packaging applications. *Reference Module in Food Science*, *1*, 1–10.

13. Seo, Y., Kang, T., Hong, S. M., & Choi, H. J. (2007). Nonisothermal crystallization behaviors of a polyolefin terpolymer and its foam. *Polymer*, *48*(13), 3844–3849.

14. Zuber, M., Zia, K. M., Noreen, A., Bukhari, S. A., Aslam, N., Sultan, N., ... & Shi, B. (2017). Algae-based polyolefins. In Khalid Mahmood Zia, Mohammad Zuber and Muhammad Ali (Eds.), *Algae Based Polymers, Blends, and Composites* (pp. 499–529). Elsevier.

15. Nevares, I., & del Alamo-Sanza, M. (2018). New materials for the aging of wines and beverages: Evaluation and comparison. In Alexandru Mihai Grumezescu and Alina Maria Holban (Eds.) *Food Packaging and Preservation* (pp. 375–407). Academic Press.

16. Aumnate, C., Rudolph, N., & Sarmadi, M. (2019). Recycling of polypropylene/polyethylene blends: Effect of chain structure on the crystallization behaviors. *Polymers*, *11*(9), 1456.

17. McCartney, I. (2019). Advantages and Disadvantages of Polyethylene. Available online: https://kempner.co.uk/2019/05/08/the-advantages-and-disadvantages-of-polyethylene-blog/ (Accessed on 12 Jan 2023).

18. McCartney, I. (2019). Advantages and Disadvantages of Polyethylene. Available online: https://kempner.co.uk/2019/04/16/advantages-and-disadvantages-of-polypropylene-blog/ (Accessed on 12 Jan 2023).

19. Ames Rubber Manufacturing (2022). Available online: https://www.amesrubberonline.com/pdf/polyisobutylene-rubber-butyl.pdf (Accessed on 12 Jan 2023).

20. Greene, J. P. (2021). Elastomers and rubbers. In Greene, J. P. (Ed). *Automotive Plastics and Composites: Materials and Processing*. William Andrew, Elsevier, Amsterdam.

21. AZO Materials (2021). Polystyrene - PS. https://www.azom.com/article.aspx?ArticleID=798 (Accessed on 12 Jan 2023).

22. Yang, H. S., Kim, H. J., Park, H. J., Lee, B. J., & Hwang, T. S. (2006). Water absorption behavior and mechanical properties of lignocellulosic filler-polyolefin bio-composites. *Composite Structures*, *72*(4), 429–437.

23. Jamróz, E., Kulawik, P., & Kopel, P. (2019). The effect of nanofillers on the functional properties of biopolymer-based films: A review. *Polymers*, *11*(4), 675.

24. Gao, Y., Wang, Q., Wang, J., Huang, L., Yan, X., Zhang, X., ... & Guo, Z. (2014). Synthesis of highly efficient flame retardant high-density polyethylene nanocomposites with inorgano-layered double hydroxides as nanofiller using solvent mixing method. *ACS Applied Materials & Interfaces*, *6*(7), 5094–5104.

25. Osman, A. F., Berhanuddin, S., Halim, K. A. A., & Haq, N. (2021 May). Tensile and tear properties of high density polyethylene/organo-fluoromica nanocomposites. In *AIP Conference Proceedings* (Vol. *2339*, No. 1, p. 020080). AIP Publishing LLC.

26. Dias, O. A. T., Konar, S., Leão, A. L., Yang, W., Tjong, J., Jaffer, S., ... & Sain, M. (2021). Clean manufacturing of nanocellulose-reinforced hydrophobic flexible substrates. *Journal of Cleaner Production*, *293*, 126141.

27. Omran, A. A. B., Mohammed, A. A., Sapuan, S. M., Ilyas, R. A., Asyraf, M. R. M., Rahimian Koloor, S. S., & Petrů, M. (2021). Micro-and nanocellulose in polymer composite materials: A review. *Polymers*, *13*(2), 231.

28. Boran, S., Kiziltas, A., Kiziltas, E., & Gardner, D. (2016). Characterization of ultrafine cellulose-filled high density polyethylene composites prepared using different compounding methods. *BioResources*, *11*(4): 8178–8199.

29. Pandey, J. K., Lee, H. T., Takagi, H., Ahn, S. H., Saini, D. R., & Misra, M. (2015). Dispersion of nanocellulose (NC) in polypropylene (PP) and polyethylene (PE) matrix. In Pandey, J., Takagi, H., Nakagaito, A., Kim, H.J. (Eds.), *Handbook of Polymer Nanocomposites. Processing, Performance and Application* (pp. 179–189). Springer, Berlin, Heidelberg.

30. Noguchi, T., Niihara, K. I., Iwamoto, R., Matsuda, G. I., Endo, M., & Isogai, A. (2021). Nanocellulose/polyethylene nanocomposite sheets prepared from an oven-dried nanocellulose by elastic kneading. *Composites Science and Technology*, *207*, 108734.

31. Alavitabari, S., Mohamadi, M., Javadi, A., & Garmabi, H. (2021). The effect of secondary nanofiller on mechanical properties and formulation optimization of HDPE/nanoclay/nanoCaCO3 hybrid nanocomposites using response surface methodology. *Journal of Vinyl and Additive Technology*, *27*(1), 54–67.

32. Siročić, A. P., Rešček, A., Ščetar, M., Krehula, L. K., & Hrnjak-Murgić, Z. (2014). Development of low density polyethylene nanocomposites films for packaging. *Polymer Bulletin*, *71*(3), 705–717.

33. Saman, N. M., Awang, N. A., Ahmad, M. H., Buntat, Z., Adzis, Z. (2021). Partial discharge characteristics of low-density polyethylene nanocomposites incorporated with plasma-treated silica and boron nitride nanofillers. In *Proceedings of the 2021 3rd International Conference on High Voltage Engineering and Power Systems (ICHVEPS), 5–6 Oct. 2021*, pp. 518–523. Institute of Electrical and Electronics Engineers (IEEE).

34. Redhwi, H. H., Siddiqui, M. N., Andrady, A. L., & Hussain, S. (2013). Durability of LDPE nanocomposites with clay, silica, and zinc oxide-Part I: Mechanical properties of the nanocomposite materials. *Journal of Nanomaterials*, *2013*: 1–6.

35. Yasim-Anuar, T. A. T., Ariffin, H., Norrrahim, M. N. F., Hassan, M. A., Andou, Y., Tsukegi, T., & Nishida, H. (2020). Well-dispersed cellulose nanofiber in low density polyethylene nanocomposite by liquid-assisted extrusion. *Polymers*, *12*(4), 927.

36. Vartiainen, J., Pasanen, S., Kenttä, E., & Vähä-Nissi, M. (2018). Mechanical recycling of nanocellulose containing multilayer packaging films. *Journal of Applied Polymer Science*, *135*(19), 46237.

37. An, J. E., Jeon, G. W., & Jeong, Y. G. (2012). Preparation and properties of polypropylene nanocomposites reinforced with exfoliated graphene. *Fibers and Polymers*, *13*(4), 507–514.

38. Liu, Y., Zhang, S., Wang, X., Pan, Y., Zhang, F., & Huang, J. (2020). Mechanical and aging resistance properties of polypropylene (PP) reinforced with nanocellulose/attapulgite composites (NCC/AT). *Composite Interfaces*, *27*(1), 73–85.

39. Bharimalla, A. K., Patil, P. G., Mukherjee, S., Yadav, V., & Prasad, V. (2019). Nanocellulose-polymer composites: Novel materials for food packaging applications. In Tomy J. Gutiérrez (Ed.), *Polymers for Agri-Food Applications* (pp. 553–599). Springer, Cham.

40. Butschli, J. (2016 January 19). Modest growth for global packaging demand through 2020. *Packaging World.* https://www.packworld.com/design/package-design/news/13369500/modest-growth-for-global-packaging-demand-through-2020.

41. Ahmadi, M., Behzad, T., & Bagheri, R. (2017). Reinforcement effect of poly (methyl methacrylate)-g-cellulose nanofibers on LDPE/thermoplastic starch composites: Preparation and characterization. *Iranian Polymer Journal, 26*(10), 733–742.

42. Fotie, G., Limbo, S., & Piergiovanni, L. (2020). Manufacturing of food packaging based on nanocellulose: Current advances and challenges. *Nanomaterials, 10*(9), 1726.

43. Garces, J. M., Moll, D. J., Bicerano, J., Fibiger, R., & McLeod, D. G. (2000). Polymeric nanocomposites for automotive applications. *Advanced Materials, 12*(23), 1835–1839.

44. Gray, N., Hamzeh, Y., Kaboorani, A., & Abdulkhani, A. (2018). Influence of cellulose nanocrystal on strength and properties of low-density polyethylene and thermoplastic starch composites. *Industrial Crops and Products, 115*, 298–305.

45. Lundin, T., Cramer, S. M., Falk, R. H., & Felton, C. (2004). Accelerated weathering of natural fiber-filled polyethylene composites. *Journal of Materials in Civil Engineering, 16*(6), 547–555.

46. Ołdak, D., Kaczmarek, H., Buffeteau, T., & Sourisseau, C. (2005). Photo- and bio-degradation processes in polyethylene, cellulose and their blends studied by ATR-FTIR and Raman spectroscopies. *Journal of Materials Science, 40*(16), 4189–4198.

47. Djordjevic, N., Marinkovic, A., Nikolic, J., Drmanic, S., Rancic, M., Brkovic, D., & Uskokovic, P. (2016). A study of the barrier properties of polyethylene coated with a nanocellulose/magnetite composite film. *Journal of the Serbian Chemical Society, 81*(5), 589–605.

48. Hubbe, M. A., Ferrer, A., Tyagi, P., Yin, Y., Salas, C., Pal, L., & Rojas, O. J. (2017). Nanocellulose in thin films, coatings, and plies for packaging applications: A review. *BioResources, 12*(1), 2143–2233.

49. Sam, S. T., Nuradibah, M. A., Ismail, H., Noriman, N. Z., & Ragunathan, S. (2014). Recent advances in polyolefins/natural polymer blends used for packaging application. *Polymer-Plastics Technology and Engineering, 53*(6), 631–644.

50. dos Santos, F. A., Iulianelli, G. C. V., & Tavares, M. I. B. (2016). The use of cellulose nanofillers in obtaining polymer nanocomposites: Properties, processing, and applications. *Materials Sciences and Applications, 07*(05), 257–294.

51. Nair, S. S., Zhu, J., Deng, Y., & Ragauskas, A. J. (2014). High performance green barriers based on nanocellulose. *Sustainable Chemical Processes, 2*(1), 23.

52. Chirayil, C. J., Joy, J., Maria, H. J., Krupa, I., & Thomas, S. (2016). Polyolefins in automotive industry. In Mariam Al-Ali AlMa'adeed and Igor Krupa (Eds.), *Polyolefin Compounds and Materials: Fundamentals and Industrial Applications.* Springer International Publishing.

53. Lyu, M.-Y., & Choi, T. G. (2015). Research trends in polymer materials for use in light-weight vehicles. *International Journal of Precision Engineering and Manufacturing, 16*(1), 213–220.

54. Sadiku, E. (2009). Automotive components composed of polyolefins. In Samuel C.O. Ugbolue (Ed.), *Polyolefin Fibres* (pp. 81–132). Elsevier.

55. Sarasini, F., Tirillò, J., Sergi, C., Seghini, M. C., Cozzarini, L., & Graupner, N. (2018). Effect of basalt fibre hybridisation and sizing removal on mechanical and thermal properties of hemp fibre reinforced HDPE composites. *Composite Structures, 188*, 394–406.

56. Kausar, A. (2021). Progress in green nanocomposites for high-performance applications. *Materials Research Innovations, 25*(1), 53–65.

57. Jayaweera, C. D., Karunaratne, D. W. T. S., Bandara, S. T. S., & Walpalage, S. (2017). Investigation of the effectiveness of nanocellulose extracted from Sri Lankan Kapok, as a filler in polypropylene polymer matrix. In *2017 Moratuwa Engineering Research Conference (MERCon)* (pp. 1–6). Institute of Electrical and Electronics Engineers (IEEE), New York City.

58. Ljungberg, N., Bonini, C., Bortolussi, F., Boisson, C., Heux, L., & Cavaillé. (2005). New nanocomposite materials reinforced with cellulose whiskers in atactic polypropylene: Effect of surface and dispersion characteristics. *Biomacromolecules, 6*(5), 2732–2739.

59. Ljungberg, N., Cavaillé, J.-Y., & Heux, L. (2006). Nanocomposites of isotactic polypropylene reinforced with rod-like cellulose whiskers. *Polymer, 47*(18), 6285–6292.

60. Al-Haik, M. Y., Aldajah, S., Siddique, W., Kabir, M. M., & Haik, Y. (2020). Mechanical and thermal characterization of polypropylene-reinforced nanocrystalline cellulose nanocomposites. *Journal of Thermoplastic Composite Materials, 35*(5), 680–691.

61. Cho, S. Y., Park, H. H., Yun, Y. S., & Jin, H.-J. (2013). Influence of cellulose nanofibers on the morphology and physical properties of poly(lactic acid) foaming by supercritical carbon dioxide. *Macromolecular Research, 21*(5), 529–533.

62. Dlouhá, J., Suryanegara, L., & Yano, H. (2014). Cellulose nanofibre-poly(lactic acid) microcellular foams exhibiting high tensile toughness. *Reactive and Functional Polymers, 85*, 201–207.

63. Huang, H.-X., & Wang, J.-K. (2007). Improving polypropylene microcellular foaming through blending and the addition of nano-calcium carbonate. *Journal of Applied Polymer Science, 106*(1), 505–513.

64. Mydin, A. O., & Soleimanzadeh, S. (2012). Effect of polypropylene fiber content on flexural strength of lightweight foamed concrete at ambient and elevated temperatures. *Advances in Applied Science Research, 3*(5), 2837–2846.

65. Ito, A., Semba, T., Kitagawa, K., Okumura, H., & Yano, H. (2019). Cell morphologies and mechanical properties of cellulose nanofiber reinforced polypropylene foams. *Journal of Cellular Plastics, 55*(4), 385–400.

66. Liu, Y., Zhang, S., Wang, X., Pan, Y., Zhang, F., & Huang, J. (2020). Mechanical and aging resistance properties of polypropylene (PP) reinforced with nanocellulose/attapulgite composites (NCC/AT). *Composite Interfaces, 27*(1), 73–85.

67. Joshy, K. S., Pothen, L. A., & Thomas, S. (2016). Biomedical applications of polyolefins. In Mariam Al-Ali AlMa'adeed and Igor Krupa (Eds.), *Polyolefin Compounds and Materials: Fundamentals and Industrial Applications*, pp. 247–264. Springer International Publishing.

68. Wang, S., Feng, Q., Sun, J., Gao, F., Fan, W., Zhang, Z., Li, X., & Jiang, X. (2016). Nanocrystalline cellulose improves the biocompatibility and reduces the wear debris of ultrahigh molecular weight polyethylene via weak binding. *ACS Nano, 10*(1), 298–306.

69. Tevet, O., Von-Huth, P., Popovitz-Biro, R., Rosentsveig, R., Wagner, H. D., & Tenne, R. (2011). Friction mechanism of individual multilayered nanoparticles. *Proceedings of the National Academy of Sciences, 108*(50), 19901–19906. https://doi.org/10.1073/pnas.1106553108.

70. Wang, L., Okada, K., Hikima, Y., Ohshima, M., Sekiguchi, T., & Yano, H. (2019). Effect of cellulose nanofiber (CNF) surface treatment on cellular structures and mechanical properties of polypropylene/CNF nanocomposite foams via core-back foam injection molding. *Polymers, 11*(2), 249.

71. Wang, L., Ando, M., Kubota, M., Ishihara, S., Hikima, Y., Ohshima, M., Sekiguchi, T., Sato, A., & Yano, H. (2017). Effects of hydrophobic-modified cellulose nanofibers (CNFs) on cell morphology and mechanical properties of high void fraction polypropylene nanocomposite foams. *Composites Part A: Applied Science and Manufacturing, 98*, 166–173.

72. Wang, L., Okada, K., Sodenaga, M., Hikima, Y., Ohshima, M., Sekiguchi, T., & Yano, H. (2018). Effect of surface modification on the dispersion, rheological behavior, crystallization kinetics, and foaming ability of polypropylene/cellulose nanofiber nanocomposites. *Composites Science and Technology, 168*, 412–419.

73. Touati, Z., Boulahia, H., Belhaneche-Bensemra, N., & Massardier, V. (2019). Modification of diss fibers for biocomposites based on recycled low-density polyethylene and polypropylene blends. *Waste and Biomass Valorization, 10*(8), 2365–2378.

74. Hao, W., Wang, M., Zhou, F., Luo, H., Xie, X., Luo, F., & Cha, R. (2020). A review on nanocellulose as a lightweight filler of polyolefin composites. *Carbohydrate Polymers, 243*, 116466.

75. Pandey, J. K., Lee, H. T., Takagi, H., Ahn, S. H., Saini, D. R., & Misra, M. (2015). Dispersion of nanocellulose (NC) in polypropylene (PP) and polyethylene (PE) matrix. In J. K. Pandey, H. Takagi, A. N. Nakagaito, & H.- J. Kim (Eds.), *Handbook of Polymer Nanocomposites. Processing, Performance and Application* (pp. 179–189). Springer, Berlin Heidelberg.

76. Saber-Samandari, S., & Saber-Samandari, S. (2017). Biocompatible nanocomposite scaffolds based on copolymer-grafted chitosan for bone tissue engineering with drug delivery capability. *Materials Science and Engineering: C, 75*, 721–732.

77. Gupta, A., Simmons, W., Schueneman, G. T., Hylton, D., & Mintz, E. A. (2017). Rheological and thermo-mechanical properties of poly(lactic acid)/lignin-coated cellulose nanocrystal composites. *ACS Sustainable Chemistry & Engineering, 5*(2), 1711–1720.

78. Huang, L., Ye, Z., & Berry, R. (2016). Modification of cellulose nanocrystals with quaternary ammonium-containing hyperbranched polyethylene ionomers by ionic assembly. *ACS Sustainable Chemistry & Engineering, 4*(9), 4937–4950.

79. Iyer, K. A., Schueneman, G. T., & Torkelson, J. M. (2015). Cellulose nanocrystal/polyolefin biocomposites prepared by solid-state shear pulverization: Superior dispersion leading to synergistic property enhancements. *Polymer, 56*, 464–475.

80. Sapkota, J., Natterodt, J. C., Shirole, A., Foster, E. J., & Weder, C. (2017). Fabrication and properties of polyethylene/cellulose nanocrystal composites. *Macromolecular Materials and Engineering, 302*(1), 1600300.

81. Ferreira, F., Pinheiro, I., de Souza, S., Mei, L., & Lona, L. (2019). Polymer composites reinforced with natural fibers and nanocellulose in the automotive industry: A short review. *Journal of Composites Science, 3*(2), 51.

82. Abitbol, T., Rivkin, A., Cao, Y., Nevo, Y., Abraham, E., Ben-Shalom, T., Lapidot, S., & Shoseyov, O. (2016). Nanocellulose, a tiny fiber with huge applications. *Current Opinion in Biotechnology, 39*, 76–88.

83. Bhutto, A. W., Qureshi, K., Harijan, K., Abro, R., Abbas, T., Bazmi, A. A., Karim, S., & Yu, G. (2017). Insight into progress in pre-treatment of lignocellulosic biomass. *Energy, 122*, 724–745.

84. Thomas, P., Duolikun, T., Rumjit, N. P., Moosavi, S., Lai, C. W., Bin Johan, M. R., & Fen, L. B. (2020). Comprehensive review on nanocellulose: Recent developments, challenges and future prospects. *Journal of the Mechanical Behavior of Biomedical Materials, 110*, 103884.

85. Mathew, A. P., & Oksman, K. (2014). Bionanomaterials: Separation processes, characterization, and properties. In K. Oksman, A. P. Mathew, A. Bismarck, O. Rojas, & M. Sain, *Materials and Energy* (Vol. 5, pp. 1–3). World Scientific.

86. Chakrabarty, A., & Teramoto, Y. (2018). Recent advances in nanocellulose composites with polymers: A guide for choosing partners and how to incorporate them. *Polymers, 10*(5), 517.

16 Nanofillers in Food Packaging

A. Nazrin, R. M. O. Syafiq, S. M. Sapuan,
M. Y. M. Zuhri, and I. S. M. A. Tawakkal
Universiti Putra Malaysia

R. A. Ilyas
Universiti Teknologi Malaysia

16.1 INTRODUCTION

Nanofillers have gained intense interest as an ideal reinforcement material since they are capable of producing high performance materials just by incorporating a small quantity of nanoparticles in a matrix. There are various types of nanofillers incorporated into polymer material to modify its properties including starch nanocrystals, cellulose nanofibers, chitin nanoparticles, silica nanoparticles, and carbon nanotubes. The outstanding capabilities of these reinforcements are well known and broadly adapted into wide range applications such as aerospace, automotive, building, construction, electronic, medical, and packaging. In the last two decades, the trend of nanocomposites in food packaging had been rising with the number of publications increased five-fold during the period 2010–2019. In its pristine form, polymers exhibit different properties factorize by their cross-linkage, degree of crystallinity, degree of branching, hydrogen bonding, and polarity (Sarfraz et al., 2020). In example, high-density polyethylene (HDPE) offers a better water vapor barrier compared to polyethylene terephthalate (PET) but provides superior oxygen barrier compared to HDPE. Hence, the introduction of these nanofillers is intended to enhance the functional aspects of food packaging materials. In compliance with the green movement, industries are shifting toward bio-based resources to minimize and if possible, to cease the production of fossil-based plastics. The focus on sustainability and biodegradability propelled the demand of plastic packaging from renewable resources. Although bio-based materials had been commercialized in the market, there are still inferior to current petroleum-based plastics in

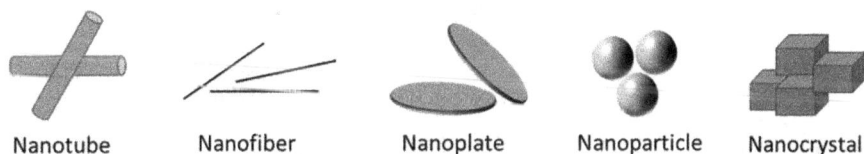

| Nanotube | Nanofiber | Nanoplate | Nanoparticle | Nanocrystal |

FIGURE 16.1 The shape and size of various nanofillers (Sarfraz et al., 2020).

DOI: 10.1201/9781003400998-16

terms of mechanical, barrier and thermal properties. Depending on the intended application, it is obligatory for food packaging to withstand high stress with deformation.
The application of nanofillers instead of micro- or macro-sized fillers had received
significant research interest due to their unique characteristics such as large surface
area (100 m²/g), high surface to volume ratio of 100, high crystallinity, lightweight, and
excellent mechanical properties, which are deemed to be promising in the characterization of superior and high-end material of multipurpose applications (Ilyas et al., 2018).

16.2 STARCH NANOCRYSTAL BASED COMPOSITES FOR PACKAGING

Starch nanocrystal (SN) can be isolated from the crystalline region of polysaccharides using acid hydrolysis in the form of platelet-like with the thickness around
6–8 nm, length of 20–40 nm, and a width of 15–30 nm (Piyada et al., 2013). Figure
16.2 elucidates microstructure of waxy maize starch and maize starch nanocrystals.
It is an excellent nanofiller for the improvement of polymeric materials particularly
the mechanical properties, thermal stability, water and gas permeability barrier due
to its high crystallinity value. Li et al. (2015) reported an improvement in the tensile
strength (72.9%), elastic modulus (305.3%), and water vapor permeability (162.4%)
of pea starch films through the reinforcement 5% of SN compared with the control
film. The well-dispersed nanofillers occupied the porous spaces in the matrix to form
a more compact structure with a tortuous path that improved the stress transfer and
decreased water vapor diffusivity. Figure 16.3 illustrates the tensile fracture surface
of PS/SN composite films. Correspondingly, the higher compatibility between pea
starch and SN results in higher thermal degradation temperature recorded through

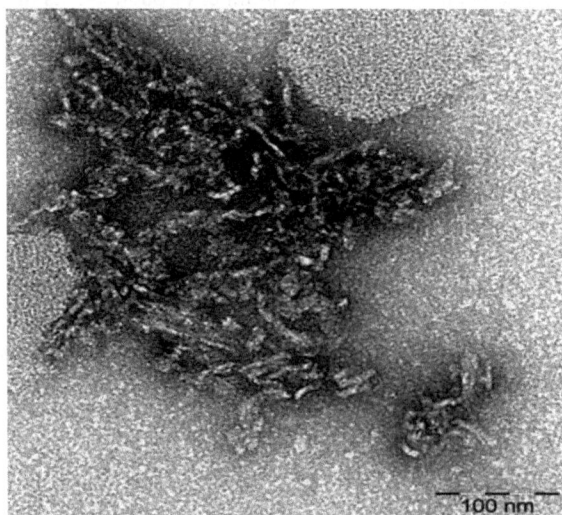

FIGURE 16.2 The transmission electron microscope (TEM) image of (a) waxy maize
starch and (b) the field emission scanning electron microscope (FESEM) image of starch
nanocrystals (García et al., 2011).

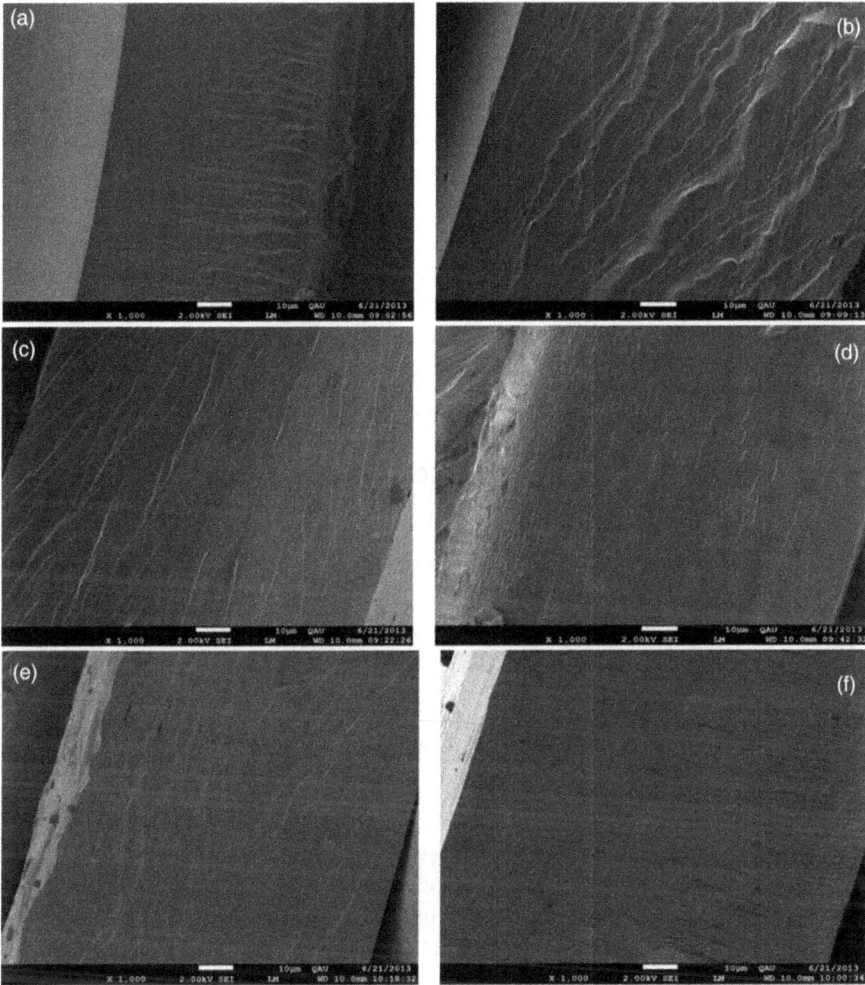

FIGURE 16.3 The SEM images of tensile fracture surface of PS/SN with different SN loading (a) 0%, (b) 1%, (c) 3%, (d) 5%, (e) 7%, (f) 9% (Li et al., 2015).

thermal gravimetric analysis. Viguié et al. (2007) found out that waxy maize starch film had its tensile strength increased from 0.38 to 0.99 MPa and elastic modulus from 17.2 to 36.6 MPa when reinforced with 5% of SN. Surprisingly, one week of aging at 88% relative humidity (RH) significantly increases both tensile strength and elastic modulus to 4.47 and 120.7 MPa, respectively. The usage of sorbitol as a plasticizer plays an important role in increasing the chain mobility if amorphous region and their ability to crystallize. At moist condition, the crystalline regions act as physical cross-links strengthening the material. Similarly, Piyada et al. (2013) stated that the tensile strength and water vapor permeability of rice starch film was improved through the reinforcement of SN. Though, both authors experienced reduction in elongation at break by utilizing sorbitol as plasticizer. Study by García et al. (2011) recorded an increment in water vapor permeability from 3.8×10^{-10} to 6.8×10^{-10} g/s m^2 Pa with the addition of 2.5% of SN into waxy maize starch film. The opposite effect of the incorporation of nanofiller is ascribed to the formation glycerol-nanocrystals bonds that established paths of high OH concentration favoring the water vapor diffusion through the film. This was also mentioned Sanyang et al. (2015) that lower water vapor permeability was recorded by using sorbitol compared to glycerol on sugar palm starch films. In case of synthetic biopolymer such as poly (butylene succinate) (PBS), Lin et al. (2011) discovered that the threshold amount of SN reinforcement in order to achieve highest tensile and elongation at break is around 5%. The melting temperature of the nanocomposites also increased from 113.3°C to 117.2°C. The nucleation of rigid nanocrystals favors the crystallinity of the nanocomposites to hinder the free motion of PBS molecular chains, which in need of additional energy for the thermal transformation.

16.3   CELLULOSE NANOFIBER-BASED COMPOSITES FOR PACKAGING

The utilization of cellulose nanostructures has been adapted in food sector especially packaging films. There are three classification of cellulose products that are currently used in food packaging namely crystalline nanocellulose (CNC), fibrillated nanocellulose (FNC), and bacterial nanocellulose (BNC). The reinforcement of cellulose nanostructures into polymer matrices primarily focuses on the tensile strength and water sensitivity as well as the effect of mechanical percolation yielded by the strong hydrogen bonding. The aim is to achieve optimum amount of filler for a well dispersion in establishing a continuous structure within the matrix. Besides that, polymer impregnated with CNC has been shown to hinder the wide spreading of foodborne pathogens by several authors. The presence of CNC in bioactive polymer films ensures a controlled release of bioactive compound such as essential oils to maintain a functionality of the films active for a longer period. Salmieri et al. (2014) reported that PLA/CNC films infused with oregano essential oil induced a quasi-total inhibition of *L. monocytogenes* in the vegetable samples. During 14 days of storage, the total phenol (TP) release was decreased from 55.4 to 47.4 µg Gallic Acid Equivalent (GAE)/mg, indicating a slow diffusion rate beneficial for long-term food storage. Bagde and Nadanathangam (2019) incorporated bacteriocins from lactic acid (*P. acidilactici* and *E. faecium*) on the surface of CNC to act as antibacterial agent in corn starch film and observed that the films incorporated with bacteriocin-immobilized CNC stayed fresh for 28 days, while the one with bacteriocin alone had fungal infection in 14 days. FNC also exhibited quite similar ability of regulating the release of loaded antibacterial agent owing to the nanoporous networks formed with the chemical reaction of the loaded antibacterial agent. It was stated by López-Rubio et al. (2007) that the incorporation of FNC in amylopectin films acts as a conventional reinforcement (strength and modulus) and promoted elasticity. The FNC capacity to retain moisture promoted the elongation of the films around 2%–7%, which was achievable for amylopectin films if 38% glycerol was added. A study by Iwatake et al. (2008) on PLA reinforced with

10% of FNC demonstrated 25% and 40% increases of tensile strength and Young's modulus, respectively, without deterioration of elongation. The percolation effect of FNC created additional entanglement structure of the nanocomposites even at low nanofiller content in restraining material deformation. As the name suggested, BC is basically synthesized through specific bacteria species in controlled environment. Similar to CNC and FNC, BNC must go through surface modification such as acetylation, oxidation, polymer grafting, and silanization to overcome incompatibility and poor dispersion when incorporated into hydrophobic matrixes. Martínez-Sanz et al. (2012) identified that melt mixing PLA/BNC masterbatch (prepared through electrospinning and solution casting) improved thermal and barrier properties of the final product compared with direct melt mixing of PLA/BNC. At 3 wt.% loading, water permeability was decreased up to 43% compared to pristine PLA. Lowering the surface hydrophilicity of BCN via acetylation is also proved to enhance its compatibility due to the reduction interfacial tension with hydrophobic matrixes. Ambrosio-Martín et al. (2015) observed a positive effect on barrier properties of PLA/BNC via melt compounding of lactic acid oligomer grafted BNC with PLA. Oligomer grafted into BNC plays an important role in occupying the free volume, which lowered the oxygen permeability rate.

16.4 CHITIN NANOPARTICLE-BASED COMPOSITES FOR PACKAGING

Chitin is a linear copolymer of N-acetyl-glucosamine and N-glucosamine, with a β-1,4 linkage commonly extracted from the exoskeletons of crustaceans (crabs, lobsters and shrimps) and considered the most abundant polysaccharide next to cellulose (Laycock et al., 2017). The utilization of chitin in food packaging application had gained numerous intentions due to its excellent features such as biocompatible, biodegradable, and non-toxic (Llanos & Tadini, 2018). In addition, chitin or chitosan is regarded as a component of packaging material with the properties to retard microorganism growth consequently preserving the food quality and shelf life. Soares et al. (2013) coated thermoplastic starch (TPS)/poly (lactic acid) (PLA) blend sheets using 0.1% (w/v) of chitosan by immersion and spraying methods. The results indicated that spraying method was more effective than immersion in improving the tensile strength from 1.7 to 3.7 MPa, Young's modulus from 60 MPa to 422 MPa, and water vapor permeability from 1.8×10^{-6} to 1.2×10^{-6} g/h m Pa. Kaisangsri et al. (2012) improved the tensile strength from 699.8 to 742. 2 kPa, elongation at break from 1.49% to 2.29%, and water solubility index from 12.93% to 6.23% of cassava starch foam by incorporating 4% (w/v) chitosan. The coagulation chitosan with starch triggered the formation of intermolecular hydrogen bonding between amino groups and hydroxyl groups, thus strengthening the tensile and elongation of the foam. Noorbakhsh-Soltani et al. (2018) indicated that enrichment of chitosan concentration leads to an enhancement in the tensile strength, Young's modulus, transparency of gelatine films and food preservation, contact angle, UV transmittance of starch films. Bie et al. (2013) explained that the infusion of chitosan in PLA/TPS blend films acted as an antimicrobial agent to inhibit the putrefying bacteria of *E. coli* and *S. aures* in fresh meat. Although the addition of TPS severely weakens the mechanical properties of the film, it activates a slow and continuous diffusion of the chitosan maintaining the long residual action of antimicrobial property. According to Gómez-Estaca et al. (2010), the antimicrobial capacity of chitosan infused polymer fluctuates depending on the preparation methods, composites formulation and bioactive compounds release. The hydrogen bonding of gelatine–chitosan films stabilizes the film structure to regulate a lesser diffusivity of the bioactive compound thus maintaining its integrity to a greater extent when exposed to bacterial strains. Pelissari et al. (2009) also mentioned that although the addition of chitosan into starch films was not effective against *B. cereus*, *E. coli*, *S. enteritidis*, and *S. aureus*, the film demonstrated an improvement in tensile strength, elongation at break, and water vapor permeability. The incorporation of hydrophobic oregano essential oil to induce antimicrobial property facilitates the chain mobility and further increases the film flexibility and water vapor permeability.

16.5   CLAY NANOPARTICLE-BASED COMPOSITES FOR PACKAGING

Clay minerals are also have been practiced as filler in polymers to induce strengthening effects. Clay mineral is visualized as a sheet-like structure of silicates. The strengthening effects of nanoclays enhance thermal and water vapor permeability of nanocomposites. Montmorillonite (MMT)-based nanoclays have been utilized for the reinforcing purpose as they are easily available, toxic-free, and environmentally friendly, suitable for food packaging application. MMT nanoparticles consist of plate-like structure with surface dimension ranging from 300 to 600 nm and a mean diameter of 1 μm (Alias et al., 2021). Ayana et al. (2014) reported that prior dispersion of sodium montmorillonite (NaMMT) into TPS stabilizes the blend structure PLA/TPS in effort to reduce the deterioration of mechanical, thermal, and barrier properties. Surface modification is often applied to improve the affinity between hydrophobic polymer and hydrophilic nanoclays. In a study by Al-Samhan et al. (2017), polypropylene (PP)/imidazolium montmorillonite (IMMT) indicated better thermal stability compared to PP/ammonium montmorillonite (AMMT). The effectiveness of surfactant is affecting the dispersion of the nanoclays in the polymer to achieve decent morphology structure. The dispersion of nanoclay can be classified into three stages namely exfoliated, intercalated, and phase separated as shown in Figure 16.4. Madaleno et al. (2010) compared the reinforcement of natural sodium montmorillonite (NaMMT) and organically modified montmorillonite (OMMT) to polyvinyl chloride (PVC). The solution blending method used in compounding the nanocomposites seemed to induced a fine dispersion of intercalated and exfoliated structures. At 2 phr, PVC/NaMMT had better tensile strength and Young's modulus compared to PVC/OMMT. Thermal stability of PVC/NaMMT showed higher thermal degradation temperature and lower decomposition rate attributed to the effect of MMT as a thermal insulation hindering the volatile decomposition rate.

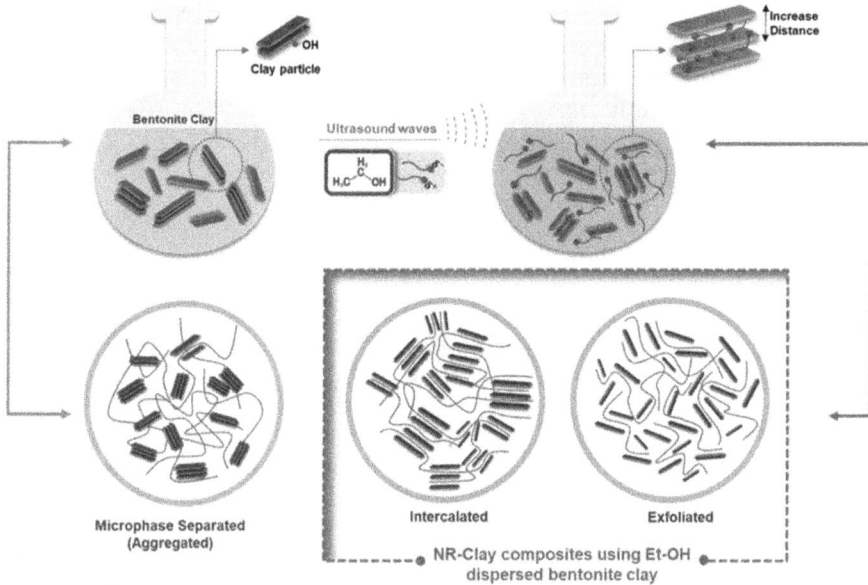

FIGURE 16.4 The dispersion stages of nanoclay in polymer (Keereerak et al., 2022).

16.6 CHALLENGES AND FUTURE RECOMMENDATIONS

The innovation of eco-friendly packaging based on bioresources promises interesting opportunities in ensuring food safety and quality. However, the hydrophilic nature of these nanocomposites tends to absorb excessive moisture and thus accelerated the deterioration of mechanical performance. Plus, the poor compatibility between hydrophilic filler and hydrophobic matrix could not achieve the desired structure in improving the mechanical strength and barrier properties. In effort to anticipate the satisfactory packaging material, it is essential to address the compatibility issue. Thus, it is essential to employ surface modification to promote fine and uniform dispersion of nanofiller into the matrix. Moreover, the introduction of supplementary component such as phenolic compound into the composites added auxiliary value in widening food packaging application.

16.7 SUMMARY

A good dispersion of nanofillers established a continuous structure of nanocomposites. Fiber with high surface area to volume had significant effect in reinforcing the matrix. Chemical treatments are applied to modify fiber crystalline structure through stripping weak components like hemicellulose and lignin. Regardless of nanofillers, nanocomposites achieved enhancement in tensile strength and thermal stability but resulted in lower elongation at break due to addition of stiff fiber. An excellent interfacial adhesion between matrix and fiber promoted efficient stress transfer and resistant against heat decomposition. The compact and complex structure can hinder the vapor diffusion to pass through it. Furthermore, nanofillers extracted from bio-resources are non-toxic, renewable, and biodegradable, thus making it suitable for packaging applications. It helps to control the release of active compound (phenols) to prolong preservative duration of the packaging film.

REFERENCES

Al-Samhan, M., Samuel, J., Al-Attar, F., & Abraham, G. (2017). Comparative effects of MMT clay modified with two different cationic surfactants on the thermal and rheological properties of polypropylene nanocomposites. *International Journal of Polymer Science*, *2017*, 1–8. https://doi.org/10.1155/2017/5717968.

Alias, A. H., Norizan, M. N., Sabaruddin, F. A., Asyraf, M. R. M., Norrrahim, M. N. F., Ilyas, A. R., Kuzmin, A. M., Rayung, M., Shazleen, S. S., Nazrin, A., Sherwani, S. F. K., Harussani, M. M., Atikah, M. S. N., Ishak, M. R., Sapuan, S. M., & Khalina, A. (2021). Hybridization of MMT/lignocellulosic fiber reinforced polymer nanocomposites for structural applications: A review. *Coatings*, 11(11), 1355. https://doi.org/10.3390/coatings11111355.

Ambrosio-Martín, J., Fabra, M. J., Lopez-Rubio, A., & Lagaron, J. M. (2015). Melt polycondensation to improve the dispersion of bacterial cellulose into polylactide via melt compounding: Enhanced barrier and mechanical properties. *Cellulose*, 22(2), 1201–1226. https://doi.org/10.1007/s10570-014-0523-9.

Ayana, B., Suin, S., & Khatua, B. B. (2014). Highly exfoliated eco-friendly thermoplastic starch (TPS)/poly (lactic acid)(PLA)/clay nanocomposites using unmodified nanoclay. *Carbohydrate Polymers*, *110*, 430–439. https://doi.org/10.1016/j.carbpol.2014.04.024.

Bagde, P., & Nadanathangam, V. (2019). Mechanical, antibacterial and biodegradable properties of starch film containing bacteriocin immobilized crystalline nanocellulose. *Carbohydrate Polymers*, 222(April), 115021. https://doi.org/10.1016/j.carbpol.2019.115021.

Bie, P., Liu, P., Yu, L., Li, X., Chen, L., & Xie, F. (2013). The properties of antimicrobial films derived from poly(lactic acid)/starch/chitosan blended matrix. *Carbohydrate Polymers*, 98(1), 959–966. https://doi.org/10.1016/j.carbpol.2013.07.004.

García, N. L., Ribba, L., Dufresne, A., Aranguren, M., & Goyanes, S. (2011). Effect of glycerol on the morphology of nanocomposites made from thermoplastic starch and starch nanocrystals. *Carbohydrate Polymers*, 84(1), 203–210. https://doi.org/10.1016/j.carbpol.2010.11.024.

Gómez-Estaca, J., López de Lacey, A., López-Caballero, M. E., Gómez-Guillén, M. C., & Montero, P. (2010). Biodegradable gelatin-chitosan films incorporated with essential oils as antimicrobial agents for fish preservation. *Food Microbiology*, 27(7), 889–896. https://doi.org/10.1016/j.fm.2010.05.012.

Ilyas, R. A., Sapuan, S. M., Ishak, M. R., & Zainudin, E. S. (2018). Development and characterization of sugar palm nanocrystalline cellulose reinforced sugar palm starch bionanocomposites. *Carbohydrate Polymers*, 202, 186–202. https://doi.org/10.1016/j.carbpol.2018.09.002.

Iwatake, A., Nogi, M., & Yano, H. (2008). Cellulose nanofiber-reinforced polylactic acid. *Composites Science and Technology*, 68(9), 2103–2106. https://doi.org/10.1016/j.compscitech.2008.03.006.

Kaisangsri, N., Kerdchoechuen, O., & Laohakunjit, N. (2012). Biodegradable foam tray from cassava starch blended with natural fiber and chitosan. *Industrial Crops and Products*, 37(1), 542–546. https://doi.org/10.1016/j.indcrop.2011.07.034.

Keereerak, A., Sukkhata, N., Lehman, N., Nakaramontri, Y., Sengloyluan, K., Johns, J. & Kalkornsurapranee, E. (2022). Development and characterization of unmodified and modified natural rubber composites filled with modified clay. *Polymers*, *14*(17), 3515. https://doi.org/10.3390/polym14173515

Laycock, B., Nikolić, M., Colwell, J. M., Gauthier, E., Halley, P., Bottle, S., & George, G. (2017). Lifetime prediction of biodegradable polymers. *Progress in Polymer Science*, *71*, 144–189. https://doi.org/10.1016/j.progpolymsci.2017.02.004.

Li, X., Qiu, C., Ji, N., Sun, C., Xiong, L., & Sun, Q. (2015). Mechanical, barrier and morphological properties of starch nanocrystals-reinforced pea starch films. *Carbohydrate Polymers*, *121*, 155–162. https://doi.org/10.1016/j.carbpol.2014.12.040.

Lin, N., Yu, J., Chang, P. R., Li, J., & Huang, J. (2011). Poly(butylene succinate)-based biocomposites filled with polysaccharide nanocrystals: Structure and properties. *Polymer Composites*, *32*(3), 472–482. https://doi.org/10.1002/pc.21066.

Llanos, J. H. R., & Tadini, C. C. (2018). Preparation and characterization of bio-nanocomposite films based on cassava starch or chitosan, reinforced with montmorillonite or bamboo nanofibers. *International Journal of Biological Macromolecules*, *107*(PartA), 371–382. https://doi.org/10.1016/j.ijbiomac.2017.09.001.

López-Rubio, A., Lagaron, J. M., Ankerfors, M., Lindström, T., Nordqvist, D., Mattozzi, A., & Hedenqvist, M. S. (2007). Enhanced film forming and film properties of amylopectin using micro-fibrillated cellulose. *Carbohydrate Polymers*, *68*(4), 718–727. https://doi.org/10.1016/j.carbpol.2006.08.008.

Madaleno, L., Schjødt-Thomsen, J., & Pinto, J. C. (2010). Morphology, thermal and mechanical properties of PVC/MMT nanocomposites prepared by solution blending and solution blending+melt compounding. *Composites Science and Technology*, *70*(5), 804–814. https://doi.org/10.1016/j.compscitech.2010.01.016.

Martínez-Sanz, M., Lopez-Rubio, A., & Lagaron, J. M. (2012). Optimization of the dispersion of unmodified bacterial cellulose nanowhiskers into polylactide via melt compounding to significantly enhance barrier and mechanical properties. *Biomacromolecules*, *13*(11), 3887–3899. https://doi.org/10.1021/bm301430j.

Noorbakhsh-Soltani, S. M., Zerafat, M. M., & Sabbaghi, S. (2018). A comparative study of gelatin and starch-based nano-composite films modified by nano-cellulose and chitosan for food packaging applications. *Carbohydrate Polymers*, *189*(February), 48–55. https://doi.org/10.1016/j.carbpol.2018.02.012.

Pelissari, F. M., Grossmann, M. V. E., Yamashita, F., & Pineda, E. A. G. (2009). Antimicrobial, mechanical, and barrier properties of cassava starch−chitosan films incorporated with oregano essential oil. *Journal of Agricultural and Food Chemistry*, *57*(16), 7499–7504. https://doi.org/10.1021/jf9002363.

Piyada, K., Waranyou, S., & Thawien, W. (2013). Mechanical, thermal and structural properties of rice starch films reinforced with rice starch nanocrystals. *International Food Research Journal*, *20*(1), 439–449.

Salmieri, S., Islam, F., Khan, R. A., Hossain, F. M., Ibrahim, H. M. M., Miao, C., Hamad, W. Y., & Lacroix, M. (2014). Antimicrobial nanocomposite films made of poly(lactic acid)-cellulose nanocrystals (PLA-CNC) in food applications-part B: Effect of oregano essential oil release on the inactivation of Listeria monocytogenes in mixed vegetables. *Cellulose*, *21*(6), 4271–4285. https://doi.org/10.1007/s10570-014-0406-0.

Sanyang, M. L., Sapuan, S. M., Jawaid, M., Ishak, M. R., & Sahari, J. (2015). Effect of plasticizer type and concentration on tensile, thermal and barrier properties of biodegradable films based on sugar palm (*Arenga pinnata*) starch. *Polymers*, *7*(6), 1106–1124. https://doi.org/10.3390/polym7061106.

Sarfraz, J., Gulin-Sarfraz, T., Nilsen-Nygaard, J., & Pettersen, M. K. (2020). Nanocomposites for food packaging applications: An overview. *Nanomaterials*, *11*(1), 10. https://doi.org/10.3390/nano11010010.

Soares, F. C., Yamashita, F., Müller, C. M. O., & Pires, A. T. N. (2013). Thermoplastic starch/poly(lactic acid) sheets coated with cross-linked chitosan. *Polymer Testing*, *32*(1), 94–98. https://doi.org/10.1016/j.polymertesting.2012.09.005.

Viguié, J., Molina-Boisseau, S., & Dufresne, A. (2007). Processing and characterization of waxy maize starch films plasticized by sorbitol and reinforced with starch nano-crystals. *Macromolecular Bioscience*, *7*(11), 1206–1216. https://doi.org/10.1002/mabi.200700136.

17 Design of Recycled Aluminium (AA 7075+ AA1050 Fine Chips)-Based Composites Reinforced with Nano-SiC Whiskers, Fine Carbon Fiber for Aeronautical Applications

Özgür Aslan
Atilim University

Olga Klinkova and Dhurata Katundi
ISAE-Supmeca

Ibrahim Miskioglu
Michigan Technological University

Emin Bayraktar
ISAE-Supmeca

17.1 INTRODUCTION

The development of recycled AA7075 based composites reinforced with Nano fillers is a useful solution for the aeronautical and/or aerospace engineering due to their exceptional properties such as low density, high stiffness, high strength, etc. In general, application of nano fillers such as GNP, SiC in the composites are structural materials that are for civil and military applications [1–8]. For security fears in the aerospace area, the application of nano SiC, carbon fibers requests a consistent manufacturing process such as diffusion bonding, with different materials to construct a new composite family [6–15].

DOI: 10.1201/9781003400998-17

In the frame of this research work, nano SiC, GNP, fine carbon fibers were used in the recycled (fresh scrap) Aluminium alloy, AA7075, for the high resistance composite production as a low cost manufacturing engineering in case of certain parts of the aircraft engine. Due to the high and reliable mechanical properties of nano SiC reinforced composites in the AA 7075 alloy with sound microstructure, a novel design of these composites were developed in the frame of a common research project with French aeronautical society. This process give a suitable chemical bonding diffusion in case of Sinter+ Forging that is essential for the hybrid composites. This process is also low-cost that give final structure is compared. This process followed by a second heat treatment to reduce the residual stresses and attain a relatively soft and ductile structure. A perfect chemical diffusion bonding was carried out at the interface between matrix and reinforcements by using this process [9,13].

At the stage of the work, comprehensive experimental tests were carried out to clarify the static and cyclic tests have been carried out for time dependent properties of these composites. Microstructural analyses were carried out using Scanning Electron microscopy.

17.2   EXPERIMENTAL CONDITIONS

As a useful and low-cost manufacturing process of these composites, two major reinforcements (received from VWR), fine nano SiC (20 nm) and very fine Carbon Fibers (1–6 μm) and TiAl intermetallic powder (2–5 μm) were added in the recycled fresh scrap AA7075 aluminium matrix. The recycled aluminium in the form of chips, were supplied by the Brazilian aeronautic company, First, recycled aluminium AA7075 chips were gas atomized and then they were mixed by high energy milling in a planetary ball mill under inert argon atmosphere to prevent oxidation of the powders (20/1 ball/powder ratio). Additionally, 3 wt% of zinc stearate was used as a lubricant during the preparation of the composite. After milling operation, thermal behaviour of the aluminium alloy (AA7075) powder was evaluated with DSC-TGA and XRD. The details of these experiments were given in the former papers [1–9,11,13].

As minor reinforcements, molybdenum and copper (Mo 1 wt%, Cu 4 wt% and GNPs 0, 15 wt%) were used. During the milling process, pure nano AA1050 (3–5 wt%) was added to homogenize the mixture of the recycled aluminium alloys. Biaxial compaction of the green compact specimens were done under 250 MPa. At the final stage, a novel composite design were carried out with a combined process, "Sinter+Forging" was used at 600°C followed a soft relaxation treatment at 200°C during the 2 hours. This second heat treatment to reduce the residual stresses and attain a relatively soft and ductile structure. The static and cyclic, time dependent properties of these composites Microstructural analyses were carried out using Scanning Electron microscopy results.

17.3   RESULTS AND DISCUSSION

Table 17.1 presents the compositions of the four composites formulated for the innovative hybrid composite designed with an aluminum alloy AA7075 as matrix. Besides the major reinforcements such as SiC, TiAl, Carbon Fibers, etc., small amounts of Mo–Cu–GNPs increase strengthening of the composites generating a strong cohesion of

TABLE 17.1 (A)
Compositions of the Four Composites (wt%)

Composite Name	AA7075	Nano SiC FillerWhiskers	Carbon Fiber 1–6 μm	TiAl 2–5 μm	GNP	Mo	Cu
ANGE-I	B	15	3	-	0,20	1	1
ANGE-II	B	20	5	-	0,20	1	1
ANGE-III	B	30	5	10	0,20	1	1
ANGE-IV	B	35	5	10	0,20	1	1

TABLE 17.1 (B)
Chemical Composition of Scrap AA 7075 Alloy (wt%)

Element	Al	Cu	Fe	Mg	Mn	Si	Ni	Zn	Cr	Zr
wt.%	Balance	1.48	0.23	2.11	0.07	0.10	0.01	5.29	0.22	0.02

the reinforcements with the matrix mainly on the grain boundaries. For the fine distribution of the reinforcements and to obtain a fine grain size, nano Cu and nano Mo (<1 nm) and nano graphene platelets (GNPs, 500 m²/g with surface particle) were added to the matrix for each composite. Additionally, the presence of Mo and GNPs in the structure increase the mechanical resistance for toughening mainly due to a strong cohesion by chemical diffusion bonding at the interface between the matrix and reinforcements. We know that fine copper particles added in the composition improves even accelerated the chemical diffusion bonding in the matrix. The formation of the chemical diffusion bonding mechanism that will be presented in the next session is only indicative and should be improved with new measurements during the course of this research project.

Figure 17.1a shows XRD diagram for AA7075 based composites reinforced with SiC TiAl and Carbon Fibers indicating the phases and an additional information was given by "EDS" chemical analysis for the composite used here. Figure 17.1b present a mapping analyses of the microstructure for showing the distribution of the reinforced elements in the microstructure. Figure 17.1c give a detailed analyses of Differential Scanning Calorimetry (DSC) diagram for AA7075 alloy and simulation of fraction of solid depending on the temperature calculated with software "Thermo-Calc" in the matrix to determine the critical transformation points during the heating and cooling stages.

It seems that X-ray diffraction patterns of the surface of the composite justifies mainly the major reinforcements which is supported by the EDS analyses of the composite structure. These analyses can justify the microstructure and mapping analyses showing the distribution of the reinforcements SiC, Carbon Fibers and other reinforced particles. It is absolutely carried out a strong chemical bonding diffusion between the matrix and reinforcements that give high toughening mechanism thank to the combined method "Sinter+Forging" process.

As for certain physical properties; Electrical conductivity levels were measured with an "Agilent 4338B Milli-ohm Meter". Three specimens were measured for each

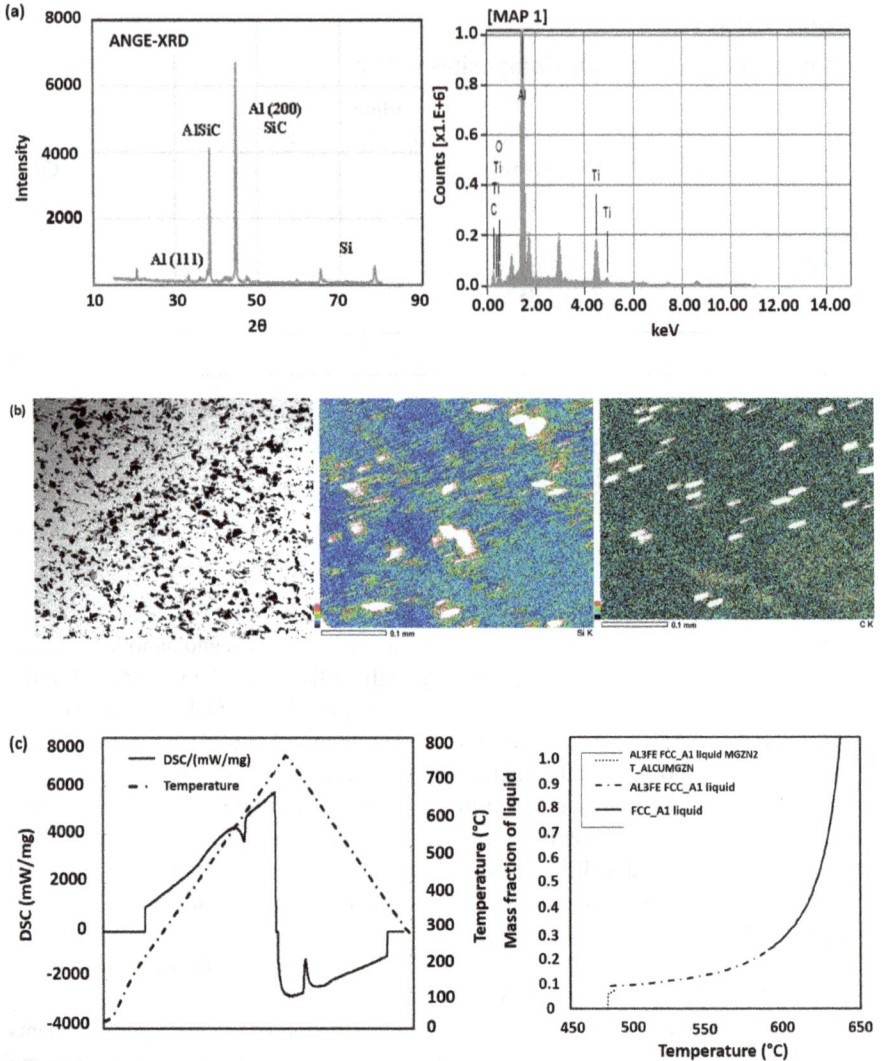

FIGURE 17.1 (a) XRD diagram for the composite of specimen ANGE III indicating the phases and an additional information was given by "EDS" chemical analysis for the composite used here, (b) Mapping analyses of the microstructure for showing the distribution of the reinforced elements in the microstructure and (c) finally Differential Scanning Calorimetry (DSC) diagram measured for A7075 alloy with a heating rate of 5°C/minute and simulation of the fraction of solid depending on the temperature for A7075 [1,3,5,6,9,10].

composite and then, the mean values are given in Table 17.2. For the measurements, DC regulated power supply voltage and current were set as 20 V and 20 A respectively. Data acquisition Card "NI9234" was connected in parallel with the output of the power to acquire the voltage data (voltage input accuracy was 24 bits). A high precision multimeter "Agilent U1253N" was connected in series to measure the current intensity (A).

For the same specimens, thermal conductivity measurements carried out in our laboratory and also micro hardness measurements taken from the samples produced under the same conditions for these composites are also presented in Table 17.2. All the data for the electrical and thermal measurements were revealed with LabVIEW program. These results obviously should be assumed an indicative data under the laboratory conditions.

In Figure 17.2, Mechanical test device adapted on the Zwick test machine (ISAE-SUPMECA/Paris) static compression test of the specimens were presented only as an example.

To understand the toughening mechanism, different type of comprehensive tests, were conducted on the mechanical properties of the new hybrid aluminium based composites. Results of these tests, static and cyclic compression at a test speed of 1 and 0.5 mm/minute, time dependent cyclic compression test with a test speed of

TABLE 17.2
Electrical and Thermal Properties Measured for Four Composites with Micro Hardness Values

Composite Name	Electrical Conductivity at Ambient (S/m)	Thermal Conductivity (W/mK)	Micro Hardness (HV$_{0,1}$)
ANGEL-I	3.25×10^9	19 ± 15	585 ± 25
ANGEL-II	3.80×10^9	30 ± 10	615 ± 30
ANGEL-III	6.40×10^9	25 ± 10	315 ± 15
ANGEL-IV	6.70×10^9	25 ± 15	285 ± 25

FIGURE 17.2   Sinter+ Forging Process, Static compression test on the cylindrical test specimens on the Zwick Test Tool. Mechanical test device adapted on the Zwick test machine (ISAE-SUPMECA/Paris).

(a)

(b)

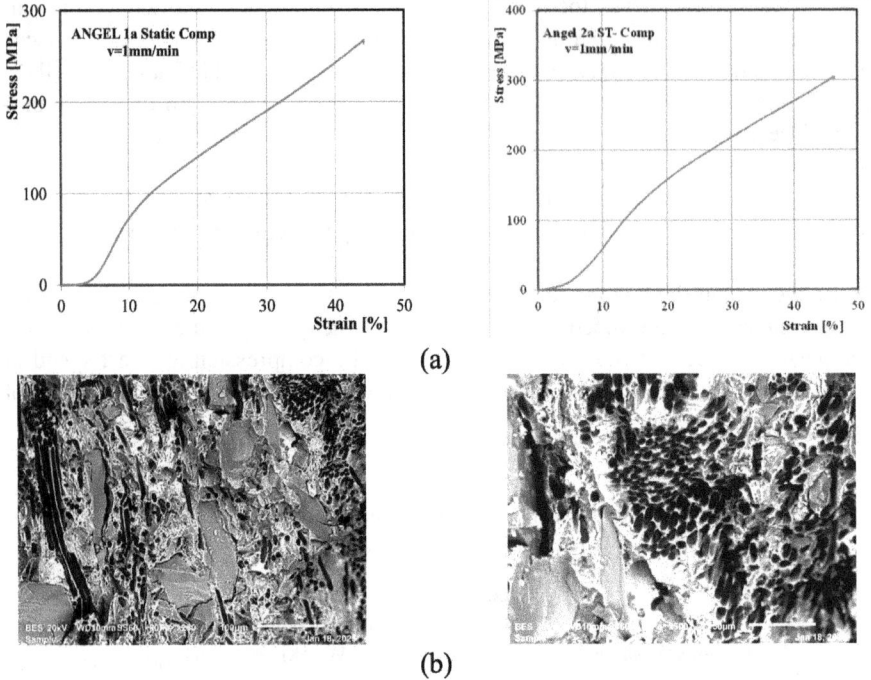

FIGURE 17.3   (a) Static Compression test carried out on the specimen ANGE 1 and ANGE II. (b) Fracture Surfaces of these tested specimens.

5 mm/second were conducted. All the results presented in the Figure 17.3a and b respectively. Only two static and cyclic compression tests (ANGE I and ANGE II) were conducted by using a cylindrical test specimen. The size of the specimen with a diameter of 20 mm and 25–30 mm in height.

The maximum stresses generated at the specimens ANGE 1 and ANGE II were about 300 ± 30 MPa carried out on the 3–4 specimens. The other test results for the specimens of ANGE III and ANGE IV were the same, we have stopped at the level of 400 ± 25 MPa. Apparent Young Modulus estimated from these tests were estimated from these compression tests are variable between 80 and 90 GPa. In case of cyclic incremental static compression test, only two compositions were exposed here such as ANGE III and ANGE IV due to having higher resistance regarding to two first compositions, ANGE I and ANGE II (Figure 17.4). Essentially this type of test can be evaluated if the composition shows a strain hardening mechanism during the cyclic loading. A detailed analyses give a small amount of strain hardening ($\Delta\sigma = \sim 0.5$–1 MPa) during the loading-unloading for the composition of ANGE III. In general way, these values can be neglected. It should be known that these results are only indicative values obtained at the laboratory scale.

These composites should undergo different variable static and cyclic bending solicitations as an interest of the aeronautical society for the aircraft engine parts. Here all the three-point bending test (3PB) results for the compositions discussed here. As indicated on the 3PB test specimens, the maximum stresses generated at

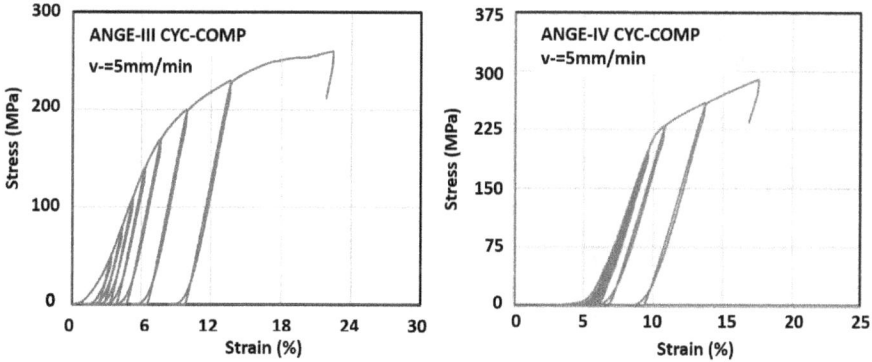

FIGURE 17.4   Cyclic incremental compression test carried out on the specimen ANGE 1II and ANGE IIV.

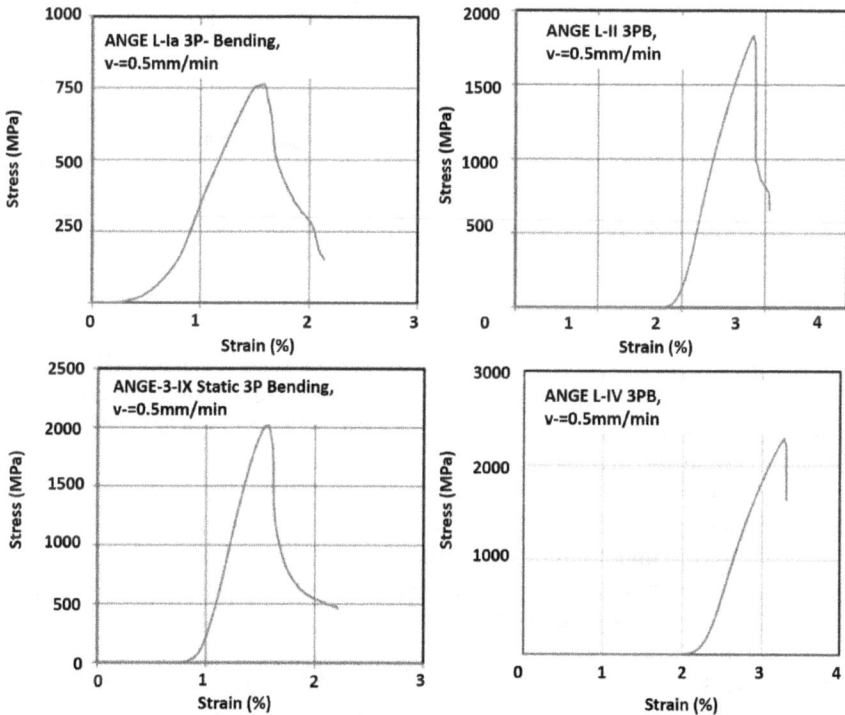

FIGURE 17.5   Static Three Point Bending (3PB) test carried out on the four composites.

the specimens ANGE 1 and ANGE II were about 750 ± 10 MPa $1,500 \pm 30$ MPa carried out on the 3–4 specimens, respectively. The other test results for the specimens of ANGE III and ANGE IV were about $2,000 \pm 15$ MPa and $2,500 \pm 30$ MPa. Once again, these results are indicative results carried out at the laboratory conditions (Figure 17.5).

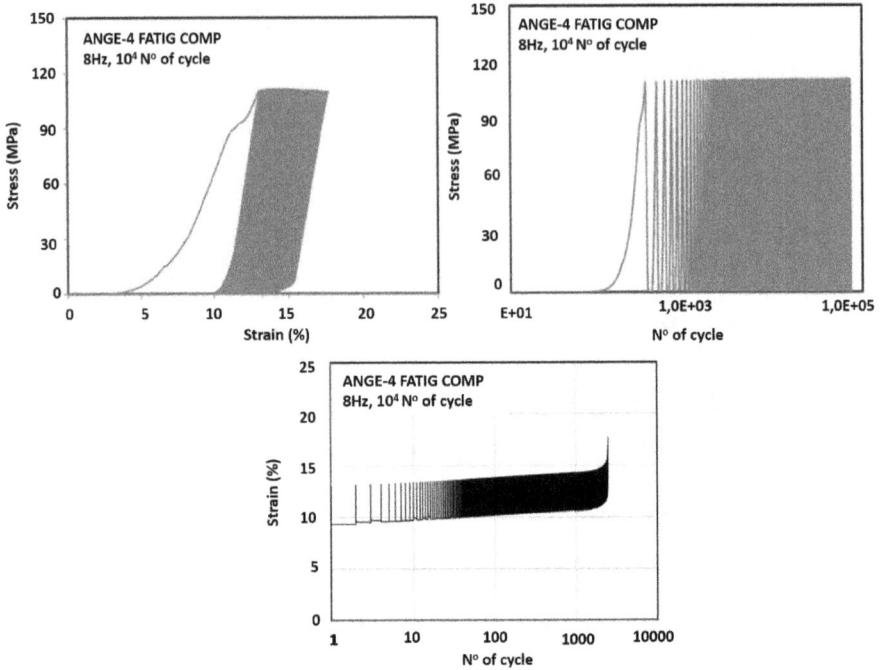

FIGURE 17.6 Low cycle Fatigue compression results evaluated on the samples ANGE IV conducted at 8 Hz for 10,000 cycle.

At the same way, these composites should expose high toughening mechanism during the cyclic (low cycle Fatigue)/time dependent test solicitations. As only the low cycle fatigue compression test results of the fourth composition, ANGE IV, has been presented here as the highest stress values (Figure 17.6). All of the tests have been conducted at 8 Hz controlled by number of cycle (10,000 cycles) at the different max stress applications. The samples of ANGE IV has shown the well accepted results up to 500–600 MPa without failure up to the 10,000 number of cycle. Essentially, the second part of this research will be target on the fatigue behaviour of these composites combining the results with FEM-modeling.

As for Time Dependent Behaviour by means of nano indentation, two compositions, ANGE III and ANGE IV were conducted with Comparative Wear Tests. Wear tests were performed using the nano-scratch testing capability of the nano-indenter. Relatively fast wear tests can be performed to compare the wear behaviour of the different samples. For the wear tests a conical tip with a 90° cone angle was used. Wear tests were run under a normal load of 50 mN applied over a linear track of 500 mm for at least 50 cycles. The wear is characterized as the area between the final and residual profiles. The results for the two samples were given in the following Table 17.3. These results have given here only as indicative values.

TABLE 17.3
General Wear Results for Two Compositions, ANGE III and ANGE IV

Values	Sample ANGE-III	Sample ANGE IV
Average (μm^2)	1010.123	920.5711
Standard deviation (μm^2)	235.1180	141.1522

17.4   CONCLUSION

This manuscript gives the partial results of our academic research project for aeronautical applications. We have developed an innovative microstructure by using high level nano filler SiC and also high level TiAl fine powder as the reinforcements in the recycled fresh scrap aluminium alloy AA7075. All of the results and interpretations are based of the partial results of our research project. A strong and high toughness hybrid composite has been developed by using a combined process called "Sinter+ Forging" This process followed by relaxation treatment and to carry out a strong chemical bonding diffusion between matrix and reinforcement. This process is much more economic and very efficient for the hybrid composites for low cost and efficient manufacturing of the small size of light hybrid composites.

ACKNOWLEDGEMENTS

This academic research has been carried out with the collaboration of ISAE-SUPMECA-Paris, Atilim University-Ankara-TR and Michigan Tech Houghton, MI-USA. Our acknowledgements are for Dr Georges Zambelis (SAFRAN-Paris) for his help and discussion of the results. For the measurements of electrical and thermal conductivity and micro hardness properties, we thank Mr Christoph Ben Brahim and Mr Matthieu SONAR from QUARTZ laboratory, ISAE-Supmeca-Paris.

REFERENCES

[1] E. Bayraktar, F. Gatamorta, H. M. Enginsoy, J. E. Polis, I. Miskioglu, New design of composites from fresh scraps of niobium for tribological applications, mechanics of composite, *Hybrid and Multifunctional Materials*, 6, 35–44, 2020.

[2] D. K. Koli, G. Agnihotri, R. Purohit, Advanced aluminium matrix composites: The critical need of automotive and aerospace engineering fields, *Materials Today: Proceedings*, 2(4–5), 3032–3041, 2015.

[3] F. Gatamorta, I. Miskioglu, D. Katundi, E. Bayraktar, Recycled Ti-Al-Cu matrix composites reinforced with silicon whiskers and γ-alumina (Al2O3) fibres through sintering+forging. In *SEM-Mechanics of Composite, Hybrid & Multi-Functional Materials*, Springer International Publishing, Vol. 5, pp. 49–54, 2023, 10.1007/978-3-031-17445-2_6.

[4] U. Kaftancıoglu, G. Zambelis, F. Gatamorta, I. Miskioglu, E. Bayraktar, Development of Ni-Al/Nb2Al/ZrO2-based composites for aircraft engine applications produced by a combined method: Sintering+forging. In *SEM-Mechanics of Composite, Hybrid & Multi-Functional Materials*, Springer International Publishing, Vol. 5, pp. 55–60, 2023, 10.1007/978-3-031-17445-2_7.

[5] I. Miskioglu, G. Zambelis, F. Gatamorta, O. Aslan, E. Bayraktar, Toughening mechanism of silicon whiskers and alumina fibers (γ-Al2O3) reinforced Ni-Al-Cu matrix composites through "sintering+forging". In *SEM-Mechanics of Composite, Hybrid & Multi-Functional Materials*, Springer International Publishing, Vol. 5, pp. 29–35, 2022, 10.1007/978-3-031-17445-2_4.

[6] F. Gatamorta, H. M. Enginsoy, E. Bayraktar, I. Miskioglu, D. Katundi, Design of recycled alumix-123 based composites reinforced with γ-Al2O3 through combined method; sinter+forging, mechanics of composite, *Hybrid and Multifunctional Materials*, 6, 9–18, 2020.

[7] D. Gu, Z. Wang, Y. Shen, Q. Li, Y. Li, In-situ tic particle reinforced Ti-Al matrix composites: Powder preparation by mechanical alloying and selective laser melting behavior, *Applied Surface Science*, 255(22), 9230–9240, 2009. https://doi.org/10.1016/j.apsusc.2009.07.008.

[8] V. A. Popov, E. V. Shelekhov, A. S. Prosviryakov, A. D. Kotov, M. G. Khomutov, Particulate metal matrix composites development on the basisof in situ synthesis of TiC reinforcing nanoparticles during mechanical alloying. *Journal of Alloys and Compounds*, 707, 365–370, 2017.

[9] F. Gatamorta, I. Miskioglu, E. Bayraktar, M. L. Melo, Recycling of aluminium-431 by high energy milling reinforced with TiC-Mo-Cu for new composites in connection applications. In *Mechanics of Composite and Multi-Functional Materials*, Vol. 5, pp. 41–46, 2019. https://doi.org/10.1007/978-3-030-30028-9_6.

[10] L. M. P. Ferreira, M. H. Robert, E. Bayraktar, D. Zaimova, New design of aluminium based composites through combined method of powder metallurgy and thixoforming, *Advanced Materials Research*, 939(1), 68–75, 2014.

[11] P. D. Srivyasa, M. S. Charoo, Role of fabrication route on the mechanical and tribological behavior of aluminium metal matrix composites – A review, *Materials Today: Proceedings*, 5, 20054–20069, 2018.

[12] L. M. P. Ferreira, E. Bayraktar, I. Miskioglu, M. H. Robert. Influence of nano particulate and fiber reinforcements on the wear response of multiferroic composites processed by powder metallurgy, *Advances in Materials and Processing Technologies*, 3(1), 23, 2016.

[13] L. F. P. Ferreira, E. Bayraktar, M. H. Robert, I. Miskioglu, Chapter 17: Particles reinforced scrap aluminium based composites by combined processing sintering+thixoforming. In *Mechanics of Composite and Multi-Functional Materials*, Springer, Berlin, pp. 145–152, 2016. 978-3-319-41766-0.

[14] L. Mihlyuzova, H.-M. Enginsoy, E. Bayraktar, S. Slavov, D. Dontchev, Tailored behaviour of scrap copper matrix composites reinforced with "Zn-Ni-Al", low cost shape memory structures, mechanics of composite, *Hybrid and Multifunctional Materials, Fracture, Fatigue, Failure and Damage Evolution*, 3, 49–54, 2021.

[15] L. Mihlyuzova, H. M. Enginsoy, D. Dontchev, E. Bayraktar, Tailored behaviour of scrap copper matrix composites reinforced with zinc and aluminium: Low cost shape memory structures, mechanics of composite, *Hybrid and Multifunctional Materials*, 6, 27–34, 2020.

18 State-of-Art Review on Nanofiller in Biodiesel Applications

M. S. M. Misenan
Yildiz Technical University

M. S. Ahmad Farabi and N. A. Zulkipli
Universiti Putra Malaysia

M. A. Mohd Saad and A. H. Shaffie
Universiti Sains Islam Malaysia

18.1 INTRODUCTION

Global warming has emerged as a significant issue as a consequence of rising greenhouse gas (GHG) emissions from the combustion of fossil fuel (Fadhli et al., 2020). Apart from global warming, population growth and global industrialization have increased energy demand and resulted in the inevitable fossil fuels depletion. For decades, environmental degradation caused by toxic fumes (i.e. SO_x and NO_x) emitted by the transportation and industrial sectors, which are powered by non-biodegradable petroleum fuel, has been a major concern (Nur et al., 2020).

Renewable energy sources have become more significant in the energy grids of several nations over the past few years. Biofuels like ethanol, biodiesel, and biogas help decrease environmental impacts by reducing the consumption of fossil fuels (Islam et al., 2020). Growing concern about environmental conservation and energy supplies over the last few decades (Misenan et al., 2022) has led to research into alternative fuels derived from non-food grade triglycerides and fats which can provide a feasible solution for these problems such as biodiesel (Farid et al., 2017). Despite the benefits to the environment and the economy, efficient disposal of the waste from the biofuel production chains is still necessary (Figure 18.1) (Costa et al., 2013).

The biorefinery concept emerged in the late 1990s. Bio-refineries are an environmentally friendly alternative to petroleum refineries. They intend to use biomass (municipal waste, agricultural waste, wood, crops, and grass) as a renewable carbon feedstock to produce primarily biofuels (biodiesel, ethanol, methane, biogasoline, and hydrogen) and chemicals such as polymers, adhesives, paper, and coatings (Alcocer-Garcia et al., 2022). The search for an alternative fuel for a diesel engine should begin before the end of the century. Sir Rudolf Diesel accomplished the world's first engine run on vegetable oil with peanut oil in the eighteenth century.

DOI: 10.1201/9781003400998-18

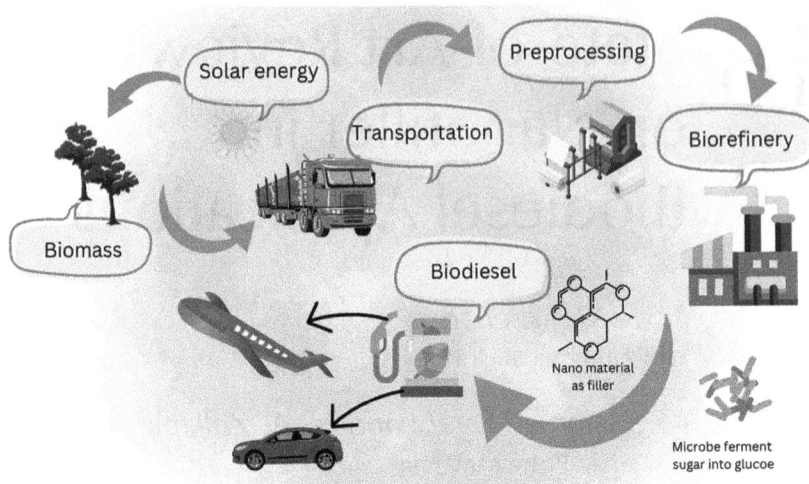

FIGURE 18.1   Utilization of renewable energy sources for daily life.

However, this invention was preceded by Patrick Duffy's idea of using vegetable oil in diesel engine (Ganesan et al., 2021).

Conventional diesel is a pure hydrocarbon with a carbon number of 18 that is derived from crude oil through the fractional distillation process. Synthesized biodiesel, on the other hand, is similar but not identical to fossil diesel. Biodiesel is a blend of long-chain hydrocarbons with fatty acid ester linkages (–COOR) that have the same carbon numbers as diesel and exhibit similar properties to hydrocarbon-rich diesel (Abed et al., 2019). Biodiesel is a green, clean-burning, and renewable fuel made from long-chain methyl esters, which are typically derived from agricultural oils (Farid et al., 2017). It is extracted from lipids/triacylglycerides, animal fats, or natural vegetable oils at mild operating conditions. Furthermore, many processes on lignocellulosic biomass have been carried out to produce bio-fuels and bio-chemicals via thermochemical conversion (Zainol et al., 2019).

The advantage of biodiesel is it comes from natural product, which is renewable, biodegradable, and can be produced according to demand (Fu et al., 2013). Unlike petroleum-based products whose underground sources and scarcely to located, very long time to produced and rapidly depleted (Taufiq-Yap et al., 2020). Other researcher also agreed that biodiesel is known as renewable source of energy, sustainable, and biodegradable (Singh et al., 2020).

Biodiesel reported content very less sulfur content hence low emission of GHGs such as CO_2, HC, SO_x and NO_x (Shahabuddin et al., 2012) and provide a complete combustion and reduced emission due to a large amount of free oxygen (Atabani et al., 2012). Using biodiesel could reduce the pollutants and movable carcinogens (Yang et al., 2013). It is practically good for human and environmentally friendly. The flashpoint of biodiesel recorded is significantly higher than regular diesel and gasoline; therefore, it is easier to store and handle, hence made it more stable and not flammable compared to other fuel.

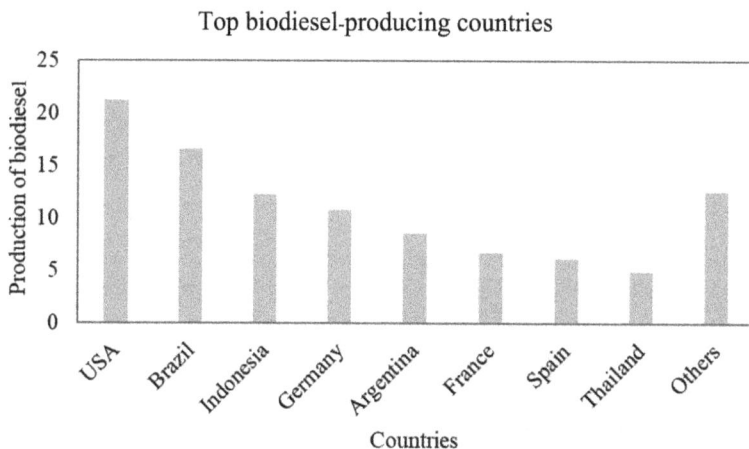

FIGURE 18.2 Top biodiesel-producing countries. (Data adapted from Abomohra et al., 2020.)

Compared to petroleum diesel, biodiesel has similar properties and also lower emissions; hence, it can be used in the transport sector as alternate solution to diesel fuel (Nguyen et al., 2010; Baskar & Aiswarya, 2016). Some researchers reported that no modification is required while using biodiesel as fuel in diesel engine (Shahabuddin et al., 2013; Takase et al., 2015). This alternative fuel can be used directly in engine either total 100% pure or formulated with diesel in various percentages to provide alternative solution to fuel in engines. Biodiesel also has better lubrication properties that can reduce the wear and increase the engine life.

With such advancements in research in recent years, biodiesel production has expanded to an industrial scale in many countries. Figure 18.2 shows the top biodiesel-producing countries in 2018. USA (21.2%) lead the countries followed by Brazil (16.6%).

The EU is now one of the world's leading biodiesel producers, accounting for 31% of total biodiesel output, which stood at 14.6 billion liters in 2019. Rapeseed, sunflower, groundnut, and soybean oil crops are the primary feedstocks for biodiesel production in the EU. Oil palms, physic nut trees, coconut palms, and Chinese tallow trees were also used in the production of biodiesel. Similarly, wheat and sugar beets have undergone extensive research for bioethanol production (Duarah et al., 2022).

The impact of the research on biodiesel production and applications has been remarkable. Based on numerous published articles as presented in Figure 18.3, there has been a growing trend in publications of this research area. Within the last 10 years, there has been an essential increase in the total number of publications related to biodiesel. Figure 18.3 shows that top topic discussed on biodiesel is energy which justifies the importance of biodiesel in energy area while Table 18.1 summarizes the previous work regarding biodiesel.

Recently, researchers from all over the world are currently interested in nanoparticle additives. Consequently, nanomaterials have emerged as promising materials

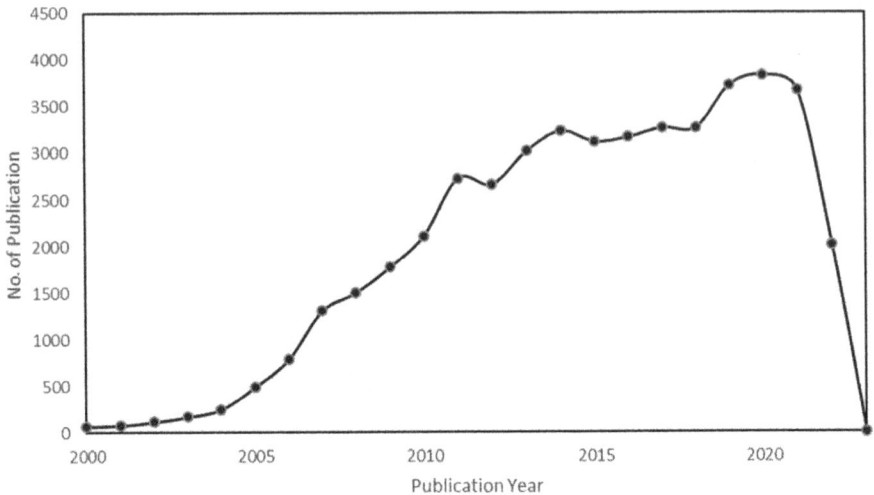

FIGURE 18.3 Trend analysis of biodiesel publications by year between 2000 and 2023.

TABLE 18.1
Previous Study Regarding Biodiesel

Year	Biodiesel Studies	Reference
2022	Production of low-emitting biomass pellets from a variety of agricultural crop leftovers and trees, including *Cedrus deodara, Pinus wallichiana, Zea mays,* and *Triticum aestivum.*	Rashedi et al. (2022)
2022	Review on developments in nanoparticles-enhanced biofuels and solar energy in Malaysian perspective.	Fayaz et al. (2022)
2020	Explored and created a native fuel using India's vast forests. For the synthesis of biodiesel, single-stage base catalyst transesterification was used. Due to their low acid content, sal seeds were employed in the manufacturing of biodiesel in India.	Pali et al. (2020)
2019	Assessment of the potential of advanced oxidative processes for the treatment of biodiesel wastewater.	Brito (2019)
2019	Alkali-catalyzed transesterification of castor oil yields batch-scale biodiesel.	Elango et al. (2019)
2018	Explored the experimental optimization of heterogeneous catalyst-accompanied ultrasound-assisted biodiesel synthesis.	Korkut and Bayramoglu (2018)
2018	CaO, MgO, and ZnO are emphasized as supporting elements on both mass and $C\text{-}Al_2O_3$, and the catalytic activity and characterization of the transesterification of soybean oil and castor oil with methanol and butanol are also discussed.	Navas et al. (2018)
2016	Reviewing the results of ultrasound-aided biodiesel transesterification utilizing diverse input materials with acid, base, and enzyme catalysts.	Ho et al. (2016)

(Continued)

TABLE 18.1 (*Continued*)
Previous Study Regarding Biodiesel

Year	Biodiesel Studies	Reference
2016	Production of a new reaction system for base-catalyzed transesterification use in the synthesis of biodiesel.	Wu et al. (2016)
2014	Production of methyl ester from corn oil, chicken oil, and other animal oils.	Alptekin et al. (2014)
2013	85% of the waste oil was converted into ester (biodiesel) when waste oil transesterification utilizing alkali (NaOH) as catalyst was studied.	Tiwari (2013)
2011	Research on the factors that determine the yield and characteristics of biodiesel made from vegetable oils.	Keera et al. (2011)

and have gained enormous demand for usage in a variety of diverse applications including sensor (Norrahim et al., 2022), optic (Norizan et al., 2022), liquid crystal (Thiruganasambanthan et al., 2022), absorbent material (Misenan et al., 2021), and energy storage (Misenan et al., 2018).

On the other hand, nanoparticles have emerged as a fresh and promising additive among the most recent additions to diesel and biodiesel fuels, improving engine performance and reducing exhaust pollutants (El-Seesy et al., 2018b). Thus, numerous researchers have concentrated their efforts on techniques for modifying fuel using nano-additives to achieve improved performance and emission characteristics. The emissions produced by compression ignition (CI) engines are subject to severe emission regulations that are in effect worldwide. Gasoline additives can change a variety of fuel characteristics, including density, sulfur concentration, and volatility, which have an impact on fuel emissions. Researchers have investigated the viability of using these modified fuels with diesel engines since the addition of a nanoparticle additive to liquid fuels as a secondary energy carrier has the potential to improve combustion characteristics (Soudagar et al., 2018).

18.2 TYPES OF BIODIESEL FEEDSTOCK

In order to ensure the continuity of the production of biodiesel, one of the most important aspects is the availability of the feedstocks. Taufiq-Yap et al. (2020) reported that the obstacle in maintaining the sustainability of the production of biodiesel in most countries is generally due to the cost and the availability of the feedstock. Nowadays, there are several types of renewable sources that have been used as a feedstock in producing the biodiesel. These feedstocks are basically classified either based on the edibility or by the generation.

Generally, the first-generation feedstock is mainly coming from the edible oils which causes the "food vs fuel" argument. In the early development of biodiesel production, edible oil feedstock that mainly comes from the vegetable's oils such as coconut oil, sunflower oil, palm oil, or corn oil has been chosen due to the easier extraction process and the availability of the crops itself. However, due to the limitation of food supply couples with constrain that comes from the environmental issues, first-generation feedstock cannot be vastly used in the production of biodiesel. Due to that concern, the second-generation feedstock comes out by utilizing non-edible oil-based feedstock in anticipating to solve the "food vs fuel" issues. The advantages of second-generation feedstock other than eradicating the food inequality also manage to reduce the production cost and more eco-friendly compared to the first-generation feedstock (Singh et al., 2020).

According to Khan et al. (2019), despite all the advantages provided by the second-generation feedstock, the GHG footprint left by the production of the non-edible oil feedstock combining with the land, water, fertilizer, and pesticides usage during the farming process has led the researchers to the third-generation feedstock which mainly comes from the waste oil and microalgae. Apart from the less land and water usage during the farming process, the primary gain of utilization of third-generation biodiesel is that it will reduce the GHG effect and the impact on the food security. Nonetheless, large amount of investment is required in purifying and extraction process of this type of feedstocks. The extraction process of microalgae shows high production cost comparable to the petroleum-derived biodiesel (Chisti, 2013). Apart from that, the sustainability of the feedstock also becomes one of the concerns when utilizing this type of sources.

Considering the disadvantages of third-generation biodiesel feedstock, fourth-generation feedstocks are coming into the surface. Aro (2016) reported that, with the fourth-generation feedstock coming into the plays, the availability of the feedstock can be increased exponentially as the production of these genetically engineered crops is principally inexhaustible, universally available, and cheap. Nonetheless, despite the generation and generation of biodiesel feedstock has been produced, Taufiq-Yap et al. (2020) reported that in Figure 18.4, there are 82% of biodiesel feedstocks that are currently being used coming from the vegetables oil. This is mostly due to the availability and cost efficiency of the feedstock despite the "food vs fuels" issues.

Different sources of feedstocks like vegetable oils, algal oils, animal fats, microbial oils, and waste oils can be used for production of biodiesel (Kaur & Ali, 2011). Production procedures for biodiesel include transesterification, pyrolysis, and supercritical fluid. From all of these methods, the most adoptive method of biodiesel production is transesterification, which produces biodiesel and glycerol as secondary product from the oil (Bet-Moushoul et al., 2016). Going through the transesterification process where glycerol is by-product, the uses of glycerol in various industrial sectors are well known, merely 100% zero waste from the biodiesel production process.

In recent decades, biodiesel has been growing swiftly as one of the alternative fuels as worlds are moving toward green energy. In addition to that, alternative fuels such as biodiesel are highly researched and utilized in order to increase the environmental benefits compared to the conventional petroleum diesel, and at the same time reduce the consumption of non-renewable sources of energy. Biodiesel or methyl ester is a type of biodegradable renewable energy which can be produced from several types of feedstock or renewable sources. This source can be from animal fats,

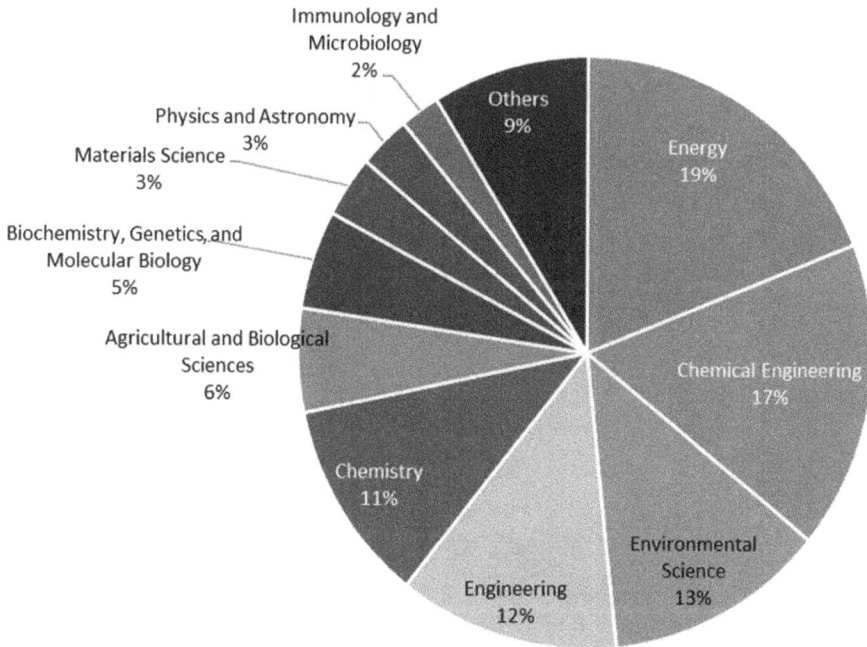

FIGURE 18.4 Trend analysis of biodiesel publications by subject area between 2000 and 2023.

vegetables oil, or waste products such as waste cooking oils (WCOs). Reportedly, there are nearly four effective methods for producing biodiesel which are pyrolysis, blending, micro-emulsions, and transesterification. All these processes have advantages and disadvantages. Production of biodiesel can be through either esterification or transesterification process depending on the acidity or basicity content of the feedstocks.

Basically, both esterification (Figure 18.5) and transesterification (Figure 18.6) processes are a reaction between feedstocks (fatty acid or triglycerides) with short-chain alcohol usually methanol to produce methyl ester. Both of these processes are reversible reaction in which the reaction needs to be derived using excess methanol to move the reaction toward the product. A large number of studies have been done in finding the best or optimum condition in increasing the methyl ester yield. This optimum condition is basically depending on the catalyst used and also the feedstock.

According to the observations made, transesterification has evolved to be the method with the greatest number of advantages, which justifies its use all over the world. Figure 18.7 shows the schematic of biodiesel production by transesterification. This is a simple process that requires animal fats or vegetable oils to promote the triglycerides, alcohol including ethanol and methanol for the esterification reaction, and a catalyst to speed up the reaction. The biodiesel produced after the reaction is raw and must be purified. Pure biodiesel is produced after quality control and purification (Esmaeili, 2022).

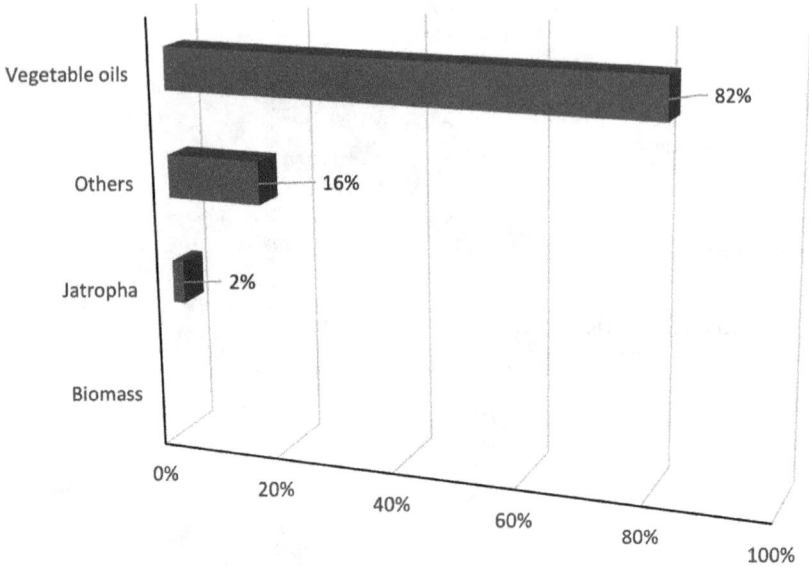

FIGURE 18.5   Percentage of biodiesel production feedstock's share.

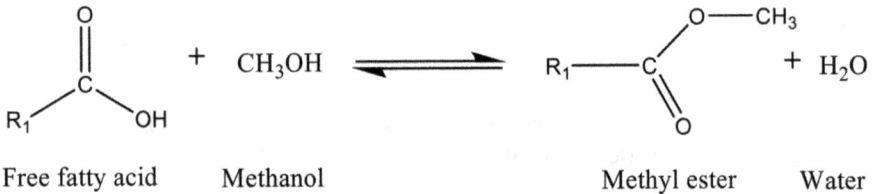

Free fatty acid   Methanol   Methyl ester   Water

FIGURE 18.6   Esterification reaction of FFA for the formation of methyl ester/biodiesel.

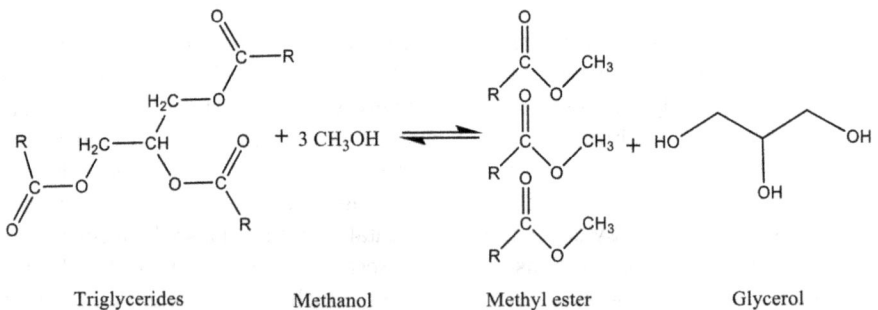

Triglycerides   Methanol   Methyl ester   Glycerol

FIGURE 18.7   Transesterification of triglycerides with methanol.

18.3   FREE FATTY ACID COMPOSITION OF FEEDSTOCK

Table 18.2 shows the classification of different types of biodiesel feedstock. Primarily, all of the listed feedstocks are having different free fatty acid (FFA) percentage. Vicentini-Polette et al. (2021) explained that determination of FFA is important in order to evaluate the raw material and also its degradation status during the storage.

TABLE 18.2

Different Classification of Biodiesel Feedstocks

Edible Oils	Non-Edible Oils	Animal Fats	Other Sources
Soybeans (Glycine max)	Jatropha curcas Mahua (*Madhuca indica*)	Pork lard	Bacteria
Rapeseed (*Brassica napus L.*)	Pongamia (*Pongamia pinnata*)	Beef tallow	Algae (Cyanobacteria)
Safflower	Camelina (*Camelina sativa*)	Poultry fat	Switchgrass
Rice bran oil (*Oryza sativa*)	Cotton seed (*Gossypium hirsutum*)	Fish oil	Terpenes Poplar
Barley	Karanja or honge (*Pongamia pinnata*)	Chicken fat	Microalgae (*Chlorella vulgaris*)
Groundnut	Cumaru	PFAD	Miscanthus
Sorghum	Cynara cardunculus		Latexes
Wheat	Abutilon muticum		Fungi
Corn	Neem (*Azadirachta indica*)		
Coconut	Jojoba (*Simmondsia chinensis*)		
Sesame (*Sesamum indicum L.*)	Passion seed (*Passiflora edulis*)		
Peanut	Moringa (*Moringa oleifera*)		
Palm and palm kernel (*Elaeis guineensis*)	Tobacco seed		
Sunflower (*Helianthus annuus*)	Rubber seed tree (*Hevea brasiliensis*)		
Canola	Salmon oil		
	Tall (*Carnegiea gigantea*)		
	Coffee ground (*Coffea arabica*)		
	Nagchampa (*Calophyllum inophyllum*)		
	Croton megalocarpus		
	Pachira glabra		
	Aleurites moluccana		
	Terminalia belerica		

Source: Adapted from Taufiq-Yap et al. (2020).

As mentioned before, the reaction to produce fatty acid methyl ester (FAME) is chosen based on the acidity of the feedstock. Meanwhile, the acidity of samples is determined from the percentage of FFA composition present in the feedstock. The FFA content in each feedstock is depended on the quality of the feedstock, location, and also the extraction process. For example, the FFA content of palm fatty acid distillate (PFAD) might differ based on the location and the extraction process. In determining the FFA content, the acid value is being measured using the alkalimetric titration technique, then calculated using Eq. 18.1 (Win & Trabold, 2018). The measured AV will then be used to calculate the FFA percentage by using Eq. 18.2 (Kim & Siang, 2022):

$$AV \left(mg\,KOH \, / \, g \right) = \frac{56.1 \times C_{KOH} \times V_{KOH}}{m} \qquad (18.1)$$

where,

56.1 = molecular weight of the solution employed for the titration
C_{KOH} = concentration of the KOH titration solution (g/mol)
V_{KOH} = volume of solution used for the titration (mL)
m = mass of the feedstock (fatty acid sample) (g)

$$FFA(\%) = acid\ value \times \frac{mol.\ wt.\ of\ oleic\ acid}{mol.\ wt.\ of\ KOH} \times \frac{10}{1,000}$$

$$= acid\ value \times \frac{282.27}{56.11} \times \frac{1}{10} \qquad (18.2)$$

$$= acid\ value \times \frac{1}{2}$$

According to Knothe and Razon (2017), the maximum FFA content in the feedstock should be less than 0.5% for the transesterification reaction to be considered in producing the FAME. If the value of FFA is higher than 0.5%, it is advisable to use bifunctional catalyst and esterification reaction before proceed with the transesterification reaction. This is to ensure that the FFA composition will be reduced before triglycerides can react with methanol to produce better FAME yields. Table 18.3 shows the different FFA percentages present in biodiesel feedstock.

18.4   NANOFILLER IN BIODIESEL APPLICATION

The metal oxide of Cu, Fe, Ce, Pt, B, Al, and Co has been widely used as additives in diesel and biodiesel fuel blends. Using cerium as nano-additives, Skillas et al. (2000) investigated the size distribution effect and composition of particulate material and reported an increase in ultrafine but a decrease in the accumulation mode. The impact of nano-additives led to an improvement in efficiency, and it also has an impact on the physicochemical characteristics and fuel emission. Numerous researchers looked into the characteristics of fuel blends with nano-additives and

TABLE 18.3

Comparison of FFA Percentage Present in Biodiesel Feedstock

Feedstock	Acid Value (mg/KOH)	FFA %	Reference
Nagchampa (*Calophyllum inophyllum*)	44	22	Atabani and César (2014)
Rubber seed tree (*Hevea brasiliensis*)	83.76	41.64	Singh et al. (2016)
Cotton seed (*Gossypium hirsutum*)	5.6	2.8	Pamuk et al. (2015)
Karanja or honge (*Pongamia pinnata*)	40	20	Naik et al. (2008)
Jatropha curcas Mahua (*Madhuca indica*)	-	20	Mekala et al. (2014)
Soybeans	121.3	60.65	Ma et al. (2018)
Corn			
Wheat	31.4	15.7	Wang and Johnson (2001)
Sunflower	2.2	1.1	Pal et al. (2015)
Rapeseed	58	29	Yuan et al. (2008)
Tobacco seed	-	>17	Veljković et al. (2006)
PFAD	-	>70	Kanjaikaew et al. (2018)
Jojoba	1.92	0.96	Gad et al. (2021)
Neem (*Azadirachta indica*)	24.4	12.2	Sathyaselvabala et al. (2010)

how they affected factors like calorific value, flash point, density, viscosity, cetane number, etc. The purpose of adding metal-based nano-additives to diesel/biodiesel is to enhance the fuel's characteristics in order to increase engine performance. The calorific value and cetane number of diesel/biodiesel blends were increased with the addition of NPs, while the sulfur level was decreased. The summary of the characteristics of the nanofuel employed in the CI engine by the most current investigations is shown in Table 18.4.

According to Sajith et al. (2010), the CeO_2 NPs have a high catalytic activity because of their huge surface area per unit volume, which improves fuel efficiency and lowers emissions. In addition, according to Iranmanesh et al. (2008), the fuel properties and combustion characteristics of Karanja biodiesel in diesel engines with addition of 5%, 10%, 15%, and 20% by volume of DEE showed improvement in combustion and cold-starting problems with enhancement in the physicochemical properties such as specific gravity, calorific value, viscosity, and liquidity with respect to the ASTM standards.

The effects of titanium oxide (TiO_2) NPs and mustard oil methyl ester (MOME) were studied by Yuvarajan et al. (2018). To increase the yield of methyl ester and eliminate any remaining methanol, the resulting mixture is warmed to 80°C.

The nano-emulsion with MOME is prepared using titanium dioxide nanoparticles. X-ray diffraction is used to evaluate TiO_2 and reveal that it has a crystalline structure made up of an anatase/rutile combination. Figure 18.8a and b exhibits images of TiO_2 nanofluid captured using SEM and TEM, respectively. These photos reveal that the TiO_2 nanofluid is clustered and has particles that are on average 50 nm in size, which is smaller than the diameter of the fuel injector nozzle. As a result, it doesn't create any obstructions in the fuel injector throughout its flow.

TABLE 18.4

Characteristics of the nanofuel employed in the CI engine.

Reference	Fuel	Kin. Viscosity (cst)	Density (kg/m³)	Sp. Gravity	Flash Point (°C)	Fire Point (°C)	Cetane Number	Calorific Value (kJ/kg)	Cloud Point (°C)	Pour Point (°C)	Total Acidity (mg of KOH/g)
Prabu (2018)	B100	4.1	873	–	85	94	–	39,500	–	1	0.46
	B20	2.58	843	–	55	63	–	41,700	–	–6	0.4
	B100A30C30	4.1	874	–	83	93	–	40,200	–	1	0.47
	B20A30C30	2.59	844	–	52	61	–	42,200	–	–5	0.4
Gardy et al. (2017)	UCO	32.91	921 at 15°C	–	289	–	–	–	–	–	4.04
	Synthesized biodiesel + TiO₂/PrSO₃H	4.8	898.1	–	171	–	–	–	–	–	0.41
Harsha Hebbar et al. (2018)	BCO	34.87 at 40°C	894	–	195	–	–	38,480	10	2	–
	BCME	4.78	875	–	155	–	–	40,320	3	–4	–
	Diesel	2.63	840	–	60	–	–	42,500	–5	–12	–
Yuvarajan et al. (2018)	MOME	4.3 at 35°C	864 at 18°C	–	–	–	52	38,108	–	–	–
	MOMET100	4.34	884	–	–	–	54	37,854	–	–	–
	MOMET200	4.38	891	–	–	–	57	37,652	–	–	–
El-Seesy et al. (2018b)	JB20	3.33 at 40°C	–	849.3 at 15.56°C		51.6		41,142	–	–	–

(Continued)

TABLE 18.4 (Continued)

Characteristics of the nanofuel employed in the CI engine.

Reference	Fuel	Kin. Viscosity (cst)	Density (kg/m³)	Sp. Gravity	Flash Point (°C)	Fire Point (°C)	Cetane Number	Calorific Value (kJ/kg)	Cloud Point (°C)	Pour Point (°C)	Total Acidity (mg of KOH/g)
	JB2025GNPs	4.05	–	850.1	–	52.3		41,160	–	–	–
	JB20100GNP	4.22	–	850.4	–	57.4		41,230	–	–	–
El-Seesy et al. (2018a)	JME (B100)	11.72	–	864.5 at 15.56°C	–	–		44,866	–	–	–
	JB20D	4.06	–	847.1	–	–	52	45,432	–	–	–
	JB20D10A	4.12	–	847.1	–	–	53.1	45,439	–	–	–
	JB20D20A	4.28	–	847.1	–	–	54.4	45,445	–	–	–
	JB20D30A	4.29	–	847.1	–	–	55.2	45,453	–	–	–
	JB20D50A	4.38	–	847.1	–	–	57	45,467	–	–	–
Sadhik Basha and Anand (2013)	JBD	5.25	895	–	85	–	53	38,880	–	–	–
	JBD50CNT	5.33	897.9	–	81	–	57	39,780	–	–	–
	JBD25AO25CNT	5.36	895.2	–	81	–	57	39,990	–	–	–
Selvan et al. (2009)	Diesel	2	830	–	50	56	46	42,300	–	–	–
	D70C10E20	2.35	820	–	11	14	44.6	39,000	–	–	–

(Continued)

TABLE 18.4 (*Continued*)

Characteristics of the nanofuel employed in the CI engine.

Reference	Fuel	Kin. Viscosity (cst)	Density (kg/m³)	Sp. Gravity	Flash Point (°C)	Fire Point (°C)	Cetane Number	Calorific Value (kJ/kg)	Cloud Point (°C)	Pour Point (°C)	Total Acidity (mg of KOH/g)
Basha and Anand (2011)	Diesel	2.1	830	–	50	–	46	42,500	–	–	–
	D2S15W100AO	5.01	859.6	–	66	–	50	39,900	–	–	–
Yang et al. (2013)	Diesel	2.8	850	–	–	–	–	45,000	–	–	–
	E10	8.8	880	–	–	–	–	38,250	–	–	–
	E15	11.4	890	–	–	–	–	36,160	–	–	–
Mehta et al. (2014)	WD	2.85	832	–	–	–	50.83	42,930	–	–	–
	WDA	2.53	834	–	–	–	49.6	42,920	–	–	–
	WDSi	2.55	832	–	–	–	51.37	42,940	–	–	–
Banapurmath et al. (2014)	HOME	5.6	880	–	170	–	–	36,016	–	–	–
	HOME25CNT	5.7	898	–	166	–	–	34,560	–	–	–
	HOME50CNT	5.8	900	–	164	–	–	34,560	–	–	–

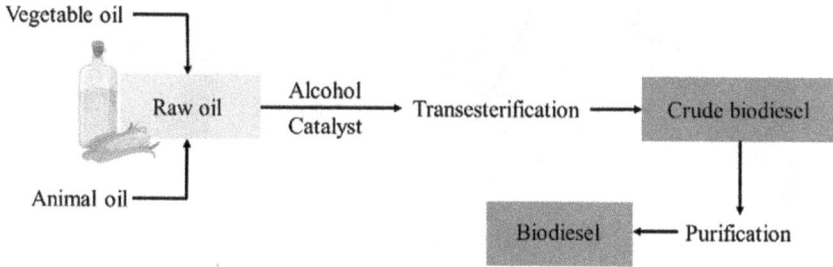

FIGURE 18.8 Schematic of biodiesel production.

Emulsion was created by adding MOME to TiO_2 nanofluid with 100 and 200 ppm of nanoparticle concentration. MOMET100 is the name of the fuel, which contains methyl ester and 100 ppm of TiO_2 nanofluid. Additionally, the gasoline known as MOMET200 contains methyl ester and 200 ppm of TiO_2 nanofluid. The results showed that because of the oxygen content in MOME, MOMET100, and MOMET200, the HC and CO were reduced. The longer delay period of biodiesel led to a higher fuel ignition temperature; therefore, NO_x emissions were higher than diesel at all loads. TiO_2 NPs with an average size of 50 nm were characterized by X-ray diffraction technique.

In other studies, the effects of zinc and calcium oxide metal additions (zinc-doped calcium oxide) on WCO biodiesel were investigated by Kataria et al. (2019). It was discovered that 12:1 was the optimal ratio for base-catalyzed transesterification of WCO. Production of biodiesel was identified by comparing its retention factor with methyl oleate ($C_{19}H_{32}O_2$), standard as shown in Figure 18.9.

Tests were performed using a CI engine with a constant speed of 1,500 rpm, a 200 bar injection pressure, a compression ratio between 15 and 17, and a range of engine loads. At full load, B40, out of all the tested fuels, displayed the highest thermal efficiency. Its CR was 17.5:1. Due to the increasing concentration of nano-additives in the blends, the cleaner and more complete combustion of the fuel led to a reduction in the emissions of HC and CO.

Researchers have found that adding carbon nanotubes (CNTs) to diesel–biodiesel fuel emulsions in doses of 100–300 ppm can reduce brake-specific fuel consumption. Balaji and Cheralathan (2014) reported that CNT nano-additive is effective in increasing the performance and controlling the NO emissions of methyl ester of neem oil-fueled diesel engines. In contrast, compared to plain diesel, the inclusion of CNTs and nano-silver particles reduces CO emissions. Soudagar et al. (2018) utilized two different diesel–biodiesel blends (B5 and B20) at three concentrations, and the hybrid nanocatalyst comprising cerium oxide on multiwall CNTs (MWCNTs) was studied (30, 60 and 90 ppm). The results showed that the combustion reaction, particularly in B20, which included 90 ppm of the catalyst B20, significantly improved due to the high surface area of the soluble nano-sized catalyst particles and their correct distribution, along with catalytic oxidation reaction (90 ppm). In comparison to catalyst-free biodiesel blends, the overall impact of the nanocatalyst inclusion at 30, 60, and 90 ppm concentrations in B5 and B20 reduced the CO, HC, soot, and NOx emissions. Significant reductions in all emissions were attained, and these reductions

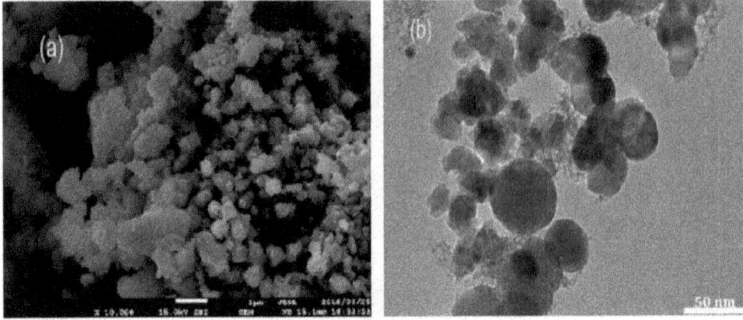

FIGURE 18.9 (a) SEM image of TiO$_2$ nanofluid and (b) TEM image of TiO$_2$ nanofluid. Modified from Yuvarajan et al. (2018).

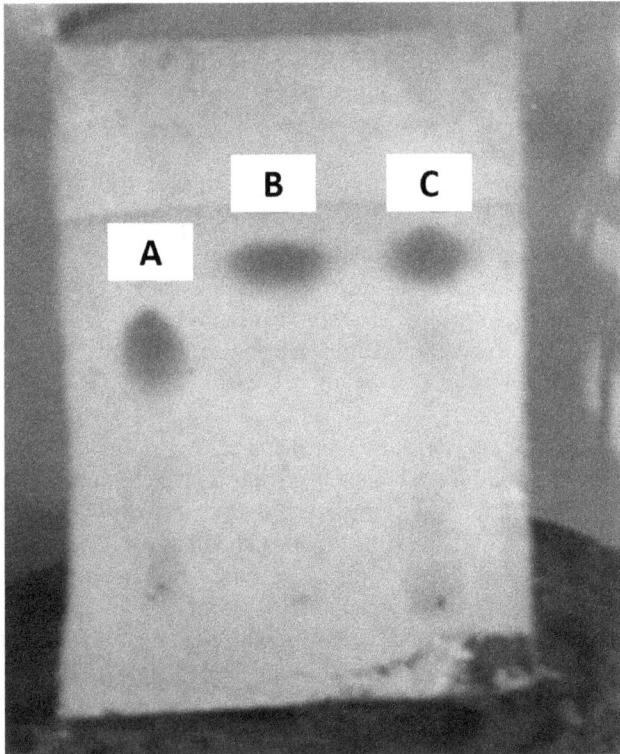

FIGURE 18.10 Thin-layer chromatographic analysis of (a) WCO, (b) WCO-derived methyl ester, and (c) methyl oleate standard. Modified from ref Kataria et al. (2019).

were directly correlated with the amount of nanocatalyst used. At 90 ppm of nano-catalyst concentration, the greatest amount of pollutants was observed to be reduced. Additionally, the CeO2-MWCNTs addition to the fuel blends had the greatest impact on HC.

18.5 LIMITATIONS OF BIODIESEL

Despite advantages mentioned above, there are some limitations to be stated. The production cost of biodiesel limits the usage of biodiesel, as well as the overlapping of food consumption demand worldwide (Kumar & Singh, 2019). Since edible oils used are high in cost, production of biodiesel creates challenges to keep profitable margin in business. Net return in this sector not always promising positive earnings resulting the industry has to go. The continuity in producing biodiesel is from government-supporting policy and the desire to protect the environment (Manaf et al., 2019).

The Ukrainian–Russian war in February 2022 has impacted fossil fuels as well as the prices of commodities including corn, palm oil, and soybean, resulting in high price due to the fear of supply shortage. Hence, some of the feedstocks used as raw materials affect the price of biodiesel produced. The price of biodiesel changes depending on the feedstock price (Marchetti, 2011). Biodiesel sources from animal-based such as beef tallow, catfish oil, chicken fat, duck tallow, waste goat tallow, and waste fish oil will surge as some of them feed on plant-based sources. Biodiesel price was reported >1.5 times higher than mineral diesel in cost (Zhang et al., 2003), while WCO significantly 2.5–3 times lower than virgin vegetable oil. (Agarwal et al., 2017).

The most significant factors that lead to degradation of biodiesel include microbial contamination, the material of the storage, types of feedstock, and nature of storage conditions like exposure to light, oxygen, and temperature. Bacteria and fungi growth in the tank is the most common reason for microbial contamination. This is possibly from interface of fuel–water interaction at the base of the tank. With proper storage management, biodiesel degradation can be hindered, thus can be stored longer. It is suggested that the tank should be always clean and dry to avoid microbial contamination. Besides that, biocides, an oxidative stability added to prevent microbial activity. The usage of tin or aluminum die casting is proven to decrease the oxidative stability of biodiesel. In addition, degradation may accelerate by storing biodiesel in copper, lead, tin, zinc, or aluminum container (Manaf et al., 2019).

Oxidation of biodiesel product may lead system deposits and cause filter plunging and fuel system malfunctions. This problem concerns vehicles user. High amount of unsaturated methyl ester hastens biodiesel degradation rate. Otherwise, research on coconut biodiesel indicates with higher concentration of saturated methyl ester which could be more stable during long-term storage (Dantas et al., 2011; Jose & Anand, 2016).

External exposures such as temperature, light, and air affect the fuel sample physically during long-term storage. Long exposure to both sunlight and air may release free radicals as sunlight, UV radiation, and oxygen drive the oxidation reaction. This reaction hastens biodiesel degradation process during long-term storage (Mittelbach & Gangl, 2001).

Furthermore, there are also microbial-derived lipids suitable for biodiesel production. Reports from Gujjala et al. mentioned that lipid with fatty acid ester linkage and FFA can be utilized as biodiesel, if they meet the specification standards (Gujjala et al., 2019). Chi et al. mentioned *Rhodosporidium toruloides, Lipomyces starkeyi, Mortierella isabellina, Cryptococcus albidus, Yarrowia lipolytica, Trichosporon fermentans, Rhodotorula glutinis,* etc., in literature as oil producer microorganism (Chi et al., 2011). In addition, yeast could be used to produce oil (Mathew et al., 2021).

Biodiesel is produced via chemical process in which catalysts are added to give best and higher output. The most common methods are by esterification or transesterification. In traditional esterification, FFA is converted into ester by adding methanol with homogeneous acid catalyst, for example, sulfuric acid. Transesterification process converts the FFA to biodiesel and glycerol with the presence of methanol and homogeneous base catalyst. Homogeneous acid catalyst may cause corrosion to reactor and slower the reaction rate. While the replacement of homogeneous catalyst may be heterogeneous catalyst, the catalyst preparation is high in cost and affects biodiesel price (Mathew et al., 2021).

Although homogeneous base catalysts such as KOH, alkali metal hydroxide, and potassium methoxide are preferable due to 4,000 times faster than acid-based catalyst, the catalyst is non-reusable for next biodiesel production (Fukuda et al., 2001). Heterogeneous catalysts such as CaO are seen to leach out during transesterification reaction. When CaO used as the catalyst in transesterification of sunflower oil, they caused carbonation of air and hydration, affecting the CaO active sites (Granados et al., 2007). Thus, biodiesel produced needs to be purified to remove the soluble fraction using ion-exchange resin (Kouzu et al., 2009).

18.6 FUTURE PROSPECTIVE

Despite the fact that the technology used to produce biodiesel is advanced and proven, the presence of highly oxygenated compounds in the biodiesel can result in poor heating values, poor storage stability, unfavorable cold flow properties, and potential engine compatibility issues. Additionally, the use of biodiesel can increase NO_x emissions. As a result, considerable research has been conducted on alternative biodiesel production routes that eliminate these oxygenated compounds (Fadhli et al., 2020).

In addition, several factors such as temperature, human labor, location and cost of land, methanol/oil ratio (MOR), catalyst dose, catalyst type such as homogeneous or heterogeneous catalysts, catalyst reusability, tax, government role, contact time, price of glycerol as a by-product, and plant production capacity are all important factors in biodiesel generation costs. As a result, the impact of all of these factors on the cost of biodiesel production should be considered (Foroutan et al., 2020). Biodiesel has been demonstrated to be a workable candidate for future energy requirements, thereby ensuring energy security. Its use as an alternative fuel will not only reduce reliance on fossil fuels, but will also provide job opportunities and economic benefits. Although biodiesel production has a major impact in addressing future energy gaps, the emphasis should be on large-scale production and cost reduction. As a result, future research should focus on making the biodiesel production process more cost-effective, with utilization of the by-product from the reaction such as glycerol. This focus may be resulting in improved economic and environmental viability.

Nanocatalysts have recently been used to produce biodiesel due to their environmental friendliness, high reactivity, high biodiesel yield, and high specific surface area (Dai et al., 2021). This type of catalyst can be appropriate alternatives to homogeneous alkaline catalysts such as potassium hydroxide or sodium hydroxide. Interestingly, the washing step can be eliminated from the factory design as the use of nanocatalyst doesn't need a washing step (Esmaeili, 2022). As a result, the impact of various catalysts must be considered when calculating the cost of biodiesel generation.

REFERENCES

Abed, K.A., Gad, M.S., El Morsi, A.K., Sayed, M.M., Elyazeed, S.A., 2019. Effect of biodiesel fuels on diesel engine emissions. *Egypt. J. Pet.* 28, 183–188. https://doi.org/10.1016/j.ejpe.2019.03.001.

Abomohra, A.E.F., Elsayed, M., Esakkimuthu, S., El-Sheekh, M., Hanelt, D., 2020. Potential of fat, oil and grease (FOG) for biodiesel production: A critical review on the recent progress and future perspectives. *Prog. Energy Combust. Sci.* 81, 100868. https://doi.org/10.1016/j.pecs.2020.100868.

Agarwal, A.K., Gupta, J.G., Dhar, A., 2017. Potential and challenges for large-scale application of biodiesel in automotive sector. *Prog. Energy Combust. Sci.* 61, 113–49.

Alcocer-Garcia, H., Segovia-Hernandez, J.G., Sanchez-Ramirez, E., Tominac, P., Zavala, V.M., 2022. Coordinated markets for furfural and levulinic acid from residual biomass: A case study in Guanajuato, Mexico. *Comput. Chem. Eng.* 156, 107568. https://doi.org/10.1016/j.compchemeng.2021.107568.

Alptekin, E., Canakci, M., Sanli, H., 2014. Biodiesel production from vegetable oil and waste animal fats in a pilot plant. *Waste Manag.* 34, 11, 2146–2154.

Aro, E.M., 2016. From first generation biofuels to advanced solar biofuels. *Ambio*, 45, 1, 24–31.

Atabani, A.E., César, A.D.S., 2014. *Calophyllum inophyllum* L. – A prospective non-edible biodiesel feedstock. Study of biodiesel production, properties, fatty acid composition, blending and engine performance. *Renew. Sustain. Energy Rev.* 37, 644–655.

Atabani, A.E., Silitonga, A.S., Badruddin, I.A., Mahlia, T.M.I., Masjuki, H.H., Mekhilef, S.A., 2012. Comprehensive review on biodiesel as an alternative energy resource and its characteristics. *Renew. Sustain. Energy Rev.* 16, 2070–2093.

Balaji, G., Cheralathan, M., 2014. Effect of CNT as additive with biodiesel on the performance and emission characteristics of a DI diesel engine. *Int. J. ChemTech Res.* 7, 3, 1230–1236.

Banapurmath, N.R., Sankaran, R., Tumbal, A.V., Naraimhalu, N.T., Hunshyal, A.M., Ayachit, N.H., 2014. Experimental investigation on direct injection diesel engine fuelled with graphene, silver and multiwalled carbon nanotubes-biodiesel blended fuels. *Int. J. Automot. Eng. Technol.* 3, 4, 129. https://doi.org/10.18245/ijaet.59113.

Basha, J.S., Anand, R.B., 2011. An experimental study in a CI engine using nanoadditive blended water-diesel emulsion fuel. *Int. J. Green Energy* 8, 3, 332–348. https://doi.org/10.1080/15435075.2011.557844.

Baskar, G., Aiswarya, R., 2016. Trends in catalytic production of biodiesel from various feedstocks. *Renew. Sustain Energy Rev.* 57, 496–504.

Bet-Moushoul, E., Farhadi, K, Mansourpanah, Y., Nikbakht, A.M., Molaei, R., Forough, M., 2016. Application of CaO-based/Au nanoparticles as heterogeneous nanocatalysts in biodiesel production. *Fuel* 164, 119–127.

Brito, S., 2019. Evaluation of advanced oxidative processes in biodiesel wastewater treatment. *J. Photochem. Photobiol. A Chem.* 375, 85–90.

Chi, Z., Zheng, Y., Jiang, A., Chen, S., 2011. Lipid production by culturing oleaginous yeast and algae with food waste and municipal wastewater in an integrated process. *Appl. Biochem. Biotechnol.* 165, 2, 442–453.

Chisti, Y., 2013. Constraints to commercialization of algal fuels. *J. Biotechnol.* 167, 3, 201–214.

Costa, A.G., Pinheiro, G.C., Pinheiro, F.G.C., Dos Santos, A.B., Santaella, S.T., Leito, R.C., 2013. Pretreatment strategies to improve anaerobic biodegradability and methane production potential of the palm oil mesocarp fibre. *Chem. Eng. J.* 230, 158–165. https://doi.org/10.1016/j.cej.2013.06.070.

Dai, Y.M., Li, Y.Y., Jia-Hao-Lin, Chen, B.Y., Chen, C.C., 2021. One-pot synthesis of acid-base bifunctional catalysts for biodiesel production. *J. Environ. Manag.* 299, 113592. https://doi.org/10.1016/j.jenvman.2021.113592.

Dantas, M.B., et al., 2011. Evaluation of the oxidative stability of corn biodiesel. *Fuel* 90, 2, 773–778.

Duarah, P., Haldar, D., Patel, A.K., Dong, C.D., Singhania, R.R., Purkait, M.K., 2022. A review on global perspectives of sustainable development in bioenergy generation. *Bioresour. Technol.* 348, 126791. https://doi.org/10.1016/j.biortech.2022.126791.

El-Seesy, A.I., Attia, A.M.A., El-Batsh, H.M., 2018a. The effect of aluminum oxide nanoparticles addition with Jojoba methyl ester-diesel fuel blend on a diesel engine performance, combustion and emission characteristics. *Fuel* 224, 147–166. https://doi.org/https://doi.org/10.1016/j.fuel.2018.03.076.

El-Seesy, A.I., Hassan, H., Ookawara, S., 2018b. Effects of graphene nanoplatelet addition to jatropha biodiesel-diesel mixture on the performance and emission characteristics of a diesel engine. *Energy* 147, 1129–1152. https://doi.org/10.1016/j.energy.2018.01.108.

Elango, R.K., Sathiasivan, K., Muthukumaran, C., Thangavelu, V., Rajesh, M., Tamilarasan, K., 2019. Transesterification of castor oil for biodiesel production: Process optimization and characterization. *Appl. Therm. Eng.* 145, 1162–1168.

Esmaeili, H., 2022. A critical review on the economic aspects and life cycle assessment of biodiesel production using heterogeneous nanocatalysts. *Fuel Process. Technol.* 230, 107224. https://doi.org/10.1016/j.fuproc.2022.107224.

Fadhli, M., Taufiq-yap, Y.H., Derawi, D., 2020. Green diesel production from palm fatty acid distillate over SBA-15-supported nickel, cobalt, and nickel/cobalt catalysts. *Biomass Bioenergy* 134, 105476. https://doi.org/10.1016/j.biombioe.2020.105476.

Farid, M.A.A., Hassan, M.A., Taufiq-Yap, Y.H., Ibrahim, M.L., Othman, M.R., Ali, A.A.M., Shirai, Y., 2017. Production of methyl esters from waste cooking oil using a heterogeneous biomass-based catalyst. *Renew. Energy* 114, 638–643. https://doi.org/10.1016/j.renene.2017.07.064.

Fayaz, H., Afghan Khan, S., AhamedSaleel, C., Shaik, S., Yusuf, A.A., Veza, I., Rizwanul Fattah, I.M., Mohammad Rawi, N.F., Asyraf, M.R.M., Alarifi, I.M., Yao, H., 2022. Developments in nanoparticles enhanced biofuels and solar energy in Malaysian perspective: A review of state of the art. *J. Nanomater.* 2022, 1–22. https://doi.org/10.1155/2022/8091576.

Foroutan, R., Mohammadi, R., Esmaeili, H., Mirzaee Bektashi, F., Tamjidi, S., 2020. Transesterification of waste edible oils to biodiesel using calcium oxide@magnesium oxide nanocatalyst. *Waste Manag.* 105, 373–383. https://doi.org/10.1016/j.wasman.2020.02.032.

Fu, X., Li, D., Chen, J., Zhang, Y., Huang, W., Zhu, Y.A., 2013. Microalgae residue based carbon solid acid catalyst for biodiesel production. *Bioresour. Technol.* 146, 767–770.

Fukuda, H., Kondo, A., Noda, H., 2001. Biodiesel fuel production by transesterification of oils. *J. Biosci. Bioeng.* 92, 5, 405–416.

Gad, H.A., Roberts, A., Hamzi, S.H., Gad, H.A., Touiss, I., Altyar, A.E., Kensara, O.A., Ashour, M.L. (2021). Jojoba oil: An updated comprehensive review on chemistry, pharmaceutical uses, and toxicity. *Polymers*, 13, 11, 1–22. https://doi.org/10.3390/polym13111711.

Ganesan, R., Manigandan, S., Shanmugam, S., Chandramohan, V.P., Sindhu, R., Kim, S.H., Brindhadevi, K., Pugazhendhi, A., 2021. A detailed scrutinize on panorama of catalysts in biodiesel synthesis. *Sci. Total Environ.* 777, 145683. https://doi.org/10.1016/j.scitotenv.2021.145683.

Gardy, J., Hassanpour, A., Lai, X., Ahmed, M.H., Rehan, M. (2017). Biodiesel production from used cooking oil using a novel surface functionalised TiO2 nano-catalyst. *Appl. Catal. B Environ.* 207, 297–310. https://doi.org/https://doi.org/10.1016/j.apcatb.2017.01.080.

Granados, M.L., et al., 2007. Biodiesel from sunflower oil by using activated calcium oxide. *Appl. Catal. B Environ.* 73, 3–4, 317–326.

Gujjala, L.K.S., et al., 2019. Biodiesel from oleaginous microbes: Opportunities and challenges. *Biofuels* 10, 1, 45–59.

Harsha Hebbar, H.R., Math, M.C., Yatish, K.V., 2018. Optimization and kinetic study of CaO nano-particles catalyzed biodiesel production from *Bombax ceiba* oil. *Energy* 143, 25–34. https://doi.org/https://doi.org/10.1016/j.energy.2017.10.118.

Ho, W.W.S., Ng, H.K., Gan, S., 2016. Advances in ultrasound-assisted transesterification for biodiesel production, *Appl. Therm. Eng.* 100, 553–563.

Iranmanesh, M., Subrahmanyam, J.P., Babu, M.K.G., 2008. Potential of diethyl ether as a blended supplementary oxygenated fuel with biodiesel to improve combustion and emission characteristics of diesel engines. *2008 SAE International Powertrains, Fuels and Lubricants Congress.* Indian Institute of Technology, New Delhi. https://doi.org/https://doi.org/10.4271/2008-01-1805.

Islam, K., Wang, H., Rehman, S., Dong, C., Hsu, H., Sze, C., Lin, K., Leu, S., 2020. Sustainability metrics of pretreatment processes in a waste derived lignocellulosic biomass biorefinery. *Bioresour. Technol.* 298, 122558.

Jose, T.K., Anand, K., 2016. Effects of biodiesel composition on its long term storage stability. *Fuel* 177, 190–196.

Kanjaikaew, U., Tongurai, C., Chongkhong, S., Prasertsit, K., 2018. Two-step esterification of palm fatty acid distillate in ethyl ester production: Optimization and sensitivity analysis. *Renew. Energy* 119, 336–344.

Kataria, J., Mohapatra, S.K., Kundu, K., 2019. Biodiesel production from waste cooking oil using heterogeneous catalysts and its operational characteristics on variable compression ratio CI engine. *J. Energy Inst.* 92, 2, 275–287. https://doi.org/10.1016/j.joei.2018.01.008.

Kaur, M, Ali, A., 2011. Lithium ion impregnated calcium oxide as nano catalyst for the biodiesel production from karanja and jatropha oils. *Renew. Energy* 36, 2866–2871.

Keera, S.T., El Sabagh, S.M., Taman, A.R., 2011. Transesterification of vegetable oil to biodiesel fuel using alkaline catalyst. *Fuel* 90, 1, 42–47.

Khan, H.M., Ali, C.H., Iqbal, T., Yasin, S., Sulaiman, M., Mahmood, H., Raashid, M., Pasha, M., Mu, B., 2019. Current scenario and potential of biodiesel production from waste cooking oil in Pakistan: An overview. *Chin. J. Chem. Eng.* 27, 10, 2238–2250.

Kim, L., Siang, C., 2022. Analysis of oils : Determination of free fatty acid (FFA), 1–2.

Knothe, G., Razon, L.F. (2017). Biodiesel fuels. *Progr. Energy Combust. Sci.* 58, 36–59.

Korkut, I., Bayramoglu, M., 2018. Selection of catalyst and reaction conditions for ultrasound assisted biodiesel production from canola oil. *Renew. Energy* 116, 543–551.

Kouzu, M., Yamanaka, S.-y., Hidaka, J.-s., Tsunomori, M., 2009. Heterogeneous catalysis of calcium oxide used for transesterification of soybean oil with refluxing methanol. *Appl. Catal. A General* 355, 1–2, 94–99.

Kumar, D., Singh, B., 2019. Algal biorefinery: An integrated approach for sustainable biodiesel production. *Biomass Bioenergy* 131, 105398.

Ma, G., Dai, L., Liu, D., Du, W., 2018. A robust two-step process for the efficient conversion of acidic soybean oil for biodiesel production. *Catalysts* 8, 11, 527.

Manaf, I.S.A., et al., 2019. A review for key challenges of the development of biodiesel industry. *Energy Convers. Manag.* 185, February, 508–517.

Marchetti, J.M., 2011. The effect of economic variables over a biodiesel production plant. *Energy Convers. Manag.* 52, 10, 3227–3233.

Mathew, G.M., et al., 2021. Recent advances in biodiesel production: Challenges and solutions. *Sci. Total Environ.* 794, 148751.

Mehta, R.N., Chakraborty, M., Parikh, P.A., 2014. Impact of hydrogen generated by splitting water with nano-silicon and nano-aluminum on diesel engine performance. *Int. J. Hydrogen Energy* 39, 15, 8098–8105. https://doi.org/10.1016/j.ijhydene.2014.03.149.

Mekala, N.K., Potumarthi, R., Baadhe, R.R., Gupta, V.K., 2014. Current bioenergy researches: Strengths and future challenges. In Vijai K. Gupta, Maria G. Tuohy, Christian P. Kubicek, Jack Saddler, and Feng Xu (Eds.), *Bioenergy Research: Advances and Applications*, pp. 1–21. Elsevier. https://doi.org/10.1016/B978-0-444-59561-4.00001-2.

Misenan, M.S.M., Ali, E.S., Sofia, A., Khiar, A., 2018. Conductivity, dielectric and modulus study of chitosan – methyl cellulose – BMIMTFSI polymer electrolyte doped with cellulose nano crystal. *AIP Conference Proceedings.* 030010. https://doi.org/10.1063/1.5041231.

Misenan, M.S.M., et al., 2021. Cellulose nanofiber as potential absorbent material for chloride ion. *Solid State Phenom.* 317, 263–269. https://doi.org/10.4028/www.scientific.net/ssp.317.263.

Misenan, M.S.M., Khiar, A.S.A., Eren, T., 2022. Polyurethane-based polymer electrolyte for lithium ion batteries: A review. *Polym. Int.* 71, 751–769. https://doi.org/10.1002/pi.6395.

Mittelbach, M., Gangl, S., 2001. Long storage stability of biodiesel made from rapeseed and used frying oil. *J. Am. Oil Chem. Soc.* 78, 6, 573–577.

Naik, M., Meher, L.C., Naik, S.N., Das, L.M., 2008. Production of biodiesel from high free fatty acid Karanja (*Pongamia pinnata*) oil. *Biomass Bioenergy* 32, 4, 354–357.

Navas, M.B., Lick, I.D., Bolla, P.A., Casella, M.L., Ruggera, J.F., 2018. Transesterification of soybean and castor oil with methanol and butanol using heterogeneous basic catalysts to obtain biodiesel. *Chem. Eng. Sci.* 187, 444–454.

Nguyen, T., Do, L., Sabatini, D.A., 2010. Biodiesel production via peanut oil extraction using diesel-based reverse-micellar microemulsions. *Fuel* 89, 2285–2291.

Norizan, M.N., Shazleen, S.S., Alias, A.H., Sabaruddin, F.A., Asyraf, M.R.M., Zainudin, E.S., Abdullah, N., Samsudin, M.S., Kamarudin, S.H., Norrrahim, M.N.F., 2022. Nanocellulose-based nanocomposites for sustainable applications: A review. *Nanomaterials* 12, 19, 3483. https://doi.org/10.3390/nano12193483.

Norrahim, M.N.F., Knight, V.F., Nurazzi, N.M., Jenol, M.A., Misenan, M.S.M., Janudin, N., Kasim, N.A.M., Shukor, M.F.A., Ilyas, R.A., Asyraf, M.R.M., Naveen, J., 2022. The frontiers of functionalized nanocellulose-based composites and their application as chemical sensors. *Polymers* 14, 4461. https://doi.org/10.3390/polym14204461.

Nur, S., Abidin, Z., Lee, H.V., Asikin-mijan, N., Juan, J.C., 2020. Ni, Zn and Fe hydrotalcite-like catalysts for catalytic biomass compound into green biofuel. *Pure Appl. Chem.* 92, 587–600.

Pal, U.S., Patra, R.K., Sahoo, N.R., Bakhara, C.K., Panda, M.K., 2015. Effect of refining on quality and composition of sunflower oil. *J. Food Sci. Technol.* 52, 7, 4613–4618.

Pali, H.S., Sharma, A., Singh, Y., Kumar, N., 2020. Sal biodiesel production using Indian abundant forest feedstock. *Fuel* 275, 117781.

Dilşat Bozdoğan Konuşkan, Murat Yilmaztekin, Mehmet Mert, Oktay Gençer. Pamuk Yağlarının Fiziko-Kimyasal Özellikleri ve Yağ Asitleri Kompozisyonu (Physico-chemical characteristic and fatty acids compositions of cottonseed oils). *Journal of Agricultural Sciences.* 23, 2, 253–259.

Prabu, A., 2018. Nanoparticles as additive in biodiesel on the working characteristics of a DI diesel engine. *Ain Shams Eng. J.* 9, 4, 2343–2349. https://doi.org/https://doi.org/10.1016/j.asej.2017.04.004.

Rashedi, A., Gul, N., Hussain, M., Hadi, R., Khan, N., et al., 2022. Life cycle environmental sustainability and cumulative energy assessment of biomass pellets biofuel derived from agroforest residues. *PLoS One* 17, 10, e0275005. https://doi.org/10.1371/journal.pone.0275005.

Sadhik Basha, J., Anand, R.B., 2013. The influence of nano additive blended biodiesel fuels on the working characteristics of a diesel engine. *J. Braz. Soc. Mech. Sci. Eng.* 35, 3, 257–264. https://doi.org/10.1007/s40430-013-0023-0.

Sajith, V., Sobhan, C.B., Peterson, G.P., 2010. Experimental investigations on the effects of cerium oxide nanoparticle fuel additives on biodiesel. *Adv. Mech. Eng.* 2010, 1–6. https://doi.org/10.1155/2010/581407.

SathyaSelvabala, V., Varathachary, T.K., Selvaraj, D.K., Ponnusamy, V., Subramanian, S., 2010. Removal of free fatty acid in *Azadirachta indica* (neem) seed oil using phosphoric acid modified mordenite for biodiesel production. *Bioresour. Technol.* 101, 15, 5897–5902. https://doi.org/10.1016/j.biortech.2010.02.092.

Selvan, V.A.M., Anand, R.B., Udayakumar, M., 2009. Effects of cerium oxide nanoparticle addition in diesel and diesel-biodiesel-ethanol blends on the performance and emission characteristics of a CI engine. *J. Eng. Appl. Sci.* 4, 7, 1–6.

Shahabuddin, M., Liaquat, A.M., Masjuki, H.H., Kalam, M.A., Mofijur, M., 2013. Ignition delay, combustion and emission characteristics of diesel engine fueled with biodiesel. *Renew. Sustain Energy Rev.* 21, 623–632.

Shahabuddin, M., Masjuki, H.H., Kalam, V., Hazrat, M.A., Mofijur, M.A., Nazira, M., Varman, A.M., Liaquat, M., 2012. Biofuel: Potential energy source in road transportation sector in Malaysia, innovation for sustainable and secure energy. *BioEnergy Res.* 15, 1371–1386.

Singh, D., Sharma, D., Soni, S.L., Sharma, S., Kumar Sharma, P., Jhalani, A., 2020. A review on feedstocks, production processes, and yield for different generations of biodiesel. *Fuel* 262, October 2019, 116553.

Singh, H.K.A.P.G., Yusup, S., Wai, C.K., 2016. Physicochemical properties of crude rubber seed oil for biogasoline production. *Proc. Eng.* 148, 426–431.

Skillas, G., Qian, Z., Baltensperger, U., Matter, U., Burtscher, H., 2000. The influence of additives on the size distribution and composition of particles produced by diesel engines. *Combust. Sci. Technol.* 154, 1, 259–273. https://doi.org/10.1080/00102200008947279.

Soudagar, M.E.M., Nik-Ghazali, N.N., Abul Kalam, M., Badruddin, I.A., Banapurmath, N.R., Akram, N., 2018. The effect of nano-additives in diesel-biodiesel fuel blends: A comprehensive review on stability, engine performance and emission characteristics. *Energy Convers. Manag.* 178, September, 146–177. https://doi.org/10.1016/j.enconman.2018.10.019.

Takase, M., Zhao, T., Zhang, M., Chen, Y., Liu, H., Yang, L., et al., 2015. An expatriate review of neem, jatropha, rubber and karanja as multipurpose non-edible biodiesel resources and comparison of their fuel, engine and emission properties. *Renew. Sustain Energy Rev.* 43, 495–520.

Taufiq-Yap, Y.H., Ahmad Farabi, M.S., Syazwani, O.N., Ibrahim, M.L., Marliza, T.S., 2020. Sustainable production of bioenergy. In Gupta, A., De, A., Aggarwal, S., Kushari, A., and Runchal, A. (Eds.), *Green Energy and Technology*, pp. 541–561. Springer, Singapore.

Thiruganasambanthan, T., Ilyas, R.A., Norrrahim, M.N.F., Kumar, T.S.M., Siengchin, S., Misenan, M.S.M., Farid, M.A.A., Nurazzi, N.M., Asyraf, M.R.M., Zakaria, S.Z.S., Razman, M.R., 2022. Emerging developments on nanocellulose as liquid crystals: A biomimetic approach. *Polymers* 14, 8, 1546. https://doi.org/10.3390/polym14081546.

Tiwari, S., 2013. Optimization of transesterification process for biodiesel production from waste oil. *Int. J. Pharm. Life Sci.* 4, 6, 2701–2704.

Veljković, V.B., Lakićević, S.H., Stamenković, O.S., Todorović, Z.B., Lazić, M.L., 2006. Biodiesel production from tobacco (*Nicotiana tabacum* L.) seed oil with a high content of free fatty acids. *Fuel* 85, 17–18, 2671–2675. https://doi.org/10.1016/j.fuel.2006.04.015.

Vicentini-Polette, C., Rodolfo Ramos, P., Bernardo Gonçalves, C., Lopes De Oliveira, A., 2021. Determination of free fatty acids in crude vegetable oil samples obtained by high-pressure processes. *Food Chem.* 12, 100166.

Wang, T. and Johnson, L.A., 2001. Refining high-free fatty acid wheat germ oil. *Journal of the American Oil Chemists' Society*, 78, 71–76.

Win, S.S., Trabold, T.A., 2018. Sustainable waste-to-energy technologies: Transesterification. In Thomas A. Trabold and Callie W. Babbitt (Eds.), *Sustainable Food Waste-to-Energy Systems*, pp. 47–67. Elsevier Inc.

Wu, L., Wei, T., Tong, Z., Zou, Y., Lin, Z., Sun, J., 2016. Bentonite-enhanced biodiesel production by NaOH-catalyzed transesterification of soybean oil with methanol. *Fuel Process. Technol.* 144, 334–340.

Yang, W.M., An, H., Chou, S.K., Vedharaji, S., Vallinagam, R., Balaji, M., Mohammad, F.E.A., Chua, K.J.E., 2013. Emulsion fuel with novel nano-organic additives for diesel engine application. *Fuel* 104, 726–731. https://doi.org/10.1016/j.fuel.2012.04.051.

Yuan, X., Liu, J., Zeng, G., Shi, J., Tong, J., Huang, G., 2008. Optimization of conversion of waste rapeseed oil with high FFA to biodiesel using response surface methodology. *Renew. Energy* 33, 7, 1678–1684.

Yuvarajan, D., Dinesh Babu, M., BeemKumar, N., Amith Kishore, P., 2018. Experimental investigation on the influence of titanium dioxide nanofluid on emission pattern of biodiesel in a diesel engine. *Atmos. Pollut. Res.* 9, 1, 47–52. https://doi.org/10.1016/j.apr.2017.06.003.

Zainol, M.M., Aishah, N., Amin, S., Asmadi, M., 2019. Kinetics and thermodynamic analysis of levulinic acid esterification using lignin-furfural carbon cryogel catalyst. *Renew. Energy* 130, 547–557. https://doi.org/10.1016/j.renene.2018.06.085.

Zhang, Y, Dubé, M.A., McLean, D.D., Kates, M., 2003. Biodiesel production from waste cooking oil: 2. Economic assessment and sensitivity analysis. *Bioresour. Technol.* 90, 3, 229–240.

19 Recent Progress of Advanced Nanomaterials in Renewable Energy

Norli Abdullah, Norherdawati Kasim, Siti Hasnawati Jamal, Intan Juliana Shamsudin, Noor Aisyah Ahmad Shah, and Siti Hasnawati Jamal
Universiti Pertahanan Nasional Malaysia

19.1 INTRODUCTION

The development of technologies for sustainable energy applications, which include energy storage, generation, conversion, collection, transport, and distribution which address major global environmental issues, greatly benefits from the use of advanced materials. (Gielen et al., 2019). One of the main goals of using advanced materials for renewable energy applications is to leverage multidisciplinary expertise to create new materials for clean energy generation, storage and conservation, and environmental remediation. Due to its fluctuating nature, renewable energy cannot completely replace fossil fuels until conversion and storage issues are resolved. Fossil fuels such as crude oil, coal, and natural gas continue to play a major role in worldwide energy systems (Romanello et al., 2022). Fossil fuels are by far the most significant contributors to global climate change, accountable for more than 75% of global greenhouse gas emissions and closely 90% of all carbon dioxide emissions (Green et al., 2022; James & Menzies, 2022). Therefore, greenhouse gas emissions must be reduced to prevent the worst effect of climate change. To accomplish this, we must reduce our reliance on fossil fuels and invest in alternative technologies that are affordable, sustainable, and reliable alternative energy sources. Renewable energy technologies capture and convert energy from the sun, wind, and the earth's core into usable forms of energy such as heat, electricity, and fuel (Li et al., 2022; Zhu et al., 2023). Agricultural products from "biomass" can also be used to generate electricity and heat when burned with coal (Oner & Dincer, 2022; She et al., 2023). Renewable energies generate electricity without producing greenhouse gases or air pollutants.

One of the most promising substitutes for fossil fuel energy is electricity. However, it was estimated that 22,000 TWh of electricity was consumed globally in 2017 and that 56.1% of that amount came from fossil fuels like coal, oil, and natural gas (Sen et al., 2021; Shehzad et al., 2022). Clean and renewable energy

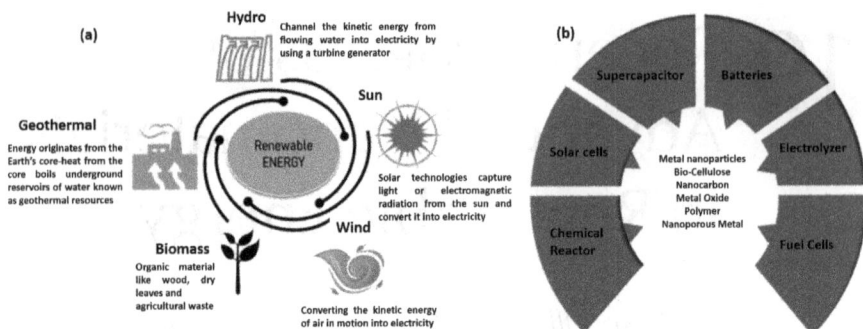

FIGURE 19.1   Type of renewable energy resources (a) and advanced materials application for the renewable energy conversion and storage devices (b).

must predominate in power generation in order to be sustainable. Energy storage and conversion technologies that are highly efficient, safe, and cost-effective are required to replace fossil fuels with clean, renewable energy. It appears that the most advanced energy storage and conversion technologies available today fall short of the aforementioned standards. Both high-density electrical storage applications and clean, renewable energy conversion with high efficiency require advanced materials. Although advanced materials have greatly improved the performance of today's sustainable energy devices, the cost and efficiency of these devices remain high, and their lifespans are short.

Therefore, the development of new and novel materials for advanced, high-capacity, and sustainable energy storage, production, and extraction technologies is critical for achieving renewable solutions for the energy grid. Nonetheless, the exploitation of these nanomaterials for scalable industrial applications still remains a challenging, and in part are lack of understanding. The chemical structure of the starting material affects the various properties and mechanisms of the results obtained. This review focuses on the current progress in the use of advanced materials for renewable energy applications, especially for supercapacitor, photovoltaic, batteries, CO_2 capture, and optoelectronic devices. Here, the performance of different types of materials such as metal-organic frameworks (MOFs) (Jahan et al., 2022; Qin et al., 2022), MXene (Ahouei et al., 2022; Cheng et al., 2022), nanocellulose (Kargupta et al., 2021; Lokhande et al., 2022), graphene oxide (GO) (Deepalakshmi et al., 2022; Li et al., 2022), reduced graphene oxide (rGO) (Kant et al., 2022; Kosukoglu et al., 2022) and conducting polymer which enable the associated advanced technologies (Figure 19.1) in various devices is summarized.

In addition, extensive discussions on their preparation method and performance will be highlighted, which will enable researchers to improve and discover suitable solutions for various industrial-related problems. In general, advanced materials perform crucial part in the development of improved, economic, and highly efficient clean energy technologies. Progress in the development of advanced materials development for renewable energy technologies and storage devices such as batteries, fuel cells, solar cells, and supercapacitors will be discussed.

19.2   ADVANCED MATERIALS FOR SUPERCAPACITORS

The two most promising options that are frequently employed as energy storage technologies are batteries and supercapacitors. Batteries may have a high energy density, but they nevertheless have a drawback: They transmit or absorb power slowly (Lipu et al., 2018; Zhang et al., 2018), while supercapacitors, often referred to as electrochemical capacitors or ultracapacitors, are a form of energy storage device that has a higher power density than batteries, a longer cycle life than regular capacitors, and a higher energy density than those batteries. Supercapacitors are suitable for circumstances where high current is needed for a brief period of time since they may be frequently charged and discharged.

A supercapacitor can only withstand a voltage of 2.7 V. High voltage is also feasible, but the supercapacitor's lifespan will be shortened. To generate higher voltage, it is possible to connect numerous supercapacitors in series, although doing so will raise their internal resistance and lower their overall capacitance. Due to their relatively high specific energy and power densities, supercapacitors are projected to bridge the gap between batteries and conventional capacitors. As a result, supercapacitors are crucial in the field of energy storage and have the potential to replace batteries (Lin et al., 2018; Raza et al., 2018). Earth-abundant and renewable carbon-based materials, such as active carbon (Sun et al., 2023), carbon nanotubes (Suganya et al., 2021; Yang et al., 2021), and graphene (Pallavolu et al., 2023), are frequently employed as the electrodes in supercapacitors. Due to its exceptional properties, graphene was shown to be a suitable electrode material among them.

Yao et al. (2022) successfully demonstrated a flexible supercapacitor built on vertical graphene (VG), carbon fabric (CF) current collector, and H_2SO_4 gel polymer electrolyte with 73% capacitance retention (Figure 19.2). The VG/CF flexible electrodes perform exceptionally electrochemically, especially at 1 M H_2SO_4 electrolyte.

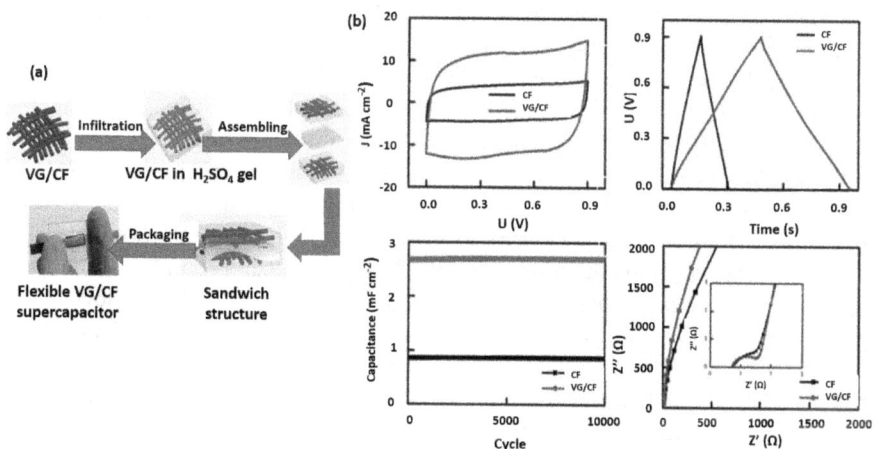

FIGURE 19.2   (a) Illustration for preparation process and (b) The Nyquist plots of electrochemical performance. Adopted from Yao et al. (2023).

A slight decrease in energy density from 119.2 to 86.6 Wh cm^2 is possible thanks to the remarkable rate performance; however, a noticeable rise in power density from 0.045 to 13.2 mW cm^2 is also possible. The flexible VG/CF supercapacitor also showed complete electrochemical stability even when mechanically bent. In the bending range of 0°–60°, the capacitance remains constant. Given by the functional performance of this material, it shows their potential as high-power energy source for the next-generation flexible electronics and optoelectronics.

Fallah et al. (2022) synthesized supercapacitor working electrode using reduced graphene oxide (rGO) and cerium metal-organic framework (Ce-MOF) nanocomposites. Ce-MOF was synthesized using a quick and simple solvothermal process over the course of 1 hour at 60°C. By physically mixing different mass fractions of 1:1, 2:1, and 1:2, the composites of MOF and rGO were created. The efficacy of all the composites was higher than that of pure Ce-MOF and rGO, according to the results, and the composite with an equal mass percentage of Ce-MOF and rGO performed best. At a scan rate of 2 mVs, the best-fabricated electrode displayed the maximum capacitance of 556.3 F g.

Amorphous cellulose nanofiber (ACF) supercapacitors with various metal carboxylate radicals such as (M: Na(I), Ca(II), and Al(III)) were studied and their charging and discharging behaviors were investigated. Comparing Na-ACF to Ca- and Al-ACFs, the storage effect of Na-ACF was higher. Based on the migration of protonic solutes, an electric double layer model in a $C_{12}H_{17}O_{11}Na$ electrolyte with an electrical resistivity of 6.8×10^3 Ω cm was used to propose the charge storage mechanism for a Na-ACF supercapacitor. The supercapacitor, which displayed quick charging upon voltage application, was able to illuminate a white LED for 7 s after charging with 10 mA at 18.5 V, according to Fukuhara et al. (2022).

Due to their superior properties such as no formatting or binder required and anisotropy, vertically aligned CNTs have received a lot of attention as potential electrode materials for supercapacitor. Aerosol-assisted catalytic chemical vapor deposition (CCVD) was employed during the process by lowering the growth temperature of the aluminum current collectors (Figure 19.3). CCVD is a cost-effective and roll-to-roll compatible process. According to Querne et al. (2023), electrodes have high areal capacitances of up to 25 F/cm^3 and 360 mF/cm^2, as well as an average gravimetric capacitance of 45 F/g. Volumetric density has little impact at high scan rates, while increasing VACNT height limits performance. With a potential difference of 2.7 V and a high energy retention capacity (90% at 13 V), a symmetrical supercapacitor based on VACNT/Al electrodes assembled in a coin cell exhibits high energy and power density.

Multivalent metal-ion hybrid supercapacitor is another promising candidate. Zinc-ion hybrid supercapacitor is a recently developed energy storage device that combines zinc-ion batteries (ZIBs) with supercapacitors. Due to high demand for quickly rechargeable batteries and supercapacitors as well as with the limited availability of lithium and cobalt, the study of sustainable energy storage using Zn-ion hybrid was investigated. Etman et al. (2021) described the application of freestanding mixed of MXene film with $Mo_{1.33}$ CT$_z$-Ti$_3$C$_2$T$_z$ in Zn-ion hybrid supercapacitors (Figure 19.4). MXenes are two-dimensional materials with outstanding properties such as superior conductivity and capacitance and enable the improvement of their performance in energy storage devices. A one-step vacuum filtration method is used

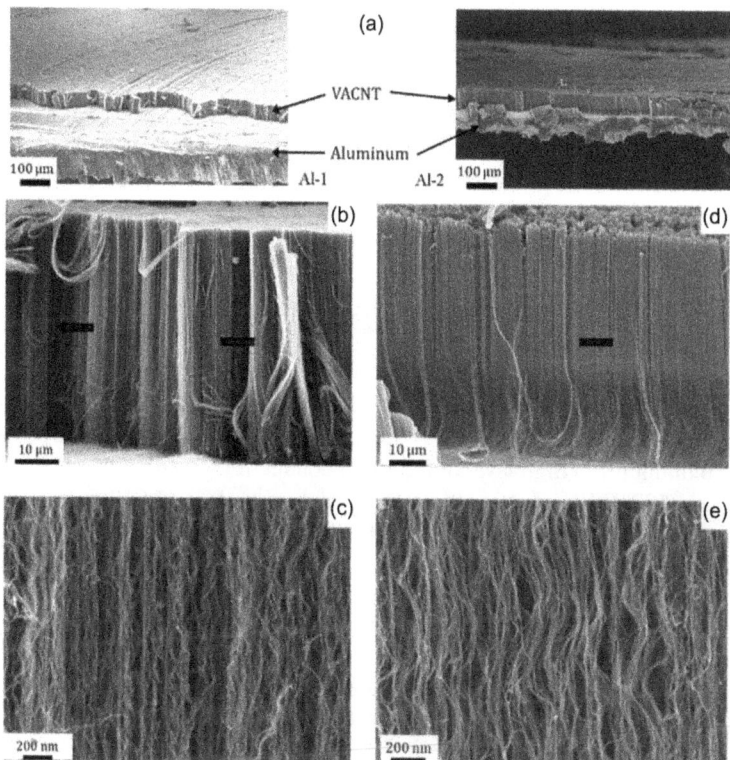

FIGURE 19.3 SEM images of VACNT (A–E). Adopted and reproduced from Querne et al. (2023).

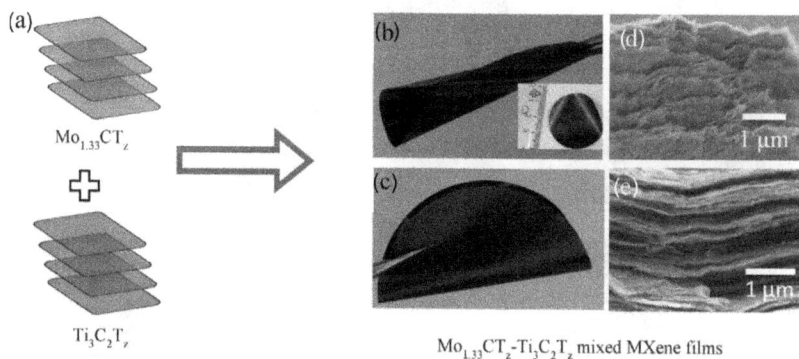

FIGURE 19.4 (a) Schematic diagram synthesis of freestanding mixed MXene film with $Mo_{1.33}$ CTz-Ti_3C_2Tz in Zn-ion hybrid supercapacitors and (b) images of MXene electrodes. Adopted from Etman et al. (2021).

to prepare mixed MXene films from pristine MXene suspensions. The capacities of mixed MXene are higher as compared to about 159 and 59 mAh/g at scan rates of 0.5 and 100 mV/s and their energy densities of approximately 103 and 38 Wh/kg, with power densities of 0.143 and 10.6 kW/kg, respectively, as compared to pristine MXene films. The effect of electrode thickness and the mechanism of charge storage on rate performance were also studied. This research paves the way for the application of Zn-ion hybrid mixed with MXene as an electrode with high energy density and power density sustainable energy storage systems.

Cellulose-derived nanostructures (CNCs, CNFs, BCNFs, CBNFs, etc.) are promisingly used in the development of electrodes for supercapacitors. Chen et al. (2018) published a method to fabricate hybrid of graphene oxide (GO) and cellulose nanocrystals (CNC) produced using spinning of non-liquid crystal and chemical reduction process. The resulting hybrid GO/CNC fibers simultaneously exhibited high capacitive performance, improved mechanical properties, and improved hydrophilicity. It also maintains a high level of electrical conductivity. With 208.2 Fcm^3 of high capacitance, 199.8 MPa of strength, 63.3° of contact angle, and 64.7 Scm of conductivity were displayed by a sample with a 100/20 GO/CNC weight ratio. Additionally, supercapacitors made from this fiber have excellent flexibility, bending stability. Their high energy and power densities (5.1 mW h cm^3 and 496.4 mW cm^3) can be used as flexible power storage (Figure 19.5).

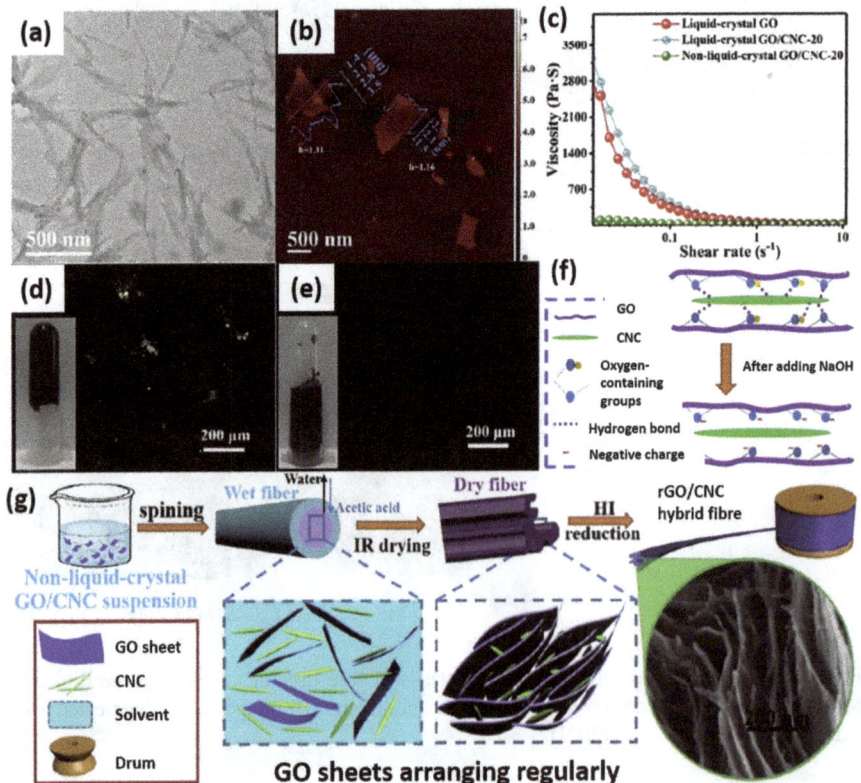

FIGURE 19.5   Images of CNC and GO (a–c). Reproduced from Chen et al. (2018).

19.3   HYBRID MATERIALS FOR SOLAR PHOTOVOLTAICS

Because of the gradual rise in the price of fossil fuels, the need for replacement energy sources has become critical. Photovoltaics are clean, abundant, and sustainable energy sources with the ability to fulfil the world's increasing energy demand. Because solar photovoltaics will play a critical role in the future, it is critical to analyze and conduct comprehensive research on materials and technology types. Solar energy has attracted a lot of interest among renewable resources due to its abundance and is one of the enormous sources of renewable energies. Despite being based on silicon, hybrid organic–inorganic materials have gained popularity in solar photovoltaics applications. Perovskites solar cells (PSCs) are promising photovoltaics material due to numerous advantages, such as high efficiency and wide applicability when compared to silicon-based solar cells. It is also have successfully attained the efficiency of power conversion greater than 25%. In addition, PSCs are also considered as a smart energy management strategy due to their minimal carbon generation during manufacture. To improve PSCs performance, hybrid halide perovskites (HHP) were introduced. The power conversion efficiency (PCE) in inorganic PSCs is still left behind the proven HHPs device efficiencies due to phase instability and low film quality. To improve phase stability, PSCs with various ratios of phenylethylammonium, PEA+ cations were incorporated into $CsPbI_3$-xBrx perovskite to achieve better crystallization growth, as organic large cations help in controlled nucleation and growth in perovskites. The perovskites films were fabricated, which mainly consist of dry method and treatment of cesium bromide (CsBr) where it facilitates significantly greater efficiency and stability due to defect passivation and stabilizing the perovskite lattice structure by the incorporation of Br-ions. There are numerous pinholes and poor morphology with no addition of PEA+ cation butpinholes were significantly reduced with the incorporation of PEA+ cation, and surface morphology was improved. According to Lee et al. (2022), with the increasing of PEA+ ratio from 0% to 10%, the crystallinity of the perovskite film that was deposited on zinc oxide (ZnO) layer was greatly improved. The result of X-ray photoelectron and UV-vis spectroscopy also depicted that the addition of PEA+ in the perovskite helps in nearly uniform distribution of Br- ion after post-annealing process.

Additionally, depth profiling research demonstrates that PEA+ interacts more with the electron transfer layer (ETL) of ZnO beneath perovskite in the intermediate phase, which facilitates the production of compact films (Lee et al., 2022). PEA+ sublimates after 300°C post-annealing treatment, as no evidence of PEA+ existence was found in post-annealed perovskite film characterizations. Perovskite solar cell (PSC) performance shows high conversion efficiency of 14.75% as well as enhanced stability, sustaining nearly 83% of its initial efficiency for 400 hours (Table 19.1).

In a different work, a novel nanocomposite of graphene–La_2CrFeW_6 perovskite compounds were prepared. A simple hydrothermal method was employed using an organic–inorganic counter electrode based on CdSe, gallic acid, urea, and sulfate in dye-sensitized solar cells (DSSCs). The nanocomposites demonstrate an outstanding electrocatalytic activity with low charge-transfer resistance. This nanocomposite's unique structure also enhances the active sites for triiodide ion reduction on the counter electrodes. The photo-conversion efficiency of the DSSCs fabricated

TABLE 19.1

Photovoltaics of PEA Assisted Perovskites Devices

Samples	J_{sc} (mA/cm^2)	PCE (%)
PEAI 0%	14.89	11.68
PEAI 2.5%	15.15	12.88
PEAI 5%	15.52	14.09
PEAI 10%	15.97	14.75
PEAI 15%	15.5	14.07

Source: Reproduced from Lee et al. (2022).

with a counter electrode based on gallic obtained a higher value of 10.40% than the conventional Pt electrode (7.49%) under standard illumination (1 sun, 100 mW/ cm^2, AM 1.5). It was also revealed that the graphene–La$_2$CrFeW$_6$ produced with an organic–inorganic base counter electrode was much more successful than those described in previous papers in which graphene was used as a counter electrode material in various ways. As a result, the current nanocomposite is regarded as an ecologically benign, low-cost, and extremely effective counter electrode material for DSSCs (Oh et al., 2021).

The electron transport layer (ETL) ZnO in perovskites (PVSK) solar cells (PSCs) also had been explored by Taukeer Khan & Khan (2022). The PSCs were fabricated on the aluminum-doped ZnO (the electron transport layer (ETL)). ZnO in perovskites (PVSK) solar cells also had been explored by Taukeer Khan & Khan (2022). Reduced graphene oxide (rGO) was used to modify the aluminum-doped ZnO (AZO) layer, and its effect on the photovoltaic performance of perovskite solar cells (PSCs) was investigated. PCE of 17.08% for the device with rGO layer compared to efficiency of 15.93% for the device without rGO layer indicates improved performance. According to steady-state photoluminescence (PL) and time-resolved photoluminescence (TRPL) spectra, the faster charge transfer at the PVSK/ETL interface is the attributes of the rGO-based device with better performance. The evaluated PV cell parameters reveal that the rGO layer restrains the recombination losses and speed up the charge transport in PSCs. Organic– inorganic hybrid PSCs are gaining interest because of their long carrier diffusion length, long lifetime, and high absorption coefficient. In other research work, copper telluride nanocrystals (NCs) have received significant interest and studied by Wu et al. (2022) due to their localized surface plasmon resonance (LSPR). They were successfully fabricated by a hot injection-like method. The TiO$_2$ precursor was spin-coated on the fluorine-doped tin oxide (FTO) to form a TiO$_2$ film as an electron transfer layer (ETL), and spiro-OMeTAD solution was spin-coated as a hole transport layer (HTL). Different concentrations of Cu$_{1.44}$ Te NCs (1%, 2%, and 3%) were used to prepare perovskite films, and NCs were successfully embedded into hybrid PSCs.

The NCs displayed a uniform cubic geometry with the average size of the Cu$_{1.44}$ Te NCs 16–17 nm, which is suitable for photovoltaic devices. A plasmonic NCs with a size smaller than 50 nm is advantageous for near-field absorption, which enhances light absorption, and small particles are advantageous for the film production of perovskite layer. The perovskite film with $Cu_{1.44}$ Te NCs doping exhibits a significant photoluminescence quenching, demonstrating that the $Cu_{1.44}$ Te NCs doping has successfully increased the carrier extraction/transport rate at the hole transport layer (HTL)/perovskite interface. The $Cu_{1.44}$ Te NCs showed strong LSPR in the NIR region, strengthening the visible light region and broadening the light absorption of perovskite films to the NIR region. This led to an increase in photocarriers and an improvement in PCE. The device modified with 2% $Cu_{1.44}$ Te NCs presented an increased PCE of 16.75%, compared to that of the reference device (13.61%). In conclusion, the performance of PSCs may be improved by copper telluride NCs with plasmonic effects, offering a new approach to enhance photovoltaic device performance (Wu et al., 2022).

Qiu et al. investigated the production of an organic–inorganic hybrid electron transport layer (ETL) using a water-soluble polyvinylpyrrolidone/tin oxide (PVP/SnO_2) perovskite film based on the one-step antisolvent approach. SnO_2 nanoparticles were employed to increase the electron mobility of the ETL. PVP has long polymer chains linked to the SnO_2 solution, causing in an organic–inorganic hybrid ELT with a SnO_2 dispersion stabilized in a solvent and controlling the arrangement of SnO_2 nanoparticles. The agglomeration of SnO_2 nanoparticles causes their irregular distribution and the presence of pinholes, prevents electrons from moving across the SnO_2/perovskite interface, and hence limits device performance. Perovskite grain development is hampered by poor wettability between SnO_2 film and the perovskite precursor solution. This method produces organic–inorganic hybrid ELT of PVP/SnO_2 with few defects and perfect wettability and the high-quality perovskite film. The resultant modified PVP-SnO_2 device displays improved operating stability, sufficient repeatability, and a high power conversion efficiency of 18.98% (Qiu et al., 2022).

The effect of carbon doping on the absorption properties of titanium dioxides (TiO_2) nanotubular (NTs) films and the photovoltaic performance of the carbon-doped TiO_2 with pulse laser deposited bismuth oxide (Bi_2O_3) was studied by Bjelajac et al. (2022). A Bi_2O_3 was applied to both the undoped and C-doped TiO_2 NTs films. Anodization was used to create a nanotubular photoanode on FTO (F-SnO_2) glass. The C-doping and crystallinity of the TiO_2 nanotubes were enhanced by further annealing in the CH_4 environment. The nanotubes were coated with Bi_2O_3, which serves as a hole transport material, using pulsed laser deposition. Results obtained from X-ray photoelectron spectroscopy analysis show a shift in the maximum position of the valence band towards lower binding energy for doped samples. Doping increases absorption by shifting the absorption edge to 567 nm. The I-V measurements made under light reveal that TiO_2 C-doping improves current density in line with the absorbance result. The samples with the highest open circuit voltage had Bi_2O_3 layers that had been deposited at 300°C, demonstrating increased p-n junction quality and better contact between Bi_2O_3 and TiO_2. This in situ annealing resulted in the formation of a close contact between Bi_2O_3 and TiO_2, allowing for faster charge transport than the contact obtained with no annealing.

19.4 ADVANCED MATERIALS FOR LITHIUM-ION, SODIUM-ION, AND LITHIUM–SULFUR BATTERIES

For lithium-ion batteries, sodium-ion batteries, and lithium–sulfur batteries, the state-of-the-art in cathode and anode nanomaterials is taken into consideration. The huge surface area and porosity of nanomaterials are the key reasons why they are currently being studied for storage devices. These properties enable the development of novel active reactions, the reduction of the ion transport path length, lower the specific surface current rate, and the enhancement of stability and specific capacity. Interest in nanomaterials has grown because of their promising electrochemical features, which include good kinetics and cycle stability (Lee & Cho, 2011; Majdi et al., 2021).

It has been widely published on the study of Fe_2O_3 as an anode electrode material for lithium-ion batteries (LIBs). Porous Fe_2O_3 nanoparticles as LIB anode materials were reported. Using a microwave-assisted template approach, the Fe-based metal-organic framework (Fe-MIL-88A) material with a spindle-like morphology was created. The creation of uniform Fe_2O_3-MW-4h nanoparticles was formed with a multicavity structure. The large pore volume and high surface area of the resulting Fe_2O_3-MW-4h nanoparticles had demonstrated the special advantages of nanomaterials. These characteristics lessen the materials' resistance while encouraging the electrolytes' flow. Most notably, the Fe_2O_3-MW-4h nanoparticles' multicavity shape might lessen the volume change that occurs during the Li+ insertion and extraction procedure. The Fe_2O_3-MW-4h nanoparticles significantly enhanced electrochemical performance as anode materials for LIBs. The microwave-assisted template is deemed viewed as a potentially effective technique for the future investigation of LIB anode electrode materials in the manufacturing of metal oxides with multicavity structures nanoparticles (Zhang et al., 2021).

Recent study on the electrochemical performance of nanocarbon separated nanosheet silicon for lithium ion battery anode material was released. Through carbonization of the mixture of nanosheet silicon (Si) and polyaniline PANI/CNT nanocomposite, nanocarbon-isolated silicon (Si) with variable CNT proportions has been effectively built. The findings implied that the CNT proportion in the electrode has a significant impact on the battery's specific capacity, rate capability, and long-term cycle stability. According to the findings, the composition of Si@CNT(M)/C/G with high CNT content has greater initial specific capacitance (1,201 mAh/g) and substantially higher capacity retention (83.68%) during the course of 500 cycles than the same composition with low CNT concentration (491.5 mAh/g, 54.85% capacity retention). The Si@CNT(M)/C/G electrode that had uniformly dispersed CNT revealed to be helpful to reduce mechanical strain, ensure good conductivity, and prevent nanosheet Si from cracking. The CNT conductive network in the electrode, which was able to retain both the structural stability of the Si and the conductivity throughout the cell cycles, has led to these observations (Sun et al., 2022).

High theoretical capacity and affordable copper sulfide (Cu_xS)-based anodes have reportedly attracted a lot of interest for modern sodium ion batteries (SIBs) (Kang et al., 2019). However, because of their unpredictable cycling performance and issues with sodium sulfide (Na_xS) electrolyte dissolving, their practical applicability was restricted. Numerous investigations on the usage of nanostructures as electrode materials in SIBs have been done. A study on nanocarbon coating layer on nano-structured copper sulfide–metal-organic framework-generated carbon was reported in the field of increased sodium-ion battery anode. Using a metal-organic framework (MOF-199) as a precursor, a two-step sulfurization and carbonization process involving H_2S gas-assisted plasma-enhanced chemical vapor deposition (PECVD) was used to create nanoporous Cu_xS with a large surface area incorporated in the MOF-derived carbon network (Cu_xS-C).

The $Cu_{1.8}S$-C was then uniformly covered with a nanocarbon layer after being hydrothermally heated and annealed. The SIB performances of the nanoporous $Cu_{1.8}S$-C/C core/shell anode materials were superior to pure $Cu_{1.8}S$-C, with capacity retention (93%) and a specific capacity of 372 mAh/g achieved during 110 cycles. The primary functions of a nanocarbon layer coated on the $Cu_{1.8}S$-C anode structure during cycling are to accommodate the volume change of the $Cu_{1.8}S$-C anode structure, enhance electrical conductivity, and hinder the dissolution of Na_xS into the electrolyte, all of which contribute to this encouraging SIB performance. Because of its physicochemical and electrochemical properties, the $Cu_{1.8}S$-C/C structure is a viable anode material for advanced and large-scale SIBs (Kang et al., 2019).

A convenient technique for converting CO_2 gas into nanostructured carbon materials was established in another study on the application of nanocarbon for SIB anode. Iron oxide was converted to a metallic phase, which then produced an iron carbide and oxide phase by adsorbing CO_2 gas on its surface. The synthesis of hollow structured nanocarbons was accomplished once the iron species were eliminated. The nanocarbon demonstrated great stability (1,200 cycles) and capacity (260 mA h/g[1]) as an anode material. Additionally, due to the hollow nanostructure and suitable carbon crystallinity, it demonstrated a high rate capability. The CO_2-H-C electrode in SIB performance is much better than that of commercial carbon materials. This iron-based metallothermic process can be employed in industrial plants that produce heat and CO_2 gas as it successfully utilizes CO_2 gas for nanomaterial synthesis and considered as new CO_2 capture and utilization (CCU) approach (Jo et al., 2019).

The next-generation flexible and wearable electronics were developed using flexible high-energy-density lithium–sulfur (Li-S) batteries based on all-fibrous sulfur cathodes and separators. Electrically conductive single-walled carbon nanotubes (CNTs) were coated with cellulose nanofibers to create flexible and free-standing sulfur cathodes (Figure 19.6). This fibrous structure produced a 3D porous electrode with a sizable surface area that can be used in high-energy-density batteries and accelerate redox kinetics and obtain high sulfur loading percentage without the need of a metal collector. The shuttle effects of lithium–polysulfide are minimized by combining these flexible sulfur cathodes with a commercial glass fiber separator coated with a CNT layer, resulting in robust cycling stability. At a charge current density of 1.57 mA/cm^2 and a temperature of 25°C, the produced Li-S batteries have high capacities of 940 mAh/g, and their Coulombic efficiency is still above 90% even after 50 charge/discharge cycles. Importantly, Li-S batteries with a high gravimetric energy density of 443 Wh/kg per cell were developed, and these batteries exhibited outstanding stability in terms of electrochemical performance even under extreme mechanical stress conditions for more than 100 cycles. Given the low cost

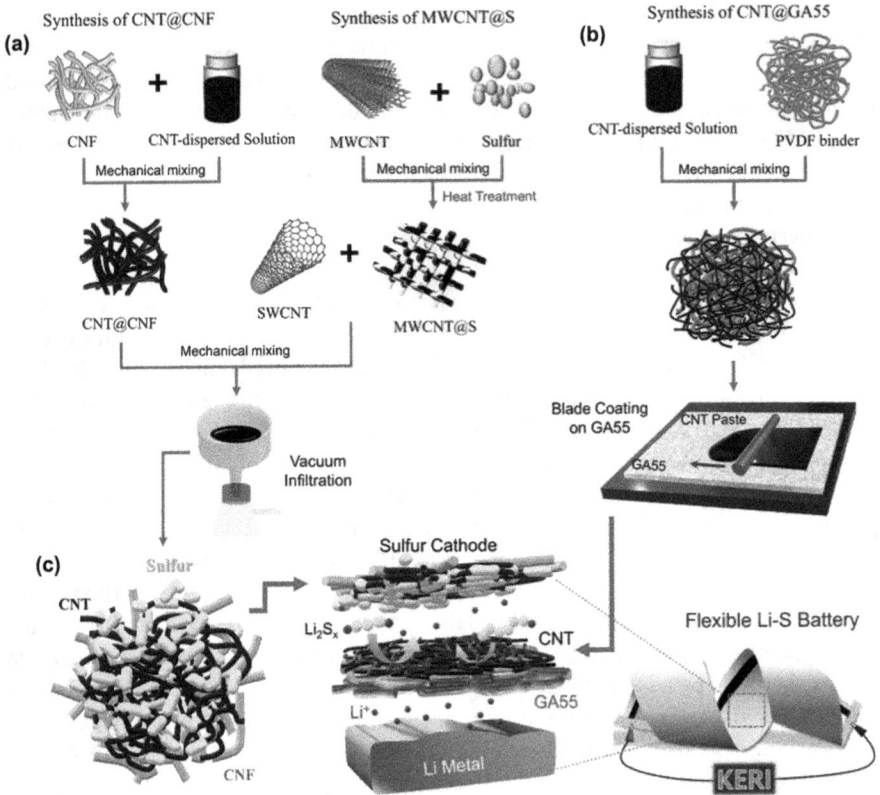

FIGURE 19.6 Fabrication procedures of the flexible Li–S battery. Adopted from Park et al. (2021).

and uncomplicated nature of the fabrication method used in this study, the materials produced are excellent candidates for commercialization. Advanced next-generation flexible batteries could be developed using the described CNF- and CNT-based composite electrodes and nanocarbon-modified separator of Li-S batteries (Park et al., 2021).

Low sulfur utilization and a high electrolyte/sulfur (E/S) ratio in lithium–sulfur batteries caused a decline in cell-based performance. A high-energy-density lithium–sulfur (Li-S) battery with a small E/S ratio and a low ratio of the negative electrode capacity to the positive electrode capacity (N/P) has been developed. It utilized a new positive electrode made of lithium polysulfides and carbon nanotubes (Li_2S_x -CNT). Li_2S_x allowed high sulfur utilization; however, given the low solubility of Li_2S_x, the cell generally has a high E/S ratio. A composite electrode made of sub-millimeter-long few-wall carbon nanotubes (CNT) and Li_2S_x ($x = 4$, 6, and 8) was developed. The Li_2S_x was deposited by solution casting and drying while the CNT served as a three-dimensional current collector and creates self-supporting sponge-like paper. Based on the total mass of a cell's interior, a full cell with Li_2S_6-CNT and Li thin foil electrodes produces 400–500 Wh kg cell-1 for E/S = 4.0 at the second and third discharge and 300 Wh kg cell-1 for E/S = 5.8 at the 97th discharge. The basis to the high energy density is the holding of the solvated Li_2S_x by the CNT sponge. The solvated Li_2S_x is held and activated by the novel design, which is compatible with various three-dimensional current collectors including metal foams and graphene foams. The design has been successfully demonstrated with a coin and is anticipated to represent a substantial advancement in electrochemical energy storage technology (Yoshie et al., 2021).

19.5 ADVANCED MATERIALS FOR CO_2 CAPTURE AND UTILIZATION

Nowadays, the emission of carbon dioxide (CO_2) gas to the atmosphere that resulted by anthropogenic activities has been among several serious global challenges to the world. Recently, an increase by 1.7% was resulted mainly by the CO_2 emission from the energy sector (Naseer et al., 2022). Numerous materials have been used to capture the emitted CO_2 because this is considered as a huge increase. The presence of an excessive amount of CO_2 in the atmosphere needs to be controlled by the use of appropriate measures, such as reducing CO_2 emissions at stationary point sources like power plants using carbon capture technologies. Therefore, one wonderful initiative by the industries is the conversion of the captured CO_2 into clean, non-polluting fuels and chemicals via photo and/or electrocatalytic routes (Karimi et al., 2022).

Recently, considerable improvements in their design and use for CO_2 capture and conversion have been made with porous materials, which have drawn a lot of attention for carbon capture (Singh et al., 2020). Using biomass, biochar, or other carbon materials as an inexpensive precursor for CO_2 adsorbent is also advantageous due to its hydrophobic nature and accessibility to resources. Due to the exponential population growth, this is also a well-known method of managing solid waste.

The intention of this study is to highlight a few advanced materials that have been utilized industrially to control or reduce the emission of CO_2 which can harm the environment. Numerous studies have been conducted on the uses of carbon materials in their varied morphologies and characteristics. Researchers are investigating the advantages of adding dopants to carbon materials, whether in situ or after treatment, and addressing any difficulties that may arise. It has been demonstrated that nitrogen doping is a highly effective method for creating improved materials for a variety of purposes, including CO_2 collection, energy conversion, and energy storage (Al-Hajri et al., 2022).

A two-dimensional substance made only of carbon (C) and nitrogen is called carbon nitride (CN) (N). It has been employed as a metal-free and visible light-active photocatalyst due to its excellent optoelectronic and physicochemical features, which include an appropriate bandgap, customizable energy-band locations, tailored surface functionalities, low cost, metal-free nature, and good thermal, chemical, and mechanical stabilities. They are connected by strong covalent bonds (Talapaneni et al., 2020). By sustainable and sporadic renewable energy sources such as sunlight and electricity through heterogeneous photo(electro)catalysis, CN and their hybrid materials have become appealing options for CO_2 capture and its reduction into clean and green low-carbon fuels and valuable chemical feedstock. Researchers have been very interested in making progress in the production of nanostructured and functionalized CN-based hybrid heterostructure materials. Studies that heavily focused on examining the mechanism of their application views have increased in recent years. The purpose and plan are to consider how materials behave in terms of light absorption, charge separation, and pathways for CO_2 movement during the reduction process.

In addition to the mentioned advanced materials, the electrochemical reduction of CO_2 (CO2RR) method holds considerable potential for the quest for efficient and suitable catalytic systems for CO_2 conversion. Promising metal-free electrocatalysts for the CO_2RR have been thought to exist in newly developed heterogeneous carbon materials. This is due to the fact of because their ample natural resources, adaptable porosity geometries, acid and basic resistance, high-temperature stability, and environmental friendliness. They have demonstrated outstanding CO_2RR qualities as catalytic activity, prolonged durability, and good selectivity. A number of carbon materials, such as carbon fibers, carbon nanotubes, graphene, diamond, nanoporous carbon, and graphene dots with heteroatom doping (N, S, and B), are highlighted that can be employed as metal-free catalysts for the CO_2RR (Duan et al., 2017). The recycling of metals in halide electrolytes, which results in metal dissolution and the generation of a highly active microstructure, is one technique for modifying the CO_2RR catalyst structure that has attracted growing interest (Samuel et al., 2020). It is interesting to note that the halide that is used affects the microstructure and CO_2RR product, offering a straightforward path to product specificity as presented in Figure 19.7.

PMOFs, or photoresponsive metal-organic frameworks, have potential for customizable CO_2 adsorption. However, because CO_2 and active sites interact weakly, modulating CO_2 adsorption on PMOFs relies on steric hindrance or structural

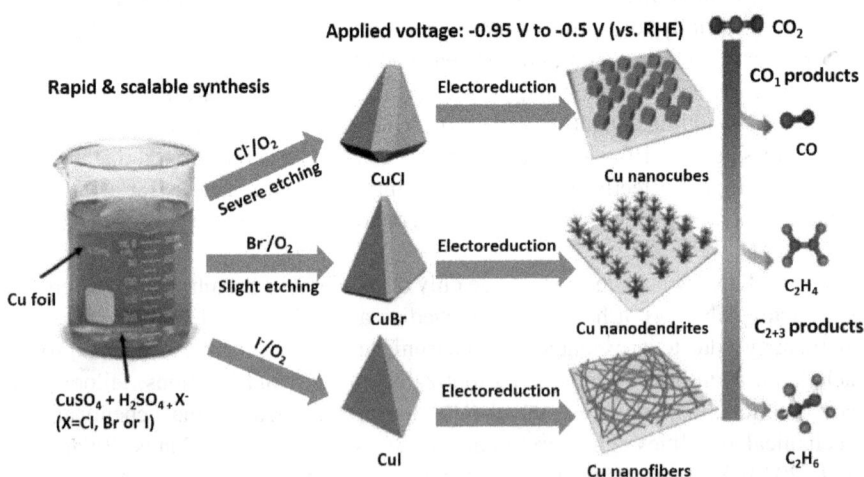

FIGURE 19.7 Schematic diagram electrochemical reduction of CO_2RR product. Reproduced from Samuel et al. (2020).

modification. Nowadays, PMOFs were used to create smart adsorbents by adding target-specific active sites which are amines (Jiang et al., 2019). The electrostatic potential of amines is considerably changed by the cis/trans isomerization of azobenzene motifs induced by UV/Vis light irradiation. This results in the exposure or shelter of amines and the effective control of CO_2 adsorption on potent active sites, which is not achievable with conventional PMOFs.

Carbon-based materials are widely used for CO_2 capture in the industrial sectors. Several researchers are looking ahead to utilize the agricultural wastes in producing green activated carbons embedded with selected nanoparticles due to enhance the efficiency of CO_2 adsorption. Bamboo waste was used to create porous carbon (BAC), which exhibits good CO_2 adsorbent characteristics, using dehydrating agent and concentrated sulfuric acid (H_2SO_4) at room temperature (Wan Isahak et al., 2015). The use of nano-CuO-supported BAC results in composite materials with a large total surface area and smaller holes of $660.8\,cm^2/g$ and $2.7\,nm$, respectively. This capture has led to the discovery of the hydroxide phase production as a carbonate intermediary and an accelerator of the CO_2 chemisorption reaction. The physical interaction of CO_2 on the surface and in the pores can be improved by the presence of BAC and metal oxide, leading to a higher adsorption capacity of $32.2\,cm^3$ of CO_2 per gram of adsorbent. The addition of nano-CuO to BAC forms an effective adsorbent that can accelerate CO_2 reduction initiatives and lower CO_2 emissions during BAC production.

Amine-based technologies are notable among CO_2 capture techniques. The reversible CO_2 reactions that amines undergo are widely recognized. As a result, numerous CO_2-containing gases, such as flue gas, can be separated from CO_2 using them. Excellent CO_2 separation properties over hydrogen (H_2) gas were demonstrated by poly(amidoamine)s (PAMAMs) added to a cross-linked polyethylene glycol (Taniguchi et al., 2017). The current CO_2 capture method at the demonstration measure uses monoethanolamine (MEA) as the CO_2 determining agent, which was easily immobilized in polyvinyl alcohol (PVA) matrix by solvent casting of an aqueous mixture of PVA and the amine. The CO_2 permeability should be enhanced for practical applications. With a thickness over $3\,nm$ and an amine fraction below $80\,wt\%$, the resulting polymeric membranes can stand on their own. The resulting polymeric membranes did, however, have some CO_2-selective gas permeability characteristics. High CO_2 separation capability required the hydroxyl group of MEA.

Other materials for CO_2 capture that is commonly explored industrially are ionic liquids. These materials are among the most growing areas in ionic liquid technology together with gas separation processes. Nowadays, its effectiveness for CO_2 capture and separation has been studied using ionic liquid polymer that has been integrated with activated carbon (Mahmood et al., 2016). Ionic liquid polymers containing activated carbon were generated by first synthesizing the ionic liquid monomers 1-vinyl-3-ethylimidazolium bromide, [veim][Br], and 1-vinyl-3-ethylimidazolium bis (trifluoromethyl-sulfonyl) imide, [veim][Tf2N]. Both monomers were found to be of a high purity by analysis using NMR and ion chromatography. According to the analytical results, activated carbon has been successfully incorporated into the lattice structure of polymers. As the generated polymer materials contain both the absorption and adsorption mechanisms for CO_2 apprehension and sequestration, it is anticipated that they will possess a greater capacity to capture CO_2.

19.6   NANOSTRUCTURED MATERIALS FOR OPTOELECTRONIC DEVICES

Optoelectronics is a field of technology that combines optics with electronics, and some examples of optoelectronic devices are sensors, emitters, and optocouplers. Optoelectronic materials were first reported in the 1910s after the studies of their optical and electronic properties. Optoelectronic devices have been manufactured in a substantial number since the first semiconductor laser demonstration in the early 1960s, and they are now widely used in computers, medical, communications, lighting, and entertainment. Perhaps their biggest contribution to humanity has been in optical-fiber communications, which has made it possible to transmit speech and data at high quality and low cost throughout the world (Sweeney & Mukherjee, 2017).

Nowadays, a lot of attention is focusing on the use of nanostructured materials for optoelectronic devices, such as solar cells, light-emitting diodes (LEDs), laser diodes, and photodetectors due to their distinctive geometry, in which their nanoscale dimensions can be perfectly integrated into a wide range of technological platforms. Some of the most studied nanostructured materials groups in this aspect generally are metal oxide NPs (Alrebdi et al., 2022; Berra et al., 2022; Jayakumar et al., 2022; Tiwari & Sahay, 2022), layered structure materials (Ganesh et al., 2022), and nanostructures of II–VI-based chalcogenides (Arshad Kamran & Alharbi, 2022) due to their distinctive physical and chemical properties.

Jayakumar et al. (2022) has synthesized Ni-doped CeO_2 NPs through hydrothermal technique and it was found to have significant effect on bandgap energy, oxygen vacancy (VO), crystallite size, photo luminescence, electrical conductivity, and magnetic and dielectric properties (Jayakumar et al., 2022). With the increase of Ni dopant concentration, the bandgap energy of UV–vis of Ni-doped CeO_2 NPs broadens, and the bandgap energy values are 3.39, 3.67, 3.74, and 3.8 eV for undoped, 5 M%, 10 M%, and 15 M% Ni-doped CeO_2 NPs, correspondingly. The values of VO increase as determined from Raman spectra. In the violet-blue-green wavelengths region, CeO_2 NPs and Ni-doped CeO_2 NPs are observed to display luminescence with emission intensity decreases with the Ni concentration increase. This result shows that oxygen vacancies increase with Ni increase which perform as non-radiative centers. All the samples exhibit paramagnetic characteristics with retentivity of the samples and saturation magnetization increases when the concentration of Ni dopant increases. The samples also demonstrate frequency dependent dielectric properties. With increase in the Ni dopant concentration and frequency, this shows that electrical conductivity increased while dielectric loss and dielectric constant values dropped for all the samples.

NiO is a wide-bandgap semiconductor material at around 4.0 eV and thus generated a lot of interest together with its other remarkable properties in enhancing optical characteristics. The decoration of NiO NPs on MWCNTs synthesized by pulsed laser ablation method has resulted in substantially enhanced nonlinearity in respect to individual NiO NPs and MWCNTs, since it took longer for photons transition and electrons to transport. The nonlinear properties were determined by Z-scan technique in which nonlinear absorption coefficient (β), nonlinear refractive index (n^2), and third nonlinear optical susceptibility (χ^3) were resolved. The β values were found to be gradually dropped as the laser beam's intensity rose. With these outstanding nonlinear optical characteristics, this NiO/MWCNTs nanocomposite has high potential being suitable for optical devices such as optical sensors and switches (Alrebdi et al., 2022).

Transition metal doped with zinc oxide (ZnO) has attracted excessive attention due to its substantial physical and chemical properties. Berra et al. (2022) has synthesized the nanostructured films of Ni-doped ZnO by the sol-gel technique which has been performed using the spin coating method on glass substrates. According to the structural findings, Ni doping caused the crystallite size to be smaller. Each sample had a polycrystalline wurtzite structure and a preference for the (002) plane orientation. Nanofiber formations can be seen on the surface of thin films as shown in their morphology micrographs. Ni-ZnO also shows materials' exceptional transparency. The photoluminescence (PL) for each film showed numerous fault emissions and a single UV emission at 387 nm. Ni doping had an influence on the PL emissions intensity, absorption, and band gap. This nanostructured thin films of Ni-doped ZnO demonstrated strong photoresponse and UV sensitivity, and thus making them a promising component for being utilized in a UV photodetector.

Lead iodide (PbI_2) is a semiconducting nanomaterial and is considered as a series of hexagonally layered structural layers of lead ions between iodide ions. PbI_2 also displays 2.5 eV of energy bandgap with high anisotropy properties. Due to these characteristics, many researchers show their interest to explore PbI_2 performance in various applications including optoelectronic applications. The synthesis of pure and Cr^{3+}-doped (1, 2.5, 5, 7, 10 wt%) PbI_2 nanosheets was effectively attained by a straightforward microwave-assisted route which is low cost by Ganesh et al. (2022). The synthesized materials were confirmed to have hexagonal phase of the aimed nanostructures with no presence of the impurity. From Figure 19.8, Cr^{3+} ions demonstrate a significant doping outcome on the PbI_2, a curved hexagonal nanosheet into clear hexagonal nanosheets with huge density and lower dimensions. For this reason, the high-density nanosheets with the presence of Cr^{3+} are seen positively prepare current electronics for electro-optical applications by providing huge rage fields. The undoped PbI_2 nanosheet is less than 100 nm in size, while in contrast, the

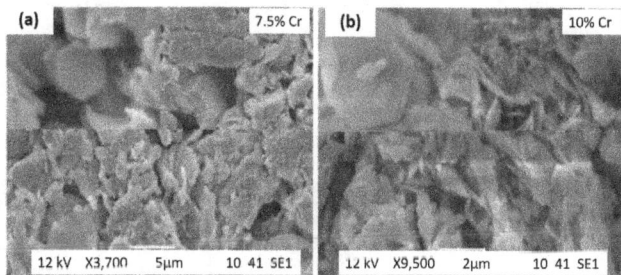

FIGURE 19.8 SEM images of Cr-doped PbI_2 nanosheets at (a) 7.5% and (b) 10% of Cr-doped. Reproduced from Ganesh et al. (2022).

obtained Cr-doped PbI_2 samples are 50–60 nm in size. Because of the Cr doping, the typical peaks of pure PbI_2 have somewhat shifted, and the energy bandgaps have changed. Both samples of pure PbI_2 and PbI_2 doped with Cr have optical bandgap values of approximately within 2.0–2.2 eV. The dielectric constant was raised to 27 from 20 with the presence of Cr- ions in the PbI_2 (Figure 19.9). This Cr-doped PbI_2 samples also show linear coefficient of Gamma absorption showing an improvement. Generally, the prepared Cr-doped PbI_2 nanostructures have the potential to be outstanding candidates for a variety of uses, such as optoelectronic detectors, with the positive improvement in the qualities of optical, radiation, and structural morphology (Ganesh et al., 2022).

Sulfides, selenides, and tellurides are nanostructures of II–VI-based chalcogenides, which have drawn a lot of scientific interest due to their wide-band-gap semiconductors' adjustable luminescence at optical emission wavelengths, because their excitons have higher stability than those of other compound types of semiconductors. In terms of preparation, these materials also simpler to be synthesized. For the first time, chemical vapor deposition (CVD) method was successfully used to synthesize one-dimensional Al-doped CdS nanobelts. High crystallinity was observed on the $Cd_{1-x}Al_xS$ nanobelts which have been synthesized vertically. From FESEM micrograph depicted in Figure 19.10, all the synthesized nanostructures have belt-like shapes, and the surfaces are very fine and smooth. It can also be seen clearly the vertical development of Al-doped CdS nanobelts on n-type Si substrates. Due to the different amounts of Al-doped into CdS, the shape and size varied somewhat. The presence of Al-tip on top of the nanobelts in Figure 19.10c confirms the vapor–liquid–solid and vapor–solid development mechanisms, and Figure 19.10d also shows that the nanobelts are around 180 nm thick. From the photoluminescence (PL) spectra, that is, one red emission band at 728 nm, and a new wide emission band in the near-infrared (NIR), they comprise of three peaks with centers at 930–961, 954–992, and 1,042 nm which correspond to the d–d transition of Al ions. 3.09% Al doping has improved sample conductivity from 1.097×10^{-6} to 2.59×10^{-5} A, proving its potential for use in optoelectronic systems, such as solar panel buffer layer components and nanoscale light sources (Arshad Kamran & Alharbi, 2022).

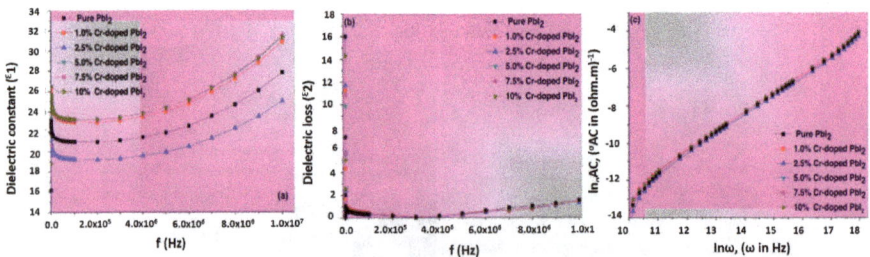

FIGURE 19.9 (a) Dielectric constant ($\varepsilon 1$), (b) dielectric loss ($\varepsilon 2$), (c) total AC electrical conductivity (σAC.Total) of undoped and Cr-doped PbI_2 nanosheets. Reproduced from Ganesh et al. (2022).

FIGURE 19.10   (a, b) Al-doped CdS nanobelt images captured using high-magnification FESEM microscopy (c) FESEM image of individual Al-doped CdS nanobelt and (d) its individual thickness. Reproduced from Arshad Kamran and Alharbi (2022).

19.7   CONCLUSIONS

In conclusion, this chapter discusses recent advancements in the use of cutting-edge materials in renewable energy technology concludes. In the academic and industrial worlds, extensive study has been performed on advanced materials for renewable energy applications such as storage and conversion, particularly for supercapacitor, photovoltaic, batteries, CO_2 capture, and optoelectronic devices. Among the materials under investigation are metal-organic frameworks (MOFs), MXene, nanocellulose, graphene oxide, reduced graphene oxide (rGO), metal oxide nanoparticles, and conducting polymer.

The key characteristics of the different kinds of supercapacitors are discussed in relation to the choice of electrode material. In order to improve advanced electrode properties like conductivity, modification of morphology, mechanical behavior, chemical and thermal stabilities, as well as specific capacitance, it is important to design hybrid or nanocomposites-based electrodes. Nanostructure material such as perovskite has been utilized in solar photovoltaics. Several approaches have been used to create organic–inorganic hybrid perovskite in order to develop nanostructures with low dimensional perovskite as promising materials. As of their high energy density, versatility, reliability, and cost advantages, lithium–sulfur and sodium ion batteries are becoming the most promising next-generation storage devices. Because of their abundant element reserves and excellent electrochemical performance, Na-ion batteries have been investigated as the most promising technology to stay competitive with Li-ion batteries in the future commercial battery market. CO_2 capture through porous solid materials is a promising method in a variety of industries. Researchers have looked into various capture routes, such as absorption and adsorption. Future membrane-based CO_2 separation will require advanced composite systems, so strategies for enhancing composite systems through alternative chemistries, material system redesign, and processing technology will be crucial in revealing the difficulties to their development. Multiple studies of optoelectrical materials have demonstrated their enormous potential for converting thermal heat into electrical energy. The basic processes for the operation of optoelectronic devices are material synthesis and physical, chemical characteristics such as band gap energy. If the anticipated full-scale practical application is to be realized, it will be necessary to further improve the quality and reproducibility of advanced materials. This includes designing the most desirable structures that can be adjusted at the nano, micro, meso, and macroscales. The techniques chosen must be capable of producing materials with manageable particle sizes that can be used for this application. Nanodimensional materials have gained widespread acceptance due to their ability to improve the capacitive performance of desired systems while maintaining high cycle life and kinetic reversibility.

REFERENCES

Ahouei, M. A., Syed, T. H., Bishop, V., Halacoglu, S., Wang, H., & Wei, W. (2022). $Ti_3C_2T_x$ MXene framework materials: Preparation, properties and applications in energy and environment. *Catalysis Today*. https://doi.org/10.1016/J.CATTOD.2022.11.001

Al-Hajri, W., De Luna, Y., & Bensalah, N. (2022). Review on recent applications of nitrogen-doped carbon materials in CO_2 Capture and energy conversion and storage. *Energy Technology*. https://doi.org/10.1002/ente.202200498

Alrebdi, T. A., Ahmed, H. A., Alkallas, F. H., Mwafy, E. A., Trabelsi, A. B. G., & Mostafa, A. M. (2022). Structural, linear and nonlinear optical properties of NiO nanoparticles–multi-walled carbon nanotubes nanocomposite for optoelectronic applications. *Radiation Physics and Chemistry*, *195*, 110088. https://doi.org/https://doi.org/10.1016/j.radphyschem.2022.110088

Arshad Kamran, M., & Alharbi, T. (2022). Effect of Al doping on photoluminescence and conductivity of 1D CdS nanobelts synthesized by CVD for optoelectronic applications. *Journal of Science: Advanced Materials and Devices*, *7*(3), 100464. https://doi.org/10.1016/j.jsamd.2022.100464.

Berra, S., Mahroug, A., Hamrit, S., Ahmad Azmin, M., Zoukel, A., Berri, S., & Selmi, N. (2022). Experimental and DFT study of structural and optical properties of Ni-doped ZnO nanofiber thin films for optoelectronic applications. *Optical Materials*, *134*, 113188. https://doi.org/https://doi.org/10.1016/j.optmat.2022.113188

Bjelajac, A., Petrović, R., Stan, G. E., Socol, G., Mihailescu, A., Mihailescu, I. N., Veltruska, K., Matolin, V., Siketić, Z., Provatas, G., Jakšić, M., & Janaćković, D. (2022). C-doped TiO_2 nanotubes with pulsed laser deposited Bi_2O_3 films for photovoltaic application. *Ceramics International*, *48*(4), 4649–4657. https://doi.org/10.1016/j.ceramint.2021.10.251.

Chen, G., Chen, T., Hou, K., Ma, W., Tebyetekerwa, M., Cheng, Y., Weng, W., & Zhu, M. (2018). Robust, hydrophilic graphene/cellulose nanocrystal fiber-based electrode with high capacitive performance and conductivity. *Carbon*, *127*, 218–227. https://doi.org/10.1016/j.carbon.2017.11.012.

Cheng, Y., Xie, Y., Yan, S., Liu, Z., Ma, Y., Yue, Y., Wang, J., Gao, Y., & Li, L. (2022). Maximizing the ion accessibility and high mechanical strength in nanoscale ion channel MXene electrodes for high-capacity zinc-ion energy storage. *Science Bulletin*, *67*(21), 2216–2224. https://doi.org/10.1016/J.SCIB.2022.10.003

Deepalakshmi, S., Revathy, M. S., Marnadu, R., & Shkir, M. (2022). Performance of sunflower petal shaped CuCo2S4 wrapping on GO asymmetric capacitor for energy storage applications. *Diamond and Related Materials*, *127*, 109133. https://doi.org/10.1016/J.DIAMOND.2022.109133

Duan, X., Xu, J., Wei, Z., Ma, J., Guo, S., Wang, S., Liu, H., & Dou, S. (2017). Metal-free carbon materials for CO_2 electrochemical reduction. *Advanced Materials*, *29*(41), 1701784. https://doi.org/https://doi.org/10.1002/adma.201701784

Etman, A. S., Halim, J., & Rosen, J. (2021). Mo1.33CTz-Ti3C2Tz mixed MXene freestanding films for zinc-ion hybrid supercapacitors. *Materials Today Energy*, *22*, 100878. https://doi.org/10.1016/J.MTENER.2021.100878.

Fallah barzoki, M., Fatemi, S., & Ganjali, M. R. (2022). Fabrication and comparison of composites of cerium metal-organic framework/reduced graphene oxide as the electrode in supercapacitor application. *Journal of Energy Storage*, *55*(PB), 105545. https://doi.org/10.1016/j.est.2022.105545

Fukuhara, M., Yokotsuka, T., Hashida, T., Miwa, T., Fujima, N., Morita, M., Nakatani, T., & Nonomura, F. (2022). Amorphous cellulose nanofiber supercapacitors with voltage-charging performance. *Scientific Reports*, *12*(1), 5619. https://doi.org/10.1038/s41598-022-09649-0

Ganesh, V., Al Abdulaal, T. H., Al-Amri, S. G. S., Zahran, H. Y., Algarni, H., Umar, A., Albargi, H. B., Yahia, I. S., & Ibrahim, M. A. (2022). Low-cost and facile synthesis of chromium doped PbI2 nanostructures for optoelectronic devices and radiation detectors: Comparative study. *Applied Surface Science Advances*, *8*. https://doi.org/10.1016/j.apsadv.2022.100226.

Gielen, D., Boshell, F., Saygin, D., Bazilian, M. D., Wagner, N., & Gorini, R. (2019). The role of renewable energy in the global energy transformation. *Energy Strategy Reviews*, *24*, 38–50. https://doi.org/10.1016/J.ESR.2019.01.006.

Green, R., Scheelbeek, P., Bentham, J., Cuevas, S., Smith, P., & Dangour, A. D. (2022). Growing health: Global linkages between patterns of food supply, sustainability, and vulnerability to climate change. *The Lancet Planetary Health*, *6*(11), e901–e908. https://doi.org/10.1016/S2542-5196(22)00223-6.

Jahan, I., Islam, M. A., Rupam, T. H., Palash, M. L., Rocky, K. A., & Saha, B. B. (2022). Enhanced water sorption onto bimetallic MOF-801 for energy conversion applications. *Sustainable Materials and Technologies*, *32*, e00442. https://doi.org/10.1016/J.SUSMAT.2022.E00442

James, N., & Menzies, M. (2022). Global and regional changes in carbon dioxide emissions: 1970–2019. *Physica A: Statistical Mechanics and Its Applications*, *608*, 128302. https://doi.org/10.1016/J.PHYSA.2022.128302.

Jayakumar, G., Albert Irudayaraj, A., Dhayal Raj, A., John Sundaram, S., & Kaviyarasu, K. (2022). Electrical and magnetic properties of nanostructured Ni doped CeO_2 for optoelectronic applications. *Journal of Physics and Chemistry of Solids*, *160*, 110369. https://doi.org/10.1016/j.jpcs.2021.110369.

Jiang, Y., Tan, P., Qi, S.-C., Liu, X.-Q., Yan, J.-H., Fan, F., & Sun, L.-B. (2019). Metal-organic frameworks with target-specific active sites switched by photoresponsive motifs: Efficient adsorbents for tailorable CO_2 capture. *Angewandte Chemie (International Ed. in English)*, *58*(20), 6600–6604. https://doi.org/10.1002/anie.201900141

Jo, C., Mun, Y., Lee, J., Lim, E., Kim, S., & Lee, J. (2019). Carbon dioxide to solid carbon at the surface of iron nanoparticle: Hollow nanocarbons for sodium ion battery anode application. *Journal of CO_2 Utilization*, *34*, 588–595. https://doi.org/10.1016/j.jcou.2019.08.003

Kang, C., Lee, Y., Kim, I., Hyun, S., Lee, T. H., Yun, S., Yoon, W. S., Moon, Y., Lee, J., Kim, S., & Lee, H. J. (2019). Highly efficient nanocarbon coating layer on the nanostructured copper sulfide-metal organic framework derived carbon for advanced sodium-ion battery anode. *Materials*, *12*(8). https://doi.org/10.3390/ma12081324.

Kant, R., Ahuja, V., Joshi, K., Gupta, H., & Bhardwaj, S. (2022). Tuning the dielectric characteristics and energy storage properties of Ni-ZnO/rGO nanocomposite. *Vacuum*, *204*, 111375. https://doi.org/10.1016/J.VACUUM.2022.111375

Kargupta, W., Seifert, R., Martinez, M., Olson, J., Tanner, J., & Batchelor, W. (2021). Sustainable production process of mechanically prepared nanocellulose from hardwood and softwood: A comparative investigation of refining energy consumption at laboratory and pilot scale. *Industrial Crops and Products*, *171*, 113868. https://doi.org/10.1016/J.INDCROP.2021.113868

Karimi, M., Shirzad, M., Silva, J. A. C., & Rodrigues, A. E. (2022). Biomass/Biochar carbon materials for CO2 capture and sequestration by cyclic adsorption processes: A review and prospects for future directions. *Journal of CO_2 Utilization*, *57*, 101890. https://doi.org/10.1016/J.JCOU.2022.101890

Kosukoglu, T., Carpan, M., Riza Tokgoz, S., & Peksoz, A. (2022). Fabrication of a new rGO@PPy/SS composite electrode with high energy storage and long cycling life for potential applications in supercapacitors. *Materials Science and Engineering: B*, *286*, 116032. https://doi.org/10.1016/J.MSEB.2022.116032

Lee, K. T., & Cho, J. (2011). Roles of nanosize in lithium reactive nanomaterials for lithium ion batteries. *Nano Today*, *6*(1), 28–41. https://doi.org/10.1016/j.nantod.2010.11.002

Li, S., Ji, W., Zou, L., Li, L., Li, Y., & Cheng, X. (2022). Crystalline TiO_2 shell microcapsules modified by Co3O4/GO nanocomposites for thermal energy storage and photocatalysis. *Materials Today Sustainability*, *19*, 100197. https://doi.org/10.1016/J.MTSUST.2022.100197

Lokhande, P. E., Singh, P. P., Vo, D. V. N., Kumar, D., Balasubramanian, K., Mubayi, A., Srivastava, A., & Sharma, A. (2022). Bacterial nanocellulose: Green polymer materials for high performance energy storage applications. *Journal of Environmental Chemical Engineering*, *10*(5), 108176. https://doi.org/10.1016/J.JECE.2022.108176

Lee, D. G., Pandey, P., Parida, B., Ryu, J., Cho, S. W., Kim, J. K., & Kang, D. W. (2022). Improving inorganic perovskite photovoltaic performance via organic cation addition for efficient solar energy utilization. *Energy*, *257*, 124640. https://doi.org/10.1016/j.energy.2022.124640.

Li, F., Liu, H., Ma, Y., Xie, X., Wang, Y., & Yang, Y. (2022). Low-carbon spatial differences of renewable energy technologies: Empirical evidence from the Yangtze River Economic Belt. *Technological Forecasting and Social Change*, *183*, 121897. https://doi.org/10.1016/J.TECHFORE.2022.121897.

Lin, Z., Goikolea, E., Balducci, A., Naoi, K., Taberna, P. L., Salanne, M., Yushin, G., & Simon, P. (2018). Materials for supercapacitors: When Li-ion battery power is not enough. *Materials Today*, *21*(4), 419–436. https://doi.org/10.1016/J.MATTOD.2018.01.035.

Lipu, M. S. H., Hannan, M. A., Hussain, A., Hoque, M. M., Ker, P. J., Saad, M. H. M., & Ayob, A. (2018). A review of state of health and remaining useful life estimation methods for lithium-ion battery in electric vehicles: Challenges and recommendations. *Journal of Cleaner Production*, *205*, 115–133. https://doi.org/10.1016/J.JCLEPRO.2018.09.065.

Mahmood, H., bin Ahmad Sayukhi, M. H. A., Moniruzzaman, M., & Yusup, S. (2016). Synthesis of ionic liquid polymer incorporating activated carbon for carbon dioxide capture and separation. *Advanced Materials Research*, *1133*, 566–570. https://doi.org/10.4028/www.scientific.net/amr.1133.566

Majdi, H. S., Latipov, Z. A., Borisov, V., Yuryevna, N. O., Kadhim, M. M., Suksatan, W., Khlewee, I. H., & Kianfar, E. (2021). Nano and battery anode: a review. *Nanoscale Research Letters*, *16*(1). https://doi.org/10.1186/s11671-021-03631-x

Naseer, M. N., Zaidi, A. A., Dutta, K., Wahab, Y. A., Jaafar, J., Nusrat, R., Ullah, I., & Kim, B. (2022). Past, present and future of materials' applications for CO_2 capture: A bibliometric analysis. *Energy Reports*, *8*, 4252–4264. https://doi.org/10.1016/J.EGYR.2022.02.301

Oh, W. C., Liu, Y., & Areerob, Y. (2021). A novel fabrication of organic-inorganic hybridized Graphene-La2CrFeW6 nanocomposite and its improved photovoltaic performance in DSSCs. *Journal of Science: Advanced Materials and Devices, 6*(2), 271–279. https://doi.org/10.1016/j.jsamd.2021.02.008.

Oner, O., & Dincer, I. (2022). Development and assessment of a hybrid biomass and wind energy-based system for cleaner production of methanol with electricity, heat and freshwater. *Journal of Cleaner Production, 367*, 132967. https://doi.org/10.1016/J.JCLEPRO.2022.132967.

Pallavolu, M. R., Prabhu, S., Nallapureddy, R. R., Kumar, A. S., Banerjee, A. N., & Joo, S. W. (2023). Bio-derived graphitic carbon quantum dot encapsulated S- and N-doped graphene sheets with unusual battery-type behavior for high-performance supercapacitor. *Carbon, 202*(P1), 93–102. https://doi.org/10.1016/j.carbon.2022.10.077.

Park, J. W., Jo, S. C., Kim, M. J., Choi, I. H., Kim, B. G., Lee, Y. J., Choi, H. Y., Kang, S., Kim, T. Y., & Baeg, K. J. (2021). Flexible high-energy-density lithium-sulfur batteries using nanocarbon-embedded fibrous sulfur cathodes and membrane separators. *NPG Asia Materials, 13*(1). https://doi.org/10.1038/s41427-021-00295-y.

Perry, S. C., Leung, P., Wang, L., & Ponce de León, C. (2020). Developments on carbon dioxide reduction: Their promise, achievements, and challenges. *Current Opinion in Electrochemistry, 20*, 88–98. https://doi.org/https://doi.org/10.1016/j.coelec.2020.04.014

Qin, M., Feaugas, O., & Zu, K. (2022). Novel metal-organic framework (MOF) based phase change material composite and its impact on building energy consumption. *Energy and Buildings, 273*, 112382. https://doi.org/10.1016/J.ENBUILD.2022.112382

Qiu, L., Mei, D., Chen, W. H., Yuan, Y., Song, L., Chen, L., Bai, B., Du, P., & Xiong, J. (2022). Organic-inorganic hybrid electron transport layer of PVP-doped SnO_2 for high-efficiency stable perovskite solar cells. *Solar Energy Materials and Solar Cells, 248*(September), 112032. https://doi.org/10.1016/j.solmat.2022.112032.

Querne, C., Vignal, T., Pinault, M., Banet, P., Mayne-L'Hermite, M., & Aubert, P. H. (2023). A comparative study of high density Vertically Aligned Carbon Nanotubes grown onto different grades of aluminum - Application to supercapacitors. *Journal of Power Sources, 553*(October 2022), 232258. https://doi.org/10.1016/j.jpowsour.2022.232258.

Raza, W., Ali, F., Raza, N., Luo, Y., Kim, K. H., Yang, J., Kumar, S., Mehmood, A., & Kwon, E. E. (2018). Recent advancements in supercapacitor technology. *Nano Energy, 52*, 441–473. https://doi.org/10.1016/J.NANOEN.2018.08.013.

Romanello, M., Di Napoli, C., Drummond, P., Green, C., Kennard, H., Lampard, P., Scamman, D., Arnell, N., Ayeb-Karlsson, S., Ford, L. B., Belesova, K., Bowen, K., Cai, W., Callaghan, M., Campbell-Lendrum, D., Chambers, J., van Daalen, K. R., Dalin, C., Dasandi, N., ... Costello, A. (2022). The 2022 report of the Lancet Countdown on health and climate change: Health at the mercy of fossil fuels. *The Lancet, 400*(10363), 1619–1654. https://doi.org/10.1016/S0140-6736(22)01540-9.

Sen, D., Tunç, K. M. M., & Günay, M. E. (2021). Forecasting electricity consumption of OECD countries: A global machine learning modeling approach. *Utilities Policy, 70*, 101222. https://doi.org/10.1016/J.JUP.2021.101222.

She, C., Zu, X., Yang, Z., Chen, L., Xie, Z., Yang, H., Yang, D., Yi, G., Qin, Y., Lin, X., Zhang, W., Dong, H., & Qiu, X. (2023). Construction of anodic electron transfer chain based on CuCl2/TiOSO4 synergetic mediators for highly efficient conversion of biomass wastes into electricity at low temperature. *Chemical Engineering Journal, 452*, 139266. https://doi.org/10.1016/J.CEJ.2022.139266.

Shehzad, K., Zaman, U., Zaman, B. U., Liu, X., & Jafri, R. A. (2022). Lithium production, electricity consumption, and greenhouse gas emissions: An imperious role of economic globalization. *Journal of Cleaner Production, 372*, 133689. https://doi.org/10.1016/J.JCLEPRO.2022.133689.

Singh, G., Lee, J., Karakoti, A., Bahadur, R., Yi, J., Zhao, D., Albahily, K., & Vinu, A. (2020). Emerging trends in porous materials for CO_2 capture and conversion. *Chemical Society Reviews*, *49*(13), 4360–4404. https://doi.org/10.1039/d0cs00075b

Suganya, B., Maruthamuthu, S., Chandrasekaran, J., Saravanakumar, B., Vijayakumar, E., Marnadu, R., Al-Enizi, A. M., & Ubaidullah, M. (2021). Design of zinc vanadate (Zn3V2O8)/nitrogen doped multiwall carbon nanotubes (N-MWCNT) towards supercapacitor electrode applications. *Journal of Electroanalytical Chemistry*, *881*, 114936. https://doi.org/10.1016/J.JELECHEM.2020.114936.

Sun, B., Fan, X., Hou, R., Zhao, G., Liu, Q., Zhou, H., & Liang, P. (2023). Electrode made of NiCo double hydroxide on oxidized activated carbon for asymmetric supercapacitors. *Chemical Engineering Journal*, *454*, 140280. https://doi.org/10.1016/J.CEJ.2022.140280.

Sun, W., Xu, L., & Zhu, A. (2022). Preparation and electrochemical performance of nanocarbon-isolated nano-sheet silicon lithium-ion battery anode material. *Journal of Solid State Electrochemistry*, *26*(11), 2585–2593. https://doi.org/10.1007/s10008-022-05275-y

Sweeney, S. J., & Mukherjee, J. (2017). Optoelectronic devices and materials BT - Springer handbook of electronic and photonic materials. In S. Kasap & P. Capper (Eds.), *Springer Handbook of Electronic and Photonic Materials, 2017* (p. 1). Springer International Publishing. https://doi.org/10.1007/978-3-319-48933-9_35

Talapaneni, S. N., Singh, G., Kim, I. Y., AlBahily, K., Al-Muhtaseb, A. H., Karakoti, A. S., Tavakkoli, E., & Vinu, A. (2020). Nanostructured carbon nitrides for CO_2 capture and conversion. *Advanced Materials*, *32*(18). https://doi.org/10.1002/adma.201904635

Taniguchi, I., Kinugasa, K., Toyoda, M., & Minezaki, K. (2017). Effect of amine structure on CO_2 capture by polymeric membranes. *Science and Technology of Advanced Materials*, *18*(1), 950–958. https://doi.org/10.1080/14686996.2017.1399045

Taukeer Khan, M., & Khan, F. (2022). Enhancement in photovoltaic performance of perovskites solar cells through modifying the electron transport layer with reduced graphene oxide. *Materials Letters*, *323*(June), 132578. https://doi.org/10.1016/j.matlet.2022.132578.

Tiwari, A., & Sahay, P. P. (2022). Impact of heterovalent cations (Ga, Co) co-doping on the physical properties of ZnO films for optoelectronic applications. *Brazilian Journal of Physics*, *52*(5), 176. https://doi.org/10.1007/s13538-022-01175-8

Wan Isahak, W. N. R., Che Ramli, Z. A., Lahuri, A. H., Yusop, M. R., Mohamed Hisham, M. W., & Yarmo, M. A. (2015). Enhancement of CO_2 capture using CuO nanoparticles supported on green activated carbon. *Advanced Materials Research*, *1087*, 111–115. https://doi.org/10.4028/www.scientific.net/amr.1087.111

Wu, Y., Juan, F., Wang, B., Sun, S., Jia, J., Wei, H., & Cao, B. (2022). Enhanced photovoltaic performance of perovskite solar cells modified with plasmonic $Cu_{1.44}Te$ nanocrystals. *Optik*, *271*(November), 170229. https://doi.org/10.1016/j.ijleo.2022.170229.

Yang, Q., Song, R., Wang, Y., Hu, X., Chen, Z., Li, Z., & Tan, W. (2021). One-pot synthesis of Zr-MOFs on MWCNTs for high-performance electrochemical supercapacitor. *Colloids and Surfaces A: Physicochemical and Engineering Aspects*, *631*, 127665. https://doi.org/10.1016/J.COLSURFA.2021.127665.

Yao, Z., Quan, B., Yang, T., Li, J., & Gu, C. (2023). Flexible supercapacitors based on vertical graphene/carbon fabric with high rate performance. *Applied Surface Science*, *610*, 155535. https://doi.org/10.1016/J.APSUSC.2022.155535

Yoshie, Y., Hori, K., Mae, T., & Noda, S. (2021). High-energy-density Li–S battery with positive electrode of lithium polysulfides held by carbon nanotube sponge. *Carbon*, *182*, 32–41. https://doi.org/10.1016/j.carbon.2021.05.046

Zhang, C., Wei, Y. L., Cao, P. F., & Lin, M. C. (2018). Energy storage system: Current studies on batteries and power condition system. *Renewable and Sustainable Energy Reviews*, *82*, 3091–3106. https://doi.org/10.1016/J.RSER.2017.10.030.

Zhang, C., Chen, Z., Wang, H., Nie, Y., & Yan, J. (2021). Porous Fe_2O_3 nanoparticles as lithium-ion battery anode materials. *ACS Applied Nano Materials*, *4*(9), 8744–8752. https://doi.org/10.1021/acsanm.1c01312

Zhu, T., Curtis, J., & Clancy, M. (2023). Modelling barriers to low-carbon technologies in energy system analysis: The example of renewable heat in Ireland. *Applied Energy*, *330*, 120314. https://doi.org/10.1016/J.APENERGY.2022.120314.

20 Emerging Development on Nanocellulose and Its Composites in Biomedical Sectors

Mohd Nor Faiz Norrrahim
Universiti Pertahanan Nasional Malaysia

Khairul Anwar Ishak
Universiti Malaya

Mohd Saiful Asmal Rani
Universiti Putra Malaysia

Nik Noorul Shakira Mohamed Shakrin,
Muhammad Faizan Abdul Shukor, Nurul Naqirah
Shukor, Nor Syaza Syahirah Amat Junaidi, Victor
Feizal Knight, and Noor Azilah Mohd Kasim
Universiti Pertahanan Nasional Malaysia

Mohd Azwan Jenol
Universiti Putra Malaysia

Mohd Nurazzi Norizan
Universiti Sains Malaysia

20.1 INTRODUCTION

The biomedical sector is expected to continue to play a major role in shaping the future of healthcare, improving patient outcomes and advancing the state of medicine. Research and development (R&D) plays a crucial role in advancing the biomedical sector. Biomedical R&D is responsible for the discovery and development of new treatments, drugs, diagnostic tools, and medical technologies that improve human health and well-being [1]. These advancements help in reducing the burden of diseases, improving patient outcomes, and extending human lifespan. Additionally,

DOI: 10.1201/9781003400998-20

R&D also plays a significant role in the development of new knowledge and understanding of the underlying mechanisms of diseases and biological systems, which is essential for developing new and more effective treatments.

In recent years, biomaterials are emerging as an important area in medical science, with a growing number of research studies focused on developing new and innovative materials for use in medical applications [2,3]. The development of new biomaterials has the potential to address a wide range of medical needs, from improving patient outcomes to advancing medical technologies. Some of the key areas where biomaterials are emerging include drug delivery, tissue engineering, wound healing, and implantable devices [4–6]. For example, researchers are developing new and improved materials for drug delivery systems, such as nanomaterials and polymeric materials, that can help to increase the efficacy of drugs and improve patient outcomes. Similarly, the use of biomaterials in tissue engineering is helping to advance regenerative medicine and create new approaches for treating damaged or diseased tissues. Biomaterials are also being used to develop wound dressings and bandages that can promote healing and reduce the risk of infection. There are several types of biomaterials that are currently being used in medical applications such as silk, collagen, chitin, cellulose, and nanocellulose [7–10]. Each type of biomaterial has its own unique properties, advantages, and limitations, and researchers are constantly working to develop new and improved biomaterials for use in medical applications.

Among those listed biomaterials, nanocellulose has been attracting increasing attention for its potential applications in biomedical fields. Nanocellulose is a biodegradable and biocompatible material made from cellulose, which is the most abundant polymer on earth. Research on nanomaterials has received much attention in recent years due to their unique and promising properties. Nanomaterials are materials with structures and properties that are determined by their size and shape at the nanoscale (typically 1–100 nm) [11–13]. One of the unique properties of nanocellulose is its high strength and stiffness, making it a promising candidate for use in medical implants and scaffolds. In addition, nanocellulose also has excellent hydrophilic properties, which make it suitable for use in wound dressings and drug delivery systems. Another advantage of nanocellulose is its biodegradability, which means that it can be safely degraded by the body after its use, reducing the risk of long-term complications. Additionally, nanocellulose is non-toxic, which further enhances its potential for use in biomedical applications.

As a result, researchers and medical professionals are actively exploring the potential of nanocellulose for use in the medical field and developing new techniques for its production and functionalization [14–16]. The ongoing research is aimed at fully realizing its potential in the medical field. Further research is needed to fully understand the potential and limitations of nanocellulose for medical applications. Nevertheless, nanocellulose is a promising material that has the potential to revolutionize the medical field with its unique properties and biocompatibility.

Therefore, this chapter covers the emerging use of nanocellulose in biomedical applications and its potential impact on the field of medicine. The chapter will discuss the properties of nanocellulose that make it suitable for use in biomedical applications, as well as the current research findings on nanocellulose in biomedical applications. The chapter will also discuss the challenges and limitations associated with the use of nanocellulose in biomedical applications, as well as future directions for research in this field. Ultimately, the goal of this chapter is to provide a comprehensive overview of the current state of nanocellulose in biomedical applications and to highlight its potential as a promising biomaterial for the future.

20.2   NANOCELLULOSE AND ITS COMPOSITES

Although nanocellulose has many advantages, the bioinert property of native cellulose limited the passive application of nanocellulose. As a result, additional biofunctionalization or modification of nanocellulose is required to extend nanocellulose's key role in a wide range of biomedical or pharmaceutical applications, which improves their physical and chemical properties. It is basically stated that the properties of nanocomposite are determined by the diffusion and dispersal of nanofillers in polymer mediums, as well as the connection formed between them. Due to the strong inter- and intramolecular hydrogen bonds between hydroxy groups, nanocellulose has a higher specific surface area and a large number of hydroxyl groups (OH) on its surface, which results in the hydrophilicity of the nanocellulose surface and its strong self-aggregation which causes the dispersing nanocellulose uniformly in most non-polar polymeric materials is difficult [17]. It should also be noted that the reactive groups on the nanocellulose surface may result in poor thermal and dimensional stability of the samples [18]. However, from another angle, the abundant reactive groups on the surface of nanocellulose provide numerous available routes for the desired modifications, which primarily aim to reduce aggregation and differences in polarity of nanoparticles. Surface modifications in nanocellulose could improve its compatibility, dispersibility, and related performance, opening the door to a wide range of high-value-added applications. In general, nanocellulose modifications include structural adjustments and functional modifications to hydroxyl groups found in its structure. Esterification, silylation, amination, oxidation, sulfonation, carboxymethylation, polymer grafting, and other chemical modification procedures are commonly used to establish covalent connections with reactive groups on the surface of nanocellulose as illustrated in Figure 20.1.

A few researchers have stated in a variety of research papers that the diffusion and dispersion of nanostuffing or support are recognized as essential conditions to energize the properties of polymer nanocomposites. Carbon nanofiber causes self-agglomeration, which affects the functionality of polymer nanocomposite. Few approaches are used to modify the exterior characteristics of nanoparticles to avoid agglomeration and improve communication between stuffing and medium. The hydrophilic or hydrophobic property of the nanoparticle can be achieved by using such modification techniques [19]. Various functionalization techniques have been used to increase its surface polarity and hydrophilicity. This can be accomplished using various surface functionalization strategies, which typically involve the chemistry of hydroxyl function. This is accomplished through covalent or noncovalent interactions between the stuffing and the polymer medium. Various studies on the use of various types of nanocellulose have been conducted in recent years.

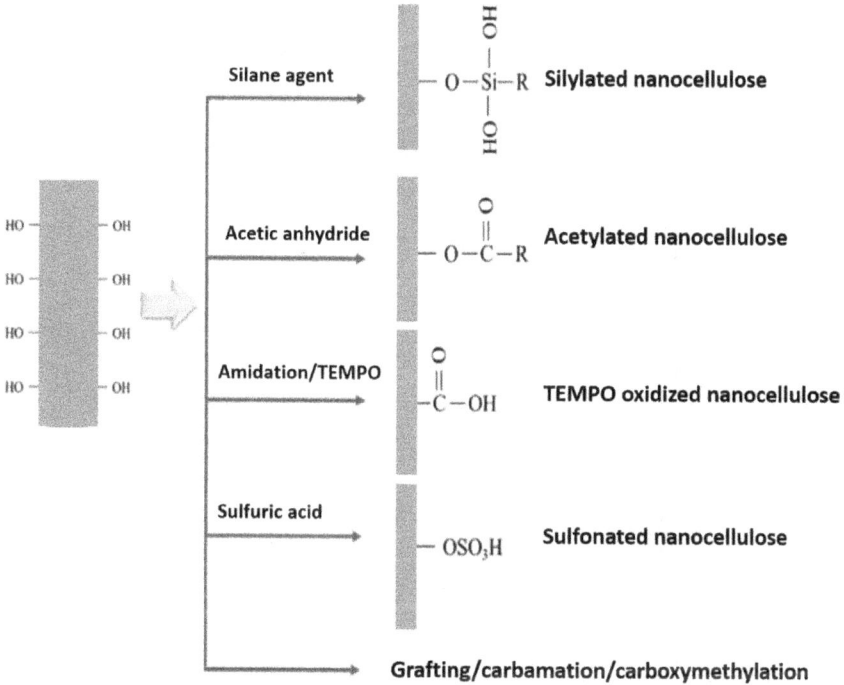

FIGURE 20.1 Schematic diagram of the most used chemical modification techniques for nanocelluloses.

The nanocellulose constituents mentioned above are used to strengthen polymer mediums [20]. Previous researchers struggled to achieve a uniform distribution of nanocellulose in the two water-solvable/insolvable polymer mediums. As a result, chemical processes are used to modify the surface with the outline of hydrophobic moieties to achieve a uniform distribution of nanocellulose in water-insoluble polymer mediums. There are numerous advantages to surface modification of nanocellulose; for example, they can improve the interfacial linkage between nanocellulose and polymer medium, resulting in the enactment of nanocomposites [21].

Depending on the extraction procedure, different types of nanocelluloses are obtained and classified by the microstructures of nanocellulose and the type of hydrogen bond. The above-mentioned microstructures differ in terms of the extraction procedure as well as the source of nanocellulose, namely cellulose nanocrystals extracted via acid hydrolysis and represented by packets of rod-like crystallites [22]. The use of various types of nanocellulose composites, particularly in polymer mediums, has become extremely common due to notable advantages such as low cost, biodegradable material, feasible extraction from sustainable sources, outstanding mechanical properties, barrier characteristics, and a high degree of surface polarization [23]. It has been discovered that cellulose and its derivatives are the most secure and safe substances used in food and biomedical applications. Table 20.1 outlines the most commonly used functionality [24].

TABLE 20.1

The Most Recent Approaches for Surface Functionalization of Nanocellulose

Reaction Type	Exposed Functionality	Reagents	Temperature	Solvent	Time	Catalyst	Reference
Thiol-Alkene	SH	MPTMS	Room temperature	H_2O/ EtOH	2 hours	-	[25]
Azido functionalization	N_3	NaN_3	100°C	DMF	24 hours	-	[26]
Carboxylation	COOH	BTCA/SHP	70°C	H_2O	2 hours	-	[27]
Epoxidation	Epoxide	ECH	60°C	H_2O	2 hours	-	[28]
Amination	NH_2	Fmoc-L-leucine/EDC	Room temperature	DMF	24 hours	DMAP	[29]
Carboxymethylation	Carboxymethyl	Chloroacetic acid	60°C	H_2O/ EtOH/ iPrOH	2 hours	-	[30]
Oxidation	COOH	NaClO/NaBr	0°C	H_2O	4 hours	TEMPO	[31]
Sulfonation	SO_3^-	H_2SO_4/HCL	Room temperature	H_2O	10 hours	-	[32]

20.3 NANOCELLULOSE AND ITS COMPOSITES APPLICATION IN BIOMEDICAL

There are several emerging applications of nanocellulose in the biomedical field such as tissue engineering, wound healing, drug delivery, biodegradable implant, and more (Figure 20.2). Several research are done to fully understand the potential and limitations of nanocellulose in this area of interest. Nevertheless, nanocellulose is a promising material that has the potential to revolutionize the medical field with its unique properties and biocompatibility. In this section, several achievements regarding nanocellulose application in biomedical were discussed.

20.3.1 TISSUE ENGINEERING

Tissue engineering (TE) is the process of isolating and inoculating specific cells in scaffolds while controlling biochemical factors to create functional substitutes. TE is often used to regenerate skin, muscles, ligaments, bone and other tissues for tissue replacement or repairing [33]. Scaffolds, regulatory factors, and seeding cells are the three key elements of TE. Exploring the application of nanocellulose in tissue engineering as supporting matrix (bioscaffold) would lead to further development and advancement in the biomedical field. Due to the abundance of OH groups in their molecular chain (three OH groups in one b-1,4-glucose), nanocellulose can be

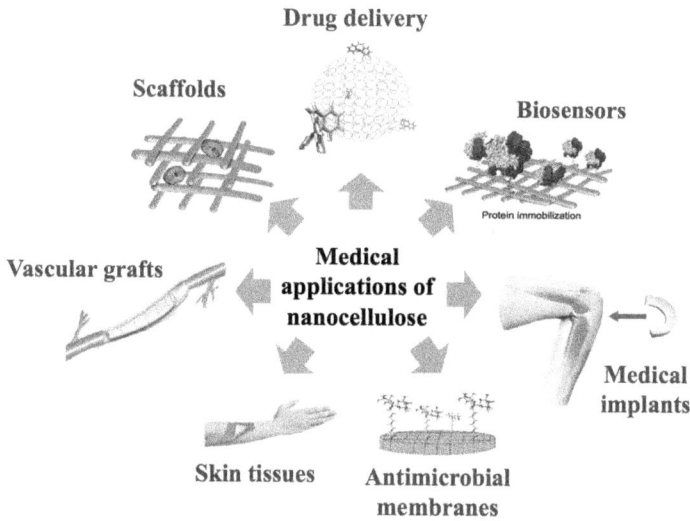

FIGURE 20.2   Medical applications of nanocellulose. (Reproduced from ref. [13].)

easily modified with other functional chemicals and also form H-bonds with other polymer matrices, which would increase their biocompatibility, cellular adaptability, etc. Here, we discuss different types of tissue that has been regenerated or engineered using a nanocellulose-based composite.

Sukul et al. [34] developed a nanocellulose-based hydrogel, loaded with vascular endothelial growth factor (VEGF) and bone morphogenic protein (BMP), to study the regeneration of bone tissue in orthopedic defects of rat bone marrow stem cells (RBMSCs). The outcome demonstrated that the cells properly adhered to the hydrogel and proliferated to form bone tissue. After 7 and 14 days of treatment, the effect of VEGF and BMP has promoted high alkaline phosphatase (ALP) protein expression in RBMSC, indicating its improved bone regeneration potential.

A patch of lysozyme nanofibers (LNF), derived from the white egg, was used for wound healing applications [35]. When incorporated with nanocellulose, the fiber patch became more thermally stable, and its mechanical properties were also improved. More importantly, the LNF/nanocellulose patch promotes better adhesion of L929 fibroblast cells with no adverse effects on the cells as compared to the pristine nanocellulose patch. The result suggests the potential of a nanocellulose composite in connective tissue regeneration.

A nanocellulose–polypropylene composite was developed by Anton-Sales and colleagues for tissue repairing in hernioplasty [36]. The nanocellulose offered better mechanical stability and anti-adhesive qualities to the composite for strengthening soft tissue of abdominal wall. The nanocellulose condense in the polymer matrix upon drying, forming strong inter- and intramolecular bonds. The tear resistance of the composite exceeded 16 N/cm, which is a threshold value for abdominal wall reinforcement applications. This demonstrates that the less stretchy dried nanocellulose is better suited for hernioplasty.

Hemostats are frequently used to stop bleeding during surgical procedures or accidents, and they are the first step toward wound healing. Silk-fibroin (SF)/TEMPO-nanocellulose scaffolds, loaded with thrombin, were developed by Shefa et al. [37] for hemostatic applications. The hemostatic potential of the developed nanocellulose composites was tested on the bleeding ear of a rabbit, and the outcome was improved blood adsorption capacity and biocompatibility. The bleeding was stopped in about 2 minutes when the thrombin-loaded scaffolds were used, as compared to 3.6 minutes for the control treatment. This suggests the potential of the developed nanocellulose composites to be used as hemostats and skin tissue repairs.

Peng et al. [38] used the coaxial electrospinning method to create PCL/gelatin nanocellulose membranes, incorporated with magnesium oxide (MgO) nanoparticles, for human periodontal ligament stem cells (hPDLSCs) regeneration. The membrane showed higher cellular compatibility and has antimicrobial properties. Higher activity of alkaline phosphatase and levels of osteogenic-associated gene markers (ColI and Runx2) were observed when MgO-incorporating PCL/gelatin/nanocellulose membrane was used as compared to the control group.

20.3.2 Medical Implants

Nanocellulose is well suited for medical implants due to its purity and exceptional mechanical performance. Nanocellulose forms a network that resembles the extracellular matrix (ECM) and has excellent wet-state integrity, making it suitable for suturing and sterilization. A study demonstrated the development of complex 3D forms by utilizing the aerobic process involved in the biogenesis of nanocellulose at the interphase of air/liquid culture medium [39]. To guide the formation of nanocellulose biofilms that easily form auxetic structures, solid supports were used. The measured mechanical strength (48–456 MPa tensile strength) is easily adjustable while ensuring shape stability (about 87% shape retention after 100 burst loading–unloading cycles). From cytotoxicity study, the nanocellulose showed negligible pro-inflammatory activation, indicating the potential use of the nanocellulose as supportive implants long-term comfort and stability.

The hemocompatibility (hemolysis and thrombogenicity) and acute and sub-chronic immune of nanocellulose were studied [40]. Ex vivo blood studies revealed that two-dimensional (2D) nanocellulose exhibited antihemolytic and antithrombogenic effects. In vivo studies, however, revealed that three-dimensional (3D) nanocellulose did not interfere with wound hemostasis and only elicited a mild acute inflammatory response. During culture treatment, increment of tissue infiltration to 91% was observed after 12 weeks, which was characterized by fibroblastic, capillary, and extracellular matrix infiltration. This suggest that 3D nanocellulose can be thought of as a possible implantable biomaterial for soft tissue augmentation or replacement.

Horbert et al. [41] investigated the possibility of using nanocellulose as blood vessel or cartilage implants. A standardized bovine cartilage punch model was used to analyze the regenerative capacity of nanocellulose implants. The cartilage-nanocellulose constructs exhibited vital chondrocytes (up to 90% until week 9; >80%

until week 12), preserved matrix integrity during culture, and restricted loss of matrix-bound proteoglycan from "host" cartilage. More importantly, the nanocellulose implant displayed visible cell colonization (chondrogenesis), and the progressive increase in collagen content suggests the beginning of cartilage regeneration. This demonstrates the high potential of nanocellulose as a cartilage replacement material for tissue engineering.

Layer-by-layer (LbL) meshes made of chitosan and cellulose nanofibrils were created by using direct ink writing through single- or multihead extrusion [39]. Given the role of electrostatic complexation between the oppositely charged components, the 3D-printed structure exhibited the best mechanical performance, that is, 683 MPa modulus and 2.5 MPa tensile strength. In neutral and light acidic conditions, the LbL structures showed high wet stability with gradual weight loss. In addition, excellent biocompatibility and non-cytotoxicity toward human monocytes and macrophages, as well as controllable shrinkage upon solvent exchange, suggest the suitability of the LbL meshes for biomedical implant usage.

20.3.3 Delivery of Drugs/Bioactive Compounds

Nanocellulose-based drug delivery system is a cutting-edge technology that is constantly evolving for more effective controlled and targeted delivery. In recent years, researchers have concentrated on modifying nanocellulose to form nanocomposites and hybrids that improve the drug loading and delivery efficiency through multiple approaches [42,43]. They are used in the form of aerogels, cryogels, hydrogels, film membrane, etc. [44]. Several drugs and bioactive molecules such as protein have benefited from the delivery systems.

20.3.3.1 Oral Delivery

In a spray-drying treatment study on tablets, nanocellulose demonstrated stronger encapsulation while decreasing powder porosity than conventional microcrystalline cellulose, indicating that a single spray-dried nanocellulose adjuvant can be used in tablet manufacturing [45]. Using a similar method, bacterial cellulose-coated tablets showed soft and flexible attributes with higher mechanical properties than standard Aquacoat ECD materials (30% aqueous ethylcellulose dispersion). The solid carriers composed of nanocellulose and other matrices provided good drug molecule confinement and drug release control.

Patil et al. [46] used nanocomposites of gelatinized corn starch and urea formaldehyde with nanocellulose to create a dimethyl phthalate controlled release system. Nanocellulose significantly hampered the initial release of dimethyl phthalate, but it was successful in providing controlled drug release. This was due to the complex network within starch matrix that prolonged the drug release (about 80%–95% for a period of a week).

A comparative study between nanocellulose/alginate and microcrystalline cellulose (MCC)/alginate as a delivery system for metformin hydrochloride was conducted by a group of researchers [47]. It was found that the cumulative release through nanocellulose/alginate beads was 10% greater than that of MCC/alginate and showed sustained release over the next 240 minutes. The nanocellulose incorporation contributed to the improved swelling and mechanical properties of alginate matrix.

Thomas et al. [48] have developed an alginate-nanocellulose hybrid polymer formulation. This formulation has a higher encapsulation efficiency and can be used for controlled oral drug delivery of rifampicin. It was suggested that the rifampicin-loaded hybrid polymer is a more effective means of treating *Mycobacterium tuberculosis.*

20.3.3.2 Transdermal Delivery

The transdermal drug delivery system (TDDS) allows drugs to be delivered through the skin in order to achieve therapeutic concentrations. In this manner, the drug avoids the digestive tract and liver metabolism, offering a therapeutic effect at lower doses and lowering the side effects in the gastrointestinal and hepatic systems.

Abba et al. developed stable nanocellulose membranes for crocin drug delivery [49]. Direct dissolution and transdermal passage revealed significant release and permeation profiles of crocin, indicating the potential use of the membrane as TDDS patches. Another study demonstrates nanofiber membranes have similar utility as drug delivery patches, whereby piroxicam was sustainably released for several hours when the loaded membranes were exposed to simulated human skin fluid. The storage stability of nanocellulose membranes can be improved by adding the hydrophilic and lipophilic active pharmaceutical ingredients (APIs) such as caffeine, lidocaine, ibuprofen, and diclofenac. The APIs-loaded nanocellulose membrane was tested in vivo for cutaneous compatibility, and the results confirmed the membranes' good storage stability as well as its actual potential as TDDS patch. Hivechi et al. [50] blended polycaprolactone (PCL) with nanocellulose, forming composites for tetracycline drug delivery. It was observed that the tetracycline drug was more sustainably released out of the PCL/nanocellulose composite as compared to pure PCL, indicating the significant role of nanocellulose in restricting the movement of tetracycline. Saidi and colleagues synthesized pH-responsive N-methacryloyl glycine (MGly)/nanocellulose composites to deliver diclofenac via transdermal and oral routes [51]. The resulting composites had better mechanical, thermal, and viscoelastic properties and increased water uptake potential. Because of the presence of nanocellulose in the composite, a more controlled delivery of diclofenac was observed than with pure N-methacryloyl glycine.

Beside transdermal delivery patches, TDDS also has a remarkable application in skin tissue healing; however, only small drugs can penetrate the skin, thus limiting its application. Erdagi et al. [52] created gelatin/diosgenin-nanocellulose hydrogels with exceptional swelling capacity for neomycin drug delivery in wound healing applications by using genipin as a cross-linker. The developed hydrogels were tested for cytocompatibility with in vitro cell cultures of human dermal fibroblasts and showed no adverse effect. Furthermore, the hydrogel demonstrated synergistic effects with the neomycin drug, with a 90% sustainable release within 24 hours. Polydopamine (PDA)/nanocellulose hydrogels were developed by Liu and colleagues for tetracycline delivery in wound healing [53]. Calcium chloride was used to chemically cross-link the hydrogels. The hydrogels composite exhibited pH responsive property with higher wound healing efficacy than the pure hydrogel. Another pH-responsive hydrogel, made of borneol/mono-6-(2-hydroxy-3-(tri-methyl ammonium) propyl)-cyclodextrin (BN/EPTAC-CD)/di-aldehyde carboxymethyl cellulose (DCMC), was also fabricated for similar application [54]. The components within the hydrogel were cross-linked with tobramycin. In an acidic environment, tobramycin imine bonds break, facilitating drug release and accelerating wound healing process.

20.3.3.3 Targeted/Local Delivery

The drug could be released by local drug delivery systems either at or near the target site, which would increase the effectiveness of the system while simultaneously lowering the required dose. Therefore, systemic exposure is limited, which leads to reduced toxicity to healthy tissue [55].

Laurén et al. [56] investigated the use of nanocellulose as an injectable hydrogel for local drug delivery and controlled release. They used a small drug (I-b-CIT) and a large protein drug (Tc-HSA) to test the effect of molecule size on release rate. From the results, the hydrogel reduced the large drug's elimination rate by twofold as compared to the small drug, which is probably due to the lack of apparent binding of small molecules to nanocellulose. Thus, it is suggested that nanocellulose hydrogel can be used as a potential matrix for controlled release or local delivery of large compounds (such as macromolecular proteins and peptide drugs).

In a study, biodegradable polymer-based films were developed by using nanocellulose and anionic nanocellulose (as film-forming substances), incorporated with various combinations of mucoadhesive components like mucin, pectin, and chitosan (as functional bioadhesion enhancers), as the delivery system for the treatment of oral diseases such as periodontitis [57]. From the results, rapid metronidazole release and the patch's inactivity after detachment were observed, which is good as high and rapid dosing is preferable for local drug delivery.

20.3.4 ANTIMICROBIAL MATERIAL FOR WOUND DRESSING

The potential use of honey-incorporated nanocellulose/polyvinylpyrrolidone (PVP) composite film for wound dressing was studied [58]. The first-order kinetic was intended to constantly release honey from the film. The maximum amount of active ingredients was released after 48 hours, indicating sustainable release. In addition to its wound-healing properties, honey also provides significant antimicrobial activity against Bacillus subtilis, Staphylococcus aureus, and Proteus mirabilis. Controlled release combined with wound healing and antimicrobial efficacy primarily demonstrate the potential of the nanocellulose film composite as a good dressing for chronic wound treatments.

To fabricate composite film dressings with antibacterial activity and improved mechanical properties, sodium alginate was efficiently incorporated into the nanocellulose matrix via vacuum suction, followed by cross-linking via immersion in separate solutions of six cations, that is, silver, zinc, cobalt manganese, and cerium [59]. The results indicated that the fabricated composites displayed not only antibacterial activities, but also pH-responsive, non-toxic to fibroblast cells and improved mechanical properties, making them suitable for use as smart dressings. From a wound healing study using rat as the model, the skin wounds covered by the composite dressings healed faster than the pure nanocellulose.

Carboxylated nanocellulose with varying oxidation levels were developed for wound dressing with a strong antibacterial effect [60]. The film's antibacterial activity was tested against *Pseudomonas aeruginosa* and *Staphylococcus aureus*. The results showed that higher oxidation levels in the nanocellulose resulted in a stronger antibacterial effect. Using a mouse model of surgical site infection, a nanocellulose sample (concentration 0.6 wt %) oxygenated to a DO level of 46.4 mg/L demonstrated a strong antibacterial effect against *S. aureus* in vivo. After 24 hours of treatment, the oxygenated nanocellulose dispersion reduced bacterial survival by 71%. According to a scanning electron microscopy analysis, the antibacterial property was due to the complex network of oxygenated nanocellulose that provides good bacteria entrapment.

In one study, an aminoalkylsilane (AaS) and nanocellulose graft copolymer membrane (AaS-g-nanocellulose) were combined with pullulan (Pul) and zinc oxide nanoparticles (ZnO-NPs) [59]. Pullulan (Pul) is a non-toxic polymer that promotes wound healing, whereas ZnO-NPs are well known for their antibacterial properties. The fabricated composite dressing outperformed pure nanocellulose in terms of blood clotting and antibacterial activity by up to 5 log values. The dressing's cytotoxicity test toward L929 fibroblast cells clearly proven safety due to fibroblast cell proliferation. In a rat model, the wounds covered by AaS-g-nanocellulose/Pul-ZnO showed faster healing and re-epithelialization, small blood vessel formation, and collagen synthesis. The findings show a low-cost method for producing functional dressings with high antibacterial activity and rapid wound healing.

Electrospinning was used to create nanofiber composites from chitosan (CS), polyethylene oxide (PEO), nanocellulose, and acacia plant-based extract [61]. The incorporation of nanocellulose improved the physical integrity, water evaporation rate, and thermal properties of the nanofibers. Acacia incorporation into the CS/PEO/nanocellulose increased its antifungal activity against some fungi while maintaining the original antibacterial activity. This demonstrates the nanofiber's potential use against a wide range of microorganisms, which is critical for preventing or treating infections. All composite samples showed no cytotoxicity in non-tumor cells, indicating good biocompatibility. Besides, a constant release of the acacia extract was observed for 24 hours, demonstrating its prolonged healing action and reducing the frequency with which dressings are replaced. Overall, the nanofibers developed are very promising as localized drug delivery systems for wound care applications.

20.4 CHALLENGES AND FUTURE RECOMMENDATIONS

Despite the promising potential of nanocellulose for various biomedical applications, there are several challenges that must be addressed in order to fully realize its potential. Some of these challenges include:

- Scalability: Currently, the production of nanocellulose is limited and can be expensive, making it difficult to scale up for commercial applications.
- Stability: Nanocellulose is a highly hydrophilic material that is prone to degradation in aqueous environments, which can limit its stability and long-term performance in biomedical applications.

- Biocompatibility: The biocompatibility of nanocellulose must be thoroughly tested and characterized before it can be widely used in medical applications.
- Regulation: The regulation of nanocellulose for medical applications is still in its early stages, and more research is needed to determine the safety and efficacy of nanocellulose for various biomedical applications.
- Cost: The cost of producing nanocellulose and functionalizing it for biomedical applications is currently high, which may limit its widespread use in the near future.

These are some of the challenges that must be overcome in order to fully realize the potential of nanocellulose for biomedical applications. Nevertheless, researchers are actively working to address these challenges and develop new techniques for the production and functionalization of nanocellulose for medical use.

20.5 CONCLUSION

In conclusion, nanocellulose is a highly promising nanomaterial for various biomedical applications due to its unique mechanical and physical properties, biocompatibility, and biodegradability. It has the potential to revolutionize the medical field by enabling the development of new therapies and medical devices. However, several challenges must be addressed before nanocellulose can be widely used in the medical field, including scalability, stability, biocompatibility, regulation, and cost. Nevertheless, ongoing research is aimed at addressing these challenges and unlocking the full potential of nanocellulose for biomedical applications.

ACKNOWLEDGEMENTS

The authors gratefully acknowledge financial support from the Universiti Pertahanan Nasional Malaysia in the preparation of this manuscript.

REFERENCES

1. Francis, P.J.J. Biomedical Applications of Polymers -An Overview. *Curr Trends Biomed Eng Biosci* **2018**, *15*, 44–45, doi:10.19080/ctbeb.2018.15.555909.
2. Parry, A. Nanocellulose and Its Composites for Biomedical Applications. *Curr Med Chem* **2017**, *24*(5), 512–528, doi:10.2174/0929867323666161014124008.
3. Sharip, N.S.; Yasim-Anuar, T.A.T.; Norrrahim, M.N.F.; Shazleen, S.S.; Nurazzi, N.M.; Sapuan, S.M.; Ilyas, R.A. A Review on Nanocellulose Composites in Biomedical Application. In *Composites in Biomedical Applications*; CRC Press, **2020**; pp. 161–190.
4. Manavitehrani, I.; Fathi, A.; Badr, H.; Daly, S.; Shirazi, A.N.; Dehghani, F. Biomedical Applications of Biodegradable Polyesters. *Polymers* **2016**, *8*, doi:10.3390/polym8010020.
5. Curvello, R,; Raghuwanshi, V.S.; Garnier, G. Engineering Nanocellulose Hydrogels for Biomedical Applications. *Adv Colloid Interface Sci* **2019**, *267*, 47–61, doi:10.1016/j.cis.2019.03.002.
6. Gumrah Dumanli, A. Nanocellulose and Its Composites for Biomedical Applications. *Curr Med Chem* **2016**, *24*, 512–528, doi:10.2174/0929867323666161014124008.

7. Misenan, S.; Shaffie, A.; Zulkipli, N.; Norrrahim, F. Nanocellulose in Sensors. **2022**; pp. 213–240 ISBN 9780323899093.

8. Jenol, M.A.; Norrrahim, M.N.F.; Nurazzi, N.M. 17 - Nanocellulose Nanocomposites in Textiles. In *Industrial Applications of Nanocellulose and Its Nanocomposites*; Sapuan, S.M., Norrrahim, M.N.F., Ilyas, R.A., Soutis, C. (Eds.); Woodhead Publishing, **2022**; pp. 397–408 ISBN 978-0-323-89909-3.

9. Norrrahim, M.N.F.; Ilyas, R.A.; Nurazzi, N.M.; Rani, M.S.A.; Atikah, M.S.N.; Shazleen, S.S. Chemical Pretreatment of Lignocellulosic Biomass for the Production of Bioproducts: An Overview. *Applied Science and Engineering Progress* **2021**.

10. Lee, C.H.; Lee, S.H.; Padzil, F.N.M.; Ainun, Z.M.A.; Norrrahim, M.N.F.; Chin, K.L. Biocomposites and Nanocomposites. In *Composite Materials*; CRC Press, **2021**; pp. 29–60.

11. Norrrahim, M.N.F.; Kasim, N.A.M.; Knight, V.F.; Misenan, M.S.M.; Janudin, N.; Shah, N.A.A.; Kasim, N.; Yusoff, W.Y.W.; Noor, S.A.M.; Jamal, S.H.; et al. Nanocellulose: A Bioadsorbent for Chemical Contaminant Remediation. *RSC Adv* **2021**, *11*, 7347–7368, doi:10.1039/D0RA08005E.

12. Norrrahim, M.N.F.; Norizan, M.N.; Jenol, M.A.; Farid, M.A.A.; Janudin, N.; Ujang, F.A.; Yasim-Anuar, T.A.T.; Najmuddin, S.U.F.S.; Ilyas, R.A. Emerging Development on Nanocellulose as Antimicrobial Material: An Overview. *Mater Adv* **2021**, doi:10.1039/D1MA00116G.

13. Norrrahim, M.N.F.; Kasim, N.A.M.; Knight, V.F.; Ujang, F.A.; Janudin, N.; Razak, M.A.I.A.; Shah, N.A.A.; Noor, S.A.M.; Jamal, S.H.; Ong, K.K.; et al. Nanocellulose: The Next Super Versatile Material for the Military. *Mater Adv* **2021**, doi:10.1039/D0MA01011A.

14. Karim, Z.; Afrin, S.; Husain, Q.; Danish, R. Necessity of Enzymatic Hydrolysis for Production and Functionalization of Nanocelluloses. *Crit Rev Biotechnol* **2017**, *37*, 355–370, doi:10.3109/07388551.2016.1163322.

15. Hoenders, D.; Guo, J.; Goldmann, A.S.; Barner-Kowollik, C.; Walther, A. Photochemical Ligation Meets Nanocellulose: A Versatile Platform for Self-Reporting Functional Materials. *Mater Horiz* **2018**, *5*, 560–568, doi:10.1039/c8mh00241j.

16. Frank, B.P.; Smith, C.; Caudill, E.R.; Lankone, R.S.; Carlin, K.; Benware, S.; Pedersen, J.A.; Fairbrother, D.H. Biodegradation of Functionalized Nanocellulose. *Environ Sci Technol* **2021**, *55*, 10744–10757, doi:10.1021/acs.est.0c07253.

17. Abdullah, N.A.; Rani, M.S.A.; Mohammad, M.; Sainorudin, M.H.; Asim, N.; Yaakob, Z.; Razali, H.; Emdadi, Z. Nanocellulose from Agricultural Waste as an Emerging Material for Nanotechnology Applications-an Overview. *Polimery* **2021**, *66*, 157–168.

18. Trache, D.; Tarchoun, A.; Derradji, M.; Hamidon, T.; Masruchin, N.; Brosse, N.; Hussin, M.H. Nanocellulose: From Fundamentals to Advanced Applications. *Front Chem* **2020**, *8*, doi:10.3389/fchem.2020.00392.

19. Abraham, J.; Sharika, T.; Mishra, R.; Thomas, S. Rheological Characteristics of Nanomaterials and Nanocomposites. In *Micro and Nano Fibrillar Composites (MFCs and NFCs) from Polymer Blends*; **2017**; pp. 331–350 ISBN 9780081019917.

20. Chirayil, C.; Mathew, L.; Thomas, S. Review of Recent Research in Nano Cellulose Preparation from Different Lignocellulosic Fibers. *Reviews on Advanced Materials Science* **2014**, *37*, 20–28.

21. Mishra, R.; Thomas, S.; Kalarikkal, N. *Micro and Nano Fibrillar Composites (MFCs and NFCs) from Polymer Blends*; **2017**; ISBN 9780081019917.

22. de France, K.; Hoare, T.; Cranston, E. Review of Hydrogels and Aerogels Containing Nanocellulose. *Chemistry of Materials* **2017**, *29*, doi:10.1021/acs.chemmater.7b00531.

23. Mariano, M.; Kissi, N.; Dufresne, A. Cellulose Nanocrystals and Related Nanocomposites: Review of Some Properties and Challenges. *J Polym Sci B Polym Phys* **2014**, *52*, doi:10.1002/polb.23490.

24. Xu, X.; Liu, F.; Jiang, L.; Zhu, J.Y.; Haagenson, D.; Wiesenborn, D.P. Cellulose Nanocrystals vs. Cellulose Nanofibrils: A Comparative Study on Their Microstructures and Effects as Polymer Reinforcing Agents. *ACS Appl Mater Interfaces* **2013**, *5*, 2999–3009, doi:10.1021/am302624t.

25. Valo, H.; Kovalainen, M.; Laaksonen, P.; Häkkinen, M.; Auriola, S.; Peltonen, L.; Linder, M.; Järvinen, K.; Hirvonen, J.; Laaksonen, T. Immobilization of Protein-Coated Drug Nanoparticles in Nanofibrillar Cellulose Matrices-Enhanced Stability and Release. *J Control Release* **2011**, *156*, 390–397, doi:10.1016/j.jconrel.2011.07.016.

26. Jackson, J.; Letchford, K.; Wasserman, B.; Ye, L.; Hamad, W.; Burt, H. The Use of Nanocrystalline Cellulose for the Binding and Controlled Release of Drugs. *Int J Nanomed* **2011**, *6*, 321–330, doi:10.2147/IJN.S16749.

27. Evdokimova, O.; Ivanov, V.; Agafonov, A.; Seisenbaeva, G.; Kessler, V. Cellulose Nanofiber-Titania Nanocomposites as Potential Drug Delivery Systems for Dermal Applications. *J Mater Chem B* **2015**, doi:10.1039/C4TB01823K.

28. Kalaskar, D.; Ulijn, R.; Gough, J.; Alexander, M.; Scurr, D.; Sampson, W.; Eichhorn, S. Characterisation of Amino Acid Modified Cellulose Surfaces Using ToF-SIMS and XPS. *Cellulose* **2010**, *17*, 747–756, doi:10.1007/s10570-010-9413-y.

29. Feese, E.; Sadeghifar, H.; Gracz, H.; Argyropoulos, D.; Ghiladi, R. Photobactericidal Porphyrin-Cellulose Nanocrystals: Synthesis, Characterization, and Antimicrobial Properties. *Biomacromolecules* **2011**, *12*, 3528–3539, doi:10.1021/bm200718s.

30. Eyholzer, C.; Borges, A.; Duc, F.; Bourban, P.-E.; Tingaut, P.; Zimmermann, T.; Månson, J.; Oksman, K. Biocomposite Hydrogels with Carboxymethylated, Nanofibrillated Cellulose Powder for Replacement of the Nucleus Pulposus. *Biomacromolecules* **2011**, *12*, 1419–1427, doi:10.1021/bm101131b.

31. Saito, T.; Nishiyama, Y.; Putaux, J.-L.; Vignon, M.; Isogai, A. Homogeneous Suspensions of Individualized Microfibrils from TEMPO-Catalyzed Oxidation of Native Cellulose. *Biomacromolecules* **2006**, *7*, 1687–1691, doi:10.1021/bm060154s.

32. Wang, N.; Ding, E.; Cheng, R. Preparation and Liquid Crystalline Properties of Spherical Cellulose Nanocrystals. *Langmuir* **2008**, *24*, 5–8, doi:10.1021/la702923w.

33. Jacob, S.; R., R.; Antony, S.; Madhavan, A.; Raveendran, S.; Awasthi, M.K.; Kuddus, M.; Pillai, S.; Varjani, S.; Pandey, A.; et al. Nanocellulose in Tissue Engineering and Bioremediation: Mechanism of Action. *Bioengineered* **2022**, *13*, doi:10.1080/21655979.2022.2074739.

34. Sukul, M.; Nguyen, L.T.; Min, Y.-K.; Lee, S.; Lee, B.-T. Effect of Local Sustainable Release of BMP2-VEGF from Nano-Cellulose Loaded in Sponge Biphasic Calcium Phosphate on Bone Regeneration. *Tissue Eng Part A* **2015**, doi:10.1089/ten.TEA.2014.0497.

35. Silva, N.; Garrido, P.; Moreirinha, C.; Almeida, A.; Palomares, T.; alonso-varona, A.; Vilela, C.; Freire, C. Multifunctional Nanofibrous Patches Composed of Nanocellulose and Lysozyme Nanofibers for Cutaneous Wound Healing. *Int J Biol Macromol* **2020**, *165*, doi:10.1016/j.ijbiomac.2020.09.249.

36. Anton-Sales, I.; Roig-Sanchez, S.; Traeger, K.; Weis, C.; Laromaine, A.; Turon, P.; Roig, A. In Vivo Soft Tissue Reinforcement with Bacterial Nanocellulose. *Biomater Sci* **2021**, *9*, doi:10.1039/D1BM00025J.

37. Shefa, A.; Taz, M.; Lee, S.; Lee, B.-T. Enhancement of Hemostatic Property of Plant Derived Oxidized Nanocellulose-Silk Fibroin Based Scaffolds by Thrombin Loading. *Carbohydr Polym* **2018**, *208*, doi:10.1016/j.carbpol.2018.12.056.

38. Peng, W.; Ren, S.; Zhang, Y.; Fan, R.; Zhou, Y.; Li, L.; Xu, X.; Xu, Y. MgO Nanoparticles-Incorporated PCL/Gelatin-Derived Coaxial Electrospinning Nanocellulose Membranes for Periodontal Tissue Regeneration. *Front Bioeng Biotechnol* **2021**, *9*, doi:10.3389/fbioe.2021.668428.

39. Ajdary, R.; Abidnejad, R.; Lehtonen, J.; Kuula, J.; Raussi-Lehto, E.; Kankuri, E.; Tardy, B.; Rojas, O. Bacterial Nanocellulose Enables Auxetic Supporting Implants. *Carbohydr Polym* **2022**, *284*, 119198, doi:10.1016/j.carbpol.2022.119198.

40. Osorio Delgado, M.; Cañas, A.; Puerta, J.; Díaz, L.; Naranjo, T.; Ortiz, I.; Castro, C. Ex Vivo and In Vivo Biocompatibility Assessment (Blood and Tissue) of Three-Dimensional Bacterial Nanocellulose Biomaterials for Soft Tissue Implants. *Sci Rep* **2019**, *9*, doi:10.1038/s41598-019-46918-x.

41. Horbert, V.; Boettcher, J.; Foehr, P.; Kramer, F.; Udhardt, U.; Bungartz, M.; Brinkmann, O.; Burgkart, R.; Klemm, D.; Kinne, R. Laser Perforation and Cell Seeding Improve Bacterial Nanocellulose as a Potential Cartilage Implant in the In Vitro Cartilage Punch Model. *Cellulose* **2019**, *26*, doi:10.1007/s10570-019-02286-3.

42. Salimi, S.; Sotudeh-Gharebagh, R.; Zarghami, R.; Chan, S.-Y.; Yuen, K.H. Production of Nanocellulose and Its Applications in Drug Delivery: A Critical Review. *ACS Sustain Chem Eng* **2019**, *7*, doi:10.1021/acssuschemeng.9b02744.

43. Raghav, N.; Sharma, M.; Kennedy, J. Nanocellulose: A Mini-Review on Types and Use in Drug Delivery Systems. *Carbohydrate Polymer Technologies and Applications* **2021**, *2*, 100031, doi:10.1016/j.carpta.2020.100031.

44. Das, S.; Ghosh, B.; Sarkar, K. Nanocellulose as Sustainable Biomaterials for Drug Delivery. *Sensors International* **2021**, *3*, 100135, doi:10.1016/j.sintl.2021.100135.

45. Kolakovic, R.; Peltonen, L.; Laaksonen, T.; Putkisto, K.; Laukkanen, A.; Hirvonen, J. Spray-Dried Cellulose Nanofibers as Novel Tablet Excipient. *AAPS PharmSciTech* **2011**, *12*, 1366–1373, doi:10.1208/s12249-011-9705-z.

46. Patil, M.; Patil, V.; Sapre, A.; Ambone, T.; Torris, A.; Shukla, P.; Shanmuganathan, K. Tuning Controlled Release Behavior of Starch Granules Using Nanofibrillated Cellulose Derived from Waste Sugarcane Bagasse. *ACS Sustain Chem Eng* **2018**, *6*, doi:10.1021/acssuschemeng.8b01545.

47. Guo, T.; Pei, Y.; Tang, K.; He, X.; Huang, J.; Wang, F. Mechanical and Drug Release Properties of Alginate Beads Reinforced with Cellulose. *J Appl Polym Sci* **2016**, *134*, doi:10.1002/app.44495.

48. Thomas, D.; Latha, S.; Thomas, K. Synthesis and in Vitro Evaluation of Alginate-Cellulose Nanocrystal Hybrid Nanoparticles for the Controlled Oral Delivery of Rifampicin. *J Drug Deliv Sci Technol* **2018**, *46*, 392–399, doi:10.1016/j.jddst.2018.06.004.

49. Abba, M.; Ibrahim, Z.; Chong, C.S.; Zawawi, N.; Abdul Kadir, M.; Yusof, A.; Razak, S.I.A. Transdermal Delivery of Crocin Using Bacterial Nanocellulose Membrane. *Fibers Polym* **2019**, *20*, 2025–2031, doi:10.1007/s12221-019-9076-8.

50. Hivechi, A.; Bahrami, H.; Siegel, R. Drug Release and Biodegradability of Electrospun Cellulose Nanocrystal Reinforced Polycaprolactone. *Mater Sci Eng C* **2018**, *94*, doi:10.1016/j.msec.2018.10.037.

51. Wasserman, L.; Papachin, A.A.; Borodina, Z.M.; Krivandin, A.; A.I., S.; Tarasov, V.F. Some Physico-Chemical and Thermodynamic Characteristics of Maize Starches Hydrolyzed by Glucoamylase. *Carbohydr Polym* **2019**, *212*, doi:10.1016/j.carbpol.2019.01.096.

52. Ilkar Erdagi, S.; Fahanwi, A.N.; Yildiz, U. Genipin Crosslinked Gelatin-Diosgenin-Nanocellulose Hydrogels for Potential Wound Dressing and Healing Applications. *Int J Biol Macromol* **2020**, *149*, doi:10.1016/j.ijbiomac.2020.01.279.

53. Liu, Y.; Sui, Y.; Liu, C.; Liu, C.; Wu, M.; Li, B.; Li, Y. A Physically Crosslinked Polydopamine/Nanocellulose Hydrogel as Potential Versatile Vehicles for Drug Delivery and Wound Healing. *Carbohydr Polym* **2018**, *188*, doi:10.1016/j.carbpol.2018.01.093.

54. Fan, X.; Yang, L.; Wang, T.; Tiedong, S.; Lu, S. PH-Responsive Cellulose-Based Dual Drug-Loaded Hydrogel for Wound Dressing. *Eur Polym J* **2019**, *121*, 109290, doi:10.1016/j.eurpolymj.2019.109290.

55. Tan, T.; Lee, H.; Dabdawb, W.; Hamid, S. A Review of Nanocellulose in the Drug-Delivery System. *Mater Biomed Eng* **2019**; 131–164 ISBN 9780128169131.
56. Laurén, P.; Lou, Y.-R.; Raki, M.; Urtti, A.; Bergström, K.; Yliperttula, M. Technetium-99m-Labeled Nanofibrillar Cellulose Hydrogel for in Vivo Drug Release. *Euro J Pharmaceut Sci* **2014**, *65*, doi:10.1016/j.ejps.2014.09.013.
57. Laurén, P.; Paukkonen, H.; Lipiäinen, T.; Dong, Y.; Oksanen, T.; Räikkönen, H.; Ehlers, H.; Laaksonen, P.; Yliperttula, M.; Laaksonen, T. Pectin and Mucin Enhance the Bioadhesion of Drug Loaded Nanofibrillated Cellulose Films. *Pharm Res* **2018**, *35*, doi:10.1007/s11095-018-2428-z.
58. Taher, M.; Zahan, K.; Rajaie, M.; Leong, C.R.; Ab Rashid, S.; Hamin, N.; Tan, W.-N.; Yenn, T. Nanocellulose as Drug Delivery System for Honey as Antimicrobial Wound Dressing. *Mater Today Proc* **2020**, *31*, doi:10.1016/j.matpr.2020.01.076.
59. Shahriari Khalaji, M.; Hong, S.; Hu, G.; Ji, Y.; Hong, F. Bacterial Nanocellulose-Enhanced Alginate Double-Network Hydrogels Cross-Linked with Six Metal Cations for Antibacterial Wound Dressing. *Polymers* **2020**, *12*, 2683, doi:10.3390/polym12112683.
60. Knutsen, M.; Agrenius, K.; Ugland, H.; Petronis, S.; Haglerød, C.; Håkansson, J.; Chinga Carrasco, G. Oxygenated Nanocellulose-A Material Platform for Antibacterial Wound Dressing Devices. *ACS Appl Bio Mater* **2021**, *XXXX*, doi:10.1021/acsabm.1c00819.
61. Ribeiro, A.; Costa, S.; Ferreira, D.; Calhelha, R.; Barros, L.; Stojković, D.; Soković, M.; Ferreira, I.; Fangueiro, R. Chitosan/Nanocellulose Electrospun Fibers with Enhanced Antibacterial and Antifungal Activity for Wound Dressing Applications. *React Funct Polym* **2020**, *159*, 104808, doi:10.1016/j.reactfunctpolym.2020.104808.

21 Cinnamon Pickering Emulsions as a Natural Disinfectant
Protection against Bacteria and SARS-CoV-2

Alana Gabrieli De Souza, Rennan Felix Da Silva Barbosa, Maurício Maruo Kato, Rafaela Reis Ferreira, and Luiz Fernando Grespan Setz
Federal University of ABC

Ivana Barros De Campos
Adolfo Lutz Institute

Derval Dos Santos Rosa
Federal University of ABC

Eliana Della Coletta Yudice
Adolfo Lutz Institute

21.1 INTRODUCTION

An emulsion is a dynamic system composed of two immiscible liquids dispersed, where one liquid is a continuous phase, and the other is usually dispersed as small droplets (Keivani Nahr et al., 2020). The spatial organization between the aqueous and oil phases may form an oil-in-water (O/W) emulsion or water-in-oil (W/O) emulsion. Products like milk, cream, mayonnaise, sauces, beverages, and others are examples of O/W emulsions, while margarine and butter are W/O emulsions. Besides these types of emulsion, multiple systems can also be prepared as oil-in-water-in-oil (O/W/O) or water-in-oil-in-water (W/O/W) emulsions (Mendes et al., 2020; Zhang et al., 2017) These systems have attracted considerable interest of pharmaceutical, cosmetic, food industry, among others since an emulsion can be used to carry an active agent and assist its application (Albert et al., 2019).

The emulsions are produced during a homogenization or emulsification process that promotes the fragmentation and dispersion of one phase inside the other,

DOI: 10.1201/9781003400998-21

resulting in the formation of small droplets (Pavoni et al., 2020). In this process, the interfacial region is defined by the contact of the two phases, and the interfacial tension (γ) plays a critical role in emulsion properties related to the energy transferred to the emulsion (McClements, 2015). Thus, to promote smaller droplets which leads to a higher interfacial area, more energy should be applied to the system.

However, emulsions are thermodynamically unstable; thus, the formed droplets tend to merge gradually, eventually leading to complete phase separation. This process is observed because the contact between oil and water molecules is thermodynamically unfavorable (Albert et al., 2019). Thus, stabilizing agents must be used to avoid this process and produce kinetically stable emulsion for a reasonable time (like months or even years) (Pavoni et al., 2020). Most emulsifiers employed in the industry are amphiphilic molecules, known as surfactants, which present the potential to interact both with the aqueous and oil phases and reduce the interfacial tension. Among the main used surfactants, it is possible to cite Tween 20®, Span 85®, Brij 97®, polysorbate 80, soy lecithin, and sodium caseinate, among others (Li et al., 2018; Paulo et al., 2020).

Of late, the use of surfactants has drawn concerns due to their non-biodegradability and possible carcinogenic and toxic effects of synthetic materials on human health and other living beings. Thus, new developments have been made to research new alternatives based on renewable resources, and the use of solid particles has been highlighted. Emulsions prepared using solid particles are known as Pickering emulsions according to the pioneering work of Pickering (1907), even though some earlier observations were made by Ramsden (1904). The network formed by the particles promotes higher deformation resistance which can promote remarkable physical stability for even longer periods than conventional emulsions.

Among the numerous solid particles that can be used in Pickering emulsion are magnetic nanoparticles, nanoclays, calcium carbonate, quantum dots, carbon nanotubes, and graphene oxide (Briggs et al., 2018; Komatsu et al., 2018; Zhai et al., 2018). However, the need for renewable and cost-effective materials increased interest in natural particles, such as nanocellulose (NC) (Albert et al., 2019). NCs usually are found parallel to the droplet surface, with the C-H groups of the cellulose interacting with the oil chains. Compared with other nanoparticles, NCs have shown superiority regarding biocompatibility, low density, biodegradability, and cost (Saffarionpour, 2020).

The Pickering emulsions are attractive systems to load with active substances to promote their stabilization, control their release kinetics, and improve their bioavailability. Moreover, the use of disinfectants has increased in recent years due to concerns associated with human health and the environment and they were compounded by the emergence of the SARS-CoV-2 coronavirus at the end of 2019, which declared a pandemic on March 11, 2020 (Goh et al., 2021). However, the most employed disinfectant products like quaternary ammonium salts, sodium hypochlorite, hydrogen peroxide, and alcohols can be considered potential skin irritants and/or sensitizers (allergens) (Fiorillo et al., 2020; Magurano et al., 2021; Zhou et al., 2020). In this context, a search for new environmentally friendly disinfectants has been observed as natural and green alternatives to traditional chemical disinfectants (Jiang et al., 2021).

Among the types of natural antimicrobial agents, essential oils (EO) are natural biocides that have stood out due to their versatility, abundance, and broad antimicrobial (Bailey et al., 2021; Jiang et al., 2021). In addition, its antioxidant and anti-inflammatory bioactivity allows its application in various products, such as in the food (packaging), pharmaceutical, public health, disinfectants, and sanitizers sectors, among others (Yan et al., 2020). The benefits of using essential oils as alternative disinfectants include their potential for application to porous surfaces that may not be effectively achieved by traditional chemical disinfectants, as well as the ability to combine essential oils in blends that affect microbes at different stages of the life cycle. To find out what has been reported in the literature, we summarize in Table 21.1 the main lines of research that have been conducted by Pickering emulsion using essential oil.

TABLE 21.1

Summarization of the Recent Works Used Pickering Emulsion Containing Essential Oil

Emulsion	Microorganisms or Pathogens	Summarization	Reference
Pickering emulsions based on sodium starch and cinnamon	*Escherichia coli, Staphylococcus aureus* and *Bacillus subtilis*	Pickering emulsions as a vehicle for delivering active biodegradable films to enhance their antimicrobial and antioxidant activities.	Sun et al. (2020)
Pickering emulsion was stabilized by cellulose nanocrystals (CNCs) and EO oregano	*Escherichia coli, Staphylococcus aureus, Bacillus subtilis,* and *Saccharomyces cerevisiae*	The investigation sought to characterize the antibacterial effects of EO Pickering emulsions stabilized by CNCs and factors that may influence the properties of the Pickering emulsion, such as concentration of CNCs, ionic strength, and pH.	Zhou et al. (2018)
Polydopamine microcapsules from cellulose nanocrystal stabilized Pickering emulsions for EO turpentine	Weeds, pathogens, and pest	Development of microcapsule and Pickering emulsion systems in the application of botanical pesticides for pest control.	Tang et al. (2019)
Cinnamon essential oil/ xanthan gum/chitosan composite microcapsules based on Pickering emulsions	*Escherichia coli* and *Staphylococcus aureus*	The authors indicated the prospects for application as antibacterial materials in textiles, coatings, biopesticides, and other fields.	Li et al. (2022)
Pickering emulsions EO melaleuca stabilized with cellulose nanofibrils	-	The authors reported that the emulsions have a steady-state viscosity suitable for dysphagic food applications, with pronounced thinning behavior suitable for the development of environmentally friendly skincare creams and lotions for the human body.	Ferreira et al. (2023)

Kulkarni et al. (2020a) reported that some monoterpenes, phenolic terpenoids, and phenylpropanoids such as cinnamaldehyde, carvacrol, geraniol, cinnamyl acetate, L-4-terpineol, and thymol isolated from essential oils are effective antiviral agents and have the potential to inhibit the viral spike protein from SARS-CoV-2 (Kulkarni et al., 2020b). In the case of the antimicrobial mechanism, bacteria are expected to be inhibited due to partitioning in the lipid layer of cell membranes, increasing their permeability, leading to leakage of ions and small molecules, and cell death (Yu et al., 2021b).

Most EOs are listed as known to be safe substances (GRAS) (Becerril et al., 2020). Cinnamon has a variety of secondary metabolites that result in antimicrobial properties (Figueiredo et al., 2017). These metabolites are considered non-essential for plant survival and defend against competing species and pathogenic microorganisms. The literature reports that, from this essential oil, more than 160 components have already been separated and identified, the main one being some terpenoids and phenylpropanoids, with cinnamaldehyde considered as a representative component (Figueiredo et al., 2017; Zhang et al., 2019). Cinnamaldehyde (cinnamic aldehyde or 3-phenyl-2-propenal) is a cyclic terpene alcohol, the main active ingredient in cinnamon essential oil (60%–75%) (Zhang et al., 2019). Le et al. (2020) extracted essential oil from the bark and leaves of *Cinnamomum cassia* and observed that 90% of the compounds were aromatic compounds such as (E)-cinnamaldehyde and cinnamyl acetate (Le et al., 2020). The extract showed an excellent inhibitory effect against *S. aureus* and *S. cerevisiae*. However, some studies have demonstrated cinnamaldehyde instability when exposed to air, as the reactive unsaturated aldehyde is easily oxidized to cinnamic acid, causing volatility and instability. Besides, decomposition can also occur before the essential oil can perform a bactericidal activity (Tung et al., 2010; Vasconcelos et al., 2018), requiring its stabilization under appropriate conditions. Among the main approaches used in the literature to obtain stable essential oils, encapsulation, nanoemulsification, and the Pickering emulsion approach stand out.

This chapter presents the development of cinnamon Pickering emulsions stabilized with nanocellulose for antimicrobial applications. The stability, physicochemical characteristics, and bioactive efficiency of emulsions stabilized with cellulose nanocrystals or cellulose nanofibers were compared to identify possible mechanisms of stabilization and release of the active compound. The emulsions were tested against bacteria and the SARS-CoV-2 virus, acting as potential active agents of new natural and clean-label disinfectant alternatives.

21.2 MATERIALS AND METHODS

21.2.1 MATERIALS

Cellulose nanocrystals (CNC) (L/D = 15) were prepared by ball milling according to the procedure described in Souza et al. (2019) and Lima et al. (2018). The average length and diameter of the crystals were $2.0 \pm 0.8\,\mu m$ and $131.1 \pm 29.6\,nm$. Cellulose nanofibrils (CNF) (L/D = 60) were a gift from Suzano Papel e Cclulose (São Paulo, Brazil), and showed an average length and diameter of $4.0 \pm 1.0\,\mu m$ and $66.3 \pm 20.5\,nm$ (Souza et al., 2020). Ferquima Indústria e Comércio Ltda. (São Paulo, Brazil) provided the cinnamon essential oil (*Cinnamomum cassia*, CAS Number 84961-46-6).

21.2.2 Emulsion's Preparation

The literature shows that during the development of Pickering emulsions using cellulose, some parameters should be investigated to ensure the development of stable systems. Among the main parameters that can be varied are the type of oil, the emulsification speed and time, the oil/water ratio employed, and the morphology and concentration of cellulose. Yu et al., worked with clove oil and cellulose nanocrystals at concentrations between 0.1 and 0.5 wt%, using rotations of about 22,000 rpm, and observed good stability of the system under different environmental conditions due to the steric barrier and electrostatic repulsion of CNC on the interface of emulsions (Yu et al., 2021a). Soo et al. also worked with cellulose nanocrystals and palm oil, employing a 10,000 rpm rotation speed during emulsification, and the developed Pickering systems showed high stability (Soo et al., 2021)

On the other hand, Wang et al. (2018) demonstrate the use of cellulose nanofibers at 1% concentration to develop Pickering using palm oil and a rotation speed of 10,000 rpm, observing high stability due to the formation of a three-dimensional network generated by nanofibers entanglement. Similar results were observed by Luo et al. (2021) when working with olive oil and cellulose nanofibers.

Based on the literature, emulsions were prepared using two nanocellulose concentrations: 0.5 and 1 wt%, and the emulsification parameters were selected in our previous work considering the physical stability and oil droplet size (Souza et al., 2021). The selected emulsions, their preparation conditions, physical stability, and droplet size are presented in Table 21.2. The nomenclatures were adopted considering the nanocellulose morphology, that is, CNC for cellulose nanocrystals and CNF for cellulose nanofibers, and the nanocellulose content (0.5 or 1 wt%). All emulsions were prepared using an Ultra-Turrax Blender, IKA T25 model (IKA Werke, Staufen, Germany) at 12,000 rpm.

21.2.3 Characterization

The characterization of Pickering emulsions seeking to understand the interfacial dynamics and stability of the developed systems is essential in active systems. In particular, the dynamics involving essential oils and the effect of emulsification can have a major impact on the properties and possible applications of these systems.

TABLE 21.2

Samples Used in this Work and Their Processing Parameters (Oil Concentration and Time) and Main Stability Characteristics (Creaming Index and Droplet Size after 30 Storage Days)

Sample	Oil Concentration (%)	Time (min)	Creaming Index	Droplet Size (μm)
0.5CNC	30	7	42.8%	36.5 ± 5.5
1CNC	30	3	34.4%	32.1 ± 8.4
0.5CNF	20	3	0%	55.1 ± 9.6
1CNF	20	7	0%	46.7 ± 12.6

21.2.3.1 Fourier Transform Raman spectroscopy

FT-Raman analysis was performed on MultiRaman equipment (Bruker Optics Inc., Massachusetts, USA), equipped with a wavelength of 1,064 nm and a laser of 500 mW. Data acquisition was performed in the range 500–4,000 cm^{-1}, 128 scans, and 4 cm^{-1} spectral resolution.

21.2.3.2 Nuclear Magnetic Resonance

1D (^{1}H & ^{13}C nuclei) and 2D (HSQC—heteronuclear single quantum coherence and HMBC—heteronuclear multiple bond correlation) experiments were done to obtain nuclear magnetic resonance spectra in the liquid phase using a 400 MHz spectrometer AVANCE III HD (Bruker, Germany) equipped with a wideband 5 mm probe and an automatic 24 sample changer. Both EO and emulsions were diluted at 10% (v/v) with D_2O and $CDCl_3$ for acquisition in both deuterated solvents. The spectra were processed and analyzed using the software TopSpin™ 3.6.3 (Academia License—Bruker, Germany).

21.2.3.3 Surface Tension

The surface tension was measured using an optical tensiometer Theta Lite (dpUnion, Brazil). A sample drop (8 mL) remained suspended at the tip of a 14-gauge plastic needle. The surface tension of the drop relative to air was measured for 30 seconds and then calculated. The experiments were conducted five times.

21.2.3.4 Rheology

Rheological measurements were performed using a rotational rheometer, model Viscotester IQ—Thermo HAAKE (Thermo Fisher Scientific, Massachusetts, USA), equipped with a double-cone rotor and a stationary plate, surrounded by a cylindrical wall. The curves were obtained in Controlled Rate (CR) mode by increasing the shear rate from 0 to 1,000 seconds^{-1}, 600 seconds, maintaining at 1,000 seconds^{-1} for 60 seconds, and returning to 0 second^{-1} in 600 seconds. All rheological parameters were obtained from the flow curves executed in Controlled Rate (CR) mode by HAAKE software RheoWin 4.63.0003.

21.2.3.5 Antimicrobial Tests

Bacterial isolates were obtained from the Center for Interdisciplinary Procedures of Microorganisms Collection Center of Adolfo Lutz Institute (Brazil). Affiliated to World Federation Culture Collections (WFCC) #282, Depository Collection—#017/09-SECEX/CGEN/MMA: *Salmonella enterica* (ATCC n° 10708); *Staphylococcus aureus* (ATCC n° 6538); *Pseudomonas aeruginosa* (ATCC n° 10145); *Escherichia coli* (ATCC n° 11229). Stock cultures were cultivated in Mueller-Hinton agar (Oxoid, Thermo Fisher Scientific, USA) and incubated at 37°C for 24 hours. Then, it was inoculated in culture broth to obtain turbidity equivalent to 0.5 Mc Farland, a concentration equivalent to 1.5×10^8 CFU/mL (Barbosa et al., 2021; Mai-Prochnow et al., 2016).

To carry out the tests, 30 mL agar was melted and cooled to 48°C ± 1°C. Then, 500 μL of the microorganisms at 1.5×10^8 CFU/mL were added, and this mixture was transferred to a petri dish. After solidification, a 0.9-cm diameter hole made in the middle of the plates was completed with 0.2–0.5 mL of emulsion. Incubation was carried out at $36°C \pm 1°C$ for 24 hours (ANVISA, 2007; Feldsine et al., 2002). The appearance of inhibition halos around the orifice with the product was evaluated. These halos have been measured, and the appearance of a zone of inhibition of any dimension indicates that the product under analysis has properties. The size of the zone of inhibition depends on the diffusion of the emulsion in the agar preparation. As a control, we use widely used sanitizers such as hypochlorite and Lysoform®.

21.2.3.6 SARS-CoV-2 Virus Inactivation Tests

SARS-CoV-2 positive human biological samples identified in the laboratory's routine diagnostic assay by the real-time RT-PCR method and stored in the Adolfo Lutz Institute—Santo André Regional Center biorepository were used. For this, 100 μL of the sample was previously analyzed by a rapid antigen test to confirm viral viability (Homza et al., 2021). After obtaining a positive result, a mixture was prepared with 100 μL of the contaminated sample and 100 μL of emulsion, and different contact times were tested. The initial testing time was 30 minutes, and if the emulsion resulted in a positive test, that is, there was no virus protein denaturation, the new contact times were higher (50 and 60 minutes). The opposite was also valid. Tests were performed in duplicate.

The tests were performed following the standard normative approved by the Research Ethics Committee of the Universidade Federal do ABC and Adolfo Lutz Institute CAAE: 49573421.2.3001.0059, and the Scientific-Technical Council of Adolfo Lutz Institute, CTC 18-N/2021. All procedures were performed following the biosafety standards of the Adolfo Lutz Institute at the Santo André Regional Center.

21.3 RESULTS AND DISCUSSIONS

21.3.1 Chemical Structure Characterization

Raman spectroscopy evaluates the interaction between the vibrating motion of particles in a material and the incident electromagnetic radiation. The region of the electromagnetic spectrum in which the Raman effect is observed depends on the energy of the incident radiation and the molecular energy levels involved, being a technique of high interest due to the little influence caused by the presence of water in the sample under study. Figure 21.1 shows the FT-Raman spectra of cinnamon essential oil and its CNC-stabilized emulsions. The pure cinnamon essential oil showed high similarity with its emulsions with characteristic peaks at $843\,cm^{-1}$ (out-of-plane C-H vibration), $1,001\,cm^{-1}$ (monosubstituted benzene ring), $1,032\,cm^{-1}$ (C-H torsional vibration of CH_2 group), $1,126\,cm^{-1}$ (C-C elongation), $1,162\,cm^{-1}$ (C-H bond), $1,180\,cm^{-1}$ (C-O bond), $1,206\,cm^{-1}$ (p-cymene), $1,253\,cm^{-1}$ (CH_2 vibration), 1,332, 1,393, 1,452, and $1,497\,cm^{-1}$ (CH_3 binding modes), $1,598\,cm^{-1}$ (monosubstituted benzene ring), $1,627\,cm^{-1}$ (C=C), and $1,673\,cm^{-1}$ (stretching vibrations C=O and C=C) (Andreev et al., 2001; Cebi et al., 2021; Nelson et al., 2020; Schulz et al., 2004; Schulz & Baranska, 2006).

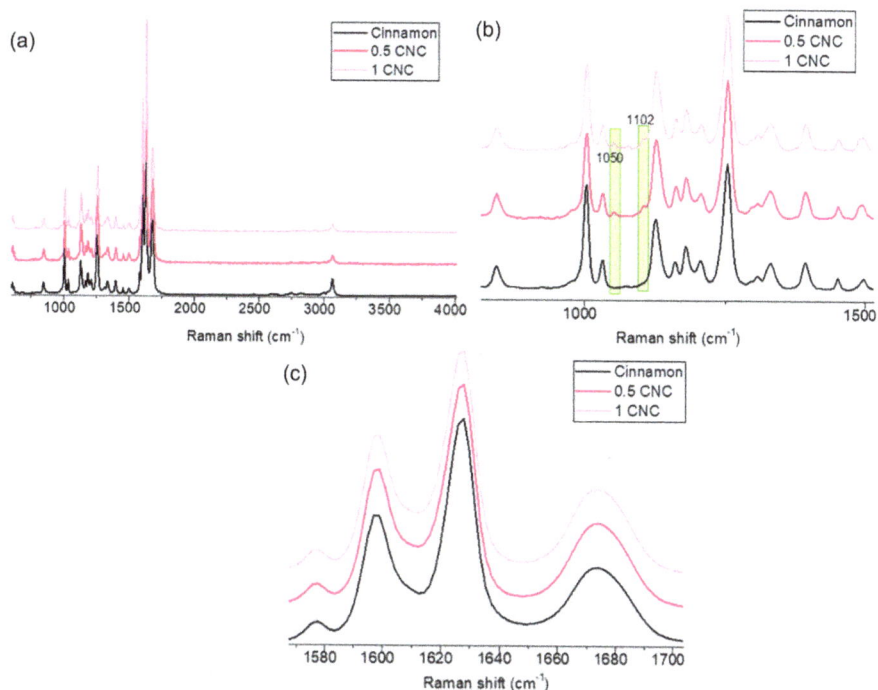

FIGURE 21.1 FT-Raman spectrum of cinnamon essential oil and its CNC-stabilized emulsions (a) full spectrum, and magnification of regions (b) 820–1,515 cm^{-1} and (c) 1,560–1,700 cm^{-1}.

After emulsification with CNCs, a low-intensity peak was observed at 1,050 cm^{-1} and a new shoulder at 1,102 cm^{-1}, associated with asymmetric C-O-C vibrations and vibrations of the C-C binding mode of the cellulose (Park et al., 2019; Sacui et al., 2014). Furthermore, the 1,497 cm^{-1} peak shifted to 1,492 cm^{-1} for both emulsions, indicating that the emulsification may have induced a disturbance in the vibration modes due to possible interactions between the components. There was no change in peaks above 1,560 cm^{-1}.

Figure 21.2 shows the FT-Raman spectra of cinnamon essential oil and its emulsions stabilized with CNF. Similar to what was observed in Figure 21.1, there were no significant changes between the spectra (Figure 21.2a), with a predominance of essential oil peaks. Figure 21.2b shows an enlargement of the region 825–1,075 cm^{-1}, making it possible to identify that after emulsification with CNF, the 1CNF sample presented a shoulder at 830 cm^{-1}, associated with C-O-C elongation of cellulose (Sacui et al., 2014). The overlapping peaks of oil and cellulose indicate the presence of both components in the mixture, and the broadening of the essential oil peak may indicate the presence of vibrations of new binding modes, being a possible indication of chemical interactions. Both emulsions showed a peak at 1,053 cm^{-1}, confirming the presence of cellulose nanofibers in the system.

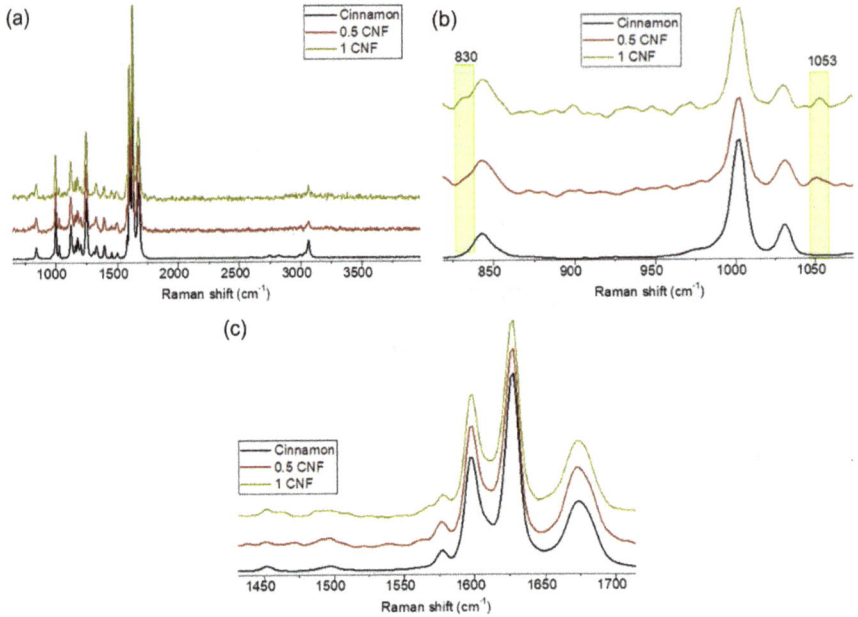

FIGURE 21.2 FT-Raman spectrum of cinnamon essential oil and its emulsions stabilized with CNF (a) full spectrum, and (b) magnification of the region 825–1,075 cm^{-1}.

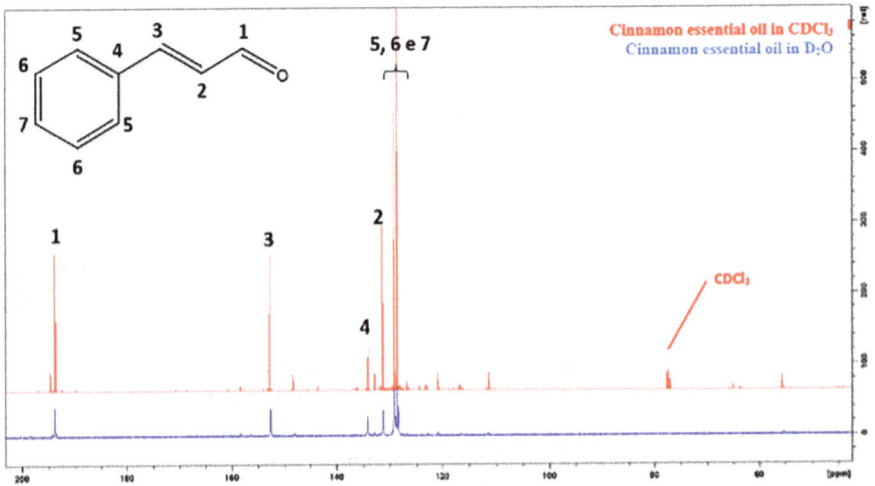

FIGURE 21.3 ^{13}C spectra of cinnamon essential oil diluted with CDCl$_3$ and D$_2$O.

Figure 21.4 illustrates the spectra of cinnamon EO in D$_2$O, and the emulsions stabilized with CNC, PECan-CNC1, and PECan-CNC2. Differences in the spectra could be seen between each other, indicating significant variation between crystalline cellulose and cinnamon EO interactions. The spectra of the CNC emulsions

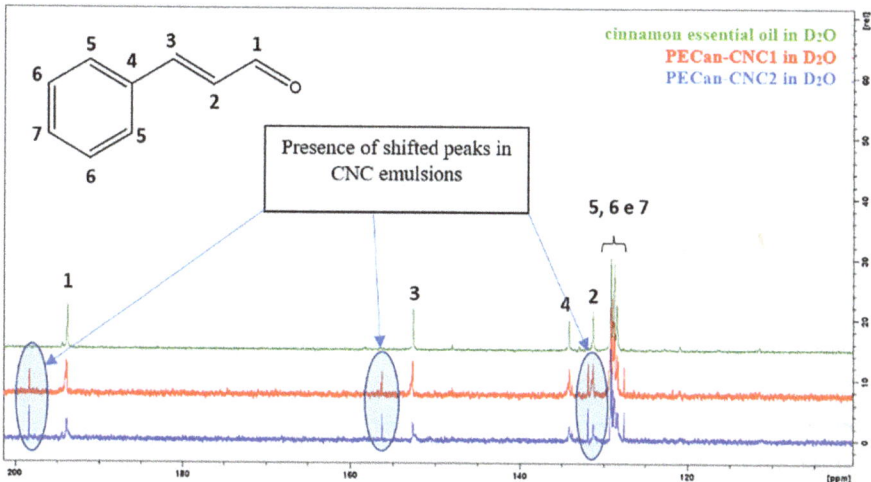

FIGURE 21.4 ^{13}C spectra of cinnamon essential oil and its CNC stabilized emulsions diluted with CDCl$_3$ and D$_2$O.

show the characteristic peaks of cinnamic aldehyde. In addition, other peaks with similar intensities and chemical shifts around 198, 156, and 133 ppm, similar to C's 1, 2, and 3, are also shown. The presence of these peaks may indicate a chemical interaction with the CNCs. The chemical environment directly influences the chemical shift of a C core, thus confirming the presence of two species of cinnamic aldehyde, a common species with characteristic chemical shifts and another species with shifted peaks.

The NMR technique was used to investigate molecular interactions and the impact on the stabilization properties of the pure oils and their Pickering emulsions. Figure 21.3 shows the ^{13}C spectrum of cinnamon essential oil in deuterated solvents CDCl$_3$ and D$_2$O. The oil demonstrated good solubility in CDCl$_3$, and the most intense peaks of the spectrum show a characteristic chemical shift, which can be attributed to cinnamic aldehyde, commonly found in cinnamon EO. On the other hand, the spectrum obtained with the diluted EO D$_2$O showed a low-resolution ^{13}C spectrum with shifted peaks and some distortion, probably caused by the heterogeneity of the magnetic field caused by a discontinuous liquid phase of the EO + D$_2$O solution. All the peaks shown in the CDCl$_3$ ^{13}C spectrum also appear in the D$_2$O spectrum, showing a partial solubility of EO in D$_2$O or H$_2$O.

Figure 21.5 illustrates the ^{13}C spectra of cinnamon essential oil in D$_2$O and PECan-CNF1 and PECan-CNF2 emulsions. The peaks in the spectra are very similar, indicating that the cinnamon EO emulsion process does not significantly alter the solubility and availability of the EO in the solution. Although the mobility of the particles may be affected, according to the NMR analyses, it is not possible to state a chemical interaction between the CNF and cinnamon essential oil structures.

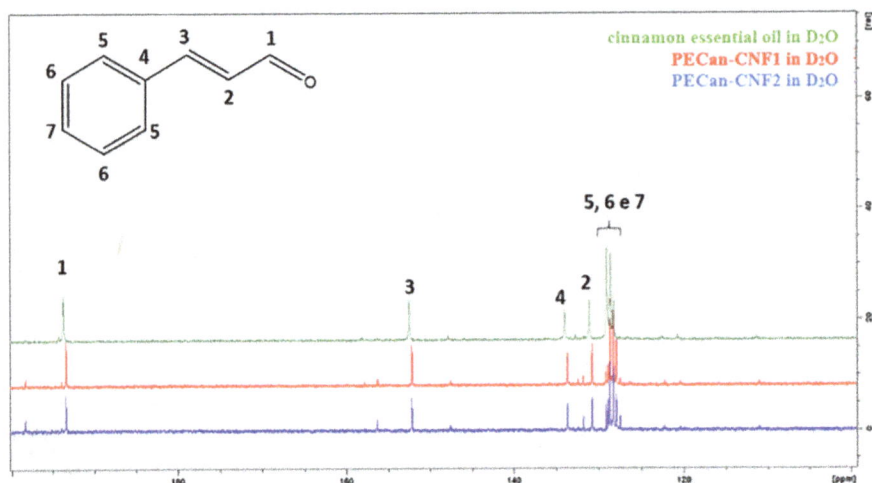

FIGURE 21.5 ^{13}C spectra of cinnamon essential oil and CNF stabilized emulsions diluted with CDCl$_3$ and D$_2$O.

21.3.2 Surface Tension

The surface tension (ST) of the Pickering emulsion samples was measured to obtain information about the surface activity of the different nanocelluloses in stabilizing the emulsion (Yuan et al., 2021). The pure cinnamon essential oil showed surface tensions (ST) of 40.2 ± 0.2 mN/m, while the mixtures of cinnamon/water without stabilizing agent showed ST of 55.6 ± 1.9 mN/m. The increase in ST values is associated with the water interaction, which has a surface tension of approximately 73 mN/m. The suspensions of CNC/water and CNF/water presented ST of 49.5 ± 1.3 and 63.9 ± 0.7 mN/m. The morphology of cellulose nanomaterials influences their contact angle and surface tension since their structures exhibit different interactions at a fluid-fluid interface, such as DLVO interactions, capillary forces, hydrophilic behavior, and monopolar and dipolar interactions (Li et al., 2019). Since the L/D of CNCs and CNFs are different, it is expected that larger particles (CNFs) may show gravity-induced capillary flotation forces due to their weight. In the case of smaller particles (CNCs), capillary dipping forces may exist depending on the surface chemistry and the position of the contact line. Anisotropic particles cause an irregular deformation of the interface around them, leading to capillary interactions. Thus, the intensity of the surface tension is dependent on the forces of interactions associated with the morphology of the nanocellulose, size, and dispersion states.

Figure 21.6 shows the surface tension values of the stabilized cinnamon essential oil emulsions, and all showed similar values, except for the sample PECan-CNC1, which can be associated with its electrostatic stability or even with the larger droplet sizes of this emulsion. The other cinnamon emulsions showed

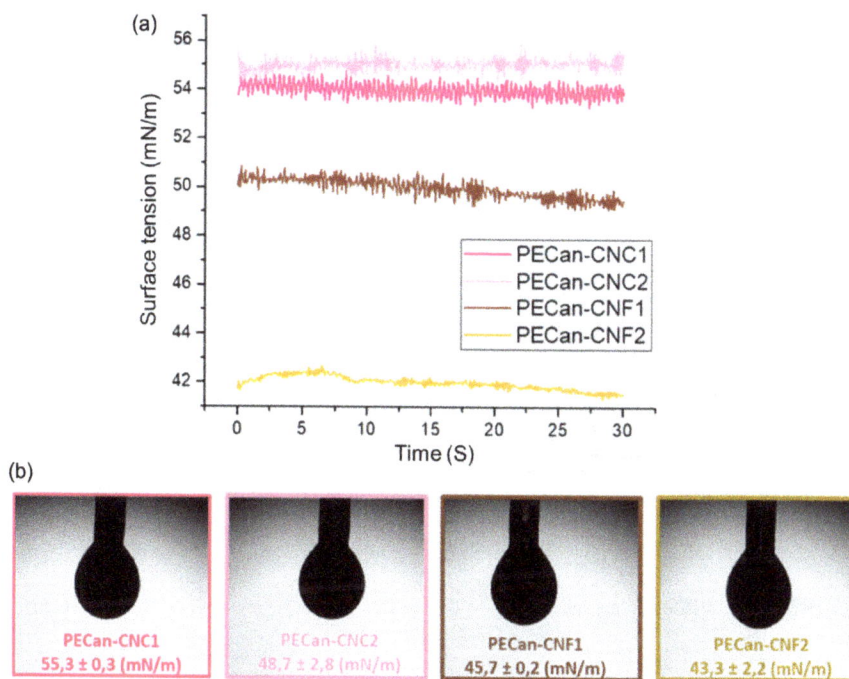

FIGURE 21.6 Surface tension data of Pickering cinnamon essential oil emulsions stabilized with CNC and CNF (a) surface tension measurements over time and (b) their respective average values.

similar values of surface tension and droplet size. Similar ST values were reported by Ozogul et al. (2017) for thymol oil emulsions stabilized with lauryl and by Yazgan (2020) in sage emulsions.

According to Zhou et al. (2018), low ST values indicate that the nanocellulose is easily adsorbed at the interface, affecting the packing of the particles at the interfaces and forming an interfacial structure, a critical feature for the physicochemical properties of Pickering emulsions. The ST values for the emulsion decreased with higher cellulose content, suggesting that it counterbalanced the oil forces related to oil aromatic structure and polarity and could result in a more stable structure.

21.3.3 RHEOLOGY

Considering a Pickering emulsion system prepared for commercial applications, its fluidity is critical for preparation, storage, and final product. Figure 21.7 presents the shear viscosity of cinnamon essential oil and its Pickering emulsions. The pure oil showed Newtonian behavior, as expected. The emulsions stabilized with CNC (PECan-CNC samples) showed similar behavior to the pure essential oil, with a slight increase in viscosity followed by a plateau. These results indicate that the addition of

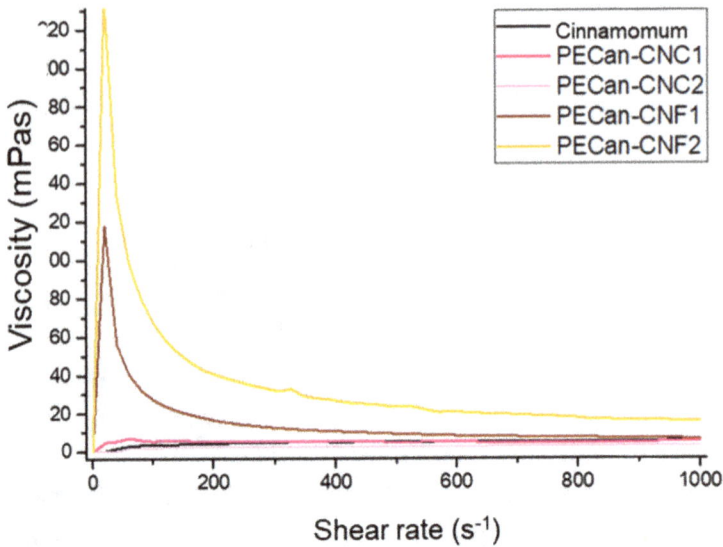

FIGURE 21.7   Viscosity versus shear rate curves of cinnamon essential oil and its Pickering emulsions.

CNC causes a disturbance in the system and possibly occurs in the effective deposition of cellulose nanocrystals in the oil droplets. Due to the system's stability, these CNCs should not be entangled/aggregated, resulting in a constant viscosity (Xu et al., 2020). In addition, the system's compatibility may assist in the homogeneous distribution of the nanoparticles throughout the emulsion, resulting in a stable continuous phase without the formation of microphases, regardless of the CNC concentration used (Dai et al., 2017; Hu et al., 2019).

The samples stabilized with CNF showed non-Newtonian behavior, with an initial increase in viscosity followed by a decrease with increasing shear rate, characteristic of pseudoplastic materials (Hosseini & Rajaei, 2020). The PECan-CNF sample's behavior indicates that with increasing shear rate, the three-dimensional network of cellulose formed became more ordered and less resistant to the flow direction, which is consistent with emulsions stabilized with polysaccharides and biopolymers (Zhu et al., 2021).

The PECan-CNF2 sample had a higher viscosity than PECan-CNF1, which may be attributed to the higher content of cellulose nanofibers used during processing or to the smaller droplet size, since the higher the viscosity of Pickering emulsions, the less possibility that the emulsion suffers from instability phenomena such as flocculation or coalescence (Xiong et al., 2018).

For samples stabilized with CNF, rheological models (Herschel-Bulkley and Ostwald-de-Waele) were applied to evaluate the internal structure and changes associated with the gel state and the influence of external forces, that is, shear. The Herschel–Bulkley model has three constants, K, n (fluid dilution), and γ (parameter related to shear stress in a nonlinear flow) (Eq. 21.1) and provides good results for solid materials up to a specific shear stress and flowing after that value (Chhabra & Richardson, 1999).

$$\tau = \tau_0 + K\gamma^n \tag{21.1}$$

where τ is the measured shear, τ_0 is the measured yield stress, K is the consistency index, n is the flow behavior index, and γ is the shear rate. The data obtained are presented in Table 21.3. The results showed a good fit with this model, and high K values were seen for PEL-CNF2, indicating that the lower essential oil content optimized the gelation effect and, consequently, there was no phase separation for this sample. Both emulsions showed $n < 1$, characteristic of non-Newtonian pseudoplastic behavior (Husin et al., 2018).

In addition, the Ostwald-de-Waele model also showed a good fit and was used to describe the time-independent flow behavior. This model relates the effective viscosity to the shear rate using three parameters, according to Eq. 21.2, where γ is the shear rate, K is the consistency index, and n is the flow behavior index.

$$\eta_a(\gamma) = K\gamma^{n-1} \tag{21.2}$$

In the models studied, the values of n were less than 1, confirming the pseudoplastic behavior of the materials. The $R^2 > 0.99$ value confirmed the reliability of the parameters obtained for the emulsions, and the K values were significantly lower than those found by Ye et al. (2017), which is related to the type of polymeric network formed during entanglement, the droplet size, and the breakup rate of this. Since there was no cross-linking agent in this work, the τ_0 and K values tend to be lower than in graphitized or cross-linked structures.

The Ostwald-de-Weale model also showed $R^2 > 0.99$, and the PECan-CNF2 sample showed the highest K and lowest n, which may be associated with the low electrostatic interaction between the droplets. This result is coherent since higher concentrations of CNF induce greater steric stability, and electrostatic charges are a secondary mechanism (Zhang et al., 2020).

TABLE 21.3

Parameters of the Herschel-Bulkley and Ostwald-de-Waele Rheological Models for PECan-CNF Samples

Sample	Model: Herschel-Bulkley			
	τ_0 (Pa)	K	n	R^2
PECan-CNF1	0.7150	0.0875	0.6038	0.9921
PECan-CNF2	0.8063	0.9825	0.3819	0.9913
	Model: Ostwald-de-Waele			
	-	K	n	R^2
PECan-CNF1	-	0.2134	0.4912	0.9912
PECan-CNF2	-	1.287	0.3509	0.9909

21.3.4 ANTIMICROBIAL TESTS

Due to the trend in the development of natural antimicrobial products, new emulsifiers based on biopolymers and essential oils have emerged in recent years. For a material to be considered antimicrobial, it must inhibit or inactivate the growth of disease-associated microorganisms that may be present in the environment, such as on surfaces, food, and packaging (Jamali et al., 2021). Due to the applicable limitations of antimicrobial materials, delivery systems can significantly improve their functionality against microorganisms, as in the case of Pickering emulsions. Initially, control trials were performed for the four bacteria studied (*E. coli, S. aureus, Pseudomonas e Salmonella*) for the validation of its feasibility, determination of the oil/emulsion concentrations to be used (standardization) (Nascimento et al., 2007; Reis et al., 2020), and analysis of the antimicrobial activity of the cellulose nanocrystals and nanofibers used in the present work.

Figure 21.8 presents the results of hypochlorite and Lysoform® against the studied microorganisms, with a set volume of 200 μL, confirming their efficacy against all

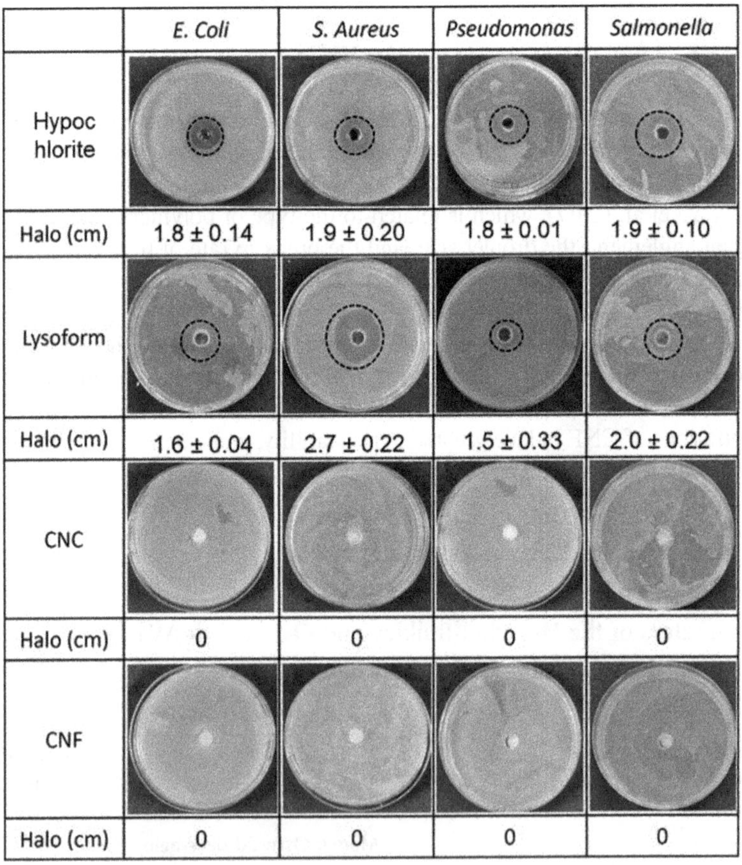

	E. Coli	*S. Aureus*	*Pseudomonas*	*Salmonella*
Hypoc hlorite				
Halo (cm)	1.8 ± 0.14	1.9 ± 0.20	1.8 ± 0.01	1.9 ± 0.10
Lysoform				
Halo (cm)	1.6 ± 0.04	2.7 ± 0.22	1.5 ± 0.33	2.0 ± 0.22
CNC				
Halo (cm)	0	0	0	0
CNF				
Halo (cm)	0	0	0	0

FIGURE 21.8 Results of antimicrobial assays obtained for standardization of the studied microorganisms using hypochlorite and Lysoform®, and antimicrobial activity of cellulose nanocrystals and nanofibers.

bacteria. Furthermore, it was found that neither CNCs nor CNFs showed bactericidal activity, which is expected since these fibers are highly biodegradable and easily attacked by bacteria and fungi (Štular et al., 2019).

Figure 21.9 presents the antimicrobial activity of cinnamon essential oil and its emulsions by the disk diffusion method. The pure essential oil showed excellent biocidal results, with larger halo diameters than the control tests for all bacteria tested. According to Zhang et al. (2016), cinnamon results in severe morphological destruction of microorganisms, increasing cell membrane permeability and causing leakage of nucleic acids and proteins.

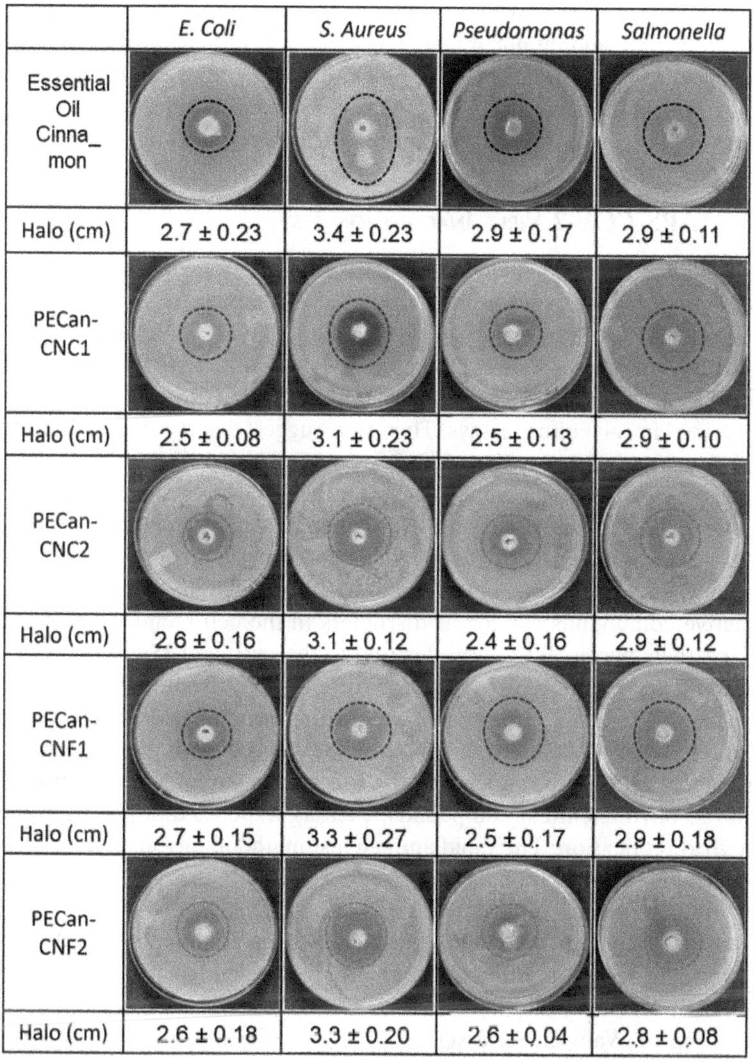

	E. Coli	S. Aureus	Pseudomonas	Salmonella
Essential Oil Cinna_ mon				
Halo (cm)	2.7 ± 0.23	3.4 ± 0.23	2.9 ± 0.17	2.9 ± 0.11
PECan-CNC1				
Halo (cm)	2.5 ± 0.08	3.1 ± 0.23	2.5 ± 0.13	2.9 ± 0.10
PECan-CNC2				
Halo (cm)	2.6 ± 0.16	3.1 ± 0.12	2.4 ± 0.16	2.9 ± 0.12
PECan-CNF1				
Halo (cm)	2.7 ± 0.15	3.3 ± 0.27	2.5 ± 0.17	2.9 ± 0.18
PECan-CNF2				
Halo (cm)	2.6 ± 0.18	3.3 ± 0.20	2.6 ± 0.04	2.8 ± 0.08

FIGURE 21.9 Results of antimicrobial assays using cinnamon essential oil and its emulsions.

The results also indicated that *E. coli* exerted more resistance to the essential oil than the other bacteria, probably due to structural differences in the outer membrane, which has a lipopolysaccharide membrane covering the cell wall. *S. aureus*, like other Gram-positive bacteria, has a single-layered peptidoglycan structure that is more susceptible to hydrophobic substances (El Atki et al., 2019).

Regarding the mechanism responsible for the antimicrobial activity, cinnamaldehyde may interfere with cell biological processes, particularly nitrogen-containing substances such as proteins and nucleic acids (Ferreira et al., 2021; Ribeiro-Santos et al., 2017).

The emulsions showed antimicrobial activity similar to pure essential oil, with no significant variations in the inhibition halos. Comparing PECan-CNC or PECan-CNF and the difference in oil concentration (30% for CNC stabilized emulsions and 20% for CNF), it could be noted that emulsions with EO content of 20% are sufficient for antibacterial action. These results highlight that the emulsions maintained the strong potential observed for the pure oil; however, the Pickering system could expand its application in several areas, including as a disinfectant material.

21.3.5 SARS-COV-2 Virus Inactivation Tests

Essential oils are natural compounds with antiviral activity against several viruses that promote human diseases, such as Influenza virus (Abou Baker et al., 2021; Najar et al., 2021; Wang et al., 2020), dengue fever (Flechas et al., 2017), Hepatitis A (Battistini et al., 2019), and others. Homza et al. (2021) presented that most rapid antigen tests that failed to detect positive samples identified by RT-qPCR were due to virus inviability verified in cell culture assays. Thus, they suggest that the rapid antigen test can be applied to check the viability of SARS-CoV-2. This work explores such a kit to check protein denaturation as an indicator of virus viability. Positive samples with the potentially viable virus were exposed to different EOs at different times. As shown in Figure 21.10, *Cinnamomum cassia* (cinnamon) oil showed great potential, suggesting that 5 minutes of exposure was sufficient to inactivate SARS-CoV-2.

Bioactive compounds can act as disrupters in the cell membranes of bacteria, increasing their permeability and destroying the cell wall (Saraiva et al., 2011). The antiviral activity of EOs is based on their lipophilic nature, which favors the entry of the oil into the viral membrane leading to membrane disruption (Wani et al., 2021). Boukhatem described that essential oils could disrupt cell membranes and displace viral envelopes, leading to virus inactivation (Yadalam et al., 2021). In addition, phytochemical compounds, such as terpenes and phenylpropanoids, disrupt viral replication. The rapid antigen test methodology applied in this study detects protein denaturation or degradation through the absence of antigen–antibody interaction, an indicator of virus inviability. Thus, if an EO disrupts the viral capsid making it non-viable, producing virus fragments, but the proteins remain in their natural conformation, the rapid antigen test will probably not be able to detect this non-viability.

According to Wani et al., essential oils have different biochemical pathways as possible targets against SARS-CoV-2 infection, such as the virus capsid, spike

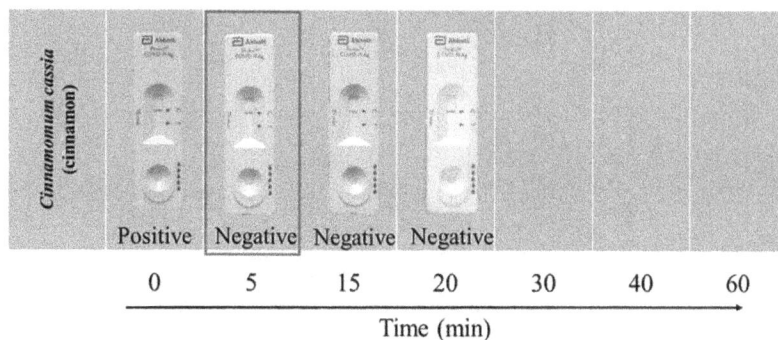

FIGURE 21.10 Results obtained for each essential oil (cinnamon, cardamom, and linalool) against SARS-CoV-2, tested by the rapid antigen test at different times.

protein, Mpro (an enzyme that assists viral replication) (Thuy et al., 2020), and RdRp (catalyzes viral replication) (da Silva et al., 2020). According to Schnitzler et al. (2007), the inactivation of the virus by EO is time-dependent due to its adsorption process to the virus envelope structures or entry into the host cells to dissolve the viral envelope.

Cinnamon EO showed strong antiviral activity. Kulkarni et al. (2020a) investigated the effect of cinnamaldehyde and cinnamyl acetate, active compounds of this oil, on the spike proteins of COVID-19. Cinnamon oil is believed to have acted based on its chemical structure and characteristics such as electronegativity, electrophilicity, chemical potential, and absolute hardness of the phytochemicals. Since cinnamaldehyde is one of the most electronegative compounds, it is expected to have excellent antiviral activity. Thus, this oil is believed to interact with the spike protein, causing a conformational change, denaturation, and degradation (Lan et al., 2020). According to Elfiky (2021), cinnamaldehyde targets the viral "SARS-CoV-2 attachment". Figure 21.11 shows a representative schematic of the expected mechanisms of action on SARS-CoV-2 after interaction with the essential oils.

Figure 21.12 shows the results obtained against SARS-CoV-2 using the cinnamon essential oil and its emulsions. The emulsions required longer times to inactivate the virus through protein destruction, and the time required for both emulsions was 40 minutes. According to Kulkarni and co-workers, cinnamaldehyde and other compounds such as carvacrol, cinnamyl acetate, geraniol, L-4-terpineol, and anethole, show good binding affinity to the SARS-CoV-2 virus via hydrogen bridges. This binding and the hydrophobic interaction between the ligands and the viral proteins may contribute to virus inactivation. Moreover, the inhibition efficiency of cinnamaldehyde is also attributed to the low electronegativity of this compound, as reported by Zhan et al. (2003), with cinnamaldehyde having electronegativity values of -4.34 (Kulkarni et al., 2020a). In emulsions, the charge balance is altered since new chemical bonds and electronic interactions occur between the components, changing the charge balance and thus resulting in an increase in the time required for the virus to be inactivated.

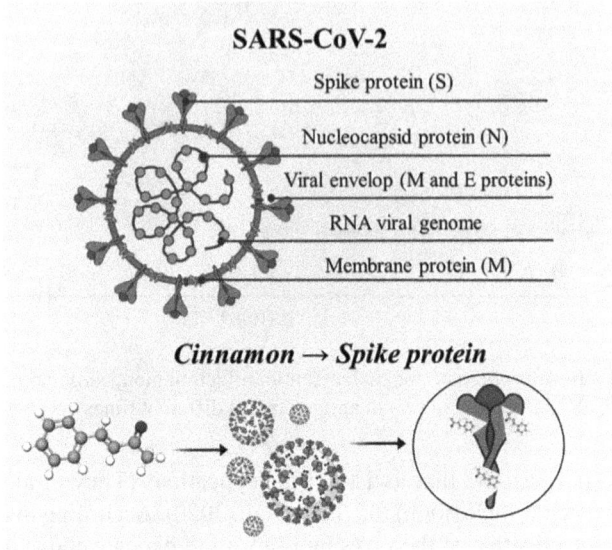

FIGURE 21.11 Diagram representing the results for each essential oil and the expected effects on SARS-CoV-2.

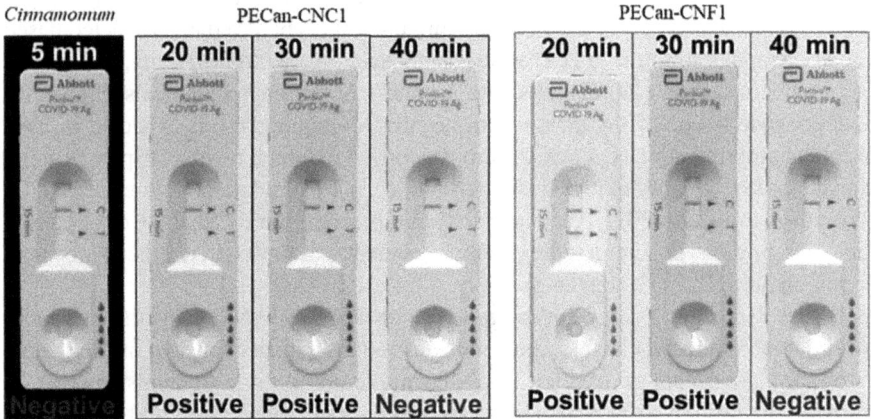

FIGURE 21.12 Results obtained for cinnamon essential oil and its emulsions against SARS-CoV-2, tested by rapid antigen test at different time points.

21.4 CONCLUSIONS

The interest in developing effective materials against bacteria and viruses has grown recently, and essential oils show potential application in the field. However, their lack of stability limits their use and requires stabilization methods, for which Pickering emulsion could assist in producing stable materials. The literature presents the use of Pickering emulsion using essential oils; meanwhile, evaluating its antibacterial and antiviral properties for future application still lacks investigation. Therefore, this work presented the use of CNC and CNF to develop cinnamon Pickering emulsions and investigated their properties with highlights for their action against different bacteria and the virus Sars-Cov-2. The cinnamon emulsions showed no chemical interaction between the oil and cellulose morphologies. In general, the emulsions stabilized with CNC showed Newtonian behavior, while the emulsions stabilized with CNF showed pseudoplastic behavior. The rheology of the emulsions indicated that two types of emulsions were prepared, solutions and gels, suitable for different products and broadening possible applications. Cinnamon emulsions showed antimicrobial activity similar to pure EO, with inhibition of all microorganisms studied and inhibition halo between 2.7 and 3.4 cm, indicating that emulsification did not alter the EO diffusivity or its bacteriostatic power. Finally, the pure oil and emulsions were tested against the SARS-CoV-2 virus by the Rapid Test method, which detects protein denaturation or degradation. The cinnamon oil needed to be in contact with the virus for 5 minutes to be negative in the rapid antigen test, while the emulsions needed 40 minutes to inactivate the virus, and its expected action was through the spike protein and the viral membrane by damaging the proteins detectable by the test. The results highlight the great stability of the developed systems with potential activity against bacteria and viruses, where the essential oil is made available slowly and can promote disinfectant properties for longer periods. Thus, new applications can be promoted which could replace conventional materials for environmentally friendly products.

ACKNOWLEDGEMENTS

This research was funded by Fundação de Amparo à Pesquisa do Estado de São Paulo (2022/01382-3, 2021/14714-1, 2021/08296-2 and 2020/13703-3) and Conselho Nacional de Desenvolvimento Científico e Tecnológico (305819/2017-8). The authors thank the CAPES (Code 001), UFABC, CEM facilities, and REVALORES Strategic Unit.

BIBLIOGRAPHY

Abou Baker, D. H., Amarowicz, R., Kandeil, A., Ali, M. A., & Ibrahim, E. A. (2021). Antiviral activity of *Lavandula angustifolia* L. and *Salvia officinalis* L. essential oils against avian influenza H5N1 virus. *Journal of Agriculture and Food Research*, *4*(December 2020), 100135. https://doi.org/10.1016/j.jafr.2021.100135.

Albert, C., Beladjine, M., Tsapis, N., Fattal, E., Agnely, F., & Huang, N. (2019). Pickering emulsions: Preparation processes, key parameters governing their properties, and potential for pharmaceutical applications. *Journal of Controlled Release*, *309*, 302–332. https://doi.org/10.1016/j.jconrel.2019.07.003.

Andreev, G. N., Schrader, B., Schulz, H., Fuchs, R., Popov, S., & Handjieva, N. (2001). Non-destructive NIR-FT-Raman analyses in practice. Part 1. Analyses of plants and historic textiles. *Analytical and Bioanalytical Chemistry*, *371*(7), 1009–1017. https://doi.org/10.1007/s00216-001-1109-6.

Anvisa, A. N. de V. S. (2007). RESOLUÇÃO-RDC No 14, DE 28 DE FEVEREIRO DE 2007-Regulamento Técnico para Produtos com Ação Antimicrobiana, harmonizado no âmbito do Mercosul, e dá outras providências.

Bailey, E. S., Curcic, M., Biros, J., Erdogmuş, H., Bac, N., & Sacco, A. (2021). Essential oil disinfectant efficacy against SARS-CoV-2 microbial surrogates. *Frontiers in Public Health, 9,* 1–6. https://doi.org/10.3389/fpubh.2021.783832.

Barbosa, R. F. da S., Yudice, E. D. C., Mitra, S. K., & Rosa, D. dos S. (2021). Characterization of rosewood and Cinnamon Cassia essential oil polymeric capsules: Stability, loading efficiency, release rate, and antimicrobial properties. *Food Control, 121*(September 2020), 107605. https://doi.org/10.1016/j.foodcont.2020.107605.

Battistini, R., Rossini, I., Ercolini, C., Goria, M., Callipo, M. R., Maurella, C., Pavoni, E., & Serracca, L. (2019). Antiviral activity of essential oils against hepatitis a virus in soft fruits. *Food and Environmental Virology, 11*(1), 90–95. https://doi.org/10.1007/s12560-019-09367-3.

Becerril, R., Nerín, C., & Silva, F. (2020). Encapsulation systems for antimicrobial food packaging components: An update. *Molecules, 25*(5), 1134. https://doi.org/10.3390/molecules25051134.

Briggs, N., Raman, A. K. Y., Barrett, L., Brown, C., Li, B., Leavitt, D., Aichele, C. P., & Crossley, S. (2018). Stable Pickering emulsions using multi-walled carbon nanotubes of varying wettability. *Colloids and Surfaces A: Physicochemical and Engineering Aspects, 537,* 227–235. https://doi.org/10.1016/j.colsurfa.2017.10.010.

Cebi, N., Arici, M., & Sagdic, O. (2021). The famous Turkish rose essential oil: Characterization and authenticity monitoring by FTIR, Raman, and GC-MS techniques combined with chemometrics. *Food Chemistry, 354*(February), 129495. https://doi.org/10.1016/j.foodchem.2021.129495.

Chhabra, R. P., & Richardson, J. F. (1999). Non-Newtonian fluid behavior. *Non-Newtonian Flow in the Process Industries, dVx,* 1–36. https://doi.org/10.1016/b978-075063770-1/50002-6.

da Silva, J. K. R., Figueiredo, P. L. B., Byler, K. G., & Setzer, W. N. (2020). Essential oils as antiviral agents. Potential of essential oils to treat sars-cov-2 infection: An in-silico investigation. *International Journal of Molecular Sciences, 21*(10), 1–35. https://doi.org/10.3390/ijms21103426.

Dai, S., Jiang, F., Shah, N. P., & Corke, H. (2017). Stability and phase behavior of konjac glucomannan-milk systems. *Food Hydrocolloids, 73,* 30–40. https://doi.org/10.1016/j.foodhyd.2017.06.025.

El Atki, Y., Aouam, I., El Kamari, F., Taroq, A., Nayme, K., Timinouni, M., Lyoussi, B., & Abdellaoui, A. (2019). Antibacterial activity of cinnamon essential oils and their synergistic potential with antibiotics. *Journal of Advanced Pharmaceutical Technology and Research, 10*(2), 63–67. https://doi.org/10.4103/japtr.JAPTR_366_18.

Elfiky, A. A. (2021). SARS-CoV-2 RNA dependent RNA polymerase (RdRp) targeting: An in silico perspective. *Journal of Biomolecular Structure and Dynamics, 39*(9), 3204–3212. https://doi.org/10.1080/07391102.2020.1761882.

Feldsine, P., Abeyta, C., & Andrews, W. H. (2002). AOAC International methods committee guidelines for validation of qualitative and quantitative food microbiological official methods of analysis. *Journal of AOAC International, 85*(5), 1187–1200. https://doi.org/10.1093/jaoac/85.5.1187.

Ferreira, G. da S., da Silva, D. J., & Rosa, D. S. (2023). Super stable Melaleuca alternifolia essential oil Pickering emulsions stabilized with cellulose nanofibrils: Rheological aspects. *Journal of Molecular Liquids, 372,* 121183. https://doi.org/10.1016/j.molliq.2022.121183.

Ferreira, R. R., Souza, A. G., & Rosa, D. S. (2021). Essential oil-loaded nanocapsules and their application on PBAT biodegradable films. *Journal of Molecular Liquids, 337,* 116488. https://doi.org/10.1016/j.molliq.2021.116488.

Figueiredo, C. S. S. e S., Viera de Oliveira, P., Felipe de Silva Saminez, W., Muniz Diniz, R., Francisco Silva Rodrigues, J., Silma Maia da Silva, M., Cláudio Nascimento da Silva, L., & Augusto Grigolin Grisotto, M. (2017). *Óleo essencial da Canela (Cinamaldeído) e suas aplicações biológicas (Cinnamon essential oil (cinnamaldehyde) and its applications), Rev. Investig, Bioméd. São Luís,* 9(2): 192–197.

Fiorillo, L., Cervino, G., Matarese, M., D'amico, C., Surace, G., Paduano, V., Fiorillo, M. T., Moschella, A., la Bruna, A., Romano, G. L., Laudicella, R., Baldari, S., & Cicciù, M. (2020). COVID-19 surface persistence: A recent data summary and its importance for medical and dental settings. *International Journal of Environmental Research and Public Health, 17*(9), 3132. MDPI AG. https://doi.org/10.3390/ijerph17093132.

Flechas, M. C., Ocazionez, R. E., & Stashenko, E. E. (2017). Evaluation of in vitro antiviral activity of essential oil compounds against dengue virus. *Pharmacognosy Journal, 10*(1), 55–59. https://doi.org/10.5530/pj.2018.1.11.

Goh, C. F., Ming, L. C., & Wong, L. C. (2021). Dermatologic reactions to disinfectant use during the COVID-19 pandemic. *Clinics in Dermatology, 39*(2), 314–322. https://doi.org/10.1016/j.clindermatol.2020.09.005.

Homza, M., Zelena, H., Janosek, J., Tomaskova, H., Jezo, E., Kloudova, A., Mrazek, J., Svagera, Z., & Prymula, R. (2021). Five antigen tests for sars-cov-2: Virus viability matters. *Viruses, 13*(4), 1–9. https://doi.org/10.3390/v13040684.

Hosseini, R. S., & Rajaei, A. (2020). Potential Pickering emulsion stabilized with chitosan-stearic acid nanogels incorporating clove essential oil to produce fish-oil-enriched mayonnaise. *Carbohydrate Polymers, 241*(April), 116340. https://doi.org/10.1016/j.carbpol.2020.116340.

Hu, Y., Tian, J., Zou, J., Yuan, X., Li, J., Liang, H., Zhan, F., & Li, B. (2019). Partial removal of acetyl groups in konjac glucomannan significantly improved the rheological properties and texture of konjac glucomannan and κ-carrageenan blends. *International Journal of Biological Macromolecules, 123*, 1165–1171. https://doi.org/10.1016/j.ijbiomac.2018.10.190.

Husin, H., Taju Ariffin, T. S., & Yahya, E. (2018). Rheological behaviour of water-in-light crude oil emulsion. *IOP Conference Series: Materials Science and Engineering, 358*(1), 1–6. https://doi.org/10.1088/1757-899X/358/1/012067.

Jamali, S. N., Assadpour, E., Feng, J., & Jafari, S. M. (2021). Natural antimicrobial-loaded nanoemulsions for the control of food spoilage/pathogenic microorganisms. *Advances in Colloid and Interface Science, 295*, 102504. https://doi.org/10.1016/j.cis.2021.102504.

Jiang, J., Ding, X., Tasoglou, A., Huber, H., Shah, A. D., Jung, N., & Boor, B. E. (2021). Real-time measurements of botanical disinfectant emissions, transformations, and multiphase inhalation exposures in buildings. *Environmental Science and Technology Letters, 8*(7), 558–566. https://doi.org/10.1021/acs.estlett.1c00390.

Keivani Nahr, F., Ghanbarzadeh, B., Samadi Kafil, H., Hamishehkar, H., & Hoseini, M. (2020). The colloidal and release properties of cardamom oil encapsulated nanostructured lipid carrier. *Journal of Dispersion Science and Technology, 42*(1), 1–9. https://doi.org/10.1080/01932691.2019.1658597.

Komatsu, S., Ikedo, Y., Asoh, T. A., Ishihara, R., & Kikuchi, A. (2018). Fabrication of hybrid capsules via CaCO3 crystallization on degradable coacervate droplets. *Langmuir, 34*(13), 3981–3986. https://doi.org/10.1021/acs.langmuir.8b00148.

Kulkarni, S. A., Nagarajan, S. K., Ramesh, V., Palaniyandi, V., Selvam, S. P., & Madhavan, T. (2020). Computational evaluation of major components from plant essential oils as potent inhibitors of SARS-CoV-2 spike protein. *Journal of Molecular Structure, 1221*, 128823. https://doi.org/10.1016/j.molstruc.2020.128823.

Lan, J., Ge, J., Yu, J., Shan, S., Zhou, H., Fan, S., Zhang, Q., Shi, X., Wang, Q., Zhang, L., & Wang, X. (2020). Structure of the SARS-CoV-2 spike receptor-binding domain bound to the ACE2 receptor. *Nature, 581*(7807), 215–220. https://doi.org/10.1038/s41586-020-2180-5.

Le, V. D., Tran, V. T., Dang, V. S., Nguyen, D. T., Dang, C. H., & Nguyen, T. D. (2020). Physicochemical characterizations, antimicrobial activity and non-isothermal decomposition kinetics of Cinnamomum cassia essential oils. *Journal of Essential Oil Research, 32*(2), 158–168. https://doi.org/10.1080/10412905.2019.1700834.

Li, Q., Wang, Y., Wu, Y., He, K., Li, Y., Luo, X., Li, B., Wang, C., & Liu, S. (2019). Flexible cellulose nanofibrils as novel pickering stabilizers: The emulsifying property and packing behavior. *Food Hydrocolloids, 88*, 180–189. https://doi.org/10.1016/j.foodhyd.2018.09.039.

Li, X., Gao, Y., Li, Y., Li, Y., Liu, H., Yang, Z., Wu, H., & Hu, Y. (2022). Formation of cinnamon essential oil/xanthan gum/chitosan composite microcapsules basing on Pickering emulsions. *Colloid and Polymer Science, 300*(10), 1187–1195. https://doi.org/10.1007/s00396-022-05019-4.

Li, Z., Wu, H., Yang, M., Xu, D., Chen, J., Feng, H., Lu, Y., Zhang, L., Yu, Y., & Kang, W. (2018). Stability mechanism of O/W Pickering emulsions stabilized with regenerated cellulose. *Carbohydrate Polymers, 181*, 224–233. https://doi.org/10.1016/j.carbpol.2017.10.080.

Lima, G. F., Souza, A. G., & Rosa, D. S. (2018). Effect of adsorption of polyethylene glycol (PEG), in aqueous media, to improve cellulose nanostructures stability. *Journal of Molecular Liquids, 268*, 415–424. https://doi.org/10.1016/j.molliq.2018.07.080.

Luo, J., Huang, K., Zhou, X., & Xu, Y. (2021). Elucidation of oil-in-water emulsions stabilized with celery cellulose. *Fuel, 291*(January), 120210. https://doi.org/10.1016/j.fuel.2021.120210.

Magurano, F., Baggieri, M., Marchi, A., Rezza, G., Nicoletti, L., Eleonora, B., Concetta, F., Stefano, F., Maedeh, K., Paola, B., Emilio, D. U., & Silvia, G. (2021). SARS-CoV-2 infection: The environmental endurance of the virus can be influenced by the increase of temperature. *Clinical Microbiology and Infection, 27*(2), 289.e5–289.e7. https://doi.org/10.1016/j.cmi.2020.10.034.

Mai-Prochnow, A., Clauson, M., Hong, J., & Murphy, A. B. (2016). Gram positive and Gram negative bacteria differ in their sensitivity to cold plasma. *Scientific Reports, 6*(December), 1–11. https://doi.org/10.1038/srep38610.

McClements, D. J. (2015). *Food Emulsions Principles, Practices, and Techniques*. CRC Press, Boca Raton. https://doi.org/10.1201/9781420039436.

Mendes, J. F., Norcino, L. B., Martins, H. H. A., Manrich, A., Otoni, C. G., Carvalho, E. E. N., Piccoli, R. H., Oliveira, J. E., Pinheiro, A. C. M., & Mattoso, L. H. C. (2020). Correlating emulsion characteristics with the properties of active starch films loaded with lemongrass essential oil. *Food Hydrocolloids, 100*. https://doi.org/10.1016/j.foodhyd.2019.105428.

Najar, B., Nardi, V., Stincarelli, M. A., Patrissi, S., Pistelli, L., & Giannecchini, S. (2021). Screening of the essential oil effects on human H1N1 influenza virus infection: An in vitro study in MDCK cells. *Natural Product Research, 0*(0), 1–4. https://doi.org/10.1080/14786419.2021.1944137.

Nascimento, P. F. C., Nascimento, A. C., Rodrigues, C. S., Antoniolli, Â. R., Santos, P. O., Barbosa Júnior, A. M., & Trindade, R. C. (2007). Atividade antimicrobiana dos óleos essenciais: Uma abordagem multifatorial dos métodos. *Revista Brasileira de Farmacognosia, 17*(1), 108–113. https://doi.org/10.1590/s0102-695x2007000100020.

Nelson, P., Adimabua, P., Wang, A., Zou, S., & Shah, N. C. (2020). Surface-enhanced Raman spectroscopy for rapid screening of cinnamon essential oils. *Applied Spectroscopy, 74*(11), 1341–1349. https://doi.org/10.1177/0003702820931154.

Ozogul, Y., Yuvka, İ., Ucar, Y., Durmus, M., Kösker, A. R., Öz, M., & Ozogul, F. (2017). Evaluation of effects of nanoemulsion based on herb essential oils (rosemary, laurel, thyme and sage) on sensory, chemical and microbiological quality of rainbow trout (Oncorhynchus mykiss) fillets during ice storage. *LWT - Food Science and Technology, 75*, 677–684. https://doi.org/10.1016/j.lwt.2016.10.009.

Park, N. M., Choi, S., Oh, J. E., & Hwang, D. Y. (2019). Facile extraction of cellulose nanocrystals. *Carbohydrate Polymers, 223*(June), 115114. https://doi.org/10.1016/j.carbpol.2019.115114.

Paulo, B. B., Alvim, I. D., Reineccius, G., & Prata, A. S. (2020). Performance of oil-in-water emulsions stabilized by different types of surface-active components. *Colloids and Surfaces B: Biointerfaces, 190*. https://doi.org/10.1016/j.colsurfb.2020.110939.

Pavoni, L., Perinelli, D. R., Bonacucina, G., Cespi, M., & Palmieri, G. F. (2020). An overview of micro-and nanoemulsions as vehicles for essential oils: Formulation, preparation and stability. In *Nanomaterials* (Vol. *10*, Issue 1). MDPI AG. https://doi.org/10.3390/nano10010135.

Pickering, S. U. (1907). Emulsions. In *Journal of the Chemical Society, Transactions* (Vol. *91*, pp. 2001–2021). https://doi.org/10.1039/CT9079102001.

Ramsden, W. (1904). Separation of solids in the surface-layers of solutions and 'suspensions' (observations on surface-membranes, bubbles, emulsions, and mechanical coagulation).- Preliminary account. *Proceedings of the Royal Society of London.*

Reis, J. B., Figueiredo, L. A., Castorani, G. M., & Veiga, S. M. O. M. (2020). Avaliação da atividade antimicrobiana dos óleos essenciais contra patógenos alimentares. *Brazilian Journal of Health Review, 3*(1), 342–363. https://doi.org/10.34119/bjhrv3n1-025.

Ribeiro-Santos, R., Andrade, M., de Melo, N. R., dos Santos, F. R., Neves, I. de A., de Carvalho, M. G., & Sanches-Silva, A. (2017). Biological activities and major components determination in essential oils intended for a biodegradable food packaging. *Industrial Crops and Products, 97*, 201–210. https://doi.org/10.1016/j.indcrop.2016.12.006.

Sacui, I. A., Nieuwendaal, R. C., Burnett, D. J., Stranick, S. J., Jorfi, M., Weder, C., Foster, E. J., Olsson, R. T., & Gilman, J. W. (2014). Comparison of the properties of cellulose nanocrystals and cellulose nanofibrils isolated from bacteria, tunicate, and wood processed using acid, enzymatic, mechanical, and oxidative methods. *ACS Applied Materials & Interfaces, 6*(9), 6127–6138. https://doi.org/10.1021/am500359f.

Saffarionpour, S. (2020). Nanocellulose for stabilization of pickering emulsions and delivery of nutraceuticals and its interfacial adsorption mechanism. https://doi.org/10.1007/s11947-020-02481-2/Published.

Saraiva, R. A., Matias, E. F. F., Coutinho, H. D. M., Costa, J. G. M., Souza, H. H. F., Fernandes, C. N., Rocha, J. B. T., & Menezes, I. R. A. (2011). Synergistic action between Caryocar coriaceum Wittm. fixed oil with aminoglycosides in vitro. *European Journal of Lipid Science and Technology, 113*(8), 967–972. https://doi.org/10.1002/ejlt.201000555.

Schnitzler, P., Koch, C., & Reichling, J. (2007). Susceptibility of drug-resistant clinical herpes simplex virus type 1 strains to essential oils of ginger, thyme, hyssop, and sandalwood. *Antimicrobial Agents and Chemotherapy, 51*(5), 1859–1862. https://doi.org/10.1128/AAC.00426-06.

Schulz, H., & Baranska, M. (2006). Rapid evaluation of quality parameters in plant products applying ATR-IR and raman spectroscopy. *Acta Horticulturae, 712*(I), 347–355. https://doi.org/10.17660/actahortic.2006.712.39.

Schulz, H., Baranska, M., Belz, H. H., Rösch, P., Strehle, M. A., & Popp, J., (2004). Chemotaxonomic characterisation of essential oil plants by vibrational spectroscopy measurements. *Vibrational Spectroscopy, 35*(1–2), 81–86. https://doi.org/10.1016/j.vibspec.2003.12.014.

Soo, Y. T., Ng, S. W., Tang, T. K., Ab Karim, N. A., Phuah, E. T., & Lee, Y. Y. (2021). Preparation of palm (*Elaeis oleifera*) pressed fibre cellulose nanocrystals via cation exchange resin: Characterisation and evaluation as Pickering emulsifier. *Journal of the Science of Food and Agriculture, 101*(10), 4161–4172. https://doi.org/10.1002/jsfa.11054.

Souza, A. G., Ferreira, R. R., Paula, L. C., Setz, L. F. G., & Rosa, D. S. (2020). The effect of essential oil chemical structures on Pickering emulsion stabilized with cellulose nanofibrils. *Journal of Molecular Liquids, 320*, 114458. https://doi.org/10.1016/j.molliq.2020.114458.

Souza, A. G., Lima, G. F., & Rosa, D. dos S. (2019). *Cellulose Nanostructures from Lignocellulosic Residues*. LAP LAMBERT Academic Publishing.

Souza, A. G. de, Ferreira, R. R., Aguilar, E. S. F., Zanata, L., & Rosa, D. dos S. (2021). Cinnamon essential oil nanocellulose-based pickering emulsions: Processing parameters effect on their formation, stabilization, and antimicrobial activity. *Polysaccharides, 2*(3), 608–625. https://doi.org/10.3390/polysaccharides2030037.

Štular, D., Golja, B., Malis, D., Jeršek, B., Tomšič, B., & Kapun, G. (2019). Antibacterial activity and biodegradation of cellulose fiber blends with incorporated ZnO. *Materials, 12*(October 2019), 3399.

Sun, H., Li, S., Chen, S., Wang, C., Liu, D., & Li, X. (2020). Antibacterial and antioxidant activities of sodium starch octenylsuccinate-based Pickering emulsion films incorporated with cinnamon essential oil. *International Journal of Biological Macromolecules*, *159*, 696–703. https://doi.org/10.1016/j.ijbiomac.2020.05.118.

Tang, C., Li, Y., Pun, J., Mohamed Osman, A. S., & Tam, K. C. (2019). Polydopamine microcapsules from cellulose nanocrystal stabilized Pickering emulsions for essential oil and pesticide encapsulation. *Colloids and Surfaces A: Physicochemical and Engineering Aspects*, *570*, 403–413. https://doi.org/10.1016/j.colsurfa.2019.03.049.

Thuy, B. T. P., My, T. T. A., Hai, N. T. T., Hieu, L. T., Hoa, T. T., Thi Phuong Loan, H., Triet, N. T., Anh, T. T. Van, Quy, P. T., Tat, P. Van, Hue, N. Van, Quang, D. T., Trung, N. T., Tung, V. T., Huynh, L. K., & Nhung, N. T. A. (2020). Investigation into SARS-CoV-2 resistance of compounds in garlic essential oil. *ACS Omega*, *5*(14), 8312–8320. https://doi.org/10.1021/acsomega.0c00772.

Tung, Y. T., Yen, P. L., Lin, C. Y., & Chang, S. T. (2010). Anti-inflammatory activities of essential oils and their constituents from different provenances of indigenous cinnamon (*Cinnamomum osmophloeum*) leaves. *Pharmaceutical Biology*, *48*(10), 1130–1136. https://doi.org/10.3109/13880200903527728.

Vasconcelos, N. G., Croda, J., & Simionatto, S. (2018). Antibacterial mechanisms of cinnamon and its constituents: A review. In *Microbial Pathogenesis* (Vol. *120*, pp. 198–203). Academic Press. https://doi.org/10.1016/j.micpath.2018.04.036.

Wang, J., Prinz, R. A., Liu, X., & Xu, X. (2020). In vitro and in vivo antiviral activity of gingerenone a on influenza a virus is mediated by targeting janus kinase 2. *Viruses*, *12*(10), 1–18. https://doi.org/10.3390/v12101141.

Wang, Y., Wang, W., Jia, H., Gao, G., Wang, X., Zhang, X., & Wang, Y. (2018). Using cellulose nanofibers and its palm oil pickering emulsion as fat substitutes in emulsified sausage. *Journal of Food Science*, *83*(6), 1740–1747. https://doi.org/10.1111/1750-3841.14164.

Wani, A. R., Yadav, K., Khursheed, A., & Rather, M. A. (2021). An updated and comprehensive review of the antiviral potential of essential oils and their chemical constituents with special focus on their mechanism of action against various influenza and coronaviruses. *Microbial Pathogenesis*, *152*(October 2020), 104620. https://doi.org/10.1016/j.micpath.2020.104620.

Xiong, W., Ren, C., Tian, M., Yang, X., Li, J., & Li, B. (2018). Emulsion stability and dilatational viscoelasticity of ovalbumin/chitosan complexes at the oil-in-water interface. *Food Chemistry*, *252*(September 2017), 181–188. https://doi.org/10.1016/j.foodchem.2018.01.067.

Xu, W., Xiong, Y., Li, Z., Luo, D., Wang, Z., Sun, Y., & Shah, B. R. (2020). Stability, microstructural and rheological properties of complex prebiotic emulsion stabilized by sodium caseinate with inulin and konjac glucomannan. *Food Hydrocolloids*, *105*(February), 105772. https://doi.org/10.1016/j.foodhyd.2020.105772.

Yadalam, P. K., Varatharajan, K., Rajapandian, K., Chopra, P., Arumuganainar, D., Nagarathnam, T., Sohn, H., & Madhavan, T. (2021). Antiviral essential oil components against SARS-CoV-2 in pre-procedural mouth rinses for dental settings during covid-19: A computational study. *Frontiers in Chemistry*, *9*(March). https://doi.org/10.3389/fchem.2021.642026.

Yan, Y., Chang, L., & Wang, L. (2020). Laboratory testing of SARS-CoV, MERS-CoV, and SARS-CoV-2 (2019-nCoV): Current status, challenges, and countermeasures. In *Reviews in Medical Virology* (Vol. *30*, Issue 3). John Wiley and Sons Ltd. https://doi.org/10.1002/rmv.2106.

Yazgan, H. (2020). Investigation of antimicrobial properties of sage essential oil and its nanoemulsion as antimicrobial agent. *LWT*, *130*(February), 109669. https://doi.org/10.1016/j.lwt.2020.109669.

Ye, F., Miao, M., Jiang, B., Campanella, O. H., Jin, Z., & Zhang, T. (2017). Elucidation of stabilizing oil-in-water Pickering emulsion with different modified maize starch-based nanoparticles. *Food Chemistry*, *229*, 152–158. https://doi.org/10.1016/j.foodchem.2017.02.062.

Yu, H., Huang, G., Ma, Y., Liu, Y., Huang, X., Zheng, Q., Yue, P., & Yang, M. (2021a). Cellulose nanocrystals based clove oil Pickering emulsion for enhanced antibacterial activity. *International Journal of Biological Macromolecules*, *170*, 24–32. https://doi.org/10.1016/j.ijbiomac.2020.12.027.

Yuan, H., Peng, J., Ren, T., Luo, Q., Luo, Y., Zhang, N., Huang, Y., Guo, X., & Wu, Y. (2021). Novel fluorescent lignin-based hydrogel with cellulose nanofibers and carbon dots for highly efficient adsorption and detection of Cr(VI). *Science of the Total Environment*, *760*. https://doi.org/10.1016/j.scitotenv.2020.143395.

Zhai, X., Gao, J., Wang, X., Mei, S., Zhao, R., Wu, Y., Hao, C., Yang, J., & Liu, Y. (2018). Inverse Pickering emulsions stabilized by carbon quantum dots: Influencing factors and their application as templates. *Chemical Engineering Journal*, *345*, 209–220. https://doi.org/10.1016/j.cej.2018.03.075.

Zhan, C.-G., Nichols, J. A., & Dixon, D. A. (2003). Ionization potential, electron affinity, electronegativity, hardness, and electron excitation energy: Molecular properties from density functional theory orbital energies. *The Journal of Physical Chemistry A*, *107*(20), 4184–4195. https://doi.org/10.1021/jp0225774.

Zhang, C., Fan, L., Fan, S., Wang, J., Luo, T., Tang, Y., Chen, Z., & Yu, L. (2019). Cinnamomum cassia Presl: A review of its traditional uses, phytochemistry, pharmacology and toxicology. *Molecules*, *24*(19). https://doi.org/10.3390/molecules24193473.

Zhang, S., Holmes, M., Ettelaie, R., & Sarkar, A. (2020). Pea protein microgel particles as Pickering stabilisers of oil-in-water emulsions: Responsiveness to pH and ionic strength. *Food Hydrocolloids*, *102*(November 2019), 105583. https://doi.org/10.1016/j.foodhyd.2019.105583.

Zhang, Y., Karimkhani, V., Makowski, B. T., Samaranayake, G., & Rowan, S. J. (2017). Nanoemulsions and nanolatexes stabilized by hydrophobically functionalized cellulose nanocrystals. *Macromolecules*, *50*(16), 6032–6042. https://doi.org/10.1021/acs.macromol.7b00982.

Zhang, Y., Liu, X., Wang, Y., Jiang, P., & Quek, S. Y. (2016). Antibacterial activity and mechanism of cinnamon essential oil against *Escherichia coli* and *Staphylococcus aureus*. *Food Control*, *59*, 282–289. https://doi.org/10.1016/j.foodcont.2015.05.032.

Zhou, F. Z., Huang, X. N., Wu, Z. L., Yin, S. W., Zhu, J. H., Tang, C. H., & Yang, X. Q. (2018). Fabrication of zein/pectin hybrid particle-stabilized pickering high internal phase emulsions with robust and ordered interface architecture. *Journal of Agricultural and Food Chemistry*, *66*(42), 11113–11123. https://doi.org/10.1021/acs.jafc.8b03714.

Zhou, J., Hu, Z., Zabihi, F., Chen, Z., & Zhu, M. (2020). Progress and perspective of antiviral protective material. In *Advanced Fiber Materials* (Vol. 2, Issue 3, pp. 123–139). Springer. https://doi.org/10.1007/s42765-020-00047-7.

Zhou, Y., Sun, S., Bei, W., Zahi, M. R., Yuan, Q., & Liang, H. (2018). Preparation and antimicrobial activity of oregano essential oil Pickering emulsion stabilized by cellulose nanocrystals. *International Journal of Biological Macromolecules*, *112*, 7–13. https://doi.org/10.1016/j.ijbiomac.2018.01.102.

Zhu, X., Chen, J., Hu, Y., Zhang, N., Fu, Y., & Chen, X. (2021). Tuning complexation of carboxymethyl cellulose/ cationic chitosan to stabilize Pickering emulsion for curcumin encapsulation. *Food Hydrocolloids*, *110*(July 2020), 106135. https://doi.org/10.1016/j.foodhyd.2020.106135.

22 Nanofillers in Automotive and Aerospace Industry

H. A. Aisyah, E. S. Zainudin, and B. F. A. Bakar
Universiti Putra Malaysia

R. A. Ilyas
Universiti Teknologi Malaysia

22.1 INTRODUCTION

Nanocomposite materials have attracted the interests of many researchers, owing to various advancements and wide range of applications. These materials can perform better in terms of electrical and thermal conductivity, mechanical strength, toughness, and stiffness, and higher flame retardancy with the addition of nanofillers. It has been discovered that incorporating minimal concentrations of these nanofillers into polymers can increase their properties without altering their processability (Li et al., 2019). Nanocomposites with various polymers have been created using a variety of nanofiller types, namely clays, carbon nanotubes (CNTs), graphene, nanocellulose (NC), and halloysite.

The application of nanofillers in nanocomposites has been widely used including in electronics components, optical devices, and automotive, aerospace, and biomedicine sectors (Shen et al., 2021; Kausar, 2020, Vashist et al., 2018). The main features of materials utilized in automotive and aerospace applications are light weight and high strength. In addition, according to Garces et al. (2020), some of the key performance characteristics of nanocomposite materials need to be improved and controlled to make the composites a viable alternative material for these industries. These characteristics include improved modulus and dimensional stability, higher heat-distortion temperature, and improved scratch and mar resistance. Different kinds of nanofillers are used to create and synthesize polymer-based nanocomposites, and they can generally be divided into two main groups: organic and inorganic nanofillers. Among these, inorganic nanofillers, particularly carbon-based nanomaterials, have recently received a lot of interest among all nanofillers as shown in Figure 22.1 (Ehsani et al., 2021).

22.2 INORGANIC NANOFILLERS IN AUTOMOTIVE AND AEROSPACE INDUSTRY

22.2.1 GRAPHENE

Graphene consists of a hexagonally arranged sheet of covalently sp^2-bonded carbon atoms that is one atom thick. There are several ways to make graphene from graphite,

DOI: 10.1201/9781003400998-22

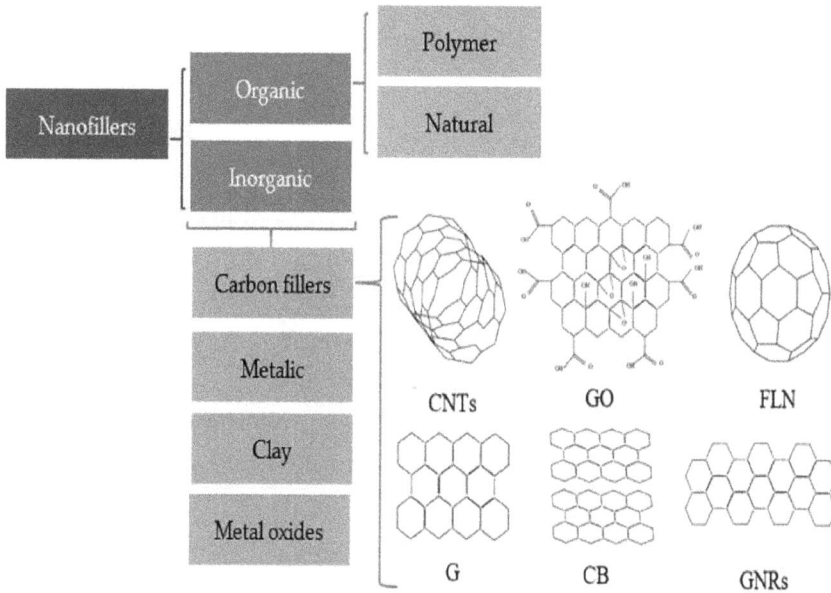

FIGURE 22.1 Various kinds of nanofillers for polymer nanocomposites. (Reproduced with a copyright permission from Ehsani et al., 2021.)

including chemical vapor deposition (CVD), chemical exfoliation (CE), mechanical exfoliation, thermal expansion of chemically intercalated graphite, and chemical reduction of graphene oxide (GO) (Bohm et al., 2021; Lim et al., 2018). In recent years, graphene has found its use in many industrial applications due to its unique properties including high optical performance, high charge carrier's mobility, and thermal conductivity (Bohm et al., 2021; Balandin et al., 2008).

Researchers and scientists have proposed, tested, and analyzed graphene materials that can be used in automotive and aerospace industries, for example, graphene nanoplatelets (GNPs) for automotive part application (Kausar, 2017). The production of GNPs is involved in three main phases: formation of polythioamide (PTA), PTA blending with polyamide 1010 (PA1010), followed by formation of PA1010/PTA and GNP nanocomposite (PA1010/PTA/GNP) as shown in Figure 22.2. The nanocomposite PA1010/PTA/GNP with varying GNP loading ranging from 0.01 to 0.03 wt% was prepared. The structure, morphology, thermal, flammability, rheology, and mechanical properties of PTA and PTA/GNP nanocomposites were compared, and they were shown to be significantly impacted by GNP as a remarkable nanofiller. The outcomes showed that the addition of GNP nanoparticles improved the flammability and thermal characteristics of processable blends and nanocomposites. The nanocomposite PA1010/PTA/GNP with 0.03 wt% of GNP loading demonstrated the best tensile strength (40 MPa), impact strength (1.9 MPa), and flexural modulus (1,373 MPa) for the production of automotive parts.

Laurenzi et al. (2020) investigated the effect of different nanofillers used on the equivalent dose absorbed by the nanocomposites in various radiation fields in space.

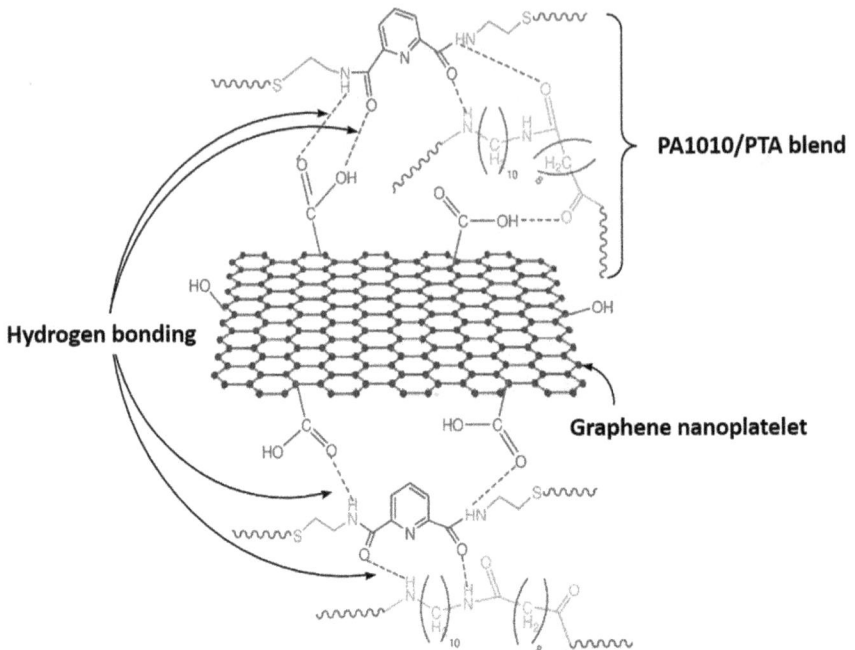

FIGURE 22.2 Preparation of PA1010/PTA/GNP nanocomposite. (Reproduced with a copyright permission from Kausar, 2017.)

The nanofillers in this study were SWCNT, GO nanoplatelets, and boron carbide (CB4). According to the results, the GO nanoplatelets were the best reinforcement for space radiation protection among the investigated fillers. It was noted that GO suspended in an epoxy matrix decreased the impact damage produced by micrometeoroid orbital debris (MMOD), and the loading and radiation shielding were improved with the addition of GO fillers. Due to the presence of functional groups rich in hydrogen that were immobilized on the planes and edges of the graphene following oxidation, the GO nanoplatelets were the most effective radiation shielding material among the examined fillers in all simulated radiation environments.

The metal–GNP nanocomposite hybrids with pure magnesium (Mg) as the matrix were successfully developed by Chen et al. (2012) using the nanoprocessing method that combines liquid-state ultrasonic processing and solid-state stirring. The discovered GNP-reinforced Mg-based metal matrix nanocomposite exhibits homogeneous GNP dispersion and significantly improved characteristics. Figure 22.3 shows that the GNP was implanted inside a Mg grain since the orientation of the Mg matrix around it is the same. Effective load transmission between the Mg matrix and GNPs and the effectiveness of the GNPs functioning as a barrier to restrict the movement of dislocation need strong interfacial bonding. Further evidence of the excellent bonding between the GNP and the Mg matrix was the absence of voids and reaction products.

A combination of poly(lactic acid) (PLA) and epoxidized palm oil (EPO) was examined for its mechanical characteristics using xGnP by Chieng et al. (2012). The melt blending process was used to effectively create PLA/EPO/xGnP green

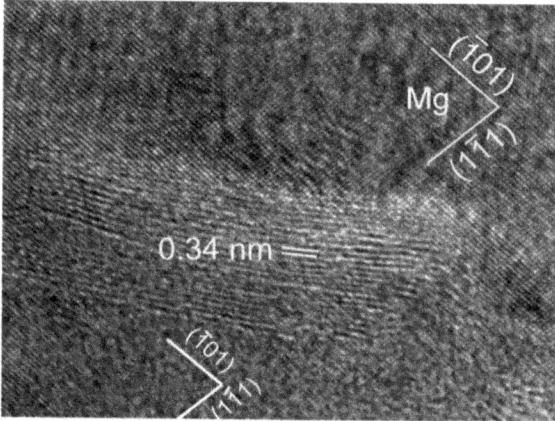

FIGURE 22.3 High-resolution transmission electron microscopy (HRTEM) image of Mg-GNP. (Reproduced with a copyright permission from Chen et al., 2012.)

nanocomposites. In comparison with PLA/EPO mix, PLA/EPO reinforced with xGnP increased the tensile strength and elongation at break of the nanocomposites by up to 26.5% and 60.6%, respectively. In PLA/EPO nanocomposites, the XRD pattern revealed the existence of a peak about 26.5°, which corresponds to the distinctive peak of GNPs. However, the flexural strength and modulus are unaffected by the addition of xGnP. The addition of 0.5 weight percent xGnP loading increased the impact strength of PLA/5 weight percent EPO by 73.6%. They concluded that a small number of GNPs (1 wt%) significantly enhanced the mechanical characteristics.

22.2.2 CARBON NANOTUBES

Carbon nanotubes (CNTs) are cylindrical molecules made of sheets of single-layer carbon atoms that have been coiled up (graphene) as shown in Figure 22.4. They usually come in two varieties: single-walled carbon nanotubes (SWCNTs), with a diameter of less than 1 nm, and multi-walled carbon nanotubes (MWCNTs), with diameters more than 100 nm and made up of multiple concentrically interconnected nanotubes. They can be as long as a millimeter or even several micrometers. According to some research, CNTs possess excellent mechanical properties in term of tensile strength and stiffness and also possess excellent electrical, thermal, and magnetic properties to be crucial materials in automotive and aerospace applications (Nurazzi et al., 2021a,b). Many researchers attempted to incorporate CNTs in these applications including developments in current technologies such as body components, electrical systems, and engine parts.

Subadra et al. (2020) investigated the effect of different concentrations of CNTs ranging from 0.05 to 0.4 wt% on the properties of CNT/fiberglass-reinforced epoxy composites. The fiberglass/CNTs/epoxy composites with higher concentrations of CNTs indicated higher strength, impact energy, and thermal stability compared with low concentrations of CNTs, which showed that the dispersion of CNTs highly depends on their existence. Furthermore, by 16% and 26%, respectively, less fuel

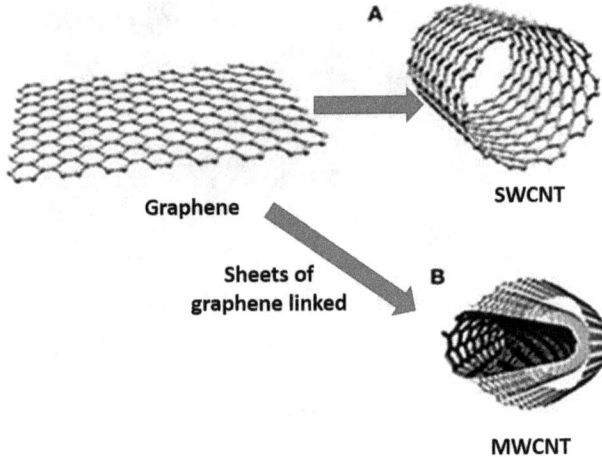

FIGURE 22.4　Roll of graphene sheets into (a) single-walled carbon nanotube (SWCNT), and (b) multi-walled carbon nanotube (MWCNT). (Reproduced with a copyright permission from David et al., 2020.)

may be consumed and greenhouse gas emissions can be reduced. Mechanical, metallurgical, and tribological behavior of magnesium/CNT composites were reported by Selvamani et al. (2017). In this study, three distinct CNT reinforcements (2%, 3%, and 4%) were added to the AZ91D-grade magnesium alloy to create stir-cast composite materials. The results showed that the hardness strength increases in the Mg/CNT composites by incorporation of 3% CNT, whereas the tensile strength, yield strength, and wear resistance increased with the incorporation of 4% CNT.

Due to CNT's outstanding mechanical and electrical properties, the application of CNTs in developing flexible batteries is favored. Many researchers attempted to incorporate CNTs to unlock the full potential of CNTs for applications in batteries. Due to their potential usage in portable and wearable electronics, this application in batteries may be adapted to a wide range of automotive uses. Zhu et al. (2021) listed research works of flexible batteries that incorporated CNTs, which play a key role in the development of high-performance flexible batteries. Additionally, CNT-based batteries have various benefits, including a distinct 1D nanoscale structure, acceptable surface chemical characteristics, and a high degree of graphitization (Song et al., 2019, Nurazzi et al., 2021b).

Andrews et al. (2018) successfully produced CNT-film transistor (CNT/TFT) for sensing environmental pressure on the tire. The performance of CNT/TFT has been shown to be comparable to or superior to that of rival technologies, such as metal oxide and organic flexible transistors (Cao et al., 2017). It was reported that CNT/TFT has been demonstrated as flexible transistors with a low-cost method for pressure sensing on non-conformal surfaces. In another study, CNTs were successfully applied in passenger tire tread compounds to improve comprehensive tire performances (Shao et al., 2018). The findings demonstrated that the addition of CNTs considerably enhanced handling and traction characteristics, making them perfect for tires used in racing and sports cars.

Composite materials reinforced with CNTs have also been reported as an excellent choice in the aerospace industry (Iqbal et al., 2021; Ramachandran et al., 2021; Joshi and Chatterjee, 2016). Islam et al. (2016) developed a chemical method to graft CNTs onto carbon fiber (CF) by direct covalent bonding to form a CNT–CF hierarchical reinforcing structure for aerospace composites and energy storage applications. A smart nanomaterial for structural health monitoring (SHM) in aerospace application using CNT and CNF was also developed by Nisha and Singh (2016). A smart coating exhibiting self-diagnostic capability made of epoxy-based CNT was created by Vertuccio et al. (2018). This conductive coating with a high sensitivity factor and a high glass transition temperature was suitable for aircrafts applications. Another study by John et al. (2022) revealed that atmospheric pressure (AP) plasma treatment, a tried-and-true direct method of dry surface modification, was used to functionalize CNTs. The functionalized CNT's high aspect ratio demonstrated a well-wetting behavior with the matrix, which resulted in a remarkable improvement in the mechanical characteristics including tensile, impact, and flexural strengths, making them suitable for prospective structural use in aerospace.

By utilizing CNTs to create metal matrix composites (MMCs) with exceptional qualities such as enhanced specific strength and wear resistance, the manufacturing of high-strength and reduced-weight advanced materials employing metal is developing quickly in the aerospace sector (Srinivasan et al., 2021; Popov et al., 2021; Ariffin et al., 2022). Numerous experimental studies on CNT-reinforced MMCs have been carried out in the aerospace industry for the manufacture of structural elements including shuttle components and engine parts. In a study by Chen et al. (2020), aluminum (Al) MMCs co-reinforced by ex-situ CNTs and in-situ alumina (Al_2O_3) nanoparticles were investigated. They found that CNTs and Al_2O_3 nanoparticles worked together to provide a protracted strain-softening stage and outstanding tensile ductility. Morphology analysis confirmed the cooperation of Al_2O_3-induced work hardening and CNT-induced load transfer resulting in a lengthy strain-softening phase, which caused the high-strength Al-CNT-Al_2O_3 composites to fracture under high strain. The findings indicated that high-performance MMCs might be created by utilizing ex-situ CNT and in-situ Al_2O_3 nanoparticles.

Kwon et al. (2011) used ball milling and hot pressing to successfully construct well-dispersed CNT-based aluminum matrix composites. In this study, they observed that the toughness of CNT-Al composites was increased by nearly seven times when compared to pure aluminum. Because of the combination of polymer resins in an aluminum–titanium–magnesium matrix, Venkatesan et al. (2018) observed reductions in coefficient of friction and wear properties in glass fiber hybrid CNT-based composites, which represent an alternative for passive thermal coverings for mating surface application. Fang et al. (2018) fabricated CNT and magnesium (Mg) matrix with different types of defects, namely monovacancy; carbon and oxygen adatoms are introduced in CNTs to investigate the effect of the defects on the interface interaction (E_{ib}) between CNT and Mg surface. They found that the impact of boron doping on E_{ib} is greater than the intermediate oxygen. In addition, the existence of holes of the boron dopant and unsaturated electrons in CNTs efficiently produces the chemical interaction between CNT and Mg matrix by revealing the micro-mechanism of the growth of E_{ib} under the action of various types of defects. In another study, Kartal et al. (2015) produced Nickel/MWCNT

(Ni/MWCNT) metal matrix nanocomposite coatings deposited by the pulse electrocode-position method. The microhardness of Ni/MWCNT hybrid nanocomposites improved as the grain size of the nickel matrix was decreased from 42 to 30.6 nm. Furthermore, the wear resistance was greatly enhanced with the addition of MWCNT. Additionally, when the CNT content and shielding plate thickness increased, the electromagnetic interference (EMI) shielding effectiveness of MWCNT/polypropylene composites improved, proving the effectiveness of the CNT nanocomposites as a heat-absorbing media in the aerospace sector (Li et al., 2016).

22.2.3 NANOCLAYS

In general, layered silicates or clay minerals containing residues of metal oxides and organic matter make up a type of materials known as nanoclays. Because of their lamellar structure and large specific surface area, clays have been proven to be efficient reinforcing fillers for polymers (Rafiee and Shahzadi, 2019). Exfoliated clay dispersion in polymers results in a notable improvement in stiffness, fire retardancy, and barrier characteristics (Oliveira and Beatrice, 2018). Example of clays are montmorillonite (MMT), saponite, laponite, hectorite, sepiolite, and vermiculite. Among the smectites, MMT is the most commonly utilized clay in polymer nanocomposites because of its availability and distinctive characteristics of intercalation, excellent exfoliation chemistry, high surface area, and reactivity (Alias et al., 2021). This makes it extremely important and effective as reinforcing fillers for polymer nanocomposite production. Basically, MMT is composed of two tetrahedral silica sheets with an alumina octahedral sheet in the center (2:1 layered structure), with hydrated exchangeable cations filling the gaps between the lattices.

Nofar et al. (2020) reported on nanocomposites made from polypropylene (PP) reinforced with 20 wt% of talc and three wt% of nanoclay (1, 3, and 5 wt%) for automotive applications using injection molding method. The reinforcement effect of the nanoclay content on the crystallization behavior, mechanical, thermal, and morphological properties of PP nanocomposites was investigated. The X-ray diffraction (XRD) results demonstrated that nanoclay had an excellent dispersion within PP, most likely as a result of its surface alteration. Adding nanoclay was found to have prolonged the melt flow index of PP–nanoclay sample as a result of PP and the nanoclay's surfactants interacted in some way. The differential scanning calorimetry (DSC) tests reveal that the addition of nanoclay enhanced the degree of crystallinity by approximately 50% due to the nucleation action of nanoclay. In term of thermal characteristics, after the addition of nanoclay, the melting temperatures (T_m) remained almost unaffected, indicating that the size of the PP crystal did not change. Additionally, stiffness, ultimate strength, and impact resistance were all improved by nanoclay incorporation.

Numerous studies on polystyrene (PS) and nanoclay-based nanocomposites for applications including in automotive and aerospace industries have been conducted (Haider et al., 2016, 2017; Samakande et al., 2007; Gul et al., 2016a). Polystyrene-organic montmorillonite (PS-OMMT) nanocomposites have been shown to decrease the heat release rate (HRR) and mass loss rate (MLR) and enhance the flame retardancy of the material (Liu et al., 2013). Other studies have reported that low loading organophilic clay platelets effectively increase the wear resistance of PS-MMT nanocomposites and enhance mechanical properties because of the improvement in the compressive and shear strength due to the strengthening effect (Liu et al., 2010). Ruamcharoen et al. (2014) studied the properties of natural rubber/polystyrene (NR/PS) composite, via the latex blending process, reinforced with bentonite clay as a nanofiller and compatibilizing filler. The Fourier transform infrared spectroscopy (FTIR) and XRD results showed that the silicate layer was intercalated by NR and PS molecular chains. The morphology of tensile fracture surfaces revealed discrete phase boundaries of PS and NR mix and their progressive disappearance with bentonite concentration, suggesting that the bentonite aided in the compatibilization of PS and NR. Furthermore, the change in the glass transition temperature (T_g) of NR to higher temperatures than those of the blends was another indicator of the bentonite clay's compatibilizing effect. Mechanical, adhesion, and morphological properties of PS and high-density polyethylene (HDPE) nanocomposites fabricated by incorporation of nano-kaolin clay were studied by George et al. (2012). The results showed that due to the excellent filler dispersion and efficient interaction of nanoparticles, the addition of kaolin clay resulted in a considerable improvement in tensile strength and tensile modulus, which appeared to enhance the stress transmission between the filler and matrix.

In another study by Chee et al. (2020), non-woven bamboo mat/woven kenaf mat-reinforced epoxy nanocomposites through hand lay-up technique were produced as shown in Figure 22.5. The nanocomposites were prepared by incorporation of nanoclays, namely MMT, OMMT, and halloysite nanotube (HNT) at 1 wt.% loading. They found that the addition of nanoclays increased the density, reduced the void content, and suppressed water uptake in all hybrid nanocomposites. Due to the evenly dispersed and effective interfacial adhesion bonding between the OMMT and epoxy matrix, the integration of OMMT was observed to demonstrate higher dimensional stability regarding water absorptions and thermal expansion as compared to MMT and HNT. Furthermore, lower water absorption resulted in less dimensional

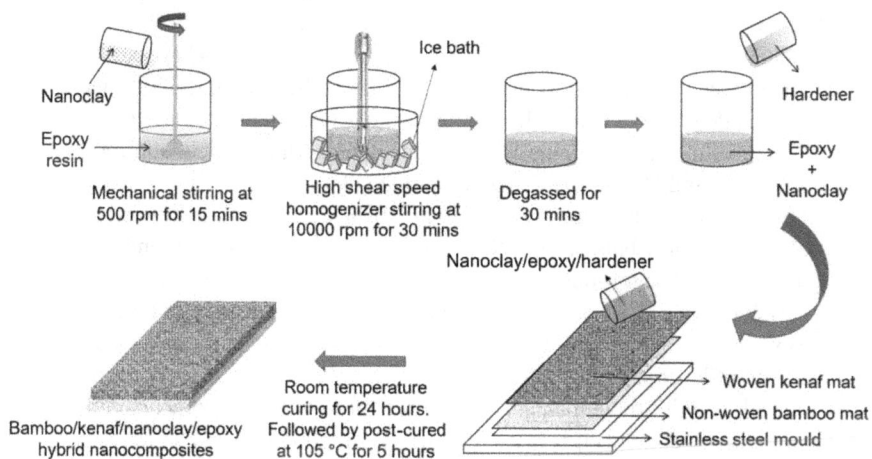

FIGURE 22.5 The preparation of bamboo/kenaf/nanoclay-reinforced epoxy hybrid nanocomposites. (Reproduced with a copyright permission from Chee et al., 2020.)

change in the hybrid composites due to the efficiency of nanoclay to fill the pores and the moisture barrier effect. In comparison with MMT and HNT, the addition of OMMT to bamboo/kenaf hybrid composites reveals a significant improvement in thermal expansion. This improvement is attributed to the high aspect ratio of the nanofillers, which prevent the thermal stress from spreading evenly within the composite's component.

Mahesh et al. (2021) investigated the mechanical properties of glycol-modified polyethylene glycol (PETG) reinforced with OMMT nanoclay and short carbon fibers (SCF) for secondary structures in aerospace and automotive applications. The tensile, compression, flexural, impact, and hardness properties of PETG/OMMT/SCF composites showed that the mechanical properties of the composites are greatly enhanced by the addition of OMMT nanoclay. Due to interstitial gaps and inadequate matrix–fiber bonding, the addition of SCF, however, has little impact on the characteristics of the composites. In the automobile industry, Gul et al. (2016b) had reported that exterior coatings made of polycarbonate (PC) and nanoclay have been utilized to withstand corrosion and abrasion without losing clarity. Poly(vinylidene fluoride) (PVDF) was used to build a composite membrane by Fang et al. (2016) by utilizing MMT nanoclays for Li-ion batteries as a separators. Liu et al. (2016) successfully developed a composite from poly-acrylic acid (PAA) with acid-treated bentonite nanoclay and untreated palygorskite that was used in air filters to remove NH_3 from contaminated air.

22.3 NATURAL NANOFILLERS IN AUTOMOTIVE AND AEROSPACE INDUSTRY

Particular attention has been given to nanofillers made from natural resources, including NC, due the growing concern about environmental sustainability. NC is a cellulosic substance that is derived from plants, animals, and microbes that have nanoscale dimensions. Generally, NC can be classified into three types: cellulose nanofiber (CNF), cellulose nanocrystals (CNC), and bacterial nanocellulose (BNC). CNC and CNF are derived mostly from lignocellulosic biomass, whereas BNC is collected from bacteria. Because of its benefits, such as lightweight, good mechanical qualities, and degradable material, the application of NC in the automotive and aerospace industries may greatly contribute to improving certain critical quality (Nurazzi et al., 2022; Ali and Hoque, 2022; Akampumuza et al., 2017).

Han et al. (2021) fabricated a lightweight, strong, and tough cellulosic structural material from natural balsawood (NW) for the application in new energy car and aerospace sectors. The top-down fabrication process through compressing rehydrated wood aerogel (WA) involved water molecules-induced hydrogen bonding between CNFs as shown in Figure 22.6. The generated WAs had a low thermal conductivity in the direction of layer stacking, which was comparable to an aerogel made entirely of cellulose. Along with the enhanced mechanical property, the compressed WAs had a higher moisture content. It was noted that the compressed WA at 18% possessed hardness of 1.60 MJ/m³ and a maximum tensile strength of 135.63 MPa, which are roughly 6 and 10 times greater, respectively, than those of NW.

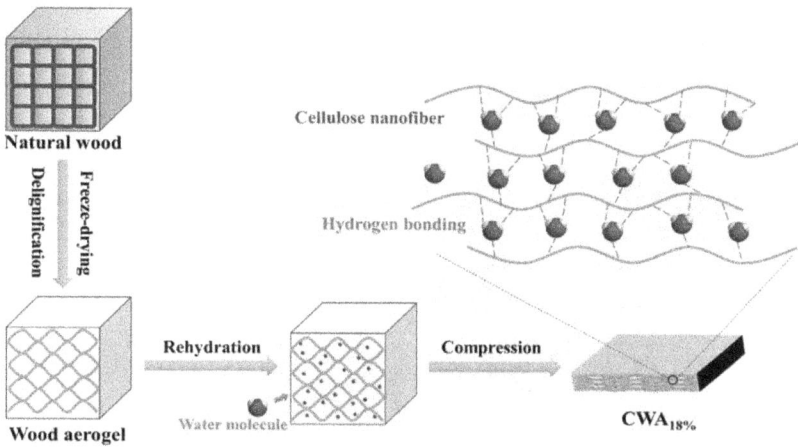

FIGURE 22.6 The preparation of compressed WA through top-down fabrication of light-weight high-performance cellulosic nanocomposites. (Reproduced with a copyright permission from Han et al., 2021.)

Tensile and flexural properties of nanocomposites of CNFs with functionalized multi-walled CNTs (f-MWCNT) as nanofiller materials were studied by Ramesh et al. (2022). The results showed that the tensile and flexural properties of the okra CNF epoxy nanocomposite with MWCNT were significantly improved when compared to the okra CNF alone by using nanoindentation techniques. The findings also showed that adding f-MWCNT nanofillers to the epoxy matrix significantly improved the tensile and flexural characteristics of the proposed CNFs. Nuruddin et al. (2017a) produced an epoxy polymer nanocomposite modified with CNFs and GNPs with different combinations of binary nanofillers. It was reported that the inclusion of CNFs improved the interaction between the graphene and polymer matrix. The superior performance of CNFs and GNPs for load transfer capacity, as well as good dispersion of CNFs and GNPs into matrix system, reduced the probability of flaw formation and resulted in good stress transfer from matrix to nanofillers, leading to improvements in flexural strength and modulus for binary nanofillers.

In another study, a lightweight composite using sheet molding compound (SMC) by added CNC in glass fiber (GF)/epoxy nanocomposite for automotive applications was successfully produced by Asadi et al. (2017). The results showed that the presence of 1% and 1.5% of CNC significantly decreased the weight of SMC composite by 7.5% while preserving its tensile and flexural capabilities. Some CNC-composite qualities, including the rubbery modulus, storage modulus, thermal, tensile, and flexural properties, were found to have been improved. The micrograph images in Figure 22.7 reveal that the addition of CNC changed the epoxy matrix characteristics of the SMC composites, resulting in rougher fracture surfaces, suggesting a strengthening effect due to stronger connections at the interfaces. Thus, the use of NC can improve the polymer matrix's mechanical characteristics while also lowering weight, which is crucial for the construction of vehicles.

FIGURE 22.7 Micrograph images of tensile fracture surface of different SMC composites; (a) and (b) 0 CNC, (c) 1 CNC, and (d) 1.5 CNC. (Reproduced with a copyright permission from Asadi et al., 2017.)

Excellent mechanical properties such as high specific strength and modulus, biodegradability, and high aspect ratio of CNF extracted from wheat straw showed its potential to improve the properties of the epoxy nanocomposite. Nuruddin et al. (2017b) investigated the different contents of CNF (1%–3%) on the epoxy polymer nanocomposite. According to the results, the CNF loading of 2% produced the greatest benefit because it allowed for the greatest cross-linking of epoxy polymers. The addition of 2% CNFs produced the greatest increases in flexure strength and modulus, of 22.5% and 31.7%, respectively. In terms of thermal stability, the first and second decomposition temperatures were increased by addition of CNF over a neat system. This increase in thermal stability is likely the result of CNFs' catalytic impact, which speeds up the cross-linking process between the polymers and the curing agent.

In a study by Martoïa et al. (2016), the CNF foam that can be used in the automotive industry as heat or sound insulation boards for interior automotive parts was produced. To produce NFC CNF foams, the NFCs CNFs were extracted from a commercial eucalyptus-bleached kraft pulp. The results revealed that under compression load, foams displayed progressive elastic, strain-hardening, and densification regimes with auxetic effects and localized strain. The yield stress and elastic modulus were also power-law functions of the foam's relative density, and their exponents reached very high values for enzymatic NFC CNF foams, possibly due of their chaotic microstructures.

22.4 CONCLUSION

Alternative nanofillers to the traditional ones include CNTs, graphene, nanoclays, and NC. Their physical, mechanical, and chemical characteristics were comparable to those of the traditional fillers, and their abundance in nature make them reliable and renewable supplies. Every nanofiller that has been mentioned significantly contributes to improving the physical and mechanical performance of polymer matrices. In conclusion, the use of nanofillers has the potential to completely replace or partially replace the synthetic nanofillers used now, which would be advantageous in many applications, particularly in the automotive and aerospace sectors.

REFERENCES

Akampumuza, O., Wambua, P. M., Ahmed, A., Li, W., & Qin, X. H. (2017). Review of the applications of biocomposites in the automotive industry. *Polymer Composites, 38*(11), 2553–2569.

Ali, N., & Hoque, M. E. (2022). Bionanocomposites in the automotive and aerospace applications. In Muthukumar, C., Thiagamani, S.M.K., Krishnasamy, S., Nagarajan, R., and Siengchin, S. (Eds.), *Polymer Based Bio-Nanocomposites* (pp. 237–253). Springer, Singapore.

Alias, A. H., Norizan, M. N., Sabaruddin, F. A., Asyraf, M. R. M., Norrrahim, M. N. F., Ilyas, A. R., … & Khalina, A. (2021). Hybridization of MMT/lignocellulosic fiber reinforced polymer nanocomposites for structural applications: A review. *Coatings, 11*(11), 1355.

Andrews, J. B., Cardenas, J. A., Lim, C. J., Noyce, S. G., Mullett, J., & Franklin, A. D. (2018). Fully printed and flexible carbon nanotube transistors for pressure sensing in automobile tires. *IEEE Sensors Journal, 18*(19), 7875–7880.

Ariffin, M. A., Muhamad, M. R., Raja, S., Jamaludin, M. F., Yusof, F., Suga, T., … & Fujii, H. (2022). Friction stir alloying of AZ61 and mild steel with Cu-CNT additive. *Journal of Materials Research and Technology, 21*, 2400–2415.

Asadi, A., Miller, M., Singh, A. V., Moon, R. J., & Kalaitzidou, K. (2017). Lightweight sheet molding compound (SMC) composites containing cellulose nanocrystals. *Composite Structures, 160*, 211–219.

Balandin, A. A., Ghosh, S., Bao, W., Calizo, I., Teweldebrhan, D., Miao, F., & Lau, C. N. (2008). Superior thermal conductivity of single-layer graphene. *Nano letters, 8*(3), 902–907.

Bohm, S., Ingle, A., Bohm, H. M., Fenech-Salerno, B., Wu, S., & Torrisi, F. (2021). Graphene production by cracking. *Philosophical Transactions of the Royal Society A, 379*(2203), 20200293.

Cao, C., Andrews, J. B., & Franklin, A. D. (2017). Completely printed, flexible, stable, and hysteresis-free carbon nanotube thin-film transistors via aerosol jet printing. *Advanced Electronic Materials, 3*(5), 1700057.

Chee, S. S., Jawaid, M., Sultan, M. T. H., Alothman, O. Y., & Abdullah, L. C. (2020). Effects of nanoclay on physical and dimensional stability of Bamboo/Kenaf/nanoclay reinforced epoxy hybrid nanocomposites. *Journal of Materials Research and Technology, 9*(3), 5871–5880.

Chen, L. Y., Konishi, H., Fehrenbacher, A., Ma, C., Xu, J. Q., Choi, H., … & Li, X. C. (2012). Novel nanoprocessing route for bulk graphene nanoplatelets reinforced metal matrix nanocomposites. *Scripta Materialia, 67*(1), 29–32.

Chen, B., Zhou, X. Y., Zhang, B., Kondoh, K., Li, J. S., & Qian, M. (2020). Microstructure, tensile properties and deformation behaviors of aluminium metal matrix composites co-reinforced by ex-situ carbon nanotubes and in-situ alumina nanoparticles. *Materials Science and Engineering: A, 795*, 139930.

Chieng, B. W., Ibrahim, N. A., Yunus, W. M. Z. W., Hussein, M. Z., & Giita Silverajah, V. S. (2012). Graphene nanoplatelets as novel reinforcement filler in poly (lactic acid)/epoxidized palm oil green nanocomposites: Mechanical properties. *International Journal of Molecular Sciences*, *13*(9), 10920–10934.

David, M. E., Ion, R. M., Grigorescu, R. M., Iancu, L., & Andrei, E. R. (2020). Nanomaterials used in conservation and restoration of cultural heritage: An up-to-date overview. *Materials*, *13*(9), 2064.

Ehsani, M., Rahimi, P., & Joseph, Y. (2021). Structure-function relationships of nanocarbon/polymer composites for chemiresistive sensing: A review. *Sensors*, *21*(9), 3291.

Fang, B., Li, J., Zhao, N., Shi, C., Ma, L., He, C., & Liu, E. (2018). Enhanced interface interaction between modified carbon nanotubes and magnesium matrix. *Composite Interfaces*, *25*(12), 1101–1114.

Fang, C., Yang, S., Zhao, X., Du, P., & Xiong, J. (2016). Electrospun montmorillonite modified poly (vinylidene fluoride) nanocomposite separators for lithium-ion batteries. *Materials Research Bulletin*, *79*, 1–7.

Garces, I. T., Aslanzadeh, S., Boluk, Y., & Ayranci, C. (2020). Cellulose nanocrystals (CNC) reinforced shape memory polyurethane ribbons for future biomedical applications and design. *Journal of Thermoplastic Composite Materials*, *33*(3), 377–392.

George, T. S., Krishnan, A., Joseph, N., Anjana, R., & George, K. E. (2012). Effect of maleic anhydride grafting on nanokaolinclay reinforced polystyrene/high density polyethylene blends. *Polymer Composites*, *33*(9), 1465–1472.

Gul, S., Kausar, A., Muhammad, B., & Jabeen, S. (2016a). Research progress on properties and applications of polymer/clay nanocomposite. *Polymer-Plastics Technology and Engineering*, *55*(7), 684–703.

Gul, S., Kausar, A., Muhammad, B., & Jabeen, S. (2016b). Technical relevance of epoxy/clay nanocomposite with organically modified montmorillonite: A review. *Polymer-Plastics Technology and Engineering*, *55*(13), 1393–1415.

Haider, S., Kausar, A., & Muhammad, B. (2016). Overview of various sorts of polymer nanocomposite reinforced with layered silicate. *Polymer-Plastics Technology and Engineering*, *55*(7), 723–743.

Haider, S., Kausar, A., & Muhammad, B. (2017). Overview on polystyrene/nanoclay composite: Physical properties and application. *Polymer-Plastics Technology and Engineering*, *56*(9), 917–931.

Han, X., Wang, Z., Ding, L., Chen, L., Wang, F., Pu, J., & Jiang, S. (2021). Water molecule-induced hydrogen bonding between cellulose nanofibers toward highly strong and tough materials from wood aerogel. *Chinese Chemical Letters*, *32*(10), 3105–3108.

Iqbal, A., Saeed, A., & Ul-Hamid, A. (2021). A review featuring the fundamentals and advancements of polymer/CNT nanocomposite application in aerospace industry. *Polymer Bulletin*, *78*(1), 539–557.

Islam, M. S., Deng, Y., Tong, L., Faisal, S. N., Roy, A. K., Minett, A. I., & Gomes, V. G. (2016). Grafting carbon nanotubes directly onto carbon fibers for superior mechanical stability: Towards next generation aerospace composites and energy storage applications. *Carbon*, *96*, 701–710.

John, V. L., Gomathi, N., Joseph, K., Mathew, D., Chandran, S. M., & Neogi, S. (2022). Plasma functionalized CNT/cyanate ester nanocomposites for aerospace structural applications. *ChemistrySelect*, *7*(39), e202201260.

Joshi, M., & Chatterjee, U. (2016). Polymer nanocomposite: An advanced material for aerospace applications. In Sohel Rana and Raul Fangueiro (Eds.), *Advanced Composite Materials for Aerospace Engineering* (pp. 241–264). Woodhead Publishing.

Kartal, M., Uysal, M., Gul, H., Alp, A., & Akbulut, H. (2015). Pulse electrocodeposition of Ni/MWCNT nanocomposite coatings. *Surface Engineering*, *31*(9), 659–665.

Kausar, A. (2017). Polyamide 1010/polythioamide blend reinforced with graphene nanoplatelet for automotive part application. *Advances in Materials Science*, *17*(3), 24.

Kausar, A. (2020). A review of high performance polymer nanocomposites for packaging applications in electronics and food industries. *Journal of Plastic Film & Sheeting*, *36*(1), 94–112.

Kwon, H., Bradbury, C. R., & Leparoux, M. (2011). Fabrication of functionally graded carbon nanotube-reinforced aluminum matrix composite. *Advanced Engineering Materials*, *13*(4), 325–329.

Laurenzi, S., de Zanet, G., & Santonicola, M. G. (2020). Numerical investigation of radiation shielding properties of polyethylene-based nanocomposite materials in different space environments. *Acta Astronautica*, *170*, 530–538.

Li, Y., Huang, X., Zeng, L., Li, R., Tian, H., Fu, X., ... & Zhong, W. H. (2019). A review of the electrical and mechanical properties of carbon nanofiller-reinforced polymer composites. *Journal of Materials Science*, *54*(2), 1036–1076.

Li, Z., Chen, S., Nambiar, S., Sun, Y., Zhang, M., Zheng, W., & Yeow, J. T. (2016). PMMA/MWCNT nanocomposite for proton radiation shielding applications. *Nanotechnology*, *27*(23), 234001.

Lim, J. Y., Mubarak, N. M., Abdullah, E. C., Nizamuddin, S., & Khalid, M. (2018). Recent trends in the synthesis of graphene and graphene oxide based nanomaterials for removal of heavy metals-A review. *Journal of Industrial and Engineering Chemistry*, *66*, 29–44.

Liu, E., Sarkar, B., Wang, L., & Naidu, R. (2016). Copper-complexed clay/poly-acrylic acid composites: Extremely efficient adsorbents of ammonia gas. *Applied Clay Science*, *121*, 154–161.

Liu, J., Fu, M., Jing, M., & Li, Q. (2013). Flame retardancy and charring behavior of polystyrene-organic montmorillonite nanocomposites. *Polymers for Advanced Technologies*, *24*(3), 273–281.

Liu, S. P., Huang, I. J., Chang, K. C., & Yeh, J. M. (2010). Mechanical properties of polystyrene-montmorillonite nanocomposites-Prepared by melt intercalation. *Journal of Applied Polymer Science*, *115*(1), 288–296.

Mahesh, V., Joseph, A. S., Mahesh, V., & Harursampath, D. (2021). Investigation on the mechanical properties of additively manufactured PETG composites reinforced with OMMT nanoclay and carbon fibers. *Polymer Composites*, *42*(5), 2380–2395.

Martoïa, F., Cochereau, T., Dumont, P. J., Orgéas, L., Terrien, M., & Belgacem, M. N. (2016). Cellulose nanofibril foams: Links between ice-templating conditions, microstructures and mechanical properties. *Materials & Design*, *104*, 376–391.

Nisha, M. S., & Singh, D. (2016). Manufacturing of smart nanomaterials for structural health monitoring (SHM) in aerospace application using CNT and CNF. *Journal of Nano Research*, *37*, 42–50. Trans Tech Publications Ltd.

Nofar, M., Ozgen, E., & Girginer, B. (2020). Injection-molded PP composites reinforced with talc and nanoclay for automotive applications. *Journal of Thermoplastic Composite Materials*, *33*(11), 1478–1498.

Nurazzi, N. M., Asyraf, M. M., Khalina, A., Abdullah, N., Sabaruddin, F. A., Kamarudin, S. H., ... & Sapuan, S. M. (2021a). Fabrication, functionalization, and application of carbon nanotube-reinforced polymer composite: An overview. *Polymers*, *13*(7), 1047.

Nurazzi, N. M., Jenol, M. A., Kamarudin, S. H., Aisyah, H. A., Hao, L. C., Yusuff, S. M., ... & Norli, A. (2022). Nanocellulose composites in the automotive industry. In S.M. Sapuan, M.N.F. Norrrahim, R.A. Ilyas & Constantinos Soutis (Eds.), *Industrial Applications of Nanocellulose and Its Nanocomposites* (pp. 439–467). Woodhead Publishing.

Nurazzi, N. M., Sabaruddin, F. A., Harussani, M. M., Kamarudin, S. H., Rayung, M., Asyraf, M. R. M., ... & Khalina, A. (2021b). Mechanical performance and applications of cnts reinforced polymer composites-A review. *Nanomaterials*, *11*(9), 2186.

Nuruddin, M., Hosur, M., Gupta, R., Hosur, G., Tcherbi-Narteh, A., & Jeelani, S. (2017a). Cellulose nanofibers-graphene nanoplatelets hybrids nanofillers as high-performance multifunctional reinforcements in epoxy composites. *Polymers and Polymer Composites*, *25*(4), 273–284.

Nuruddin, M., Hosur, M., Mahdi, T., & Jeelani, S. (2017b). Flexural, viscoelastic and thermal properties of epoxy polymer composites modified with cellulose nanofibers extracted from wheat straw. *Sensors & Transducers, 210*(3), 1.

Oliveira, A. D., & Beatrice, C. A. G. (2018). Polymer nanocomposites with different types of nanofiller. In Subbarayan Sivasankaran (Eds.) *Nanocomposites-Recent Evolutions*, 103–104.

Popov, V. V., Pismenny, A., Larianovsky, N., Lapteva, A., & Safranchik, D. (2021). Corrosion resistance of Al-CNT metal matrix composites. *Materials, 14*(13), 3530.

Rafiee, R., & Shahzadi, R. (2019). Mechanical properties of nanoclay and nanoclay reinforced polymers: A review. *Polymer Composites, 40*(2), 431–445.

Ramachandran, K., Boopalan, V., Bear, J. C., & Subramani, R. (2021). Multi-walled carbon nanotubes (MWCNTs)-reinforced ceramic nanocomposites for aerospace applications: A review. *Journal of Materials Science, 57*(6), 1–31.

Ramesh, A., Srinivasulu, N. V., & Rani, M. I. (2022). Influences of functionalized multiwalled carbon nanotube on the tensile and flexural properties of okra cellulose nanofibers/epoxy nanocomposites. In Narasimham, G.S.V.L., Babu, A.V., Reddy, S.S., and Dhanasekaran, R. (Eds.), *Innovations in Mechanical Engineering* (pp. 639–651). Springer, Singapore.

Ruamcharoen, J., Ratana, T., & Ruamcharoen, P. (2014). Bentonite as a reinforcing and compatibilizing filler for natural rubber and polystyrene blends in latex stage. *Polymer Engineering & Science, 54*(6), 1436–1443.

Samakande, A., Hartmann, P. C., Cloete, V., & Sanderson, R. D. (2007). Use of acrylic based surfmers for the preparation of exfoliated polystyrene-clay nanocomposites. *Polymer, 48*(6), 1490–1499.

Selvamani, S. T., Premkumar, S., Vigneshwar, M., Hariprasath, P., & Palanikumar, K. (2017). Influence of carbon nano tubes on mechanical, metallurgical and tribological behavior of magnesium nanocomposites. *Journal of Magnesium and Alloys, 5*(3), 326–335.

Shao, H. Q., Wei, H., & He, J. H. (2018). Dynamic properties and tire performances of composites filled with carbon nanotubes. *Rubber Chemistry and Technology, 91*(3), 609–620.

Shen, X., Zheng, Q., & Kim, J. K. (2021). Rational design of two-dimensional nanofillers for polymer nanocomposites toward multifunctional applications. *Progress in Materials Science, 115*, 100708.

Song, W. J., Yoo, S., Song, G., Lee, S., Kong, M., Rim, J., … & Park, S. (2019). Recent progress in stretchable batteries for wearable electronics. *Batteries & Supercaps, 2*(3), 181–199.

Srinivasan, V., Kunjiappan, S., & Palanisamy, P. (2021). A brief review of carbon nanotube reinforced metal matrix composites for aerospace and defense applications. *International Nano Letters, 11*(4), 321–345.

Subadra, S. P., Yousef, S., Griskevicius, P., & Makarevicius, V. (2020). High-performance fiberglass/epoxy reinforced by functionalized CNTs for vehicle applications with less fuel consumption and greenhouse gas emissions. *Polymer Testing, 86*, 106480.

Vashist, A., Kaushik, A., Ghosal, A., Bala, J., Nikkhah-Moshaie, R., A. Wani, W., … & Nair, M. (2018). Nanocomposite hydrogels: Advances in nanofillers used for nanomedicine. *Gels, 4*(3), 75.

Venkatesan, M., Palanikumar, K., & Boopathy, S. R. (2018). Experimental investigation and analysis on the wear properties of glass fiber and CNT reinforced hybrid polymer composites. *Science and Engineering of Composite Materials, 25*(5), 963–974.

Vertuccio, L., Guadagno, L., Spinelli, G., Lamberti, P., Zarrelli, M., Russo, S., & Iannuzzo, G. (2018). Smart coatings of epoxy based CNTs designed to meet practical expectations in aeronautics. *Composites Part B: Engineering, 147*, 42–46.

Zhu, S., Huang, A., Wang, Q., & Xu, Y. (2021). MOF-derived porous carbon nanofibers wrapping Sn nanoparticles as flexible anodes for lithium/sodium ion batteries. *Nanotechnology, 32*(16), 165401.

Index

Note: **Bold** page numbers refer to tables, *Italic* page numbers refer to figures.

For Product Safety Concerns and Information please contact our EU
representative GPSR@taylorandfrancis.com
Taylor & Francis Verlag GmbH, Kaufingerstraße 24, 80331 München, Germany

www.ingramcontent.com/pod-product-compliance
Lightning Source LLC
Chambersburg PA
CBHW060742220326
41598CB00022B/2303